Encyclopedia of Food Science

Encyclopedia of Food Science

Edited by **Johann Wells**

R CALLISTO REFERENCE

New York

Published by Callisto Reference,
106 Park Avenue, Suite 200,
New York, NY 10016, USA
www.callistoreference.com

Encyclopedia of Food Science
Edited by Johann Wells

International Standard Book Number: 978-1-63239-248-0 (Hardback)

Printed in the United States of America.

Contents

Preface

Food science is a multidisciplinary field of science, which includes different areas of scientific studies which converge at the study of food. It includes aspects of biochemistry, microbiology and chemical engineering. According to the Institute of Food Technologists, food science is defined as "the discipline in which the engineering, biological, and physical sciences are used to study the nature of foods, its causes of deterioration, principles underlying food processing and the improvement of foods for the consuming public". Food scientists play an important role in developing new food products, designing procedures for production of these food products, selection of materials used for packaging. Preservatives and shelf-life studies are a very important part of microbiological and chemical studies. Some of the related sub-disciplinary fields included in the study of food science are-:

1) Food chemistry
2) Food engineering
3) Food microbiology
4) Food packaging
5) Food preservation
6) Food technology
7) Molecular gastronomy
8) Sensory analysis, etc.

This book provides some of the key topics on Food Science, compiled together in a single text. The book would be a significant source of information to a broad spectrum of readers.

At the end of the preface, I wish to thank the editorial team, the project manager, and all the contributors, for making this book come to life. I also extend gratitude to my family and friends for being a constant source of encouragement at every step of my research.

Editor

Angiotensin I-Converting Enzyme Inhibitory Peptides of Chia (*Salvia hispanica*) Produced by Enzymatic Hydrolysis

**Maira Rubi Segura Campos, Fanny Peralta González,
Luis Chel Guerrero, and David Betancur Ancona**

*Facultad de Ingeniería Química, Universidad Autónoma de Yucatán, Periférico Norte. Km 33.5, Tablaje Catastral 13615,
Colonia Chuburná de Hidalgo Inn, 97203 Mérida, YUC, Mexico*

Correspondence should be addressed to David Betancur Ancona; bancona@uady.mx

Academic Editor: Fabienne Remize

Synthetic angiotensin I-converting enzyme (ACE-I) inhibitors can have undesirable side effects, while natural inhibitors have no side effects and are potential nutraceuticals. A protein-rich fraction from chia (*Salvia hispanica* L.) seed was hydrolyzed with an Alcalase-Flavourzyme sequential system and the hydrolysate ultrafiltered through four molecular weight cut-off membranes (1 kDa, 3 kDa, 5 kDa, and 10 kDa). ACE-I inhibitory activity was quantified in the hydrolysate and ultrafiltered fractions. The hydrolysate was extensive (DH = 51.64%) and had 58.46% ACE-inhibitory activity. Inhibition ranged from 53.84% to 69.31% in the five ultrafiltered fractions and was highest in the <1 kDa fraction (69.31%). This fraction's amino acid composition was identified and then it was purified by gel filtration chromatography and ACE-I inhibition measured in the purified fractions. Amino acid composition suggested that hydrophobic residues contributed substantially to chia peptide ACE-I inhibitory strength, probably by blocking angiotensin II production. Inhibitory activity ranged from 48.41% to 62.58% in the purified fractions, but fraction F1 (1.5–2.5 kDa) exhibited the highest inhibition (IC$_{50}$ = 3.97 μg/mL; 427–455 mL elution volume). The results point out the possibility of obtaining bioactive peptides from chia proteins by means of a controlled protein hydrolysis using Alcalase-Flavourzyme sequential system.

1. Introduction

Cardiovascular disease (CVD) affects the heart and blood vessels and is the principal cause of death worldwide. Considered the primary risk factor for CVD, high blood pressure, or hypertension, consists of a sustained increase in blood pressure levels. In 2000, >25% of the population worldwide (approximately 1 billion) suffered from hypertension, a figure predicted to increase to 1.56 billion by 2025 [1].

Angiotensin I-converting enzyme (peptidyl carboxy peptidase, EC 3.4.15.1, ACE) belongs to the class of zinc proteases that require zinc and chloride for activation. ACE plays an important role in blood pressure regulation via the renin-angiotensin system (RAS) and the kallikrein-kinnin system (KKS). In the KKS, ACE inactivates the vasodilator bradykinin, while in the RAS, ACE acts as an exopeptidase cleaving His-Leu from the C-terminal of decapeptide angiotensin I and producing the potent vasoconstrictor octapeptide angiotensin II [2].

Use of enzyme technologies for protein recovery and modification has led to production of a broad spectrum of food ingredients and industrial products. Hydrolysis selectivity is commonly manipulated by employing proteases from different sources due to their specificity for peptide bonds adjacent to certain amino acids. Enzymatic hydrolysis of food proteins is an efficient way to recover potent bioactive peptides [3]. ACE-I inhibitory peptides derived from food proteins have attracted particular attention for their ability to prevent hypertension. Compared with chemosynthetic drugs, peptides from food proteins may have reduced toxic effects in humans. They therefore hold promise as potent functional food additives and represent a healthier and more natural alternative to ACE inhibitor drugs. Dietary ACE-I inhibitory peptides may be bioavailable. Some ACE-I inhibitory peptides resist digestion, can be absorbed in the intestine, and are stable in the blood, suggesting they may produce an acute blood pressure-lowering effect after oral administration. The first ACE-I inhibitory peptide was isolated from snake venom [4], and since then many others have been discovered in enzymatic hydrolysates of different food proteins. Food protein sources used to date include casein, whey protein, fish protein, chicken eggs, and wheat germ. Inhibitory activities and sequences are remarkably different in many of these peptides [5]. The soybean proteins are other novel sources to production of the ACE inhibitory peptides. β-conglycinin and glycinin were hydrolysed by an acid proteinase from Monascus purpureus. The degree of hydrolysis and inhibitory activities of angiotensin I-converting enzyme (ACE) increased with increasing proteolysis time [6].

Population growth and shrinking food resources are ongoing challenges in developing countries, while excessive animal protein intake and associated unhealthy levels of saturated fats exposure are increasingly common in developed countries. In response, research interest has steadily grown in the search for new sources of proteins from nonconventional raw materials. The genus Salvia L. belongs to the Lamiaceae family and includes about 900 species found worldwide, most mainly in the Mediterranean, Southeast Africa, and Central and South America. Cultivated for culinary, medicinal, and ornamental uses, Salvia species form part of ethnopharmacological traditions and are an important crop, especially for small farmers [7]. Recent research has addressed the chemical composition, biological properties, and possible applications of its essential oils, which may be sources for economically promising natural products with uses in the food, pharmaceutical, and cosmetic industries. The present study objective was to identify and quantify ACE-I inhibitory activity in protein hydrolysates from a Salvia hispanica protein rich fraction hydrolyzed with an Alcalase-Flavourzyme sequential system, and in ultrafiltered fractions from this hydrolysate.

2. Materials and Methods

2.1. Materials. Chia (S. hispanica, L.) seeds were obtained in Yucatan state, Mexico. Reagents were of analytical grade and purchased from J. T. Baker (Phillipsburg, NJ, USA), Sigma (Sigma Chemical Co., St. Louis, MO, USA), Merck (Darmstadt, Germany), and Bio-Rad (Bio-Rad Laboratories Inc., Hercules, CA, USA). Angiotensin-converting enzyme from rabbit lung (2 units/mg protein) was purchased by Sigma (A6778 Sigma Chemical Co., St. Louis, MO, USA). The Alcalase 2.4 L and Flavourzyme 500 MG enzymes were purchased from Novo Laboratories (Copenhagen, Denmark).

2.2. Protein Rich Fraction. Flour was produced from 6 kg chia seed by first removing all impurities and damaged seeds, crushing the remaining sound seeds (Moulinex DPA139, Zapopan, Mexico), and milling them (Krups 203 Mill, Mexico City, Mexico). Oil extraction from the milled seeds was done with hexane in a Soxhlet system for 2 h. The remaining fraction was milled with 0.5 mm screen (Thomas Model 4 Wiley,

Swedesboro, NJ, USA). The defatted chia flour was dried in a Labline stove at 60°C for 24 h. Extraction of the protein-rich fraction was done by dry fractionation of the defatted flour according to Vázquez-Ovando et al. [8]. Briefly, 500 g flour was sifted for 20 min using a Tyler 100 mesh (140 μm screen) and a Ro-Tap agitation system.

2.3. Enzymatic Hydrolysis. The chia protein-rich fraction was hydrolyzed in batches by sequential treatment with Alcalase and Flavourzyme. A predigestion with Alcalase for 60 min was followed by incubation with Flavourzyme for 90 min. Hydrolysis conditions were substrate concentration, 2 g/100 g; enzyme/substrate ratio, 0.3 AU g^{-1} for Alcalase and 50 LAPU g^{-1} for Flavourzyme; pH, 7 for Alcalase and 8 for Flavourzyme and temperature, 50°C. Hydrolysis was done in a reaction vessel equipped with a stirrer, thermometer, and pH electrode. In all treatments, the reaction was stopped by heating to 85°C for 15 min, followed by centrifuging at 9880 ×g for 20 min to remove the insoluble portion [9].

2.4. Degree of Hydrolysis. Degree of hydrolysis (DH) was calculated by determining free amino groups with o-phthaldialdehyde following the methodology described by Nielsen et al. [10] as follows: DH = $h/h_{tot} * 100$, where h_{tot} is the total number of peptide bonds per protein equivalent, and h is the number of hydrolyzed bonds. The h_{tot} factor is dependent on the raw material amino acid composition and was determined by reverse-phase high performance liquid chromatography (RP-HPLC) [11]. Samples (2–4 mg protein) were treated with 4 mL of 6 mol equivalent to L^{-1} HCl, placed in hydrolysis tubes, and gassed with nitrogen at 110°C for 24 h. They were then dried in a rotavapor and suspended in 1 mol L^{-1} sodium borate buffer at pH 9.0. Amino acid derivatization was performed at 50°C using diethyl ethoxymethylenemalonate. Amino acids were separated using HPLC with a reversed-phase column (300 × 3.9 mm, Nova-Pak C18, 4 mm; Waters) and a binary gradient system with 25 mmol L^{-1} sodium acetate containing (A) 0.02 g L^{-1} sodium azide at pH 6.0 and (B) acetonitrile as solvent. Flow rate was 0.9 mL min^{-1}, and elution gradient was time 0.0–3.0 min, linear gradient A : B (91 : 9) to A–B (86 : 14); time 3.0–13.0 min, elution with A–B (86–14); time 13.0–30.0 min, linear gradient A–B (86 : 14) to A–B (69 : 31); time 30.0–35.0 min, elution with A–B (69 : 31).

2.5. Hydrolysate Fractionation by Ultrafiltration. The hydrolysate was fractionated by ultrafiltration [12], using a high performance ultrafiltration cell (Model 2000, Millipore). Five fractions were prepared using four molecular weight cutoff (MWCO) membranes: 1 kDa, 3 kDa, 5 kDa, and 10 kDa. Soluble fractions prepared by centrifugation (9880 ×g for 20 min) were passed through the membrane starting with the largest MWCO membrane cartridge (10 kDa). Retentate and permeate were collected separately, and the retentate recirculated into the feed until maximum permeate yield was reached at this size, as indicated by a decrease in permeate flow rate. Permeate from the 10 kDa membrane was then filtered through the 5 kDa membrane with recirculation until maximum permeate yield was reached. The 5 kDa permeate was then recirculated through the 3 kDa membrane and the 3 kDa permeate through the 1 kDa membrane. This process minimized contamination of the larger molecular weight fractions with smaller molecular weight fractions, while producing enough retentates and permeates for the following analyses. The five ultrafiltered peptide fractions (UPF) were prepared and designated as >10 kDa (10 kDa retentate); 5–10 kDa (10 kDa permeate–5 kDa retentate); 3–5 kDa (5 kDa permeate–3 kDa retentate); 1–3 kDa (3 kDa permeate–1 kDa retentate); <1 kDa (1 kDa permeate).

2.6. ACE-I Inhibitory Activity. ACE-I inhibitory activity in the hydrolysate and its UPF was analyzed following the method of Hayakari et al. [13] which is based on the fact that ACE-I hydrolyzes hippuryl-L-histidyl-L-leucine (HHL) yielding hippuric acid and L-histidyl-L-leucine. This method relies on the colorimetric reaction of hippuric acid with 2,4,6-trichloro-S-triazine (TT) in a 0.5 mL incubation mixture containing 40 μmol potassium phosphate buffer (pH 8.3), 300 μmol sodium chloride, 40 μmol 3% HHL in potassium phosphate buffer (pH 8.3), and 100 mU/mL ACE-I. This mixture was incubated at 37°C/45 min and the reaction terminated by addition of TT (3% v/v) in dioxane and 3 mL 0.2 M potassium phosphate buffer (pH 8.3). After centrifuging the reaction mixture at 10,000 ×g for 10 min, enzymatic activity was determined in the supernatant by measuring absorbance at 382 nm. All runs were done in triplicate. ACE-I inhibitory activity was quantified by a regression analysis of ACE-I inhibitory activity (%) versus peptide concentration, and IC$_{50}$ values (i.e., the peptide concentration in μg protein/mL required to produce 50% ACE-I inhibition under the described conditions) was defined and calculated as follows:

$$\text{ACE-I inhibitory activity } (\%) = \frac{(A - B)}{(A - C)} \times 100, \quad (1)$$

where A represents absorbance in the presence of ACE-I sample, B absorbance of the control, and C absorbance of the reaction blank.

Consider the following:

$$\text{IC}_{50} = \frac{(50 - b)}{m}, \quad (2)$$

where b is the intersection and m is the slope.

2.7. Amino Acid Composition. Amino acid composition was determined in the UPF with the highest biological activity, according to the method of Alaiz et al. [11] Samples (2–4 mg protein) were treated with 4 mL of 6 mol equivalent L^{-1} HCl, placed in hydrolysis tubes, and gassed with nitrogen at 110°C for 24 h. They were then dried in a rotavapor and suspended in 1 mol L^{-1} sodium borate buffer at pH 9.0. Amino acid derivatization was performed at 50°C using diethyl ethoxymethylenemalonate. Amino acids were separated using HPLC with a reversed-phase column (300 × 3.9 mm, Nova-Pak C18, 4 mm; Waters) and a binary gradient system with 25 mmol L^{-1} sodium acetate containing (A) 0.02 g L^{-1} sodium azide at pH 6.0 and (B) acetonitrile as solvent. Flow

rate was 0.9 mL min^{-1}, and the elution gradient was time 0.0–3.0 min, linear gradient A : B (91 : 9) to A–B (86 : 14); time 3.0–13.0 min, elution with A–B (86–14); time 13.0–30.0 min, linear gradient A–B (86 : 14) to A–B (69 : 31); time 30.0–35.0 min, elution with A–B (69 : 31).

2.8. G-50 Gel Filtration Chromatography. After filtration through 10, 5, 3, and 1 kDa membranes in a high performance ultrafiltration cell, 10 mL of the fraction with highest ACE-I inhibitory activity was injected into a Sephadex G-50 gel filtration column (3 cm × 79 cm) at a flow rate of 25 mL/h of 50 mM ammonium bicarbonate (pH 9.1). The resulting fractions were collected to assay ACE-I inhibitory activity [13]. Peptide molecular masses were determined by referring to a calibration curve running molecular mass markers on the Sephadex G-50 under identical conditions and those used for the test samples. Molecular mass standards were thyroglobulin (670 kDa), bovine gamma globulin (158 kDa), equine myoglobin (17 kDa), vitamin B$_{12}$ (1.35 kDa), and Thr-Gln (0.25 kDa). Fractions selected for further peptide purification were pooled and lyophilized before RP-HPLC.

2.9. Statistical Analysis. All results were analyzed in triplicate using descriptive statistics to estimate means and variation. One-way ANOVAs were run to evaluate *in vitro* ACE-I inhibitory activity and a Duncan multiple range done to identify differences between treatments. All analyses were done according to Montgomery [14] and processed using the Statgraphics Plus version 5.1 software.

3. Results and Discussion

3.1. Enzymatic Hydrolysis of Protein-Rich Fraction. With a protein content of 46.7%, the chia protein-rich fraction proved to be good starter material for hydrolysis. Production of extensive (i.e., >50% DH) hydrolysates requires use of more than one protease because a single enzyme cannot achieve such high DHs within a reasonable time period. For this reason, an Alcalase-Flavourzyme sequential system was used in the present study to produce an extensive hydrolysate. Alcalase (EC 3.4.21.62) is a proteinase from *Bacillus licheniformis* and the Flavourzyme (EC 3.4.11.1) is a fungal protease from *Aspergillus oryzae* with both endo- and exopeptidase activities. The bacterial endoprotease Alcalase is limited by its specificity, resulting in DHs no higher than 20%–25%, depending on substrate, but can attain these DHs in a relatively short time under moderate conditions. When Alcalase is used to hydrolyze protein it tends to produce peptides whose C-terminals are amino acids with large side chains and no charge (aromatic and aliphatic amino acids), such as Ile, Leu, Val, Met, Phe, Tyr, and Trp. The hydrolysis process is accelerated when peptide N-terminals have hydrophobic amino acids [15]. The fungal protease Flavourzyme has broader specificity, which, when combined with its exopeptidase activity, can generate DH values as high as 50%. Flavourzyme is recommended for production of hydrolysates or peptides with biological activity and low bitterness. Both Alcalase and Flavourzyme tend to generate peptides with hydrophobic amino acid C-terminals, and QSAR analyses of ACE-I inhibitory peptides have shown that peptides with hydrophobic amino acid C-terminals exhibit potentially strong ACE-I inhibition. Therefore, Alcalase and Flavourzyme are probably suitable for preparing high-activity ACE-I inhibiting peptides. In addition, both proteases are suitable for industrial applications and are microbial enzymes, meaning that they are easily obtained and relatively low cost compared to other enzymes such as proteinase K and chymotrypsin C [15].

Chia hydrolysates obtained with the Alcalase-Flavourzyme sequential system had clear biological activity and are promising prospects for use in new product development. The highest DH in the present study (51.6%) was attained with Flavourzyme at 150 min. However, it was made possible by Alcalase predigestion, which increases the number of N-terminal sites, thus facilitating hydrolysis by Flavourzyme. This DH was higher than reported for hard-to-cook bean hydrolyzed with Alcalase-Flavourzyme (43%) or pepsin-pancreatin (26.2%) for 90 min [16]. But it was lower than reported for chickpea (65%) hydrolyzed with Alcalase-Flavourzyme at 150 min [17] and for rapeseed (60%) hydrolyzed with the same system at 3 h [18]. Variations in DH values are probably the result of protein source and hydrolytic specificity. When hydrolyzed sequentially with Alcalase and Flavourzyme, chia S. hispanica is an appropriate substrate for producing extensive hydrolysates (DH higher than 10%), and a natural source of peptides with potential bioactivity.

3.2. ACE-I Inhibitory Activity.

A number of natural ACE-I inhibitors have been isolated from different organism proteins, including peptides extracted by enzymatic hydrolysis. When added to food systems, enzymatic hydrolysates have exhibited advantages such as improved water-binding capacity, emulsifying stability, protein solubility, and nutritional quality. Enzymatic hydrolysis has become a valuable tool for modifying protein functionality [19]. During hydrolysis, hydrophobicity of amino-acid side chains is normally due to relatively small peptides with molecular weights between 1,000 and 6,000 Da. Enzymatic hydrolysis is an effective way of producing bioactive peptides, which are short peptides released from food proteins by hydrolysis and with biological activities that may be beneficial to the organism [20]. Bioactive peptides usually contain 3–20 amino acids per molecule and are inactive within the parent protein molecule sequence [21].

The protein hydrolysate obtained from the chia protein-rich fraction had 58.5% ACE-I inhibition. This is lower than the 79.5% reported for yak milk casein hydrolyzed with Alcalase for 240 min [22], but higher than the 5%–50% obtained for protein hydrolysates from amaranth (Amaranthus hypochondriacus) albumin 1 and globulin [23].

Ultrafiltration of the hydrolysate produced UPFs with increased biological activity. ACE-I inhibition ranged from 53.8% to 69.3% (Figure 1), with clear increases ($P < 0.05$) in activity in progressively smaller fractions; that is, the >10 kDa fraction had the lowest activity and the <1 kDa had the

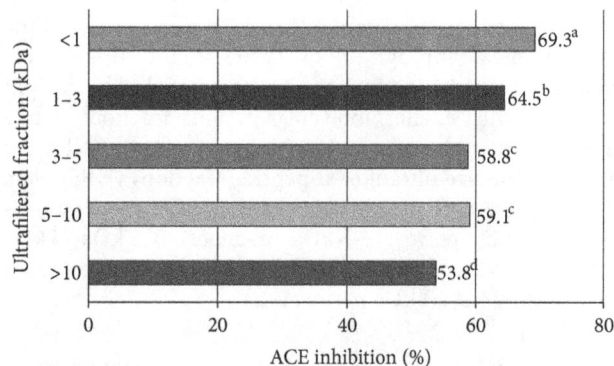

FIGURE 1: ACE-I inhibition percentage of peptide fractions obtained by ultrafiltration of a S. hispanica protein hydrolysate. [a–d]Different superscript letters indicate statistical difference ($P < 0.05$). Data are the mean of three replicates.

highest. Use of ultrafiltration membranes clearly helped to enrich specific peptide fractions. This coincides with a study in which the <1 kDa fraction of an Alaska pollock (Theragra chalcogramma) frame protein hydrolysate had the highest ACE-inhibitory activity (87.6%, $IC_{50} = 457 \mu g/mL$) [24]. In another study, protein hydrolysate from Chinese soft-shelled turtle had a lower ACE-I inhibitory effect ($IC_{50} = 280 \mu g/mL$) than its corresponding <5000 kDa ultrafiltered fraction ($IC_{50} = 190 \pm 5 \mu g/mL$) [25]. A greater inhibitory effect at smaller molecular weights was also observed in a yak casein hydrolysate fraction, in which the <6 kDa fraction was the most effective (85.4%) [22]. When taken in conjunction, these results support the suggestion that ultrafiltration is an effective way of enriching ACE-I inhibitory peptides from chia proteins.

3.3. Amino Acid Composition.

An amino acid profile was generated for the <1 kDa UPF because it had the highest ACE-I inhibitory activity. During hydrolysis, asparagine and glutamine partially converted to aspartic acid and glutamic acid, respectively; the data for asparagine and/or aspartic acid were therefore reported as Asx while those for glutamine and/or glutamic acid were reported as Glx. The high inhibitory activity (69.3%) exhibited by the <1 kDa UPF was probably due to its high concentration of hydrophobic amino acids (41.68 g/100 g), including Pro (6.11 g/100 g), Phe (11.03 g/100 g), Leu (10.23 g/100 g), and Ile (6.57 g/100 g) (Table 1).

Amino acid C-terminal hydrophobicity has the greatest influence on ACE-I inhibitory activity, and the higher the hydrophobicity the higher the inhibitory activity [15]. This coincides with a number of previous studies. Cheung et al. [26] found that dipeptides could have high ACE-I inhibitory activity if C-terminals were aromatic amino acids and proline, and N-terminals were aliphatic amino acid branches. Hellberg et al. [27] measured Cheung's peptides samples in the same laboratory, modeled the results with a QSAR, and found that dipeptides with positively-charged amino acids at the N-terminal and bulky hydrophobic amino acids at the C-terminal had stronger ACE-I inhibitory activity. Using Z

TABLE 1: Amino acid contents (g/100 g) of the <1 kDa ultrafiltered fraction from a *S. hispanica* protein hydrolysate.

Amino acid	$F < 1$ kDa
Asx	7.17
Glx	6.47
Ser	5.14
His	4.04
Gly	4.48
Thr	4.34
Arg	12.36
Ala	3.34
Pro	6.11
Tyr	2.46
Val	1.73
Met	2.06
Cys	4.24
Ile	6.57
Trp	0.61
Leu	10.23
Phe	11.03
Lys	7.61

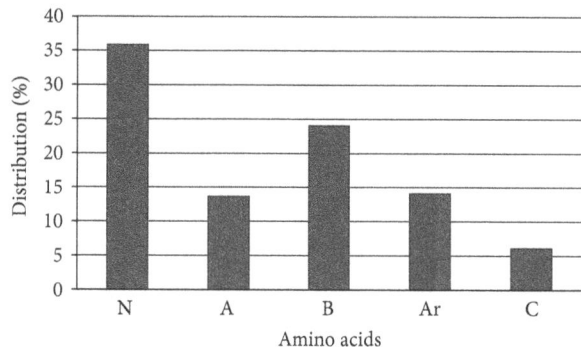

FIGURE 2: Amino acid distribution in the <1 kDa ultrafiltered fraction of a *S. hispanica* protein hydrolysate. N: neutral amino acids (including Gly, Ala, Ser, Thr, Val, Leu, and Ile). A: acid amino acids (including Asp and Glu). B: basic amino acids (including Lys, His, and Arg). Ar: aromatic amino acids (including Phe, Tyr, and Trp). C: cyclic amino acids (including Pro).

descriptors to investigate the quantitative structure-activity relationship of ACE-inhibitory dipeptides, Wu et al. [28] found that ACE-inhibitory activity was strongly affected by the three-dimensional chemical properties and hydrophobicity of C-terminal amino acids; that is, the higher the volume and the greater the hydrophobicity of the amino acids, the higher the ACE-I inhibitory activity. Therefore, some dipeptides with hydrophobic amino acids at the C-terminal, such as phenylalanine, tryptophan, and tyrosine, will have high ACE-I inhibitory activity. Wu et al. [28] also reported that strongly hydrophobic and small N-terminal amino acids, such as valine, leucine, and isoleucine, were more suitable for high-activity tripeptides. Low charge, large size, and weak hydrophobicity in the second amino acid from the N-terminal were more suitable high activity. Finally, for the C-terminal, high charge, large volume, and strongly hydrophobic residues (e.g., aromatic amino acids) were more suitable. In an analysis of ACE-I inhibitory peptides from milk sources, Pripp et al. [29] found that for peptides with ≤6 amino acids at the C-terminal, ACE-I inhibition was strongly affected by hydrophobicity, positive charge, and volume of amino acids adjacent to the C-terminal, whereas the N-terminal amino acid had no direct relationship. Hydrophobicity and C-terminal amino acid size are apparently the principal aspects affecting ACE-inhibitory activity. It stands to reason that hydrophobic amino acids, aromatic amino acids, and branched-chain amino acids are important components in high-activity peptides, and that proteins with high contents of these amino acid types (especially aromatic amino acids) have more potential to produce high activity ACE-I inhibitory peptides. Therefore, digestion of proteins to produce peptides with hydrophobic amino acids at the C-terminal will tend to increase ACE-I inhibitory activity in any derived hydrolysates [20].

Although the structure-activity relationship of ACE-I inhibitory peptides has not yet been established, these peptides show some common features. Studies of the structure-activity relationships in different ACE-I inhibitory peptides indicate that binding to ACE is strongly influenced by substrate C-terminal tripeptide sequence. Ondetti and Cushman [30] proposed a binding model for interactions between the substrate and active ACE site. C-terminal tripeptide residues may interact with the S1, S1', and S2' subsites at the active ACE site. ACE appears to prefer substrates or competitive inhibitors that contain hydrophobic amino acid residues at the three C-terminal positions. Captopril owes its potency and selectivity to chemical design guided by a hypothetical active site model based on the observed properties of ACE and on an analogy to the known active site of a related zinc-containing peptidase. ACE's zinc ion is appropriately located between S1 and S1', allowing it to participate in hydrolytic cleavage of the substrate peptide bond and resulting in release of a dipeptide product. Studies of the structure-activity relationship in ACE-I inhibitory peptides have shown that those with potent inhibitory activity have proline, phenylalanine, or tyrosine at the C-terminal, as well as hydrophobic amino acids in their sequence [31]. In light of this previous research, the ACE-I inhibitory activity observed here in the chia hydrolysate was probably due to amino acid composition (Figure 2).

3.4. Gel Filtration Chromatography of the <1 kDa Ultrafiltered Fraction. Of the UPFs, the <1 kDa fraction exhibited the highest ACE-I inhibitory activity and was selected for further fractionation. A molecular weight profile was generated of this UPF using gel filtration chromatography (Sephadex G-50 column). This profile was typical of a protein hydrolysate formed by a pool of peptides, with gradually decreasing molecular masses. Elution volumes between 406 and 518 mL included free amino acids and peptides with molecular masses ranging from 0.4 to 3.6 kDa. This range was fractionated into three fractions and ACE-I inhibitory activity determined for each. Fractions with elution volumes smaller than 406 mL and greater than 518 mL were not analyzed because they largely included peptides with high molecular weights, as well as free amino acids. ACE-I inhibitory activity (%) in the <1 kDa UPF ranged from 48.4% to 62.6% (Figure 3). The highest ACE-I inhibitory activity was observed in fraction F1 (62.6%; 427–455 mL elution volume). Its molecular mass was approximately 1.5–2.5 kDa, indicative of 7–12 amino acid residues. The IC_{50} value for F1 (3.97 μg/mL) was lower than those of gel filtration (Sephadex G-25) peptide fractions from tuna broth hydrolysate (210 to 25,260 μg/mL) [32] or from buckwheat *Fagopyrum esculentum* Moench (Sephadex C-25 = 25,715.1 μg/mL; Sephadex G-10 = 21,315.1 μg/mL). Other studies suggest that ACE inhibition by hydrolysates depends on the source species and hydrolysate purity level. For instance, production of a 1–3 kDa protein hydrolysate from Alaska pollock frame using an ultrafiltration membrane bioreactor system resulted in high ACE-I inhibition (IC_{50} = 110 μg/mL), but further purification using consecutive chromatographic methods in a SP-Sephadex C-25 column and HPLC in an octadecylsilane column resulted in a still stronger effect (IC_{50} = 14.7 μg/mL) [24]. In another example, an enzymatic hydrolysate from cuttlefish (*Sepia officinalis*) muscle protein was found to have high ACE-I inhibitory activity (87.1 ± 0.9% at 200 μg/mL). However, size exclusion chromatography with a Sephadex G-25 produced a fraction (P6) with yet higher ACE-I inhibition (IC_{50} = 11.6 μmol/L), which, when fractionated by reverse-phase (RP)-HPLC, was found to contain Ala-His-Ser-Tyr, Gly-Asp-Ala-Pro, Ala-Gly-Ser-Pro, and Asp-Phe-Gly [33].

In ACE-I inhibitory peptides from chia, the protein-rich fractions are not as potent as hypertension treatment drugs but hold promise as a safe, natural therapeutic agent without adverse side effects. The potential of chia protein-derived peptides as antihypertension agents depends on the ability of these peptides to reach their target site without suffering degradation and consequent inactivation by intestinal or plasma peptidases. Resistance to peptidase degradation is a probable prerequisite for any ACE inhibitory hydrolysates/peptides to exercise an antihypertensive effect after oral or intravenous administration. Proline-containing peptides are generally resistant to degradation by digestive enzymes [34], and tripeptides containing a Pro-Pro C-terminal are resistant to proline-specific peptidases [35]. However, peptide degradation or fragmentation results in smaller peptides and therefore in potentially more potent ACE-I inhibitory activity. Clearly, *in vivo* studies are needed to confirm the effect of these peptides since it is both difficult and unwise to extrapolate directly from *in vitro* to *in vivo* activity. The main challenges in doing this are determining bioavailability of ACE-I inhibitory peptides after oral administration and the fact that peptides may influence blood pressure by mechanisms other than ACE-I inhibition. To exert an antihypertensive effect after oral ingestion, ACE-I inhibitory peptides must reach the cardiovascular system in an active form, meaning that they need to remain active

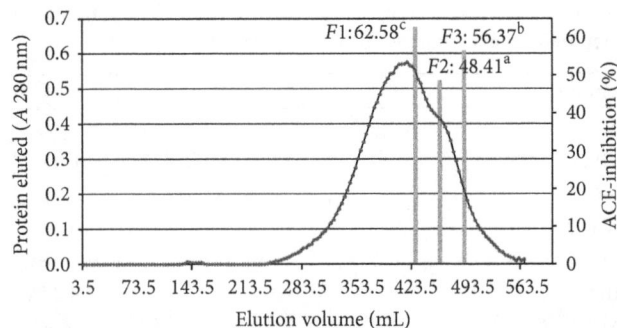

FIGURE 3: Elution profile of the <1 kDa ultrafiltration fraction of the *S. hispanica* protein hydrolysate purified in a Sephadex G-50 gel filtration column. [a–e] Different superscript letters indicate statistical difference ($P < 0.05$). Data are the mean of three replicates.

during digestion by human proteases and transport through the intestinal wall into the blood. Bioavailability has been studied for some ACE-I inhibitory peptides, and it is known that proline-containing peptides are generally resistant to degradation by digestive enzymes. Peptides can be absorbed intact through the intestine by paracellular and transcellular routes, although postabsorption bioactivity potency is inversely correlated to chain length [36].

All the chia derivatives studied here had IC_{50} values far higher than the 0.0013 μg/mL of Captopril, a synthetic ACE-I inhibitor [32]. Nonetheless, the chia purified peptides have biological potential, and the $F < 1$ kDa fraction had high ACE-I inhibitory activity, suggesting that ACE-I inhibitory peptides are rich in hydrophobic amino acids (aromatic or branched chains) and in proline. ACE-I inhibiting peptides from food sources have garnered increasing attention in recent years as promising natural biofunctional alternatives to synthetic drugs. Many of these peptides have been discovered in enzymatic hydrolysates of different food-source proteins and subsequently applied in the prevention of hypertension and initial treatment of mildly hypertensive individuals. The ongoing search for natural ACE inhibitors may eventually help to create safer and less costly alternatives to synthetic pharmaceutical treatments [25].

4. Conclusions

Chia proteins hydrolyzed with the Alcalase-Flavourzyme sequential enzyme system resulted in hydrolysates with ACE-I inhibitory activity. Ultrafiltration produced a very low molecular weight fraction (<1 kDa) which had the highest activity. The hydrolysates and ultrafiltered fractions are potential ingredients for development of functional foods. Enzymatic hydrolysis and ultrafiltration are promising bioprocesses for production of new bioactive food ingredients such as ACE inhibitory peptides purified from chia hydrolysate.

Conflict of Interests

The authors declare no direct financial relationship with trademarks mentioned in this paper.

Acknowledgments

This research was supported by the Consejo Nacional de Ciencia y Tecnología (CONACYT) de México and the Programa de Mejoramiento al Profesorado (PROMEP) de la Secretaría de Educación Pública (SEP-México).

References

[1] C. L. Jao, S. L. Huang, and K. C. Hsu, "Angiotensin I-converting enzyme inhibitory peptides: inhibition mode, bioavailability, and antihypertensive effects," *BioMedicine*, vol. 2, no. 4, pp. 130–136, 2012.

[2] S. C. Ko, N. Kang, E. A. Kim et al., "A novel angiotensin I-converting enzyme (ACE) inhibitory peptide from a marine Chlorella ellipsoidea and its antihypertensive effect in spontaneously hypertensive rats," *Process Biochemistry*, vol. 47, pp. 2005–2011, 2012.

[3] S. Nalinanon, S. Benjakul, H. Kishimura, and F. Shahidi, "Functionalities and antioxidant properties of protein hydrolysates from the muscle of ornate threadfin bream treated with pepsin from skipjack tuna," *Food Chemistry*, vol. 124, no. 4, pp. 1354–1362, 2011.

[4] S. H. Ferreira, D. C. Bartelt, and L. J. Greene, "Isolation of bradykinin-potentiating peptides from bothrops jararaca venom," *Biochemistry*, vol. 9, no. 13, pp. 2583–2593, 1970.

[5] H. Ni, L. Li, G. Liu, and S. Q. Hu, "Inhibition Mechanism and Model of an Angiotensin I-Converting Enzyme (ACE) Inhibitory hexapeptide from yeast (*Saccharomyces cerevisiae*)," *PLOS ONE*, vol. 7, no. 5, Article ID e37077, 2012.

[6] M. Kuba, C. Tana, S. Tawata, and M. Yasuda, "Production of angiotensin I-converting enzyme inhibitory peptides from soybean protein with Monascus purpureus acid proteinase," *Process Biochemistry*, vol. 40, no. 6, pp. 2191–2196, 2005.

[7] A. J. Mossi, R. L. Cansian, N. Paroul et al., "Morphological characterisation and agronomical parameters of different species of Salvia sp. (Lamiaceae)," *Brazilian Journal of Biology*, vol. 71, no. 1, pp. 121–129, 2011.

[8] J. A. Vázquez-Ovando, J. G. Rosado-Rubio, L. A. Chel-Guerrero, and D. A. Betancur-Ancona, "Dry processing of chia (*Salvia hispanica* L.) flour: chemical and characterization of fiber and protein," *CYTA Journal of Food*, vol. 8, no. 2, pp. 117–127, 2010.

[9] M. R. Segura-Campos, I. M. Salazar-Vega, L. A. Chel-Guerrero, and D. A. Betancur-Ancona, "Biological potential of chia (*Salvia hispanica* L.) protein hydrolysates and their incorporation into functional foods," *LWT- Food Science and Technology*, vol. 50, pp. 723–731, 2013.

[10] P. M. Nielsen, D. Petersen, and C. Dambmann, "Improved method for determining food protein degree of hydrolysis," *Journal of Food Science*, vol. 66, no. 5, pp. 642–646, 2001.

[11] M. Alaiz, J. L. Navarro, J. Giron, and E. Vioque, "Amino acid analysis of high-performance liquid chromatography after derivatization with diethyl ethoxymethylenemalonate," *Journal of Chromatography*, vol. 591, no. 1-2, pp. 181–186, 1992.

[12] M. J. Cho, N. Unklesbay, F. H. Hsieh, and A. D. Clarke, "Hydrophobicity of bitter peptides from soy protein hydrolysates," *Journal of Agricultural and Food Chemistry*, vol. 52, no. 19, pp. 5895–5901, 2004.

[13] M. Hayakari, Y. Kondo, and H. Izumi, "A rapid and simple spectrophotometric assay of angiotensin-converting enzyme," *Analytical Biochemistry*, vol. 84, no. 2, pp. 361–369, 1978.

[14] D. Montgomery, *Diseño Y Análisis De Experimentos*, Limusa-Wiley, Mexico City, Mexico, 2004.

[15] R. He, H. Ma, W. Zhao et al., "Modeling the QSAR of ACE-Inhibitory peptides with ANN and its applied illustration," *International Journal of Peptides*, vol. 2012, Article ID 620609, 9 pages, 2012.

[16] J. Ruiz-Ruiz, G. Dávila-Ortíz, L. Chel-Guerrero, and D. Betancur-Ancona, "Angiotensin I-converting enzyme inhibitory and antioxidant peptide fractions from hard-to-cook bean enzymatic hydrolysates," *Journal of Food Biochemistry*, vol. 37, no. 1, pp. 26–35, 2013.

[17] J. Pedroche, M. M. Yust, J. Girón-Calle, M. Alaiz, F. Millán, and J. Vioque, "Utilisation of chickpea protein isolates for production of peptides with angiotensin I-converting enzyme (ACE)-inhibitory activity," *Journal of the Science of Food and Agriculture*, vol. 82, no. 9, pp. 960–965, 2002.

[18] J. Vioque, R. Sánchez-Vioque, A. Clemente, J. Pedroche, J. Bautista, and F. Millan, "Production and characterization of an extensive rapeseed protein hydrolysate," *Journal of the American Oil Chemists' Society*, vol. 76, no. 7, pp. 819–823, 1999.

[19] H. Korhonen, A. Pihlanto-Leppälä, P. Rantamäki, and T. Tupasela, "Impact of processing on bioactive proteins and peptides," *Trends in Food Science and Technology*, vol. 9, no. 8-9, pp. 307–319, 1998.

[20] J. Y. Je, J. Y. Park, W. K. Jung, P. J. Park, and S. K. Kim, "Isolation of angiotensin I converting enzyme (ACE) inhibitor from fermented oyster sauce, *Crassostrea gigas*," *Food Chemistry*, vol. 90, no. 4, pp. 809–814, 2005.

[21] S. R. Kim and H. G. Byun, "The novel angiotensin I-converting enzyme inhibitory peptides from rainbow trout muscle hydrolysate," *Fisheries and Aquatic Sciences*, vol. 15, no. 3, pp. 183–190, 2012.

[22] X. Y. Mao, J. R. Ni, W. L. Sun, P. P. Hao, and L. Fan, "Value-added utilization of yak milk casein for the production of angiotensin-I-converting enzyme inhibitory peptides," *Food Chemistry*, vol. 103, no. 4, pp. 1282–1287, 2007.

[23] E. G. Tovar-Pérez, I. Guerrero-Legarreta, A. Farrés-González, and J. Soriano-Santos, "Angiotensin I-converting enzyme-inhibitory peptide fractions from albumin 1 and globulin as obtained of amaranth grain," *Food Chemistry*, vol. 116, no. 2, pp. 437–444, 2009.

[24] J. Y. Je, P. J. Park, J. Y. Kwon, and S. K. Kim, "A novel angiotensin I converting enzyme inhibitory peptide from Alaska pollack (*Theragra chalcogramma*) frame protein hydrolysate," *Journal of Agricultural and Food Chemistry*, vol. 52, no. 26, pp. 7842–7845, 2004.

[25] L. Liu, B. Lu, L. Gong, L. Liu, X. Wu, and Y. Zhang, "Studies on bioactive peptides from Chinese soft-shelled turtle (*Pelodiscus sinensis*) with functionalities of ACE inhibition and antioxidation," *African Journal of Biotechnology*, vol. 11, no. 25, pp. 6723–6729, 2012.

[26] H. S. Cheung, F. L. Wang, M. A. Ondetti, E. F. Sabo, and D. W. Cushman, "Binding of peptide substrates and inhibitors of angiotensin-converting enzyme. Importance of the COOH-terminal dipeptide sequence," *Journal of Biological Chemistry*, vol. 255, no. 2, pp. 401–407, 1980.

[27] S. Hellberg, L. Eriksson, and J. Jonsson, "Minimum analogue peptide sets (MAPS) for quantitative structure-activity relationships," *International Journal of Peptide and Protein Research*, vol. 37, no. 5, pp. 414–424, 1991.

[28] J. Wu, R. E. Aluko, and S. Nakai, "Structural requirements of angiotensin I-converting enzyme inhibitory peptides: quantitative structure-activity relationship study of Di- and tripeptides," *Journal of Agricultural and Food Chemistry*, vol. 54, no. 3, pp. 732–738, 2006.

[29] A. H. Pripp, T. Isaksson, L. Stepaniak, T. Sørhaug, and Y. Ardö, "Quantitative structure activity relationship modelling of peptides and proteins as a tool in food science," *Trends in Food Science and Technology*, vol. 16, no. 11, pp. 484–494, 2005.

[30] M. A. Ondetti and D. W. Cushman, "Enzymes of the renin-angiotensin system and their inhibitors," *Annual Review of Biochemistry*, vol. 51, pp. 283–308, 1982.

[31] H. G. Byun and S. K. Kim, "Structure and activity of angiotensin I converting enzyme inhibitory peptides derived from alaskan pollack skin," *Journal of Biochemistry and Molecular Biology*, vol. 35, no. 2, pp. 239–243, 2002.

[32] J. S. Hwang and W. C. Ko, "Angiotensin I-converting enzyme inhibitory activity of protein hydrolysates from tuna broth," *Journal of Food and Drug Analysis*, vol. 12, no. 3, pp. 232–237, 2004.

[33] R. Balti, N. Nedjar-Arroume, E. Y. Adjé, D. Guillochon, and M. Nasri, "Analysis of novel angiotensin I-converting enzyme inhibitory peptides from enzymatic hydrolysates of cuttlefish (*Sepia officinalis*) muscle proteins," *Journal of Agricultural and Food Chemistry*, vol. 58, no. 6, pp. 3840–3846, 2010.

[34] G. N. Kim, H. D. Jang, and C. I. Kim, "Antioxidant capacity of caseinophosphopeptides prepared from sodium caseinate using Alcalase," *Food Chemistry*, vol. 104, no. 4, pp. 1359–1365, 2007.

[35] W. L. Mock, P. C. Green, and K. D. Boyer, "Specificity and pH dependence for acylproline cleavage by prolidase," *Journal of Biological Chemistry*, vol. 265, no. 32, pp. 19600–19605, 1990.

[36] S. Sharma, R. Singh, and S. Rana, "Boactive peptides: a review," *International Journal Bioautomation*, vol. 15, no. 4, pp. 223–250, 2011.

Effect of Cholesterol Removal Processing Using β-Cyclodextrin on Main Components of Milk

A. M. Maskooki,[1] **S. H. R. Beheshti,**[2] **S. Valibeigi,**[1] **and J. Feizi**[2]

[1] *Food Processing Department, Research Institute of Food Science and Technology (RIFST), Km. 12 Asian Road, Mashhad, Iran*
[2] *TESTA Quality Control Laboratory, North-East Food Technology and Biotechnology Zone, Km. 12 Asian Road, Mashhad, Iran*

Correspondence should be addressed to A. M. Maskooki; a.maskooki@rifst.ac.ir

Academic Editor: Philip Cox

Various concentrations (0%, 0.5%, 1% and 1.5%) of β-CD were mixed with different fat contents (1%, 2.5% and 3%) of raw (unhomogenized) and homogenized milk at two mixing temperatures of 8 and 20°C. The cholesterol residue, fat, protein, lactose, solid nonfat (SNF), density, and ash content of milk were measured for each treatment. The results statistically analysed and showed that the cholesterol content of milk remarkably decreased as the β-CD was increased particularly in homogenized milk at 20°C. However, the reduction rate of cholesterol was decreased when extra β-CD was added due to its intermolecular reactions. The maximum cholesterol reduction was achieved at the level of 1% β-CD. The fat content, SNF, protein, lactose, and density content were decreased with increasing β-CD whereas it did not affect ash content.

1. Introduction

According to the World Health Organization, 2010 report, cardiovascular diseases are the first among the top 10 causes of death. An estimated 16.7 million people or 29.2% of total global deaths are due to the various forms of cardiovascular disease (CVD) [1, 2]. The direct relation between high blood cholesterol and CVD has been proved. Cholesterol is a typical animal sterol; for example, its content in milk fat is 95%–98% [3]. The main source of cholesterol in food comes from animal origin, such as meat, milk, and eggs. The other major source of cholesterol is produced by the liver in the body [4]. Milk and dairy products contain relatively high level cholesterol that can elevate the blood cholesterol [5]. First attempts on reducing cholesterol in food go back to early twentieth century. Denis and Minot [6] determined the cholesterol content of animal and human milks and they suggested the relation between blood plasma cholesterol and food intake. Although so many papers were published about the relation of food intake and increasing of blood plasma cholesterol, no serious attempt took place on reducing cholesterol in food until 1960. Since the 1960s, large number of physicochemical methods was recommended to

reduce cholesterol in food as well as blood cholesterol [7]. Cholesterol could be removed in an efficient manner from milk fat up to 90% using supercritical CO_2 technology [8]. Bobby and Joseph Jr. [9] developed and patented a process for the production of cholesterol-free milk based on extraction of cholesterol from the milk fat globule membranes using an organic polar solvent and without substantial loss of solid milk fat. Solvent-free and low cholesterol products were recovered from the reseparated and washed cream fractions. Cholesterol removal from dairy product using saponin and diatomaceous earth also was patented by Richardson et al. [10]. They suggested that the process could be useful particularly for raw and pasteurized milks, cream, and butter. In spite of effectiveness of the mentioned methods, some complications exist because of the organic solvents and saponin residues in food and safety problems for human body. The worries about solvent residue and harmful saponin consumption for human health motivated the investigator to find nontoxic and effective substances instead of unsafe materials. In the last years, several studies describing the use of beta cyclodextrin (β-CD) and food applications have been published [11]. It has been proved that the β-CD molecule can be used as nontoxic and nondigestible molecule to remove cholesterol effectively from milk and animal products for improving their nutritional characteristics [12]. β-CD is a cyclic oligosaccharide consisting of seven glucopyranose molecules that are linked together with α 1–4 bonds. β-CD is frequently used as building blocks. More than 20 substituents have been linked to β-CD by a regioselective manner [13]. The most widespread use of β-CD is in cholesterol removal from animal products, such as eggs and dairy products. β-CD treated materials show 80% removal of cholesterol in various food containing cholesterol [14]. Chiu et al. [15] removed more than 85% cholesterol content of egg yolk by immobilization of β-CD on chitosan beads. The process for preparing low cholesterol dairy product using β-CD has been patented by Graille et al. [16]. Lee et al. [17] suggested that the 94.3% of cholesterol content of homogenized milk with 3.6% fat could be removed by addition of 1.5% β-CD at mixing temperature of 10°C for 10 min of mixing time. The method for removing cholesterol from milk and cream was developed and patented by Kwak et al. [18]. Optimization of cholesterol removal in cream using β-CD by response surface method was investigated by Ahn and Kwak [19]. Kim and coworkers [20–22] developed a cross-linked β-CD and epichlorohydrin to recycle the cholesterol removal process. They concluded that cross-linked β-CD would result in almost 100% effective recycling efficiency. Further, this method was applied and optimized in order to remove cholesterol from milk and dairy product and recover the cross-linked β-CD by Kwak et al. [23]. Recently, Dias et al. [24] investigated butter cholesterol removal using different combinatorial methods with β-CD. Although many investigations were carried out to demonstrate the feasibility of β-CD as excellent substance for removing cholesterol from food including milk, few investigations have been reported on the effect of β-CD on physicochemical properties of milk with various fat content. Ha et al. [25] measured the amount of nutrients (lactose, short-chain FFA, FAA, and water-soluble vitamins) that were entrapped during cholesterol removal from cream by cross-linked β-CD. They concluded that the amount of entrapped nutrients was negligible during cholesterol removal from cream by cross-linked β-CD. In spite of this study, the cholesterol removal process by β-CD may affect main components and physicochemical properties of milk. The aims of this study are evaluation of the feasibility of β-CD in cholesterol removal from raw and homogenized milks with various fat contents and the effect of cholesterol reduction process on main components of milk.

2. Materials and Methods

2.1. Raw Materials and Reagents. Commercial raw (unhomogenized) and homogenized milks with 1%, 2.5%, and 3% fat content were purchased from a Pegah Dairy Co. Industry as needed (located in Mashhad Iran).

β-CD, cholesterol, and 5α-cholestane were purchased from Sigma Chemical Co. (St. Louis, MO, USA). All reagents and solvents were gas-chromatographic grade.

2.2. Cholesterol Removal Processing. Overall, 100 mL of raw or homogenized milk with various fat contents (1%, 2.5%, and 3%) were separately placed in 500 mL beakers and different concentrations of β-CD (0%, 0.5%, 1%, and 1.5%) were added to each of samples. The mixture was stirred vigorously by VELP stirrer 700 rpm at two temperature conditions of 8 and 20°C for 10 and 15 min. The mixtures were centrifuged with Eppendorf centrifuge model 5810 at 112–448 ×g for 15 min. The supernatant containing cholesterol reduced milk was decanted and removed for further measurements. The protein, lactose, fat, solid nonfat (SNF), ash, and density were measured using MilkoScan device MCC model (Milkotronic CO., Bulgaria).

2.3. Modified Cholesterol Analysis. The determination of cholesterol was carried out using HPLC-SYKAM, S1122 (Germany) with capillary column Symmetry C_{18}, length of 25 cm, diameter 4.6, vol. 5 μm, and UV detector 205 nm. The temperature of column was adjusted to 40°C using column oven S4011. The preparation of solutions and reagents method was described by Klatt [26]. We have used this method with some modifications. Accurately 30 mL milk with constant fat content was poured in 250 mL Erlenmeyer flask. Also, 40 mL of 95% ethanol and 8 mL of 50% KOH (W/W) were added to flask. Then, the flask was placed on a magnetic heater-stirrer, connected to condenser, and stirred vigorously for 70 min to ensure the saponification reaction was completed. Once more, 60 mL of 95% ethanol were added and stirring procedure was continued without heating. The condenser was disconnected and the flask was removed after 15 min. The saponified solution flask was topped with stopper and rested to be cooled at room temperature for 24 h. The extraction procedure was carried out by adding 100 mL toluene to saponified sample while stirring for 30 min. The solution was poured into 500 mL decantation funnel without rinsing and then 110 mL of KOH were added. The decantation funnel was shaked up vigorously for 10 sec. We let layers be separated and then we discarded the aqueous lower layer. The extraction stage was repeated by adding 40 mL of 0.5 N KOH solutions. The aqueous solution was discarded and the toluene was washed with 40 mL H_2O for three times. The toluene residue was evaporated by rotary vacuum evaporator. The remained cholesterol was dissolved in 5 mL methanol HPLC grade and was injected to HPLC device.

2.4. Determination of Main Components of Milk. The determination of main components and some physicochemical properties of milk was carried out by ultrasonic milk analysis device. Table 1 shows the capability and accuracy of ultrasonic milk analysis.

2.5. Statistical Procedure. Each experiment was repeated at least three times. Multifactor Randomized Complete Block Design (RCBD) was used to analyse the acquired raw data statistically. The obtained means were evaluated by Duncan Multiple Range Test (D.M.R.T). Statistical analysis was performed using Sigma Stat 3.1, MiniTab15, and Microsoft Excel softwares.

TABLE 1: Capability and accuracy of milk by ultrasonic milk analysis for determination of physicochemical properties of milk.

Parameter	Measuring range	Accuracy
Fat	0,01%–45%	±0,06%
Solids nonfat (SNF)	3%–40%	±0,15%
Density	1000–1160 kg/m	±0,3
Protein	2%–15%	±0,15%
Lactose	0,01%–20%	±0,2%
Added water	0%–70%	±3
Milk sample temperature	5°C–40°C	±1%
Freezing point	(−0,4°C)–(−0,7°C)	±0,005%
Salts	0,4%–4%	±0,05%
pH	0–14	±0,05%
Conductivity	2–14 (mS·cm^{-1})	±0,05%

3. Results and Discussions

3.1. Effect of β-CD Concentration. The cholesterol content of both raw and homogenized milks was significantly ($P \leq 0.05$) reduced with increasing the β-CD. Figure 1 shows the cholesterol residues and percentage reduction of cholesterol in treated raw and homogenized milks. The reduction effect is more obvious in homogenized milk than in the raw one. However, there were no remarkable differences between the reduction rates of raw and homogenized milks at 0.5% and 1% β-CD concentrations ($P \geq 0.05$). The absorption of cholesterol was reduced when using 1.5% β-CD particularly in raw milk. The obtained results in this study showed that cholesterol is effectively removed from milk and dairy product by β-CD which is in agreement with other investigators [16, 17, 27, 28]. The lipid complexes in homogenized milk are significantly smaller than unhomogenized (raw), and therefore the possibility of cholesterol molecule inclusion by internal cavity of β-CD is higher in homogenized milk than in raw one. The cholesterol removal trend was slightly decreased when 1.5% β-CD was applied particularly in raw milk. This phenomenon can be due to intermolecular reaction effects of extra β-CD. Kim et al. (1999) have suggested that an excess of β-CD could compete with itself to bind to cholesterol molecules, and consequently cholesterol adsorption is decreased. Therefore, it seems that the concentration of 1% β-CD may be sufficiently effective to remove greater than 90% of cholesterol from homogenized milk as shown in Figure 1.

3.2. Effect of Operation Temperature. The β-CD was added and mixed with milk samples at two temperatures 8 and 20°C for 15 min. The adsorption of cholesterol was significantly increased when β-CD was added and mixed at 20°C. Figure 2 shows the effect of mixing temperature on reduction of cholesterol in treated milks. The differences were more obvious in higher concentration of β-CD. However, the diversity of opinions exists among investigators on this result. Lee et al. [17] reported that no difference was found in cholesterol removal at mixing temperature of 4, 10, 15, or 20°C, while Oakenfull and Sihdu [29] disagreed with this result. They indicated that the removal of cholesterol

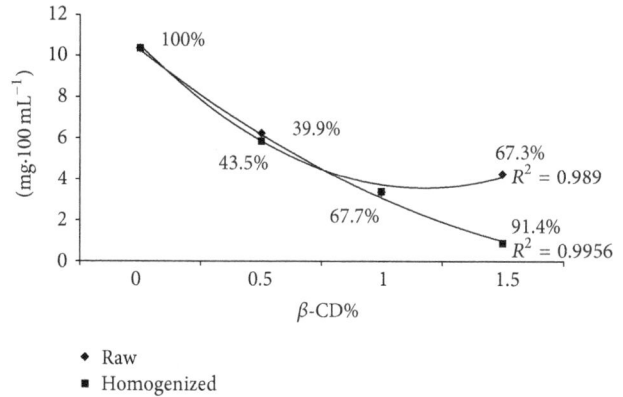

FIGURE 1: Cholesterol reduction in both raw and homogenized milks by various amount of β-CD.

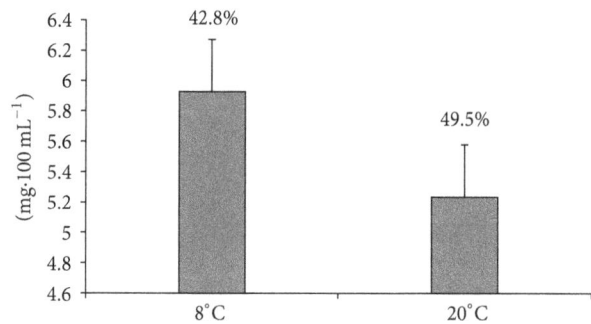

FIGURE 2: The cholesterol reduction in milk for both mixing temperature conditions.

using β-CD significantly is influenced by temperature. They reported that 77%, 63%, and 62% cholesterol were removed when milk was treated with 1.0% β-CD at 4, 8, and 40°C, respectively, during 10 min of mixing. Yen and Tsui [30], and Kim et al. [22] also confirmed that the removal of cholesterol with β-CD from lard which was stirred at 50°C is greater in comparison with the temperature condition of 27 or 40°C. Actually, our overall results showed that the higher cholesterol removed when mixing temperature was increased as demonstrated in Figure 2. This result is in agreement with investigations concerning the effect of mixing temperature on the reduction of cholesterol by β-CD that were carried out by other investigators [22, 29, 30]. Naturally, the aggregations of milk lipoproteins at the mixing temperature of 20°C are lower than 8°C due to reaching of lipids to their melting points, and consequently there is more chance for the entrapment of molecules in β-CD cavities. Therefore, we can expect more cholesterol molecules to be trapped in β-CD cavities at higher temperatures.

Figure 3 shows the amount of cholesterol reduction in both raw and homogenized milks in various concentrations of β-CD at both temperature conditions of 8 and 20°C. We observed that the increase of cholesterol removal depends on amount of β-CD in higher temperatures. There were no remarkable statistical differences between cholesterol reduction in both raw and homogenized milk samples

TABLE 2: Effect of various concentrations of β-CD on main constituents and density of raw and homogenized milk.

Milk	β-CD (%)	Fat (g/100 g) SEM	Protein (g/100 g) SEM	SNF (g/100 g) SEM	Lactose (g/100 g) SEM	Density (g/100 mL) SEM	Ash (g/100 g) SEM
Raw	0	3.133 ± 0,002	3.1633 ± 0,011	8.633 ± 0.019	4.751 ± 0.08	1.0313 ± 0.003	0.714 ± 0.008
	0.5	3.111 ± 0,002	3.1025 ± 0,015	8.564 ± 0.03	4.733 ± 0.06	1.0311 ± 0.003	0.711 ± 0.008
	1	3.0908 ± 0,02	3.075 ± 0,016	8.535 ± 0.03	4.724 ± 0.05	1.0309 ± 0.003	0.711 ± 0.011
	1.5	3.0663 ± 0,003	2.9696 ± 0,014	8.424 ± 0.04	4.711 ± 0.02	1.0306 ± 0.004	0.713 ± 0.013
Homogenized	0	3.131 ± 0,002	3.1633 ± 0,011	8.633 ± 0.019	4.751 ± 0.08	1.0313 ± 0.003	0.714 ± 0.008
	0.5	2.993 ± 0,001	3.0575 ± 0,016	8.436 ± 0.04	4.662 ± 0.04	1.0285 ± 0.003	0.709 ± 0.008
	1	2.882 ± 0,003	3.0271 ± 0,015	8.406 ± 0.04	4.643 ± 0.04	1.0283 ± 0.007	0.712 ± 0.011
	1.5	2.805 ± 0,006	2.9475 ± 0,014	8.319 ± 0.05	4.642 ± 0.08	1.0280 ± 0.003	0.713 ± 0.013

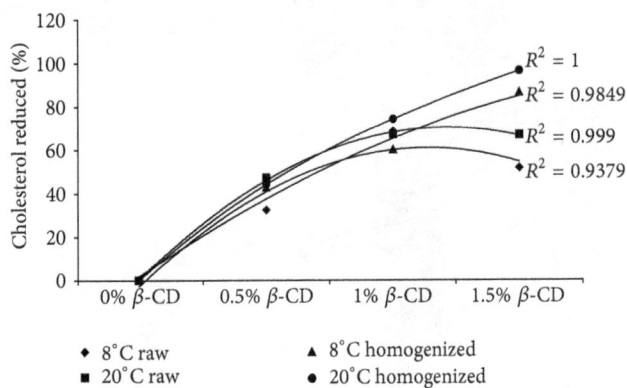

FIGURE 3: The cholesterol reduction in both raw and homogenized milks using various concentrations of β-CD for both temperature conditions.

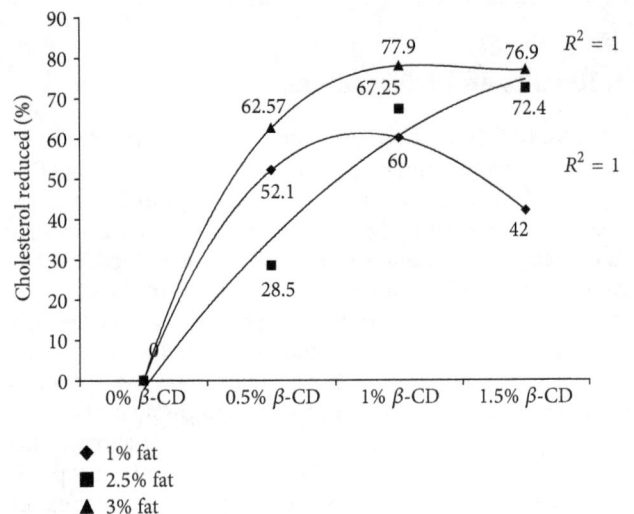

FIGURE 4: Cholesterol reduction (percentage) by various concentrations of β-CD in milk with different fat contents.

at two mixing temperatures of 8 and 20°C treated with low concentrations of 0.5% and 1% of β-CD while the differences were significant when we applied higher concentration of 1.5% β-CD as shown in Figure 4. Kim et al. [22] confirmed this result by investigation on cholesterol reduction of lard using high concentration of β-CD.

3.3. Interaction Effects of Fat Content and Concentration of β-CD.

More concentration of β-CD was needed with increasing the fat content in milk. The percentage of cholesterol, which was reduced by various concentrations of β-CD in different milk fat contents, is shown in Figure 4. The highest cholesterol reduction was achieved by 1% β-CD followed by 1.5% and 0.5% in 3% fat content of milk. The entrapped cholesterol molecules in β-CD cavities were reduced when we applied excess β-CD for milk samples with low fat content. The percentage of cholesterol reduction was only 42% when we applied 1.5% β-CD, while we achieved 52.1% and 62% with 0.5% and 1% β-CD, respectively. The best results were achieved when we applied 1% β-CD in various fat contents. The cholesterol reductions by 1% β-CD were 60%, 67.25%, and 77.9% for 1%, 2.5% and 3% of milk fat, respectively. The molecules of β-CD are linked together, therefore, the ability

of cholesterol absorption is reduced. Dias et al. [24], Ahn and Kwak [19], and Kim et al. [22] have confirmed our results.

3.4. Effect of Cholesterol Reduction by β-CD on Physicochemical Properties of Milk

3.4.1. Effect on Fat Content.
The cholesterol removal processing by β-CD on total fat content of milk was investigated as shown in Table 2. Obviously, the fat content of milk was reduced when cholesterol was removed by β-CD because cholesterol molecules are a part of milk lipids. This reduction of fat in homogenized milk samples was more obvious than in raw milk. The fat content of homogenized milk was significantly decreased with increasing the amount of β-CD due to smaller size of fat globules in homogenized milk compared with raw milk. The trend of fat decrease in raw milk samples was slower than homogenized with increasing the concentration of β-CD. Aggregation of fat molecules in raw milk inhibits the inclusion of cholesterol in β-CD molecules. Ha et al. [25, 31] suggested that the free fatty acids in food may be reduced because they are trapped in β-CD molecule

cavities. Kwak et al. [32] reported that cheddar cheese fat content made by milk treated by β-CD was lower than control milk samples. The cholesterol reduction processing such as stirring, separation, and centrifugation during separation of "β-CD + cholesterol" may affect fat reduction particularly in homogenized milk.

3.4.2. Effect on Milk Protein. The milk protein content was decreased with increasing the cholesterol absorption by β-CD. The amount of protein was significantly decreased in high concentrations of β-CD particularly in homogenized, milk, and there is a direct relation between increasing the concentration of β-CD and decreasing the protein content of milk as shown in Table 2. Kwak et al. [32] and Kim et al. [21] have already reported the protein reduction in milk after removal of cholesterol. Suzuki [33] suggested that the bitter taste due to protein hydrolyzate can be removed by 10% β-CD. The protein molecules and protein-lipid networks are smaller in homogenized milk than in raw milk. Ha et al. [25] confirmed that the protein content of cream was reduced during reduction of cholesterol by 4%–10% β-CD due to entrapment of amino acids in β-CD cavities. On the other hand, the outer surface of β-CD macromolecules has affinity to absorb the negative charges that cover the protein surfaces. Consequently, parts of protein together with "β-CD + cholesterol" leave the environment during separation of β-CD. The cholesterol reduction processing such as mixing, separation, and centrifugation may affect decrease of milk protein.

3.4.3. Effect on Nonfat Solid Content. The nonfat solid content of both raw and homogenized milks was slightly decreased after cholesterol reduction processing. Although cholesterol molecules are lipids and are independent to SNF, but presence of β-CD in milk may influence constituents such as proteins, ions, and lactose. Moreover, cholesterol removal operations may have effect on the reduction of nonfat solid content as shown in Table 2. The SNF was reduced with increasing the amounts of β-CD, and it was proved that some constituents of milk may entrapped in the cavity of β-CD and leave the environment after separation of β-CD. Ha et al. [31] confirmed the decreasing of some nutritional materials during cholesterol removal processing of cream using β-CD.

3.4.4. Effect on Lactose. The same effect was observed on lactose. The lactose content in both raw and homogenized milks was decreased after cholesterol removal processing. This phenomenon was more obvious when the amount of β-CD increased particularly in homogenized milk as illustrated in Table 2. Ha et al. [25, 31] reported the effect of cholesterol removal using β-CD on lactose previously, and they noted that the reduction of lactose is very low and negligible. However, we guess that the decrease of lactose is due to cholesterol reduction processing and the ability to bind lactose to β-CD and subsequently exit the milk suspension in association with "β-CD + cholesterol" complex.

3.4.5. Effect on Density. The density of milk strongly depends on existing of material in milk and is generally decreased when the main constituents of milk are reduced. The density of milk was significantly decreased after removing the cholesterol by β-CD because of losing some components of milk after cholesterol removal treatment. Decrease of milk density in homogenized milk is more than raw milk as illustrated in Table 2. The density of homogenized milk samples was significantly changed from 1.0313 (control) to 1.028 (treated samples) when 1.5% β-CD was applied to remove the cholesterol of milk. We could not find any published report about this phenomenon.

3.4.6. Effect on Ash Content. The effect of cholesterol minimization treating by β-CD had no remarkable effect on ash content of treated milk samples. Generally, the minerals of milk are sodium, potassium, calcium, and magnesium. These minerals could not be affected by β-CD as well as cholesterol removal operation and there was no remarkable effect on ash content of milk.

4. Conclusions

This study depicts the ability of various concentrations of β-CD to remove cholesterol from both raw and homogenized fresh milks and their effects on main constituents of milk. Many investigations have been carried out on cholesterol removal using β-CD. Undoubtedly, the ability of β-CD to exclusively remove cholesterol from milk has been proved and we confirmed these results. However, a few studies reported the effects of cholesterol reduction on nutritional compounds such as protein, lactose, and fat contents of milk. Although the β-CD molecules are edible and nontoxic and as a results they can be used safely as cholesterol removal agent from milk and dairy products, the effects of cholesterol removal processing are very important. This operations decreased the fat, density, SNF, and lactose content of milk. This phenomenon may be due to cohesiveness of these compounds with β-CD or existence some excess operations such as filtration and centrifugation during separation of "β-CD + cholesterol." Further investigation needs to demonstrate the quality of milk after cholesterol removal processing and to develop prevention methods of leaving nutritional materials of milk in association with β-CD.

Highlights

(i) The ability of β-CD to remove cholesterol from milk has been proved.

(ii) The β-CD molecules are safe and can be used as cholesterol removal agent.

(iii) The effects of cholesterol removal processing are very important.

(iv) The cholesterol removal operations can be decreased some main components of milk.

Acknowledgments

The authors acknowledge the Iran Research Institute (RIFST) and Iran National Science Foundation (INSF) for providing the financial support. The authors would like to thank the manager of TESTA quality control laboratory for providing the laboratory facilities. Their special thanks are dedicated to Dr. Mohsen Fathi Najafi (Vaccine and Serum Research Institute of Razi) for technical and scientific advices and Dr. Rasoul Kadkhodaee assistant professor and director of Research Institute of Food Science and Technology (RIFST). Authors also thank Mr. Mohammad Asadi and Mrs. Samira Yeganehparast experts of TESTA Co. for laboratory and technical support.

References

[1] WHO, "Cardiovascular disease: prevention and control," *Worldwide Missions Statistics*, 2010, http://www.who.org/.

[2] World Health Organization (WHO) World Health Statistics, "Global health indicator tables and footnotes part 2," 2010, http://www.who.int/whosis/whostat/2010/en/.

[3] M. Careri, L. Elviri, and A. Mangia, "Liquid chromatography—UV determination and liquid chromatography—atmospheric pressure chemical ionization mass spectrometric characterization of sitosterol and stigmasterol in soybean oil," *Journal of Chromatography A*, vol. 935, no. 1-2, pp. 249–257, 2001.

[4] S. Snider and A. Wehrman, "Cholesterol, Food and Nutrition Specialist," 1997, http://ag.udel.edu/extension/fnutri/pdf/Nutrition/fnf-18.pdf.

[5] B. Nataf, O. Milckelson, A. L. Keys, and W. E. Peterson, "The cholesterol content of cows' milk," *Journal of Nutrition*, vol. 3, pp. 13–18, 2009.

[6] W. Denis and A. S. Minot, "Constituents of milk non protein nitrogen," *Journal of Biological Chemistry*, vol. 37, pp. 353–366, 1919.

[7] M. I. Gurr, "Dietary lipids and coronary disease: old evidence, new perspectives and progress," *Lipid Research*, vol. 31, pp. 195–243, 1992.

[8] R. Bradely and L. Jir, "Removal of cholesterol from milk fat using supercritical carbon dioxide," *Journal Dairy Science*, vol. 72, pp. 2834–2840, 1989.

[9] R. J. Bobby and A. C. Joseph Jr., "Production of low cholesterol milk by organic solvent extraction," US patent. 4,997,668, 1991.

[10] T. Richardson, D. Calif, R. Jimenez, and F. Champaign, "Process to remove cholesterol from dairy products," US Patent No: 5,326,579, 1994.

[11] I. C. Munro, P. M. Newberne, V. R. Young, and A. Bär, "Safety assessment of γ-cyclodextrin," *Regulatory Toxicology and Pharmacology*, vol. 39, pp. S3–S13, 2004.

[12] G. Astray, C. Gonzalez-Barreiro, J. C. Mejuto, R. Rial-Otero, and J. Simal-Gándara, "A review on the use of cyclodextrins in foods," *Food Hydrocolloids*, vol. 23, no. 7, pp. 1631–1640, 2009.

[13] E. M. M. Del Valle, "Cyclodextrins and their uses: a review," *Process Biochemistry*, vol. 39, no. 9, pp. 1033–1046, 2004.

[14] A. R. Hedges, "Industrial applications of cyclodextrins," *Chemical Reviews*, vol. 98, no. 5, pp. 2035–2044, 1998.

[15] S.-H. Chiu, T.-W. Chung, R. Giridhar, and W.-T. Wu, "Immobilization of β-cyclodextrin in chitosan beads for separation of cholesterol from egg yolk," *Food Research International*, vol. 37, no. 3, pp. 217–223, 2004.

[16] J. Graille, D. Pioch, M. Serpelloni, and L. Mentink, "Process for preparing dairy products with a low content of sterols, particularly of cholesterol," US. Patent. 5,264,226, 1993.

[17] D. K. Lee, J. Ahn, and H. S. Kwak, "Cholesterol removal from homogenized milk with β-cyclodextrin," *Journal of Dairy Science*, vol. 82, no. 11, pp. 2327–2330, 1999.

[18] H. S. Kwak, S. H. Kim, J. H. Kim, H. J. Choi, and J. Kang, "Immobilized β-cyclodextrin as a simple and recyclable method for cholesterol removal in milk," *Archives of Pharmacal Research*, vol. 27, no. 8, pp. 873–877, 2004.

[19] J. Ahn and H. S. Kwak, "Optimizing cholesterol removal in cream using β-cyclodextrin and response surface methodology," *Journal of Food Science*, vol. 64, no. 4, pp. 629–632, 1999.

[20] S. H. Kim, J. Ahn, and H. S. Kwak, "Crosslinking of beta-cyclodextrin on cholesterol removal from milk," *Archives of Pharmacal Research*, vol. 27, no. 11, pp. 1183–1187, 2004.

[21] S.-H. Kim, E.-M. Han, J. Ahn, and H.-S. Kwak, "Effect of crosslinked β-cyclodextrin on quality of cholesterol-reduced cream cheese," *Asian-Australasian Journal of Animal Sciences*, vol. 18, no. 4, pp. 584–589, 2005.

[22] S. H. Kim, H. Y. Kim, and H. S. Kwak, "Cholesterol removal from lard with cross-linked β-cyclodextrin," *Asian-Australasian Journal of Animal Sciences*, vol. 20, no. 9, pp. 1468–1472, 2007.

[23] H. S. Kwak, S. H. Kim, and J. Kang, "Methods for cross linking betacyclodextrin for the cholesterol trapping and regeneration there of," US Patent No: 2007/0093447, 2007.

[24] H. M. A. M. Dias, F. Berbicz, F. Pedrochi, M. L. Baesso, and G. Matioli, "Butter cholesterol removal using different complexation methods with beta-cyclodextrin, and the contribution of photoacoustic spectroscopy to the evaluation of the complex," *Food Research International*, vol. 43, no. 4, pp. 1104–1110, 2010.

[25] H. J. Ha, J. E. Lee, Y. H. Chang, and H.-S. Kwak, "Entrapment of nutrients during cholesterol removal from cream by crosslinked β-cyclodextrin," *International Journal of Dairy Technology*, vol. 63, no. 1, pp. 119–126, 2010.

[26] L. V. Klatt, "Cholesterol analysis in food by direct saponification—gas chromatographic method: collaborative study," *Journal of AOAC International*, vol. 78, pp. 75–79, 1995.

[27] H. S. Kwak, J. J. Ahn, and D. K. Lee, "Method for removing cholesterol from milk and cream," US Patent No. 6110517, 2000.

[28] H. S. Kwak, H. M. Sun, J. Ahn, and H. J. Kwon, "Optimization of β- CD recycling process for cholesterol removal in cream," *Asian-Australasian Journal of Animal Sciences*, vol. 14, no. 4, pp. 548–552, 2001.

[29] D. G. Oakenfull and G. S. Sihdu, "Cholesterol reduction," International Patent 91/11114, 1991.

[30] G. C. Yen and L. T. Tsui, "Cholesterol removal from a lard mixture with β- CD," *Journal of Food Science*, vol. 60, pp. 561–564, 1995.

[31] H. J. Ha, S. S. Jeon, Y. H. Chang, and H. S. Kwak, "Entrapment of milk nutrients during cholesterol removal from milk by crosslinked β-cyclodextrin," *Korean Journal for Food Science of Animal Resources*, vol. 29, no. 5, pp. 566–572, 2009.

[32] H. S. Kwak, C. S. Jung, S. Y. Shim, and J. Ahn, "Removal of cholesterol from cheddar cheese by β-cyclodextrin," *Journal of Agricultural and Food Chemistry*, vol. 50, no. 25, pp. 7293–7298, 2002.

[33] J. Suzuki, "Molecular encapsulation of volatile, easily oxydizable flavour substances by cyclodextrins," *Acta Chimica Academiae Scientiarum Hungaricae*, vol. 101, pp. 27–46, 1975.

3

Polyphenol Bioaccessibility and Sugar Reducing Capacity of Black, Green, and White Teas

Shelly Coe, Ann Fraser, and Lisa Ryan

Functional Food Centre, Oxford Brookes University, Gipsy Lane, Oxford OX3 0BP, UK

Correspondence should be addressed to Lisa Ryan; lisaryan@brookes.ac.uk

Academic Editor: Jose M. Prieto-Garcia

Tea (*Camellia sinensis*) is a widely consumed beverage and recognised for its potential enhancing effect on human health due to its rich polyphenol content. While a number of studies have investigated the quantity and type of polyphenols present in different tea samples, no study has reported the potential effect of digestive enzymes on the availability of tea polyphenols for human absorption or the subsequent impact on glycaemic response. The objectives of the present study were to assess the total polyphenol content of different teas, to assess the bioaccessibility of polyphenols in whole and bagged teas, and to determine the effect of black, white, and green tea infusions on sugar release. All of the teas were a significant source of polyphenols (10–116 mg Gallic acid equivalents/g). There was an overall increase in the release of polyphenols from both the bagged and the whole teas following *in vitro* digestion. Bagged green tea significantly ($P < 0.05$) reduced rapidly digestible starch from white bread samples compared to control and black and white bagged teas. The present study confirms that tea is a rich source of polyphenols and highlights the potential benefits it may have on modulating glycaemic response in humans.

1. Introduction

One of the most widely consumed beverages throughout the world is tea produced from the tea plant (*Camellia sinensis*). Tea for consumption is classified according to the methods used in its production. Geographical consumption patterns of the different teas vary greatly with green tea consumed mainly in Asia and the Middle East and black tea consumed mostly in western countries.

Tea has been found to be a rich source of polyphenols and antioxidants [1]. This together with evidence from epidemiological studies [2] and high consumption rates worldwide has led to growing interest in tea as a product that may significantly contribute to human health.

The polyphenol profile of the different teas is affected by their different methods of production. Black tea is produced by wilting, crushing, and partial oxidation and consequently is rich in theaflavins and thearubigins [3]. In green tea production oxidation is minimised resulting in catechins being dominant [4], particularly epigallocatechin-gallate (EGCG), white tea undergoes the least processing and is produced

from young leaves and buds resulting in high levels of EGCG [5], though generally lower than the levels found in green tea.

Reduced risk of coronary heart disease [6], stroke incidence [7], chronic inflammation [8], and cancer incidence [9] is associated with black and green tea consumption. Recent studies have focused on the impact of tea and tea polyphenols on blood glucose (BG) and insulin sensitivity. Black tea has been shown to decrease plasma glucose and enhance insulin concentrations after consumption in comparison to a control and a caffeine drink [10]. Aldughpassi and Wolever [11] showed that 250 mL of black tea with test meals actually increased overall mean peak BG compared to water though a reduction in the standard error might indicate the ability of tea compounds to improve the precision of the BG response. A study of green tea catechins in insulin resistant induced obese rats suggested that they may impact glucose control through several pathways [12].

The objectives of this study were as follows:

(1) to assess the total polyphenol content of different commercial teas;

(2) to assess the bioaccessibility of polyphenols from whole and bagged teas after *in vitro* digestion;

(3) to determine the effect of black, white, and green tea infusions on sugar release from bread after an *in vitro* digestion model.

2. Materials and Methods

2.1. Chemicals. All chemicals and reagents were of analytical grade and purchased from Sigma-Aldrich (Poole, UK). The different teas were sourced directly from a specialist supplier as whole teas or purchased in bag form from a local Tesco supermarket.

2.2. Study Protocol. The weight of each whole tea used was approximately 3 g except for the "Flowering Osmanthus" and "Jasmine & Lily" teas where an entire bulb was used (the weight of each bulb was recorded). For teas in bagged format, one tea bag (approximately 2.5 g) was used. All tea samples were prepared using a standard protocol. Each tea was infused in 200 mL of boiling water (90°C unless otherwise specified on the manufacturer's instructions) for three minutes and then stirred six times before the tea was removed. The resultant sample was then left at room temperature to cool for an additional 17 minutes before testing commenced. All tests were carried out on a minimum of three separate occasions and samples were analysed in triplicate for each test.

2.3. Analysis of Polyphenol Content

2.3.1. Folin-Ciocalteu (FCR). Samples from nineteen brands of commercially available tea were chosen. An aliquot (200 μL) of each tea sample was added to 1.5 mL of freshly prepared Folin-Ciocalteu reagent (1 : 10 v/v with water). The mixture was allowed to equilibrate for 5 min and then mixed with 1.5 mL of 60 g/L sodium carbonate solution. After incubation at room temperature for 90 min, the absorbance of the mixture was read at 725 nm using the respective solvent as blank. The results were expressed as mg of gallic acid equivalents (GAEs) per gram of tea.

2.4. Bioaccessibility of Tea Polyphenols. Samples from nineteen brands of commercially available tea were analysed using an *in vitro* digestion model adapted from Ryan and others [13]. A total of 4 mL of each 200 mL tea infusion was added to an amber vial and made up to a volume of 15 mL with saline. A 1 mL baseline aliquot was taken from each sample. The samples were acidified to pH 2 by the addition of 1 mL of a porcine pepsin preparation (0.04 g pepsin in 1 mL 0.1 M HCl) and then incubated at 37°C in a shaking water bath at 95 rpm for 1 hour. Gastric aliquots were taken. The pH was increased to 5.3 with 0.9 M sodium bicarbonate, followed by the addition of 200 μL of the bile salts glycodeoxycholate (0.04 g in 1 mL saline), taurodeoxycholate (0.025 g in 1 mL saline) and taurocholate (0.04 g in 1 mL saline), and 100 μL of pancreatin (0.04 g in 500 μL saline) to each sample. The pH was adjusted to 7.4 using 1 M NaOH and overlaid with

nitrogen. The samples were then incubated in a shaking water bath for two hours at 37°C. Duodenal aliquots were taken and samples were frozen until analysis.

2.5. Measurement of Sugar Release. The relative glycaemic impact (RGI) of one black tea, one green tea, and one white tea sample from the Clipper brand was measured in order to assess the effect of the tea polyphenols on the inhibition of starch breakdown. Bread was used as the starch source. This was achieved by subjecting samples of tea combined with bread to an *in vitro* digestion procedure and measuring the resultant reducing sugars released.

2.5.1. Bread Preparation. White bread dough was made to a recipe of 190 g warm tap water, 1 tbsp virgin olive oil, 1 tsp salt, 1 tbsp sugar, 1 tbsp dried milk powder, 350 g strong white flour, and 1.5 tsp of yeast. The dough was then baked in a *Russell Hobbs* bread maker (model no: 18036, Manchester, UK) for a total of 3 hours and 20 minutes. Samples of the bread were then prepared by weighing 2.5 g samples and placing each into 60 mL specimen pots. The pots were inserted into an aluminium heating block and covered with an insulating sheet in readiness for testing.

2.5.2. In Vitro Digestion. An *in vitro* digestion procedure was used to test the tea samples. This consisted of a simulated gastric digestion phase followed by an ileal digestion phase with timed sampling at the end of the gastric phase and during the ileal phase [14]. A volume of 30 mL of each tea infusion (1 tea bag/infusion) was added to its own individual bread sample. A 250 μL baseline sample was extracted for each sample at $t = 0$ min and added to a test tube in a ratio of 1 : 4 in ethanol. This was followed by the addition to each sample of 0.1 mL 10% α-amylase, 0.8 mL 1 M HCl, and 1 mL 10% pepsin solution in 0.05 M HCl to each. The resultant mixture was stirred slowly at 130 rpm every 15 s for 30 min at 37°C to complete the gastric digestion phase, and then gastric aliquots were taken. The ileal phase was initiated by the addition of 2 mL 1 M NaHCO$_3$ and 5 mL 0.2 M Na maleate buffer (pH 6) to each sample, and the volume was increased to 55 mL with dH$_2$O. In quick succession, 0.1 mL of amyloglucosidase and 1 mL of 2% pancreatin solution (in maleate buffer, pH 6) were added to each sample. Samples were then incubated for 120 minutes with constant slow mixing, and aliquots were taken at 20, 60, and 120 minutes during ileal digestion. The tubes were centrifuged (1000 $\times$g, 2 min) in a Biofuge Primo Centrifuge (Heraeus Instruments, Kendro Laboratory Products, Germany) and an aliquot of the supernatant was removed for analysis of reducing sugars.

2.5.3. Analysis of Reducing Sugars Released during Digestion. Sugar released from the bread during digestion was measured by a colourimetric method adapted from Englyst and Hudson [15] designed to measure monosaccharides after an amyloglucosidase secondary digestion to complete depolymerisation of starch fragments. A total of 0.05 mL of 10 mg/mL glucose standard or sample from the *in vitro* digestion was added to 0.25 mL of enzyme solution A (1% amyloglucosidase in acetate buffer, pH 5.2). Each sample was incubated for 10 minutes at 25°C and then 0.75 mL of 3,5-Dinitrosalicylic acid (DNS) mixture (0.5 mg/mL glucose : 4 M NaOH : DNS reagent mixed in ratio 1 : 1 : 5) was added. The resultant sample was then heated for 15 minutes at 95°C in a water bath. Following this, 3 mL of water was added to each sample which was then left to cool for 20 min in a cold water bath. Absorbance was read at 530 nm on a Shimadzu UV-1201 spectrophotometer (Shimadzu Corporation, Australia) and sugar release was measured in mg per g of bread sample. Slowly digestible starch (SDS) was extrapolated by subtracting the rapidly digestible starch (RDS) measurement at 20 min from the reducing sugars measurement at 120 min during ileal digestion [14].

2.6. Statistical Analysis. All experiments were carried out in triplicate and each had a minimum of three replicates for each tea. The data are presented as means (±SEM) and comparisons between samples were carried out by an ANOVA and Tukey's multiple comparison test (SPSS, version 17; SPSS Inc., Chicago, IL, USA). A probability of 5% or less was considered statistically significant.

3. Results

3.1. Polyphenol Content. Table 1 illustrates that all teas were a significant source of polyphenols. Of the whole teas, "Kagoshima Sencha" brand had a significantly higher polyphenol content ($P < 0.05$) compared to the other commercial whole teas analysed. Of the bagged varieties, green tea infusion had a significantly higher polyphenol content than both white and black teas, as measured by FCR ($P < 0.05$).

3.2. Polyphenol Bioaccessibility. The *in vitro* digestion model enables the measurement of polyphenols potentially available for absorption after the gastric and duodenal phases of digestion. Bioaccessibility refers to the proportion of polyphenols which are presented to the brush border for absorption after digestion and gives some indication as to their potential bioavailability *in vivo*.

Table 2 illustrates that the polyphenol content of all tea infusions was enhanced following the gastric digestion phase. This enhancement continued into the duodenal phase, although some tea polyphenols became less bioaccessible relative to the gastric phase.

3.3. Sugar Release. At 20 minutes into the duodenal phase of digestion, green tea significantly suppressed RDS release in white bread to 253.83 mg/g bread sample when compared to white tea and the control bread ($P < 0.05$; Figure 1). Black tea showed no significant effect on sugar release at this time point. In all teas, there was a nonsignificant trend to increase SDS release compared to the control.

4. Discussion

4.1. Polyphenol Content. Of the bagged teas, the black tea had the lowest polyphenol content. The different methods of

TABLE 1: Polyphenol content (expressed as gallic acid equivalents (GAEs) per gram and per serving (3 g in 200 mL water)).

Tea	GAE (mg/g Tea)	GAE (mg/serving)
Jing Assam Breakfast[1]	48.6	145.9
Organic Jade Sword[1]	42.6	127.9
Organic Dragon Well[1]	54.9	164.6
Jing Earl Grey[1]	62.2	186.7
Jasmine Pearls[1]	23.3	69.9
Flowering Osmanthus[1]	10.3	75.2
Flowering Jasmine and Lily[1]	13.8	96.3
Tieguanyin[1]	28.5	85.5
Moroccan Mint[1]	48.7	146.2
Jing Ceylon[1]	58.7	176.0
Jing Darjeeling 2nd Flush[1]	47.3	141.8
Jasmine Silver Needle[1]	20.4	61.1
Yellow Gold Oolong[1]	23.5	70.5
Jun Shan Silver Needle[1]	38.6	115.7
Kagoshima Sencha[1]	95.3[a]	285.8
Taiwan Ali Shan Oolong[1]	20.1	60.3
Clipper Black Tea[2]	87.9	263.7
Clipper Green Tea[2]	115.5[b]	346.5
Clipper White Tea[2]	102.8	308.4

[1]Whole teas, [2]bagged teas.
[a]significantly ($P < 0.05$) greater than all other whole tea samples.
[b]significantly ($P < 0.05$) greater than all other bagged tea samples.

processing and production impact the polyphenol content of the resultant teas. Total black tea polyphenols decrease during fermentation, and the longer tea is subjected to processing, the lower the polyphenol content [16]. Turkmen and others [17] found that black tea polyphenol content as measured by FCR reached a maximum of 131.9 mg GAE/g tea extract compared to the 87.9 GAE/g in the current study indicating that the polyphenol content of the same tea can vary widely.

The green tea infusion was shown to have more overall reducing power than both black tea and white tea. The production process used in black tea results in the formation of theaflavins. Theaflavins in black tea are dimers of catechins and contain more hydroxyl groups in their structure. In green tea, catechins remain dominant. This could in part explain the differences in reducing potential between black tea and green tea [16].

There has been very little research to date on the polyphenol content and health effects of white tea. Rusak and others [3] found that green tea was a richer source of polyphenols than white tea and the current study supports this. They also found that the concentration of catechins was significantly higher in green tea leaves than white tea leaves.

Ryan and Carolan [18] found that green teas varied in their polyphenol content, ranging from 250 to 750 mg GAE/tea bag. They also found that both the structure of the tea bag and the infusion time influenced the polyphenol content of the teas. FCR values in the current study were

TABLE 2: % Bioaccessibility of the polyphenol content after the gastric and duodenal phases of digestion.

Tea	Gastric (%)	Duodenal (%)
Jing Assam Breakfast[1]	140.7	121.2
Organic Jade Sword[1]	133.2	131.0
Organic Dragon Well[1]	123.2	128.1
Jing Earl Grey[1]	121.0	127.7
Jasmine Pearls[1]	172.2	204.0
Flowering Osmanthus[1]	176.8	189.3
Flowering Jasmine and Lily[1]	160.5	174.6
Tieguanyin[1]	161.7	185.2
Moroccan Mint[1]	128.0	124.3
Jing Ceylon[1]	124.6	127.4
Jing Darjeeling 2nd flush[1]	133.5	142.4
Jasmine Silver Needle[1]	204.1	233.6
Yellow Gold Oolong[1]	194.4	233.5
Jun Shan Silver Needle[1]	153.9	147.1
Kagoshima Sencha[1]	143.7	134.7
Taiwan Ali Shan Oolong[1]	207.9	231.1
Clipper Black Tea[2]	136.6	126.6
Clipper Green Tea[2]	132.0	125.3
Clipper White Tea[2]	165.1	176.6

[1] Whole teas, [2] bagged teas.

slightly lower than those found by Ryan and Carolan [18], averaging 115.5 mg GAE/g tea bag. Rusak and others [3] found that extraction of catechins from green tea was affected by the form of tea used, with extraction from loose green tea leaves being more effective than from refined bagged tea leaves. However, the form of the tea did not affect white tea catechins. Greater tea bag size results in tea solids diffusing into solution faster because of the larger surface area available in which the contents can diffuse [16]. The material of the tea bag can also have an impact on polyphenol diffusion. In the current study, tea bag infusion time and stirring/squeezing of the tea bags were kept constant for all three teas. Tea bags were different weights, the black tea bag weighing more than both white tea and green tea. Weights were, however, corrected for upon calculations.

Of the whole teas analysed, "Kagoshima Sencha" had a significantly higher polyphenol content compared to the other teas. This was the only whole tea that had a higher polyphenol content than any of the bagged teas. Whole teas tend to be more compacted and have undergone much less processing than the bagged tea. Unpublished data from our laboratory indicate that the grinding of leaves to form the bagged teas has the effect of releasing polyphenols and this may in part explain the higher polyphenol content in the bagged teas.

Overall, all teas were shown to be good sources of polyphenols.

4.2. Polyphenol Bioaccessibility. In the current study, polyphenol bioaccessibility increased in all teas from baseline to

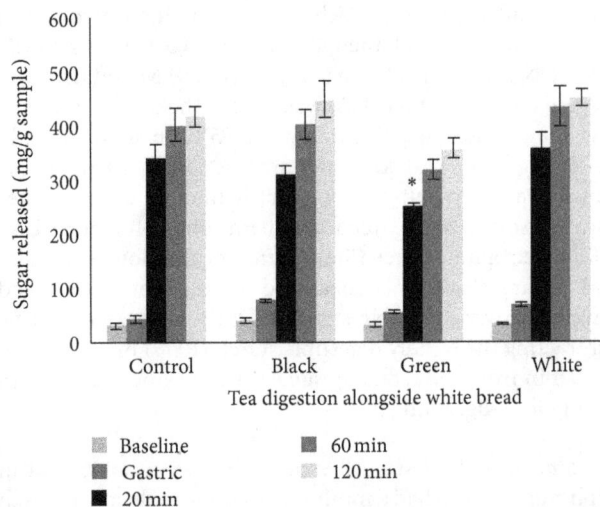

FIGURE 1: Sugar release from bread samples, when digested alongside either no tea (control), black, green, or white tea. * = significant reduction in sugar release compared to control white bread, $P <$ 0.05. Values reported as mg sugar released/g bread sample. Values represent mean ± standard error of the means (SEM). 20, 60, and 120 min = stages of intestinal digestion.

gastric phases suggesting that the polyphenols in these teas may become more available in humans after consumption.

The bioaccessibility increase from baseline to gastric phase was similar for both bagged and whole teas. The compact and relatively unprocessed nature of the whole teas resulted in a lower polyphenol content at baseline. However, the bioaccessibility increases suggest that the digestive enzymes can further release polyphenols from the tea infusion. Green and others [19] looked at the effect of in vitro digestion specifically on catechins in green tea and found that catechins had less than 20% recovery after in vitro digestion. However, the current study is the first to report and compare the bioaccessibility of total polyphenols from green, white, and black tea infusions. In the current study, with the exception of white bagged tea, the bioaccessibility decreased slightly from gastric to duodenal phases in bagged teas. However, at the duodenal stage values still remained above baseline for all tea varieties. This decrease may be due to the increase in pH during the duodenal phase. Rusak and others [3] showed that the form of tea, either loose or bagged did not affect white tea catechin stability. Therefore, compounds in white tea may be more stable, and thus less susceptible to degradation compared to those in the other teas.

4.3. Starch Digestion and Sugar Release. Green tea was the only tea shown to significantly reduce sugar release from white bread. This is a promising finding in that green tea may reduce the RDS of starch rich foods such as bread. However, tea polyphenols have been shown to reduce starch retrogradation. Wu and others [20] found that increased levels of added purified polyphenols (50% EGCG) resulted in decreased retrogradation for resistant starch in rice. From these results, it was predicted that the hydroxyl radical of tea polyphenols combined with rice starch to form hydrogen bonds, preventing the reassociation of the starch chains.

In the current study, white tea had no significant effect on sugar release, although there was a slight increase compared to the control bread. Certain polyphenols or other compounds in white tea may be responsible for interfering with the natural chemical bonds in the bread, therefore rendering the starch more susceptible to degradation. White tea polyphenols were shown to be the most stable throughout digestion, and therefore they may have a greater impact on starch than those polyphenols which are degraded more readily.

Different teas contain different polyphenols, and therefore it is plausible that each tea may have a different effect on sugar release. Teas were used at low concentrations in this study, that is, one tea bag per infusion. Therefore, green tea at low concentrations reduces sugar release from starch samples, whereas white tea at a low dose seems to have the opposite effect. Teas were made as infusions and then added to bread samples at the baseline phase of digestion, whereas other studies have baked tea into breads or looked at the effect of purified tea polyphenol rich extracts. Therefore, different study designs may account for variability in study results. However, what can be seen is that different types of tea do have an effect on starch breakdown and sugar release from breads.

Unpublished data from our laboratory found that glycaemic response (GR) increased over 180 minutes following the consumption of white bread with added black tea extract in comparison to eating white bread alone. A more pronounced effect was also seen when a tea infusion was consumed alongside the white bread and the time taken for this combination to exert its peak GR was prolonged compared to either the control bread or the bread with tea extract added. However, Koh and others [21] found black tea to reduce starch digestion, yet found no effect with green or oolong tea. In the current study, each tea was presented as an infusion alongside bread during digestion and the black tea infusion had no significant effect on sugar release from bread. The form in which the black tea is used, the preparation method, and the concentration used may all be factors to account for the variation in results between studies.

Deshpande and Salunkhe [22] found that in isolation tannic acid and catechins decreased *in vitro* digestibility of various types of starch sources. Also, Bjork and Nyman [23] found that phytic acid and tannic acid reduced starch hydrolysis in the digestive tract. However, isolated polyphenols may have a different effect on starch digestion compared to phenolics in combination. For example, green tea is one of the richest sources of phenolic compounds and includes catechins such as EGCG, procyanidins, and quercetin [24]. Therefore, a combination of the phenolics in tea may have synergistic effects, either enhancing or reducing the degree of starch breakdown and sugar release compared to isolated catechins or tannins [25, 26]. The reason for the reduction in sugar release seen in the current study may be because of the structural bonding of the green tea polyphenols with starch molecules. Both black tea and green tea contain conjugate forms of catechins, and these compounds may interfere with

the way in which starch is broken down. Therefore, future research could look into accessing the polyphenol profile of the different teas to determine which polyphenols have an effect on reducing starch breakdown.

Tea polyphenols may have an inhibitory effect on digestive enzymes such as α-amylase and α-glucosidase. Black tea and to a lesser extent green tea were shown to inhibit α-amylase in human saliva and removal of tea tannins resulted in loss of the inhibitory activity [27]. Hara and Honda [28] found that both catechins and theaflavins inhibited salivary α-amylase and whilst, various teas differed in the extent of inhibitory activity, black tea showed consistently greater inhibitory activity. Kwon and others [29] found that black tea and white tea showed almost a 40% inhibition of α-amylase, with green tea showing only a 30% inhibition. They also found black and white tea to have a higher α-glucosidase inhibitory activity than green and oolong teas. At higher concentrations of polyphenols, {50 μg GAE/mL, 100 μg GAE/mL and 200 μg GAE/mL} α-glucosidase inhibition was increased in all teas. Although enzyme inhibition was not tested in the current study, a future area of study is to investigate the effect of tea polyphenols on digestive enzymes.

Finally, it should be noted that many studies evaluating the GR to foods contain the addition of tea or coffee in their test meals. Based on the current results and on previous studies, the polyphenols in tea may affect BG, thus affecting the results of the test foods studied. It is important that when testing foods for different effects on disease parameters, the combination of ingredients is considered. A single food in isolation may have different effects on GR than that food with an additional tea beverage.

Conflict of Interests

The authors report no conflict of interests.

Acknowledgment

The authors would like to thank JING Tea Ltd for the generous gift of all the whole tea samples.

References

[1] C. Lakenbrink, S. Lapczynski, B. Maiwald, and U. H. Engelhardt, "Flavonoids and other polyphenols in consumer brews of tea and other caffeinated beverages," *Journal of Agricultural and Food Chemistry*, vol. 48, no. 7, pp. 2848–2852, 2000.

[2] N. Khan and H. Mukhtar, "Tea polyphenols for health promotion," *Life Sciences*, vol. 81, no. 7, pp. 519–533, 2007.

[3] G. Rusak, D. Komes, S. Likić, D. Horzic, and M. Kovač, "Phenolic content and antioxidative capacity of green and white tea extracts depending on extraction conditions and the solvent used," *Food Chemistry*, vol. 110, no. 4, pp. 852–858, 2008.

[4] C. Cabrera, R. Artacho, and R. Giménez, "Beneficial effects of green tea: a review," *Journal of the American College of Nutrition*, vol. 25, no. 2, pp. 79–99, 2006.

[5] G. Santana-Rios, G. A. Orner, A. Amantana, C. Provost, S. Y. Wu, and R. H. Dashwood, "Potent antimutagenic activity of

white tea in comparison with green tea in the Salmonella assay," *Mutation Research*, vol. 495, no. 1-2, pp. 61–74, 2001.

[6] K. J. Mukamal, K. MacDermott, J. A. Vinson, N. Oyama, W. J. Manning, and M. A. Mittleman, "A 6-month randomized pilot study of black tea and cardiovascular risk factors," *The American Heart Journal*, vol. 154, no. 4, pp. 724.e1–726.e1, 2007.

[7] S. O. Keli, M. G. L. Hertog, E. J. M. Feskens, and D. Kromhout, "Flavonoids, antioxidant vitamins and risk of stroke: the Zutphen Study," *Archives of Internal Medicine*, vol. 154, pp. 637–642, 1995.

[8] V. Sharma and L. J. M. Rao, "A thought on the biological activities of black tea," *Critical Reviews in Food Science and Nutrition*, vol. 49, no. 5, pp. 379–404, 2009.

[9] T. Kuzuhara, M. Suganuma, and H. Fujiki, "Green tea catechin as a chemical chaperone in cancer prevention," *Cancer Letters*, vol. 261, no. 1, pp. 12–20, 2008.

[10] J. A. Bryans, P. A. Judd, and P. R. Ellis, "The effect of consuming instant black tea on postprandial plasma glucose and insulin concentrations in healthy humans," *Journal of the American College of Nutrition*, vol. 26, no. 5, pp. 471–477, 2007.

[11] A. Aldughpassi and T. M. S. Wolever, "Effect of coffee and tea on the glycaemic index of foods: no effect on mean but reduced variability," *The British Journal of Nutrition*, vol. 101, no. 9, pp. 1282–1285, 2009.

[12] J. Yan, Y. Zhao, S. Suo, Y. Liu, and B. Zhao, "Green tea catechins ameliorate adipose insulin resistance by improving oxidative stress," *Free Radical Biology and Medicine*, vol. 52, no. 9, pp. 1648–1657, 2012.

[13] L. Ryan, O. O' Connell, L. O' Sullivan, S. A. Aherne, and N. M. O' Brien, "Micellarisation of carotenoids from raw and cooked vegetables," *Plant Foods for Human Nutrition*, vol. 63, no. 3, pp. 127–133, 2008.

[14] S. Mishra and J. A. Monro, "Digestibility of starch fractions in wholegrain rolled oats," *Journal of Cereal Science*, vol. 50, no. 1, pp. 61–66, 2009.

[15] H. N. Englyst and G. J. Hudson, "Colorimetric method for routine measurement of dietary fibre as non-starch polysaccharides: a comparison with gas-liquid chromatography," *Food Chemistry*, vol. 24, no. 1, pp. 63–76, 1987.

[16] C. Astill, M. R. Birch, C. Dacombe, P. G. Humphrey, and P. T. Martin, "Factors affecting the caffeine and polyphenol contents of black and green tea infusions," *Journal of Agricultural and Food Chemistry*, vol. 49, no. 11, pp. 5340–5347, 2001.

[17] N. Turkmen, F. Sari, and Y. S. Velioglu, "Effects of extraction solvents on concentration and antioxidant activity of black and black mate tea polyphenols determined by ferrous tartrate and Folin-Ciocalteu methods," *Food Chemistry*, vol. 99, no. 4, pp. 835–841, 2006.

[18] L. Ryan and S. Carolan, "Determination of the total antioxidant capacity and total polyphenol content of commercially available green tea," *Proceedings of the Nutrition Society*, vol. 70, p. E139, 2011.

[19] R. J. Green, A. S. Murphy, B. Schulz, B. A. Watkins, and M. G. Ferruzzi, "Common tea formulations modulate in vitro digestive recovery of green tea catechins," *Molecular Nutrition and Food Research*, vol. 51, no. 9, pp. 1152–1162, 2007.

[20] Y. Wu, Z. Chen, X. Li, and M. Li, "Effect of tea polyphenols on the retrogradation of rice starch," *Food Research International*, vol. 42, no. 2, pp. 221–225, 2009.

[21] L. W. Koh, L. L. Wong, Y. Y. Loo, S. Kasapis, and D. Huang, "Evaluation of different teas against starch digestibility by mammalian glycosidases," *Journal of Agricultural and Food Chemistry*, vol. 58, no. 1, pp. 148–154, 2010.

[22] S. S. Deshpande and D. K. Salunkhe, "Interactions of tannic acid and catechin with legume starches," *Journal of Food Science*, vol. 47, pp. 2080–2081, 1982.

[23] I. M. Bjork and M. A. Nyman, "In vitro effects of phytic acid and polyphenols on starch digestion and fiber degradation," *Journal of Food Science*, vol. 52, no. 6, pp. 1588–1594, 1987.

[24] V. Neveu, J. Perez-Jimenez, F. Vos et al., "Phenol-Explorer: an online comprehensive database on polyphenol contents in foods," *Database*, vol. 2010, 2010.

[25] A. Conte, S. Pellegrini, and D. Tagliazucchi, "Effect of resveratrol and catechin on PC12 tyrosine kinase activities and their synergistic protection from β-amyloid toxicity," *Drugs under Experimental and Clinical Research*, vol. 29, no. 5-6, pp. 243–255, 2003.

[26] D. M. Morré and D. J. Morré, "Anticancer activity of grape and grape skin extracts alone and combined with green tea infusions," *Cancer Letters*, vol. 238, no. 2, pp. 202–209, 2006.

[27] J. Zhang and S. Kashket, "Inhibition of salivary amylase by black and green teas and their effects on the intraoral hydrolysis of starch," *Caries Research*, vol. 32, no. 3, pp. 233–238, 1998.

[28] Y. Hara and M. Honda, "The inhibition of alpha-amylase by tea phenols," *Agricultural and Biological Chemistry*, vol. 54, pp. 1939–1945, 1990.

[29] Y. I. Kwon, E. Apostolidis, and K. Shetty, "Inhibitory potential of wine and tea against α-amylase and α-glucosidase for management of hyperglycemia linked to type 2 diabetes," *Journal of Food Biochemistry*, vol. 32, no. 1, pp. 15–31, 2008.

4

Development of Low-Fat Soft Dough Biscuits Using Carbohydrate-Based Fat Replacers

Bhawna Chugh,[1] **Gurmukh Singh,**[1] **and B. K. Kumbhar**[2]

[1] *Department of Food Science & Technology, G. B. Pant University of Agriculture and Technology, Uttarakhand, Pantnagar 263145, India*
[2] *Department of Post Harvest Process and Food Engineering, G. B. Pant University of Agriculture and Technology, Uttarakhand, Pantnagar 263145, India*

Correspondence should be addressed to Bhawna Chugh; bhawna84chugh@rediffmail.com

Academic Editor: Marie Walsh

Experiments were conducted to develop low-fat soft dough biscuits using carbohydrate-based fat replacers (maltodextrin and guar gum). A central composite rotatable design was used to optimise the level of sugar 24–36%, composite fat (fat 10.5–24.5%, maltodextrin 10.4–24%, and guar gum 0.1–0.5%), ammonium bicarbonate 0.5–2.5%, and water 20–24% for production of low-fat biscuits. Diameter ($P < 0.01$) and stress-strain ratio ($P < 0.05$) decreased significantly with increase in the amount of sugar. There was a significant decrease ($P < 0.01$) in spread ratio at high amount of water. Hardness was significantly affected by the interactions of ammonium bicarbonate with sugar ($P < 0.05$) and fat ($P < 0.1$). The optimum level of ingredients obtained for low-fat biscuits was sugar 31.7 g, fat 13.55 g, maltodextrin 21.15 g, guar gum 0.3 g, ammonium bicarbonate 2.21 g, and water 21 mL based on 100 g flour. The fat level in the optimised low-fat biscuit formulation was found to be 8.48% as compared to 22.65% in control; therefore, the reduction in fat was 62.5%.

1. Introduction

The Indian bakery industry is the largest of the food processing industries, estimated to be over $1,400 million. The major products within this industry include bread, cakes, pastries, and biscuits. Short dough biscuits are products made from soft and weak wheat flours and are characterised by a formula high in sugar and shortening. Fat in a biscuit formulation has a multifaceted function. It is the principal ingredient responsible for tenderness, keeping quality, grain, and texture, and adding richness to biscuits [1].

The food industry is primarily driven by consumer health trends. A present day dietary concern is the consumption of a large amount of fat and sugar. With the growing incidence of obesity and diabetes, low calorie foods have gained immense popularity. Most well-maintained strategies in terms of fat reduction diets involve either the use of low-fat foods or fat substitutes or modifications such as trimming of fat from foods. So, the use of fat mimetics instead of conventional fats and oils helps in reducing calorie intake, whereas fat substitutes are either resistant to digestive lipases or partially digested [2, 3]. Fat replacers are grouped broadly into either lipid-, carbohydrate-, or protein-based materials. Carbohydrate-based replacers incorporate water into a gel type structure, resulting in lubricant or flow properties similar to those of fats in some food systems. It is likely that desirable textures can be achieved using these types of substitutes, and there are few regulatory obstacles regarding any toxicological potential [2]. Maltodextrin can be used in a gram-for-gram fat replacement in bakery goods that provides 4 kcal or 16.8 kJ/g [4]. Maltodextrins generally have a DE (dextrose equivalent) between 3 and 20. The higher the DE value, the higher the solubility and sweetness. Gums are referred to as hydrophilic colloids which can fulfil some of the bulking properties of sucrose and fat. The replacement of 50% of fat by soluble β-glucan and amylodextrins derived from oat flour resulted in cookies not significantly different from the full-fat ones, but at higher substitution levels moistness and overall quality were decreased [5]. Also tenderness of biscuits decreased with the increase of fat substitution by pectin, gum, or oat-based fat mimetics [6]. The previous work on fat replacers indicated that polydextrose, maltodextrins, and Simplesse are the most appropriate as far as cookies properties are concerned, but the main problem noted is the high hardness of the biscuits [7]. Gallagher et al. [8] developed low-fat biscuits using sugar and fat replacers and reported their effect on biscuit dimensions, colour, and texture. Sudha et al. [9] also reported the effect of fat replacers, namely, maltodextrin and polydextrose, on the biscuit hardness.

Flour, sugar, fat, water, and salt are the main components in a biscuit formulation. Changes made to these principal components have significant effects on final biscuit quality [10, 11]. Fat level and type have a significant effect on the rheological characteristics of biscuit dough and on the properties of the baked biscuits [12]. Replacement of the sensory properties of fat is difficult in low moisture bakery foods like cookies with final moisture between 3 and 4% [13]. Sugar delivers sweetness, influences the structural and textural properties of cookies, and enhances incorporation of air into the fat during cookie dough preparation. Increasing water quantity produced a reduction of consistency and an increase in fluidity and in adhesiveness of dough. The quantity of water affected the behaviour of the dough after baking. A slight increase in biscuit length was observed when the water content was increased particularly from 21% upwards [14].

The main objective of this study was to develop low-fat biscuits using combinations of carbohydrate-based fat replacers. Moreover, to produce acceptable quality low-fat biscuit, the level of other ingredients, namely, sugar, ammonium bicarbonate, and water, was varied to take into account the synergetic effect on the physical and sensory parameters. The response surface methodology was used to minimise the number of baking trials while gathering all information relating to ingredient interactions and quality characteristics.

2. Materials and Methods

2.1. Materials. Refined wheat flour, whole wheat flour, sugar, sodium bicarbonate, ammonium bicarbonate, skim milk powder, vanilla essence, and hydrogenated fat were procured from local market. Liquid glucose was collected from Uttarakhand maize processing unit, SIDCUL, Rudrapur, Uttarakhand, India. Carbohydrate based fat replacers, maltodextrin (DE = 16) and guar gum, were procured from M/s Ensigns Healthcare Private Ltd., Pune, Maharashtra and Hindustan Gum and Chemicals Ltd., Bhiwani, Haryana, India.

2.2. Experimental Plan. Response surface methodology which involves design of experiments, selection of levels of variables in experimental runs, fitting mathematical models, and finally selecting variables' level by optimizing the response was employed in the study [15]. A central composite rotatable design (CCRD) was used to design the experiments comprising of four independent variables (Table 1). The parameters that influence the product quality and acceptability were taken as responses. The statistical

software package Design-Expert 8.0.6 (Trial version), Stat-Ease Inc., Minneapolis, USA (http://www.statease.com/), was used to construct the experimental design as well as analyze the data. The experimental design and the codes for the processing variables have been reported in Table 1. A total of 32 combinations were generated for the four independent variables, and the experiments at centre point were repeated eight times. The magnitude of the effect of independent variables on the responses was based on magnitude of regression coefficients.

2.3. Preparation of Biscuits. Biscuits were prepared using traditional creamery method given by Whitley [16]. For preparation of low-fat biscuits, sugar, fat, maltodextrin, guar gum, ammonium bicarbonate, and water were mixed in the quantities on 100 g flour (63.7 g white flour and 36.3 g whole wheat flour) basis as per the experimental design to form different formulations. In these formulations, fixed amounts of liquid glucose (3 g), skim milk powder (3 g), sodium bicarbonate (1.18 g), and vanilla essence (4 drops) were mixed.

The control biscuit formulation contained the following ingredients at the indicated level: flour, 100 g (63.7 g white flour and 36.3 g whole wheat flour); liquid glucose, 3 g; skim milk powder, 3 g; sodium bicarbonate, 1.18 g; vanilla essence (4 drops); sugar, 26.5 g; fat, 35 g; ammonium bicarbonate, 1.22 g; water, 23 mL [16].

The dough was sheeted to 4 mm height biscuits which were baked in an air circulation oven at 190 ± 2°C for 8 min. The biscuits were cooled for 30 min at room temperature and stored in low density polyethylene packs before further analysis.

2.4. Physical and Chemical Evaluation of the Biscuits. The biscuit diameter and thickness were determined by AACC [17] methods. Spread ratio was then calculated as diameter divided by thickness of the biscuit. Crude fat was determined using standard AOAC [18] method.

2.5. Texture Analysis. Hardness of biscuits was measured by Stable Micro Systems Texture Analyzer (TAXT 2i). It was measured in terms of maximum force used to break the biscuit sample. The biscuits were placed under sharp blade cutting probe, 70 mm long and 0.4 mm thick. The analyzer was set at a "return to start" cycle, a speed of 1 mm/s and a distance of 3 mm, and pretest speed 5 mm/s and posttest speed 10 mm/s. The maximum force was expressed in N. Stress was calculated by dividing the maximum force by area of blade, and strain was expressed as the maximum distance travelled by probe to break the biscuit. All measurements were replicated six times, and the mean values are reported.

2.6. Sensory Evaluation. Ten semitrained panelists carried sensory evaluation of low-fat soft dough biscuits and compared with the control samples. Three 1-hour preliminary sessions were conducted to train the panelists so as to familiarize themselves with the samples. In the first session, the subjects described two very different biscuits (control and low-fat biscuit) and mainly focussed on the texture change.

TABLE 1: Experimental design matrix for manufacture of low-fat biscuits.

Expt. No.	Coded form				Physical parameters					Sensory parameters				
	X_1	X_2	X_3	X_4	Diameter (cm)	Thickness (cm)	Spread ratio	Hardness (N)	Stress-strain ratio	Colour	Texture	Taste	Flavour	Overall acceptability
1	−1	−1	−1	−1	6.28	0.64	9.81	49.32	1.660	3.72	3.89	**4.25**	4.03	4.14
2	+1	−1	−1	−1	6.42	0.64	10.03	36.59	0.760	3.50	4.08	3.75	3.83	3.79
3	−1	+1	−1	−1	6.65	*0.57*	11.67	*25.07*	*0.574*	3.20	4.00	3.40	3.60	3.30
4	+1	+1	−1	−1	6.55	**0.73**	8.97	39.82	1.067	4.25	3.75	3.25	3.25	3.38
5	−1	−1	+1	−1	6.63	0.63	10.52	35.53	0.829	3.75	4.00	3.13	3.13	3.13
6	+1	−1	+1	−1	6.82	*0.57*	**11.96**	31.62	0.796	3.40	3.50	3.40	3.40	3.40
7	−1	+1	+1	−1	6.42	0.67	9.58	**84.02**	**3.193**	3.20	3.30	3.47	3.36	3.31
8	+1	+1	+1	−1	6.88	0.59	11.66	31.85	0.669	3.17	3.40	3.70	3.40	3.16
9	−1	−1	−1	+1	6.28	0.62	10.13	27.92	0.578	4.00	**4.25**	4.13	**4.13**	**4.25**
10	+1	−1	−1	+1	*6.08*	0.72	*8.44*	50.66	1.508	3.75	3.95	3.60	3.70	3.97
11	−1	+1	−1	+1	6.36	0.65	9.78	53.30	1.417	4.17	3.75	3.92	4.00	4.04
12	+1	+1	−1	+1	6.70	0.67	10.00	40.16	1.035	**4.83**	3.45	2.92	3.17	3.04
13	−1	−1	+1	+1	6.28	0.72	8.72	42.92	1.312	3.10	3.60	3.35	3.20	3.45
14	+1	−1	+1	+1	7.02	0.72	9.75	32.23	0.586	3.50	3.50	3.25	3.42	3.33
15	−1	+1	+1	+1	6.88	0.64	10.75	61.90	1.995	3.10	3.53	3.33	3.39	3.00
16	+1	+1	+1	+1	6.83	0.67	10.19	42.67	1.022	3.75	3.50	3.25	3.00	3.25
17	−α	0	0	0	6.12	0.65	9.42	48.08	1.396	3.92	3.92	3.58	3.50	3.71
18	+α	0	0	0	**7.03**	0.71	9.90	34.12	0.906	3.60	3.75	3.50	3.58	3.25
19	0	−α	0	0	6.32	0.72	8.78	55.70	1.531	4.10	3.60	3.70	3.50	3.70
20	0	+α	0	0	6.70	0.65	10.31	59.22	1.734	3.20	*3.25*	*2.88*	2.88	2.88
21	0	0	−α	0	6.23	0.63	9.89	62.01	1.698	3.70	3.71	3.21	3.12	3.12
22	0	0	+α	0	6.90	0.58	11.90	33.50	0.864	*3.00*	3.40	3.10	3.10	*2.36*
23	0	0	0	−α	6.67	0.64	10.42	38.38	0.756	3.40	4.00	3.50	3.40	3.50
24	0	0	0	+α	6.75	**0.73**	9.25	45.64	1.475	3.30	3.33	3.17	3.25	3.40
25	0	0	0	0	6.50	0.58	11.21	68.04	1.587	3.50	3.60	3.50	3.50	3.50
26	0	0	0	0	6.57	0.60	10.95	70.11	1.647	3.60	3.50	3.40	3.40	3.40
27	0	0	0	0	6.50	0.60	10.83	67.04	1.555	3.60	3.50	3.55	3.30	3.45
28	0	0	0	0	6.60	0.62	10.65	67.77	1.582	3.50	3.55	3.50	3.40	3.50
29	0	0	0	0	6.50	0.60	10.83	69.14	1.657	3.50	3.60	3.40	3.40	3.44
30	0	0	0	0	6.60	0.60	11.00	71.54	1.648	3.60	3.53	3.35	3.35	3.40
31	0	0	0	0	6.58	0.62	10.61	67.98	1.624	3.55	3.50	3.56	3.31	3.50
32	0	0	0	0	6.58	0.62	10.61	67.20	1.611	3.58	3.57	3.40	3.40	3.43

X_1: sugar, X_2: composite fat, X_3: ammonium bicarbonate, X_4: water, MD: maltodextrin, GG: guar gum, A.B.: ammonium bicarbonate; figures in bold are maximum values, and figures in italics are minimum values.

In the second session, the most frequently cited attributes were selected, and their definitions and the protocols scoring them were developed. During the third session, the panel lists were able to understand the test and were given a score sheet to evaluate sensory attributes, namely, colour, taste, texture, flavor, and overall acceptability (OAA), and asked to score samples on 5-point scale where scores 1, 2, 3, 4, and 5 represented poor, fair, satisfactory, good, and excellent, respectively [19]. Panelists were instructed to cleanse their palate with tap water before tasting each sample. Product characterization was carried out under "day light" illumination and in isolated booths within a sensory laboratory.

2.7. Data Analysis. The experimental data were analysed using second-order model given below:

$$y = \beta_0 + \sum_{i=1}^{4} \beta_i x_i + \sum_{i=1}^{4} \beta_{ii} x_i^2 + \sum_{i=1}^{3} \sum_{j=i+1}^{4} \beta_{ij} x_i x_j, \quad (1)$$

where y = response, x_i, x_j = coded processing parameters, and β_0, β_i, β_{ii}, β_{ij} = regression coefficients.

Adequacy of the model was determined using coefficient of determination (R^2), F-value, and adequacy of precision. The effect of variables at linear, quadratic and interactive levels on the response was described using various levels of significance.

Numerical optimization technique of the Design-Expert (8.0.6) software was used for simultaneous optimization of the multiple responses, and for this some constraints had to be decided. These constraints set the guidelines to get the desired results. The goal seeking begins at a random starting point and proceeds up or down the steepest stop on the response surface for a maximum or minimum value of a response, respectively. The response values and the analysis of the models gave the valuable information in deciding constraints for independent variables and responses. Therefore, all the independent variables were kept within experimental range except composite fat which was kept at minimum as our main goal is to produce low-fat biscuit. The multiple responses, namely, spread ratio, hardness, stress-strain ratio, and overall acceptability (OAA), were considered for optimization as they represent quality attributes adequately. The numerical optimization finds a point that maximizes the desirability function.

3. Results and Discussion

The experimental results for physical and sensory parameters are reported in Table 1. The product with varied formulations had diameter, thickness, spread ratio, hardness, and stress-strain ratio in the ranges of 6.08–7.03 cm, 0.57–0.73 cm, 8.44–11.96, 25.07–84.02 N, and 0.574–3.19, respectively. The sensory score ranged 3.00–4.83, 3.25–4.25, 2.88–4.25, 2.88–4.13, and 2.36–4.25 for colour, texture, taste, flavor, and OAA, respectively. The corresponding values of physical parameters for control biscuits were 5.98 cm, 0.62 cm, 9.64, 32.52 N, and 0.73, respectively, while for sensory parameters were 4.5, 4.0, 3.33, 3.83, and 4.02, respectively. Most of the combinations were better than the control formulation.

The Design-Expert software was used to fit the second-order response surface model (1) into the experimental data of all responses using regression analysis, and the resulting predictive equations are given below. All models have adequate precision ratio of more than 4, thus indicative of the fact that the experiments were carried out with adequate precision, and moreover, the F-value was found significant in all models:

$$\text{diameter } (y) = 6.5537 + 0.1391 * X_1$$
$$+ 0.0925 * X_2 + 0.1575 * X_3$$
$$- 0.0025 * X_4 - 0.0137 * X_1 * X_2$$
$$+ 0.0725 * X_1 * X_3 + 0.0087 * X_1 * X_4$$
$$- 0.058 * X_2 * X_3 + 0.0475 * X_2 * X_4$$
$$+ 0.0462 * X_3 * X_4 + 0.0015 * X_1 * X_1$$
$$- 0.0146 * X_2 * X_2 - 0.0009 * X_3 * X_3$$
$$+ 0.0353 * X_4 * X_4 \quad \left(R^2 = 79.84\%\right),$$

$$\text{thickness } (y) = 0.605 + 0.0120 * X_1$$
$$- 0.0087 * X_2 - 0.0054 * X_3$$
$$+ 0.0229 * X_4 + 0.0056 * X_1 * X_2$$
$$- 0.0243 * X_1 * X_3 + 0.0081 * X_1 * X_4$$
$$- 0.0043 * X_2 * X_3 - 0.0143 * X_2 * X_4$$
$$+ 0.0131 * X_3 * X_4 + 0.0169 * X_1 * X_1$$
$$+ 0.0182 * X_2 * X_2 - 0.0017 * X_3 * X_3$$
$$+ 0.0182 * X_4 * X_4 \quad \left(R^2 = 79.72\%\right),$$

$$\text{spread ratio } (y)$$
$$= 10.8368 + 0.0424 * X_1$$
$$+ 0.2621 * X_2 + 0.3467 * X_3$$
$$- 0.366 * X_4 - 0.1224 * X_1 * X_2$$
$$+ 0.4960 * X_1 * X_3 - 0.1276 * X_1 * X_4$$
$$- 0.0487 * X_2 * X_3 + 0.2583 * X_2 * X_4$$
$$- 0.1369 * X_3 * X_4 - 0.2708 * X_1 * X_1$$
$$- 0.2998 * X_2 * X_2 + 0.0376 * X_3 * X_3$$
$$- 0.2269 * X_4 * X_4 \quad \left(R^2 = 75.66\%\right),$$

$$\text{hardness } (y) = 68.6025 - 4.2627 * X_1$$
$$+ 3.2931 * X_2 - 0.7124 * X_3$$
$$+ 1.3522 * X_4 - 4.075 * X_1 * X_2$$
$$- 6.1027 * X_1 * X_3 + 2.1093 * X_1 * X_4$$
$$+ 5.2673 * X_2 * X_3 + 1.037 * X_2 * X_4$$
$$- 1.5337 * X_3 * X_4 - 7.5804 * X_1 * X_1$$
$$- 3.4916 * X_2 * X_2 - 5.9171 * X_3 * X_3$$
$$- 7.3353 * X_4 * X_4 \quad \left(R^2 = 77.71\%\right),$$

$$\text{stress-strain ratio } (y)$$
$$= 1.6139 - 0.2123 * X_1$$
$$+ 0.1394 * X_2 + 0.0056 * X_3$$
$$+ 0.0559 * X_4 - 0.1659 * X_1 * X_2$$
$$- 0.2748 * X_1 * X_3 + 0.1132 * X_1 * X_4$$
$$+ 0.2355 * X_2 * X_3 + 0.0017 * X_2 * X_4$$
$$- 0.0656 * X_3 * X_4 - 0.1336 * X_1 * X_1$$
$$- 0.0132 * X_2 * X_2 - 0.1012 * X_3 * X_3$$
$$- 0.1424 * X_4 * X_4 \quad \left(R^2 = 62.87\%\right),$$

$$\text{colour}(y) = 3.5537 + 0.0529 * X_1$$
$$+ 0.0354 * X_2 - 0.2437 * X_3$$
$$+ 0.0754 * X_4 + 0.1718 * X_1 * X_2$$
$$- 0.0356 * X_1 * X_3 + 0.0631 * X_1 * X_4$$
$$- 0.1256 * X_2 * X_3 + 0.1281 * X_2 * X_4$$
$$- 0.1343 * X_3 * X_4 + 0.0718 * X_1 * X_1$$
$$+ 0.0443 * X_2 * X_2 - 0.0306 * X_3 * X_3$$
$$- 0.0306 * X_4 * X_4 \quad (R^2 = 72.0\%),$$

$$\text{texture}(y) = 3.5437 - 0.0637 * X_1$$
$$- 0.1162 * X_2 - 0.142 * X_3$$
$$- 0.072 * X_4 + 0.0143 * X_1 * X_2$$
$$+ 0.0081 * X_1 * X_3 - 0.0168 * X_1 * X_4$$
$$+ 0.0218 * X_2 * X_3 - 0.0031 * X_2 * X_4$$
$$+ 0.0156 * X_3 * X_4 + 0.0887 * X_1 * X_1$$
$$- 0.0137 * X_2 * X_2 + 0.0187 * X_3 * X_3$$
$$+ 0.0462 * X_4 * X_4 \quad (R^2 = 71.0\%),$$

$$\text{taste}(y) = 3.4575 - 0.08417 * X_1$$
$$- 0.1358 * X_2 - 0.1066 * X_3$$
$$- 0.0525 * X_4 - 0.0087 * X_1 * X_2$$
$$+ 0.1562 * X_1 * X_3 - 0.0975 * X_1 * X_4$$
$$+ 0.1787 * X_2 * X_3 - 0.0125 * X_2 * X_4$$
$$- 0.0275 * X_3 * X_4 + 0.05 * X_1 * X_1$$
$$- 0.0125 * X_2 * X_2 - 0.0462 * X_3 * X_3$$
$$- 0.0012 * X_4 * X_4 \quad (R^2 = 79.27\%),$$

$$\text{flavour}(y) = 33.3825 - 0.0629 * X_1$$
$$- 0.1212 * X_2 - 0.1437 * X_3$$
$$- 0.012 * X_4 - 0.0868 * X_1 * X_2$$
$$+ 0.1218 * X_1 * X_3 - 0.0743 * X_1 * X_4$$
$$+ 0.1043 * X_2 * X_3 - 0.0068 * X_2 * X_4$$
$$- 0.0356 * X_3 * X_4 + 0.0742 * X_1 * X_1$$
$$- 0.0132 * X_2 * X_2 - 0.0332 * X_3 * X_3$$
$$+ 0.0205 * X_4 * X_4 \quad (R^2 = 72.97\%),$$

$$\text{overall acceptability}(y)$$
$$= 3.4525 - 0.0925 * X_1$$
$$- 0.1925 * X_2 - 0.225 * X_3$$

$$+ 0.0216 * X_4 - 0.0212 * X_1 * X_2$$
$$+ 0.1125 * X_1 * X_3 - 0.0625 * X_1 * X_4$$
$$+ 0.1125 * X_2 * X_3 - 0.0225 * X_2 * X_4$$
$$- 0.0412 * X_3 * X_4 + 0.0495 * X_1 * X_1$$
$$+ 0.002 * X_2 * X_2 - 0.1354 * X_3 * X_3$$
$$+ 0.042 * X_4 * X_4 \quad (R^2 = 82.65\%).$$
$$(2)$$

Based on the regression analysis, the results are discussed below.

3.1. Effect of Independent Variables on Different Responses

3.1.1. Effect of Sugar. The level of sugar had a significant effect on all the responses except spread ratio. The effect of sugar was significant on diameter ($P < 0.01$), thickness ($P < 0.1$), hardness ($P < 0.1$), and stress-strain ratio ($P < 0.05$) at linear level. It affected thickness ($P < 0.01$), spread ratio ($P < 0.05$), and hardness ($P < 0.01$) at quadratic level also. Diameter and thickness increased with increase in sugar level, while hardness and stress-strain ratio decreased. Similar findings were also reported by Pareyt et al. [20] who found increase in spread of biscuits with increase in the level of sugar. Higher sucrose levels in the cookie dough recipe lead to increased sucrose dissolution during baking. This results in higher quantities of solvent phase, and as a consequence, spread increases.

It was observed that sugar had a significant effect on texture ($P < 0.1$), taste ($P < 0.05$), and OAA ($P < 0.05$) scores. Moreover, the sugar also affected significantly texture ($P < 0.05$) and flavour ($P < 0.1$) quadratically.

The overall effect of sugar was found significant on all physical characteristics as it affected taste, flavor, and OAA significantly at $P < 0.01$, $P < 0.05$, and $P < 0.1$, respectively.

3.1.2. Effect of Composite Fat (Fat and Fat Replacers). It was found that combination of fat and fat replacers significantly affected the diameter ($P < 0.01$) and spread ratio ($P < 0.05$). Diameter and spread ratio increased with increase in the level of fat replacer (maltodextrin and guar gum). Sudha et al. [9] also reported that replacement of fat with maltodextrin at different levels had improving effect on the spread and texture of the biscuits. The quadratic term of fat was significant for thickness ($P < 0.01$), spread ratio ($P < 0.05$), and hardness ($P < 0.1$) while insignificant for other responses. There was a significant ($P < 0.01$) effect of composite fat on all the sensory responses except colour.

The overall effect of composite fat was significant on all physical responses of biscuit, namely, diameter, thickness, spread rati,o and hardness, except stress-strain ratio. The overall effect of composite fat was significant on all the sensory parameters except texture. The overall effect of fat was observed to be more pronounced on OAA ($P < 0.01$)

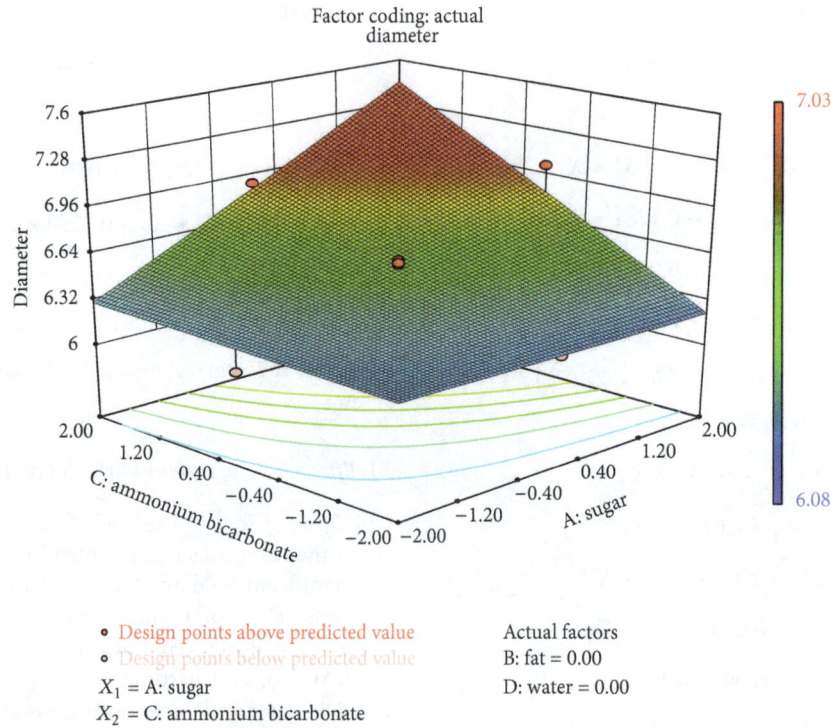

FIGURE 1: Surface plot representing the effect of sugar (X_1) and ammonium bicarbonate (X_3) on diameter of the biscuits.

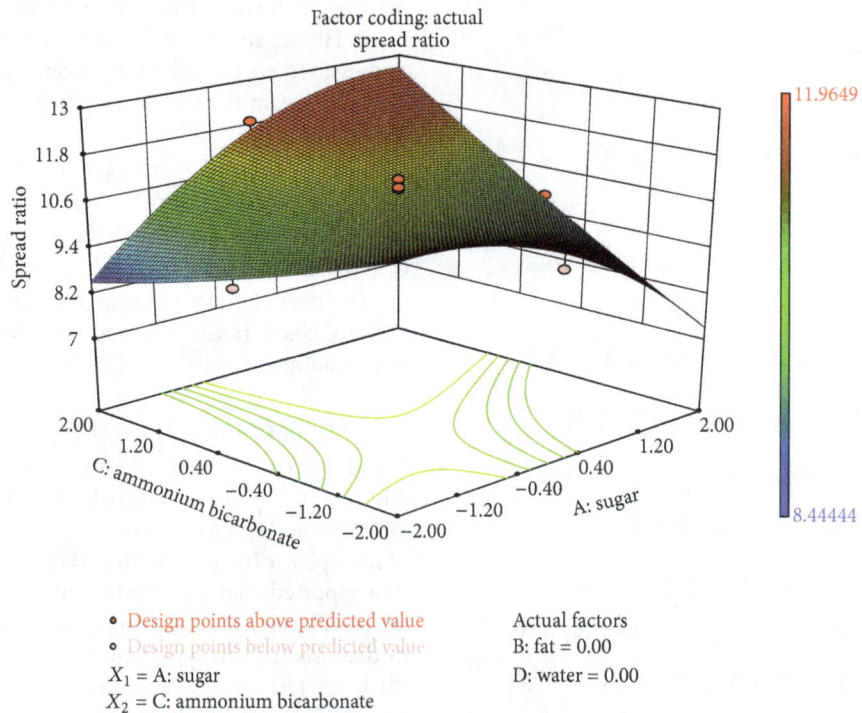

FIGURE 2: Surface plot representing the effect of sugar (X_1) and ammonium bicarbonate (X_3) on spread ratio of the biscuits.

and taste ($P < 0.01$) than on flavour ($P < 0.05$) followed by colour ($P < 0.1$).

3.1.3. Effect of Ammonium Bicarbonate.
Ammonium bicarbonate had significant effect on diameter ($P < 0.01$) and spread ratio ($P < 0.05$). With increase in the level of ammonium bicarbonate, diameter and spread ratio of product increased. Similar findings were reported by Finney et al. [21]. However, the quadratic term of ammonium bicarbonate had significant effect on hardness ($P < 0.1$).

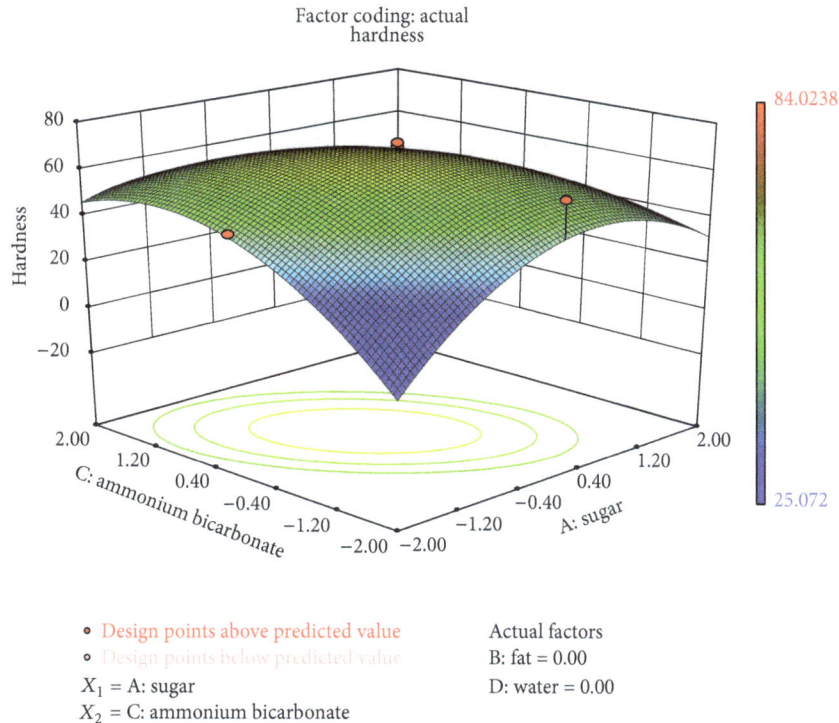

FIGURE 3: Surface plot representing the effect of sugar (X_1) and ammonium bicarbonate (X_3) on hardness of the biscuits.

The total effect of ammonium bicarbonate on all the sensory parameters was significant at $P < 0.01$ except taste ($P < 0.05$). The ammonium bicarbonate affected the OAA significantly ($P < 0.01$) at quadratic level.

The overall effect of ammonium bicarbonate was found significant on all the physical and textural parameters, and it was maximum on diameter ($P < 0.01$). The overall effect of ammonium bicarbonate on all sensory parameters was found to be significant at $P < 0.01$, except on texture, where the level of significance was $P < 0.05$.

3.1.4. Effect of Water. Water had a significant ($P < 0.01$) effect on thickness and spread ratio both at linear and quadratic levels. Maache-Rezzoug et al. [14] reported the reduction in thickness and weight of biscuits with increase in the water concentration. It also affected thickness ($P < 0.01$), hardness ($P < 0.01$), spread ratio ($P < 0.1$), and stress-strain ratio ($P < 0.1$) quadratically.

Amongst the sensory parameters, only texture was significantly ($P < 0.1$) affected by water. It decreased with increase in water level. The overall effect of water on thickness was significant at $P < 0.01$, while it was significant at $P < 0.05$ on spread ratio and hardness. It was found that the overall effect of water on all the sensory responses was insignificant.

3.1.5. Synergistic Effect of Independent Variables. The interaction of sugar and ammonium bicarbonate affected all physical characteristics significantly, while interaction of composite fat and water had a significant effect on thickness ($P < 0.1$) and spread ratio ($P < 0.1$). The interaction

between ammonium bicarbonate and water was significant on thickness. Figures 1–3 showed the effect of sugar and ammonium bicarbonate on diameter, spread ratio, and hardness, respectively. Diameter and spread ratio increased with increase in sugar and ammonium bicarbonate levels as shown in Figures 1 and 2, respectively. A similar finding was also reported by Finney et al. [21]. Maache-Rezzoug et al. [14] also showed the positive correlation between sugar content and length. Hardness was maximum around the centre level of both sugar and ammonium bicarbonate (Figure 3). Stress-strain ratio was found to be significantly ($P < 0.05$) affected by the interaction of ammonium bicarbonate with sugar. The interaction of sugar and ammonium bicarbonate was found to be more significant for taste ($P < 0.01$) than flavour and OAA ($P < 0.05$). The interactive effect of water and sugar was found significant on taste ($P < 0.05$).

Hardness ($P < 0.1$) and stress-strain ratio ($P < 0.05$) were significantly affected by ammonium bicarbonate and fat as shown in Figures 4 and 5. Figure 5 shows that the hardness increased as the level of fat in the formulation decreased. It can also be concluded that the hardness increased at higher values of fat replacer (maltodextrin and guar gum). Sudha et al. [9] also demonstrated the effect of maltodextrin on the breaking strength of biscuit. They reported that force required to break biscuits containing 70% less fat was almost three times more than that required to break the control biscuits. Mamat et al. [22] also reported higher hardness for a biscuit with lower-fat content than a biscuit with normal-fat content. The stress-strain ratio ($\sigma_{max}/\varepsilon_{max}$) is related to brittleness of the sample [23, 24]. Zoulias et al. [25] also reported the increase in stress-strain ratio of cookies by replacement of fat

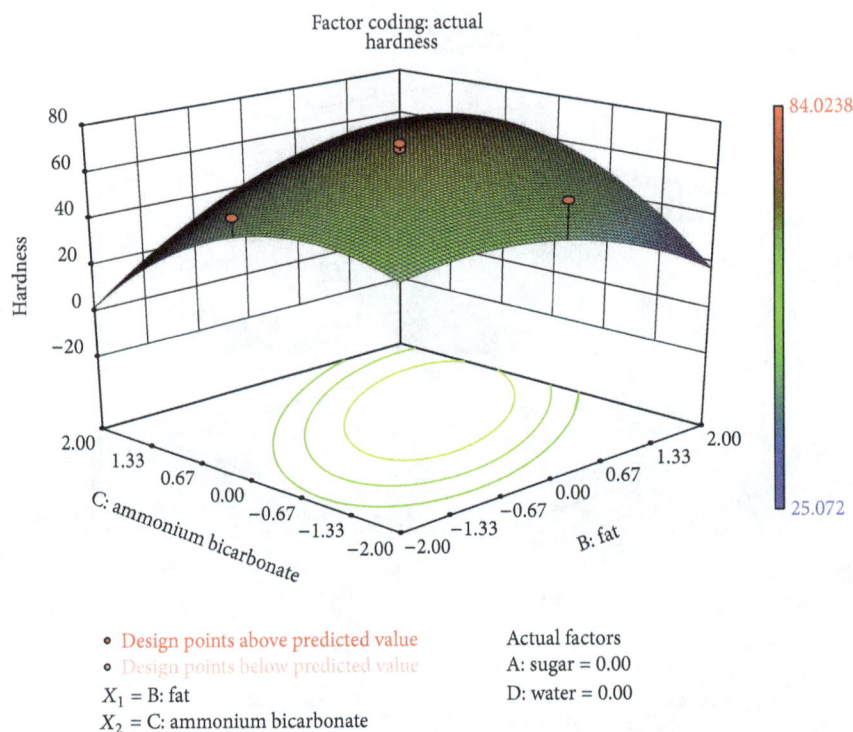

Factor coding: actual
hardness

FIGURE 4: Surface plot representing the effect of fat (X_2) and ammonium bicarbonate (X_3) on hardness of the biscuits.

with fat mimetics which resulted in the production of brittle cookies. Cookies that present a high ratio of $\sigma_{max}/\varepsilon_{max}$ are less compressible, more brittle and break easily. Brittleness can be considered a pleasant sensorial characteristic for the cookies as far as it does not become extremely great.

The level of ammonium bicarbonate and fat also had a significant effect on OAA (Figure 6). There was a decrease in OAA at higher levels of fat replacers. A similar effect of fat replacers on cookies was reported by Zoulias et al. [7]. They found that the low-fat cookies had significantly lower flavour and general acceptance scores than the control cookies. It was observed that the effect of interaction of ammonium bicarbonate and composite fat was significant on colour, taste, and flavour.

The effect of interaction of fat and water was significant on thickness ($P < 0.1$) and spread ratio ($P < 0.1$). Spread ratio decreased with increase in water and fat replacer level (Figure 7). It was noticed that colour score decreased significantly ($P < 0.1$) by increase in the levels of fat replacers and water. Low-fat biscuits were found to be darker than the control biscuits because of the higher degree of the Maillard browning reactions which might be the result of carbohydrate nature of these fat replacers. A similar finding was also reported by Sanchez et al. [26] who found high colour intensity in biscuits made from carbohydrate-based fat replacers. It was also observed that the interaction of sugar and fat significantly affected colour ($P < 0.05$) and flavour ($P < 0.1$).

The interaction between ammonium bicarbonate and water significantly affected thickness. Goldstein and Seetharaman [27] correlated the increase in cookie height

with increasing moisture content in the samples. Also, colour was found to be significantly ($P < 0.1$) affected by the interaction of ammonium bicarbonate with water.

3.1.6. Combined Effect of Independent Variables. ANOVA (Tables 2 and 3) is used to show the total effect of variables individually and combination of all variables at linear, quadratic and interactive levels. It was found that all independent variables had significant effect on diameter ($P < 0.01$), thickness ($P < 0.01$), and spread ratio ($P < 0.01$) at linear level. They affected thickness and spread ratio quadratically at $P < 0.01$ and $P < 0.05$, respectively. At interactive level, they affected spread ratio ($P < 0.1$), hardness ($P < 0.1$), stress-strain ratio ($P < 0.1$), and thickness ($P < 0.05$).

The combined effect of independent variables was found on all sensory parameters at linear, quadratic, and interactive levels. They affected all parameters linearly at $P < 0.01$. At quadratic level, they affected texture ($P < 0.1$) and OAA ($P < 0.05$). At interactive level, they significantly affected colour ($P < 0.05$), flavour ($P < 0.05$), and taste ($P < 0.01$).

3.2. Optimization of Independent Variables for Low-Fat Biscuits. Design-Expert (8.0.6 trial version) software was employed to optimise ingredient level based on maximum spread ratio and OAA and minimum stress-strain ratio and composite fat (fat, maltodextrin, and guar gum) of biscuits using numerical methods of optimization. The optimum condition for different parameters obtained was sugar of 31.74 g, fat of 13.55 g, maltodextrin of 21.15 g, guar gum of 0.3 g, ammonium bicarbonate of 2.21 g, and water of 21 mL.

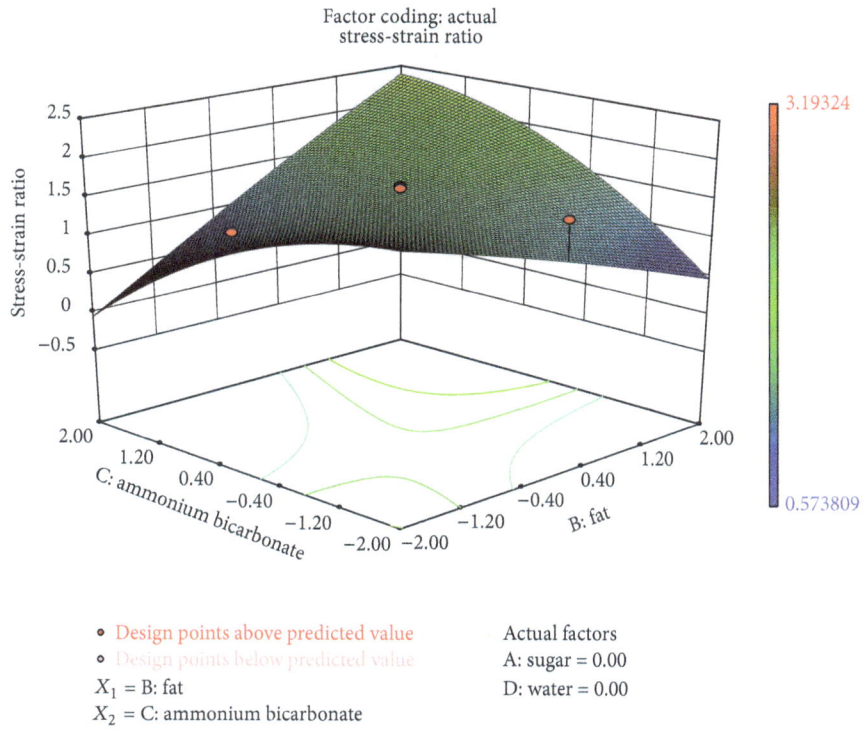

FIGURE 5: Surface plots representing the effect of fat (X_2) and ammonium bicarbonate (X_3) on stress-strain ratio of the biscuits.

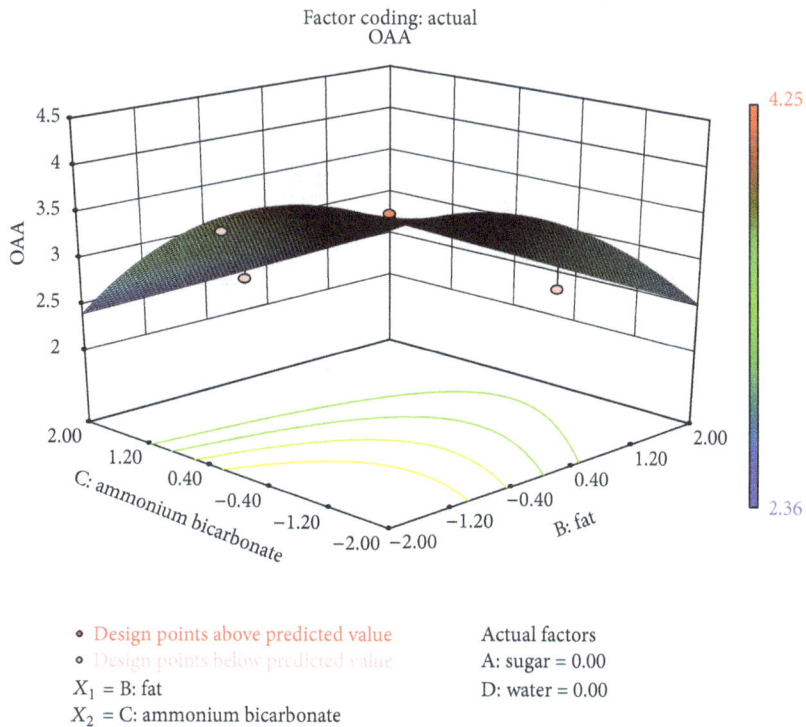

FIGURE 6: Surface plot representing the effect of fat (X_2) and ammonium bicarbonate (X_3) on overall acceptability of the biscuits.

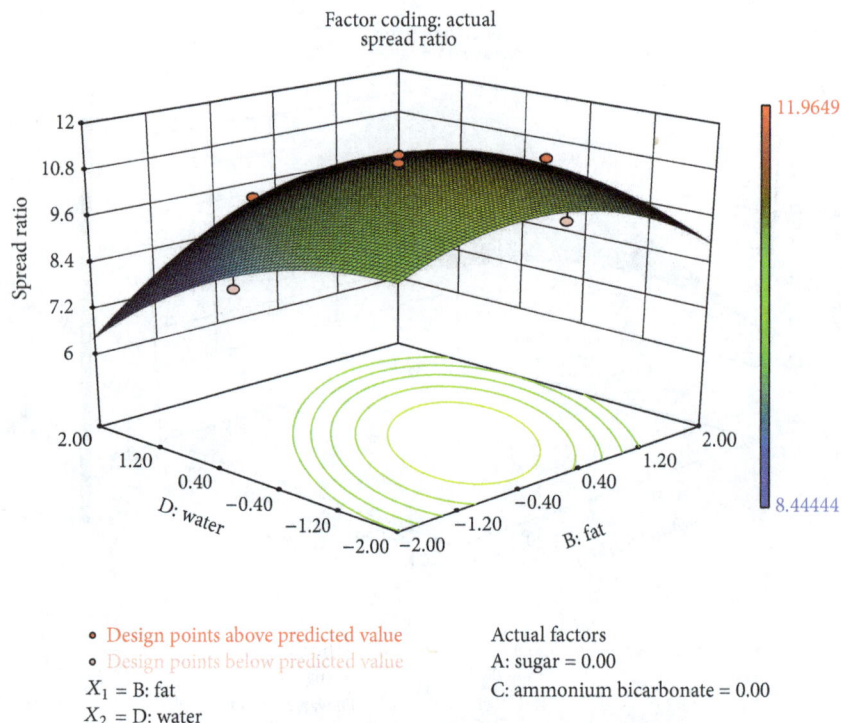

FIGURE 7: Surface plot representing the effect of fat (X_2) and water (X_4) on spread ratio of the biscuits.

TABLE 2: ANOVA for the overall effect of processing parameters on the physical and textural responses[a].

Responses	Mean squares				
	Diameter (cm)	Thickness (cm)	Spread ratio	Hardness (N)	Stress-strain ratio
Total individual effect of processing parameters					
Sugar	0.1106[***]	0.0046[***]	1.3295[**]	613.2477[***]	0.6928[**]
Fat	0.0612[*]	0.0031[**]	1.1300[**]	269.4439[*]	0.3601
Ammonium bicarbonate	0.1537[***]	0.0026[**]	1.4404[**]	424.7674[**]	0.4937[*]
Water	0.0217	0.0059[***]	1.2738[**]	353.4363[**]	0.1897
Combined effect of all processing parameters					
Linear level	0.3164[***]	0.0046[***]	1.9498[***]	188.1162	0.4062
Quadratic level	0.0108	0.0070[***]	1.5964[**]	171.7865	0.3586
Interactive level	0.0356	0.0029[**]	0.9737[*]	238.5898[*]	0.4685[*]

[a]Significant at [*]10%, [**]5%, and [***]1%.

TABLE 3: ANOVA for the overall effect of processing parameters on the sensory responses[a].

Responses	Mean squares				
	Colour	Texture	Taste	Flavour	Overall acceptability
Total individual effect of processing parameters					
Sugar	0.1553	0.0678	0.1575[***]	0.1409[**]	0.1100[*]
Fat	0.2152[*]	0.0681	0.1924[***]	0.1307[**]	0.2214[***]
Ammonium bicarbonate	0.4030[***]	0.1015[**]	0.25[***]	0.1921[***]	0.4377[***]
Water	0.1559	0.0393	0.0465	0.0251	0.0322
Combined effect of all processing parameters					
Linear level	0.4149[***]	0.2577[***]	0.238[***]	0.2368[***]	0.5802[***]
Quadratic level	0.0665	0.0779[*]	0.0354	0.0532	0.1666[**]
Interactive level	0.2267[**]	0.0034	0.1782[***]	0.107[**]	0.0850

[a]Significant at [*]10%, [**]5%, and [***]1%.

TABLE 4: Verification of the models by comparing the experimental values with the predicted values.

Response	Predicted value	Experimental value[*]
Spread ratio	11.92	10.93 ± 0.06
Hardness, N	25.07	31.47 ± 1.02
Stress-strain ratio	0.409	0.599 ± 0.017
Overall acceptability	4.53	4.17 ± 0.41

[*] Average of ten experiments.

The levels were based on 100 g flour. The comparison between predicted and experimental results is shown in Table 4. It shows a good agreement between predicted and experimental values.

4. Conclusions

RSM was successfully used for optimizing different ingredients for the manufacture of low-fat soft dough biscuits. Results of the study indicate that hardness increases with increase in level of sugar and fat replacers and decrease in fat level. With increase in ammonium bicarbonate, diameter and spread ratio increase. Interactive effect of increased fat and water level decreases spread ratio. The optimized product had 62.5% replacement of fat with carbohydrate-based fat replacers, maltodextrin and guar gum.

Acknowledgment

The authors are grateful to Dr. S. K. Jha, a Senior Scientist at the Division of Post Harvest Technology, Indian Agricultural Research Institute, New Delhi, for permitting the use of texture analyzer.

References

[1] R. D. O'brien, *Fats and Oils: Formulating and Processing for Applications*, CRC Press, Boca Raton, Fla, USA, 2003.

[2] C. A. Hassell, "Nutritional implications of fat substitutes," *Cereal Foods World*, vol. 38, pp. 142–144, 1993.

[3] C. C. Akoh, "Fat replacers," *Food Technology*, vol. 52, no. 3, pp. 47–53, 1998.

[4] A. M. Altschul, "Low-calorie foods—a scientific status summary by the Institute of Food Technologists expert panel on food safety and nutrition," *Food Technology*, vol. 43, no. 4, p. 113, 1989.

[5] G. E. Inglett, K. Warner, and R. K. Newman, "Sensory and nutritional evaluations of oatrim," *Cereal Foods World*, vol. 39, pp. 755–759, 1994.

[6] F. D. Conforti, S. A. Charles, and S. E. Dunkan, "Sensory evaluation and consumer acceptance of carbohydrate-based fat replacers in biscuits," *Journal of Consumer Studies and Home Economics*, vol. 20, pp. 285–296, 1996.

[7] E. I. Zoulias, V. Oreopoulou, and E. Kounalaki, "Effect of fat and sugar replacement on cookie properties," *Journal of the Science of Food and Agriculture*, vol. 82, no. 14, pp. 1637–1644, 2002.

[8] E. Gallagher, C. M. O'Brien, A. G. M. Scannell, and E. K. Arendt, "Use of response surface methodology to produce functional short dough biscuits," *Journal of Food Engineering*, vol. 56, no. 2-3, pp. 269–271, 2003.

[9] M. L. Sudha, A. K. Srivastava, R. Vetrimani, and K. Leelavathi, "Fat replacement in soft dough biscuits: its implications on dough rheology and biscuit quality," *Journal of Food Engineering*, vol. 80, no. 3, pp. 922–930, 2007.

[10] D. Manley, *Technology of Biscuits, Crackers and Cookies*, Woodhead Publishing, Cambridge, UK, 2nd edition, 1996.

[11] K. Wehrle, E. Gallagher, D. P. Neville, M. K. Keogh, and E. K. Arendt, "Microencapsulated high-fat powders in biscuit production," *European Food Research and Technology*, vol. 208, no. 5-6, pp. 388–393, 1999.

[12] R. S. Manohar and P. H. Rao, "Effect of emulsifiers, fat level and type on the rheological characteristics of biscuit dough and quality of biscuits," *Journal of the Science of Food and Agriculture*, vol. 79, no. 10, pp. 1223–1231, 1999.

[13] J. L. Vetter, "Calorie and fat modified bakery products," *American Institute of Baking. Research Department Technical Bulletin*, vol. 13, no. 5, 1991.

[14] Z. Maache-Rezzoug, J. M. Bouvier, K. Allaf, and C. Patras, "Effect of principal ingredients on rheological behaviour of biscuit dough and on quality of biscuits," *Journal of Food Engineering*, vol. 35, no. 1, pp. 23–42, 1998.

[15] A. I. Khuri and J. A. Cornell, *Response Surfaces, Designs and Analysis*, Marcel Dekker, New York, NY, USA, 1987.

[16] P. R. Whitley, *Biscuit Manufacture*, Applied Science Publisher, London, UK, 1970.

[17] AACC, *Approved Methods of American Association of Cereal Chemists*, Cereal Laboratory Methods., St. Paul. Minn, USA, 1967.

[18] AOAC, *Official Methods of Analysis*, Association of Official Analytical Chemists, Washington, DC, USA, 10th edition, 1984.

[19] E. Larmond, *Laboratory Methods for Sensory Evaluation of Foods*, Canada Department of Agriculture, Ottawa, Canada, 1937.

[20] B. Pareyt, F. Talhaoui, G. Kerckhofs et al., "The role of sugar and fat in sugar-snap cookies: structural and textural properties," *Journal of Food Engineering*, vol. 90, no. 3, pp. 400–408, 2009.

[21] K. F. Finney, W. T. Yamazaki, and V. H. Morris, "Effects of varying quantities of sugar, shortening and ammonium bicarbonate on the spreading and top grain of sugar-snap cookies," *Cereal Chemistry*, vol. 27, pp. 30–41, 1950.

[22] H. Mamat, M. O. Abu Hardan, and S. E. Hill, "Physicochemical properties of commercial semi-sweet biscuit," *Food Chemistry*, vol. 121, no. 4, pp. 1029–1038, 2010.

[23] C. S. Gaines, "Instrumental measurement of the hardness of cookies and crackers," *Cereal Foods World*, vol. 36, pp. 989–996, 1991.

[24] J. C. Jackson, M. C. Bourne, and J. Barnard, "Optimization of blanching for crispness of banana chips using response surface methodology," *Journal of Food Science*, vol. 61, no. 1, pp. 165–166, 1996.

[25] E. I. Zoulias, V. Oreopoulou, and C. Tzia, "Textural properties of low-fat cookies containing carbohydrate- or protein-based fat replacers," *Journal of Food Engineering*, vol. 55, no. 4, pp. 337–342, 2002.

[26] C. Sanchez, C. F. Klopfenstein, and C. E. Walker, "Use of carbohydrate-based fat substitutes and emulsifying agents in reduced-fat shortbread cookies," *Cereal Chemistry*, vol. 72, no. 1, pp. 25–29, 1995.

[27] A. Goldstein and K. Seetharaman, "Effect of a novel mono-
glyceride stabilized oil in water emulsion shortening on cookie
properties," *Food Research International*, vol. 44, no. 5, pp. 1476–
1481, 2011.

5

Comparative Effect of Crude and Commercial Enzyme on the Juice Recovery from Bael Fruit (*Aegle marmelos* Correa) Using Principal Component Analysis

Anurag Singh,[1] **H. K. Sharma,**[2] **Sanjay Kumar,**[2] **Ashutosh Upadhyay,**[3] **and K. P. Mishra**[4]

[1] *Department of Food Technology, FET, RBS Engineering Technical Campus, Bichpuri, Agra 283105, India*

[2] *Food Engineering and Technology Department, Sant Longowal Institute of Engineering and Technology (SLIET), Longowal, Sangrur, Punjab 148106, India*

[3] *National Institute of Food Technology Entrepreneurship and Management, Kundli, Sonipat, Haryana 131028, India*

[4] *Faculty of Engineering and Technology, Mahatma Gandhi Chitrakoot Gramodaya Vishwavidyalaya, Chitrakoot, Satna, Madhya Pradesh 485331, India*

Correspondence should be addressed to Anurag Singh; rbsanurag@gmail.com

Academic Editor: Marie Walsh

The effect of incubation time, incubation temperature, and crude enzyme concentration was observed on the yield, viscosity, and clarity of the juice obtained from bael fruit pulp. The recommended enzymatic treatment conditions from the study were incubation time 475 min, incubation temperature 45°C, and crude enzyme concentration 0.20 mL/25 g bael fruit pulp. The recovery, viscosity, and clarity of the juice under these conditions were 82.9%, 1.41 cps, and 21.32%T, respectively. The variables, clarity, and yield were found as principal components for comparing different samples of the juice treated with enzyme.

1. Introduction

The bael fruit (*Aegle Marmelos* Correa) has been attributed with various nutritional and therapeutic properties. The fruit has excellent aroma which is not destroyed even during processing [1]. The bael fruit pulp contains many functional and bioactive compounds such as carotenoids, phenolics, alkaloids, coumarins, flavonoids, terpenoids, and other antioxidants which may protect against chronic diseases [2]. It has been surmised that the psoralen in the pulp increases tolerance of sunlight and aids in the maintaining of normal skin color and is considered fruitful in the treatment of leucoderma. The marmelosin ($C_{13}H_{12}O_3$) content, found in the bael fruit, is considered as panacea of various stomach ailments [3]. Bael fruit, because of its hard shell, mucilaginous texture, and numerous seeds, is not popular as fresh fruit. However, the excellent flavor and nutritive and therapeutic value of bael fruits show potential for processing into values added products. Bael is commercially considered as an important fruit, but the potential of the fruit is not fully tapped.

The edible pulp, 100 g of bael fruit contains 61.5 g water, 1.8 g protein, 0.39 g fat, 1.7 g minerals, 31.8 g carbohydrate, 55 mg carotene, 0.13 mg thiamine, 1.19 mg riboflavin, 1.1 mg niacin, and 7 to 21 mg ascorbic acid [4]. Generally, three methods of juice extraction are employed, namely, cold, hot, and enzymatic methods. The use of fungal enzyme in fruit juice extraction had shown significant increase in juice recovery as compared to cold and hot extraction methods. The enzymes, mainly pectinases, and cellulases assist in pectin and cellulolytic hydrolysis, respectively, which cause a reduction in pulp viscosity and a significant increase in juice yield [5].

The extraction of bael juice on large scale has not been explored for its commercial scale viability and exploitation but conventionally, the extraction includes addition of water to pulp, boiling and pressing of juice from the mixture. The residual pulp remaining after juice extraction still contains valuable extractable material such as particulate, flavour, and soluble solids which may further improve the final quality of the juice. By adding cell wall liquefying enzymes, it is possible to further extract valuable juice components from pulp.

The area requires wider research in terms of utilization of residue, enhanced juice yield with optimum overall acceptability. The application of commercial enzyme for the different juice clarification is reported by several researchers [6–8]. It is more economical to use crude enzyme (spore free and produced from GRAS fermentation) for the improvement of juice yield and clarity. Therefore, the present study was undertaken to use crude enzyme from *A. Niger* for the treatment of the bael pulp to improve the juice yield with optimum overall acceptability and examine the comparative effect of enzymes in crude and purified form by using Principal Component Analysis (PCA).

2. Materials and Methods

2.1. Materials. Fully ripe fresh bael fruits (*Aegle marmelos* Correa) of Kagazi variety, without any visual defects, were procured from Agricultural farm of RBS College, Bichpuri, Agra (India). The bael fruits were broken by hammering, and the pulp was scooped out with the help of stainless steel spoon. The scooped pulp was homogenized by blending manually. This pulp was used to extract juice.

2.2. Crude Enzyme Preparation. The strain of *Aspergillus niger* NCIM 548 was obtained from the National Chemical Laboratory, Pune. The strain was cultured on potato dextrose agar slant and subcultured after every 6–8 weeks. This strain was used for the production of crude enzyme under solid state fermentation (SSF) using wheat bran, corn bran, and kinnow peel (in 2 : 1 : 2 ratio) as substrate as per the method reported by Kumar et al. [9]. The enzyme contained 50 U/mL of the pectinase and 20 U/mL of the cellulase and was used for the treatment of bael fruit pulp to improve the juice yield and quality.

2.3. Experimental Design and Statistical Analysis. Response surface methodology (RSM) was used in designing the experiments as it provides the modeling and analysis of the problem in which several variables influence the output parameter, and the objective is to optimize this parameter [10]. RSM provides a reduced number of experimental runs needed to obtain sufficient information for statistically acceptable results. A five-level three-factor central composite rotatable design was employed. The independent variables were the temperature of enzyme treatment (X_1), time of treatment (X_2), and used enzyme concentration (X_3). The variables and their levels were chosen based on the limited literature available on enzymatic hydrolysis of fruits [6–8]. These were the temperature (X_1; 35–55°C), time (X_2; 210–540 min) of the enzymatic treatment, and concentration of enzyme (X_3; 0.06–0.20 mL/25 g pulp). The experimental design matrix in coded (x) form and at the actual level (X) of variables is given in Table 1. A total of 20 experiments

TABLE 1: Experimental range and levels of the independent variables.

Variables	Range and levels				
	−1.68	−1	0	+1	+1.68
Temp. (X_1, °C)	28.18	35	45	55	61.82
Time (X_2, min)	97.5	210	375	540	652.5
Conc. of crude enzyme (X_3, mL)	0.01	0.06	0.13	0.20	0.25

were carried out by using crude enzyme under different experimental conditions as given in Table 2. The response function (Y) was related to the coded variables by a second degree polynomial equation (1) as follows:

$$y = b_0 + b_1 x_1 + b_2 x_2 + b_3 x_3$$
$$+ b_{12} x_1 x_2 + b_{13} x_1 x_3 + b_{23} x_2 x_3 \qquad (1)$$
$$+ b_{11} x^2_1 + b_{22} x^2_2 + b_{33} x^2_3 + \varepsilon.$$

The coefficients of the polynomial were represented by b_0 (constant), b_1, b_2, b_3 (linear effects), b_{12}, b_{13}, b_{23} (interaction effects), b_{11}, b_{22}, b_{33} (quadratic effects), and ε (random error).

2.4. Commercial Enzyme Treatment of Bael Pulp under Optimized Conditions. The pretreatment conditions based on the application of commercial enzyme on the juice recovery from bael fruit were also optimized in our laboratory using CCRD design of Response Surface Methodology. The optimum conditions observed were concentration of pectinase 5 mg/25 g of pulp (1.64 IU/mg), time 425 min, and temperature 47°C as given in Table 4 [7].

2.5. Analysis of Response Variables

2.5.1. Enzymatic Treatment and Juice Yield. The bael pulp for the treatment of enzymes was prepared as per the procedure adopted by Singh et al. [7]. The temperature was adjusted to the required level by using a high precision water bath (Seco, Model 129, India) for all enzymatic treatment combinations. At the completion of the enzyme treatment, the treated mixture was filtered by using 6-fold cheese cloth, and the extract was heated at 90°C for 5 min to inactivate the enzyme. The extract thus obtained was considered as clear juice for determining the juice yield.

2.5.2. Clarity and Viscosity. The clarity was determined by the method given by Krop and Pilnik [11], and viscosity of the juice was determined as per the method reported by Ranganna [12]. The juice was shaken, and 10 mL portion of juice was centrifuged at 3000 rpm for 10 min to remove pulp coarse cloud particles. The clarity of the juice obtained was determined by measuring the transmittance at a wavelength of 590 nm using UV-VIS spectrophotometer (UV 5704SS, Electronics Corporation of India Ltd.).

Time required to flow through the capillary section of the Oswald viscometer was noted using a stopwatch for the reference and the sample at 20 ± 2°C.

Comparative Effect of Crude and Commercial Enzyme on the Juice Recovery from Bael Fruit (Aegle marmelos Correa) Using Principal Component Analysis

35

TABLE 2: The central composite rotatable experimental design employed for enzymatic hydrolysis pretreatment of bael pulp.

Exp. no.	Coded variables			Uncoded variables			Responses		
	X_1	X_2	X_3	Temp. (°C)	Time (min.)	Conc. of crude enzyme (mL)/25 g pulp	% Age yield	Viscosity (cps)	Clarity (%T)
1	−1	−1	−1	35	210	0.06	70.3	1.61	18.7
2	1	−1	−1	55	210	0.06	69.8	1.69	17.6
3	−1	1	−1	35	540	0.06	71.1	1.51	19.3
4	1	1	−1	55	540	0.06	72.8	1.49	18.5
5	−1	−1	1	35	210	0.2	77.1	1.44	18.3
6	1	−1	1	55	210	0.2	75.8	1.61	17.6
7	−1	1	1	35	540	0.2	79.6	1.42	20.5
8	1	1	1	55	540	0.2	80.1	1.45	20.1
9	−1.68	0	0	28.18	375	0.13	70.4	1.52	17.7
10	1.68	0	0	61.81	375	0.13	72.4	1.65	16.6
11	0	−1.68	0	45	97.50	0.13	73.2	1.59	18.7
12	0	1.68	0	45	652.5	0.13	80.1	1.46	20.5
13	0	0	−1.68	45	375	0.01	70.1	1.6	19.1
14	0	0	1.68	45	375	0.25	83.7	1.4	21
15	0	0	0	45	375	0.13	81.4	1.45	20.8
16	0	0	0	45	375	0.13	81.7	1.42	20.5
17	0	0	0	45	375	0.13	80.2	1.42	21.1
18	0	0	0	45	375	0.13	80.6	1.43	21
19	0	0	0	45	375	0.13	78.5	1.45	20.6
20	0	0	0	45	375	0.13	81	1.43	20.6

2.6. Optimization and Verification. The optimal level of three independent variables, namely, temperature (X_1), time (X_2) of the enzymatic treatment, and concentration of enzyme (X_3), was established with the help of graphical and numerical optimization procedures resulting to desirable responses which were maximum yield, maximum clarity, and minimum viscosity. For graphical optimization, a three-dimensional response surface was plotted by varying the two variables in the experimental range while keeping rest one variable constant at the centre point. The exact optimum value of individual and multiple responses was determined by a numerical optimization process using Design Expert software (DE–6) (Trial version; STAT-EASE Inc., Minneapolis, MN, USA). To verify the predicted results, the experiments were conducted at the optimized conditions. The predictive models were validated on the basis of R^2, adjusted R^2, F-value, Lack of fit, and so forth obtained from the analysis of the experimental data by using Design Expert software.

2.7. Comparative Study Using Principal Component Analysis (PCA). For exploring the feasibility of the application of the crude enzyme in place of commercial enzymes, the usages of commercial enzymes were compared with the crude enzymes under the optimized conditions, on the same variety of bael fruit. Principal component analysis was carried out by using software Minitab version 16.1 (Trial version; Minitab Inc., USA) to form a smaller number of uncorrelated variables from a large set of data. A large number of variables are

reduced to a few variables called principal components (PCs) that describe the greatest variance in the data analyzed [13]. The technique helps to understand the similarities and differences between the samples treated with crude and commercial enzymes separately and also establishes an interrelationship between the measured properties of the samples.

3. Results and Discussion

3.1. Optimization of Pretreatment Conditions by Using Crude Enzyme. The juice extracted from enzyme treated and untreated (control) pulp was evaluated for the juice yield (%), viscosity, and clarity. Table 2 shows the juice yield, apparent viscosity, and clarity under the different experimental conditions by using crude enzyme. It is clear from the data obtained that the juice has been improved significantly with respect to quantity as well as quality by the enzymatic treatment.

3.1.1. Adequacy of the Models for the Different Responses. The model was judged for its fitness and adequacy by the coefficient of determination (R^2), which is the ratio of the explained variation to the total variation. The closer the R^2 value to unity, the better the empirical model fits the actual data. The coefficients of determination, R^2, in the model were 0.9764, 0.9738, and 0.9806 (Table 3) for the regressed models predicting the juice yield, viscosity, and clarity, respectively, suggesting a good fit for the models. The adjusted R^2 was

TABLE 3: Regression coefficients of predicted quadratic polynomial models for the responses for the model.

Coefficients	Juice yield	Viscosity	Clarity
Intercept	80.57[a]	1.43[a]	20.77[a]
Linear			
A	0.28	0.035[a]	−0.36[b]
B	1.63[b]	−0.05[a]	0.68[a]
C	3.77[a]	−0.05[a]	0.41[b]
Quadratic			
A^2	−3.26[a]	0.05[a]	−1.28[a]
B^2	−1.40[b]	0.03[b]	−0.41[b]
C^2	−1.31[b]	0.02[b]	−0.25[b]
Cross product			
$A*B$	0.5	−0.03[b]	0.08
$A*C$	−0.25	0.02[b]	0.1
$B*C$	0.38	0.029[c]	0.4[b]
$R^{2(d)}$	0.9764	0.9738	0.9806
Adj. $R^{2(e)}$	0.9551	0.9503	0.9631
CV[f]	1.31	1.33	1.34
F value	45.92	41.38	56.04

Statistically significant at [a]$P < 0.001$, [b]$P < 0.05$, and [c]$P < 0.10$; [d]coefficient of multiple determination; [e]adjusted R^2; [f]coefficient of variance.

a corrected value for R^2 after elimination of the unnecessary model terms, which was very close to their corresponding R^2 values for all the responses. Higher values of adjusted R^2 also confirm the significance of the models.

The F value of 45.92, 41.38, and 56.04 for juice yield, viscosity, and juice clarity, respectively, also inferred that the models were significant ($P < 0.001$). The model for the juice yield, viscosity, and clarity can be derived by using the coefficients (Table 3) for the predictions of the data. The predicted models seemed to reasonably represent the observed values. Thus, the responses were adequately explained by the model.

The coefficient of variation (CV) describes to which extent the data are dispersed and is defined as a measure of residual variation of the data relative to the size of the mean; the small values of CV give better reproducibility. The small CV values 1.31, 1.33, and 1.34 (Table 3) of juice yield, viscosity, and clarity, respectively, suggested that the experimental results were precise and reliable.

3.1.2. Response Surface Analysis

(1) *Juice Yield.* The juice yield ranged from 69.8% to 83.7% (Table 2). The yield was minimum when crude pectinase enzyme 0.06 mL/25 g pulp was used for 210 min at 55°C, whereas the maximum juice yield was at 0.25 mL/25 g crude pectinase enzyme concentration for 375 min at 45°C.

To understand the interaction of different variables and to find the optimum level of each variable, the response surface curves were plotted. The response surface curves for juice yield are shown in Figures 1(a) and 1(b). Each response surface curve explains the effect of two variables while keeping the third variable at middle level. Figure 1(a)

represents the interactive effect of incubation temperature (X_1) and incubation time (X_2) on the juice yield, whereas the concentration of crude enzyme (X_3) was kept at middle level, that is, 0.13 mL/25 g of bael pulp. The juice yield increased with the increase in both time and temperature up to 473 min of time and 45.8°C temperature. With further increase in temperature, the yield slightly decreased but remains unaffected of increase in incubation time. The decrease in juice yield with increasing temperature beyond 45.8°C may be due to denaturation of protein which may lead to decrease in enzyme activity at higher temperature. The results are supported by the findings of Kaur et al. [6], who reported that the maximum juice yield from guava is obtained by pectinolytic enzyme treatment of pulp at 43.3°C temperature for 447 min of time.

Figure 1(b) presents the interaction effect of incubation temperature (X_1) and crude enzyme concentration (X_3) to juice yield. At higher temperature and enzyme concentration, the juice yield followed a linear behaviour which reflects that with increase in enzyme concentration and temperature, juice yield increased up to maximum concentration of enzyme, that is, 0.20 mL/25 g pulp and 45°C temperature, respectively. Thereafter, the juice yield decreased slowly beyond 45°C, which may be due to decrease in enzyme activity at higher temperature. The increase in plum juice yield with pectinase enzyme is also supported by Singh and Das Gupta [14] who reported that pectinases increase the juice yield from 48% to 77.5%.

(2) *Viscosity.* The viscosity of the bael juice ranged from 1.40 to 1.69 cps under the different experimental conditions (Table 2). The minimum viscosity was observed at 0.25 mL/25 g crude enzyme concentration for 375 min at 45°C. The corresponding condition for maximum viscosity was 0.06 mL/25 g crude enzyme concentration, for 210 min at 55°C. It is clear from Figure 2(a) that viscosity decreased with increase in both time and temperature up to 510 min of time and 44°C temperature. The findings are in accordance with Kumar and Sharma [8] who reported that the viscosity of the juice decreased with increase in temperature up to 47°C. The temperature increases the rate of enzymatic reactions. Upon enzyme treatment, degradation of pectin leads to a reduction of water holding capacity, and consequently free water was released to the system thus reducing the viscosity of the juice.

Figure 2(b) presents the interaction effect of incubation temperature (X_1) and crude enzyme concentration (X_3) to viscosity. The viscosity decreased with the increase in both concentration of enzyme and incubation temperature up to 0.20 mL/25 g of pulp of crude enzyme concentration and 40°C temperature. Lee et al. [15] observed that the viscosity of the juice decreases when the enzyme concentration is increased from 0.01% to its maximum value (0.1%).

(3) *Juice Clarity.* The clarity of the extracted juice was in a range from 16.60 to 21.10 %T. The minimum clarity was observed when 0.13 mL/25 g pulp, crude enzyme concentration was used for 375 min at 61.81°C. The rate of clarification

TABLE 4: Optimization of process variables with respect to juice yield, viscosity, and juice clarity.

		Commercial enzymes			Crude enzyme	
		Optimum value (In the range)	Optimum value (Targeted)		Optimum value (In the range)	Optimum value (Targeted)
Variables	Temperature (°C)	46.57	47	Temperature (°C)	44.97	45
	Time (min)	425.21	425	Time (min)	474.50	475
	Conc. of pectinase (mg/25 g pulp)	4.96	5.00	Conc. of crude enzyme (mL/25 g pulp)	0.20	0.20
		Predicted value	Experimental value		Predicted value	Experimental value
Responses	Juice yield (%)	85.19	84.5	Juice yield (%)	83.73	82.9
	Viscosity (cps)	1.37	1.35	Viscosity (cps)	1.39	1.41
	Juice clarity (%T)	23.25	22.43	Juice clarity (%T)	21.40	21.32

TABLE 5: Comparison of crude and commercial enzymes for the improvement of juice recovery from bael.

Parameter	Control[*]	Commercial enzyme treatment		Crude enzyme treatment	
		Treated	Difference	Treated	Difference
Juice yield (%)	69.1	84.5	15.40	82.9	13.8
Viscosity (cps)	1.69	1.35	0.34	1.41	0.28
Juice clarity (%T)	17.0	22.43	5.43	21.32	4.32

[*]The control sample was prepared by mixing 25 g of pulp with 62.5 mL water and then filtered through cheese cloth without any enzymatic treatment.

increases with an increase in enzyme concentration may be due to the exposure of positively charged protein beneath, which reduces electrostatic repulsion between cloud particles causing these particles to aggregate into larger particles and eventually settled out [16]. Multiple regression technique was used to develop a response surface analysis of juice clarity as a function of enzymatic hydrolysis process variables. All the variables affected the juice clarity significantly (Table 2). The time (X_2) had the most significant effect.

The response surface curves were plotted to explain the interaction of the variables and to determine the optimum level of each variable (Figures 3(a) and 3(b)). Figure 3(a) shows the effect of incubation temperature (X_1) and time (X_2) on juice clarity keeping the third at its middle level. It is clear from the figure that the clarity of the juice increased with both incubation time and temperature up to 509.60 min and 43.81°C temperature, respectively. Karangwa et al. [17] observed that the clarity of the blended carrot-orange juice decreased when the temperature was increased above 50°C.

Figure 3(b) reveals the effect of incubation temperature (X_1) and crude enzyme concentration (X_3) on the clarity of juice. It was evident that the clarity of juice increased with increase in temperature up to 43.90°C and crude enzyme concentration of 0.19 mL/per 25 g of pulp. Degradation of the polysaccharides like pectin leads to a reduction in water holding capacity, and consequently, free water is released to the system which increases the yield and clarity of juice [18]. With further increase in the incubation temperature, the clarity of juice decreased.

3.1.3. Optimization and Verification of Process Variables. The process was optimized by keeping the main constraints

as maximum possible juice yield, maximum clarity, and minimum viscosity of juice. Under these constraints, the optimum treatment conditions were found to be incubation temperature 44.97°C, time 474.50 min, and concentration of crude enzyme 0.20 mL/25 g of pulp (Table 4). But practically, it is difficult to maintain the recommended conditions during processing, and it is expected that there may be some deviation. Hence, the optimum conditions were targeted as temperature 45°C, time 475 min, and concentration of enzyme 0.20 mL/25 g of pulp. The experiments were conducted to check the variation in juice yield, viscosity, and clarity of juice under the optimum conditions while keeping the target constraints. The results indicate a high fit degree between the observed and predicted values from the regression model.

3.2. Optimization of Pretreatment Conditions Using Commercial Enzymes. The optimized conditions for the pretreatment of bael pulp using commercial enzymes were concentration of pectinase 5 mg/25 g of pulp (1.64 IU/mg), time 425 min, and temperature 47°C (Table 4), and the responses, juice recovery, viscosity, and clarity, were 84.5%, 1.35 cps, and 22.43%T, respectively, under the optimized conditions [7].

3.3. Comparative Effect of Crude and Commercial Enzymes for the Improvement of Juice Yield, Viscosity, and Clarity Using Principal Component Analysis (PCA). The juice yield, viscosity, and clarity from the crude enzyme treated pulp were 82.9%, 1.41 cps, and 21.32%T, respectively, under the optimized conditions. It is evident from the data (Table 5) that the crude enzyme treatment is equally competitive to the commercial enzymes. The possible cause of crude

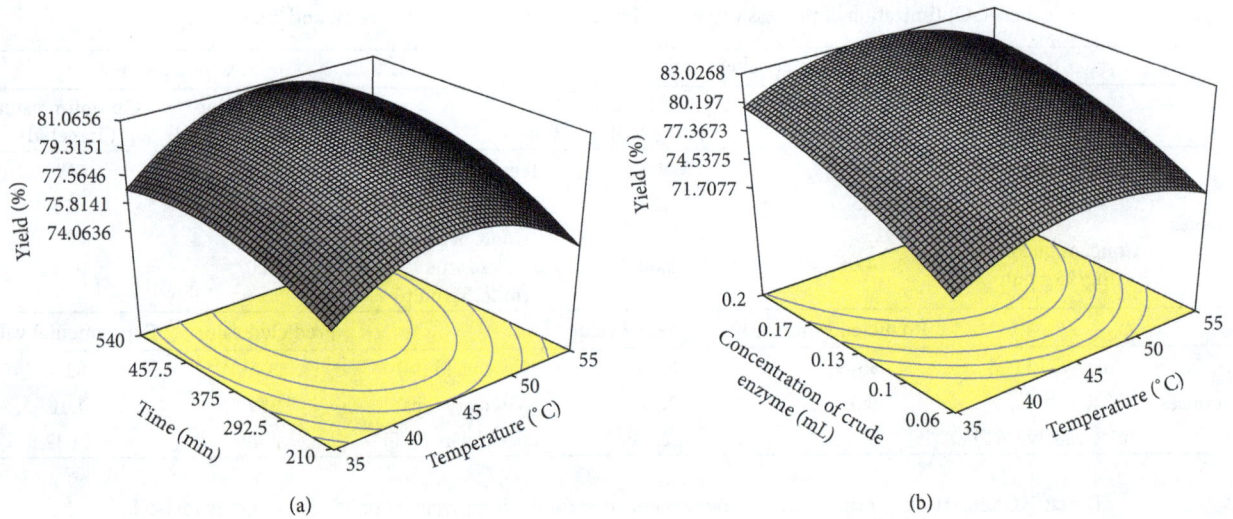

FIGURE 1: Response surface plots of juice yield as a function of (a) time and temperature and (b) concentration of crude pectinase enzyme and temperature.

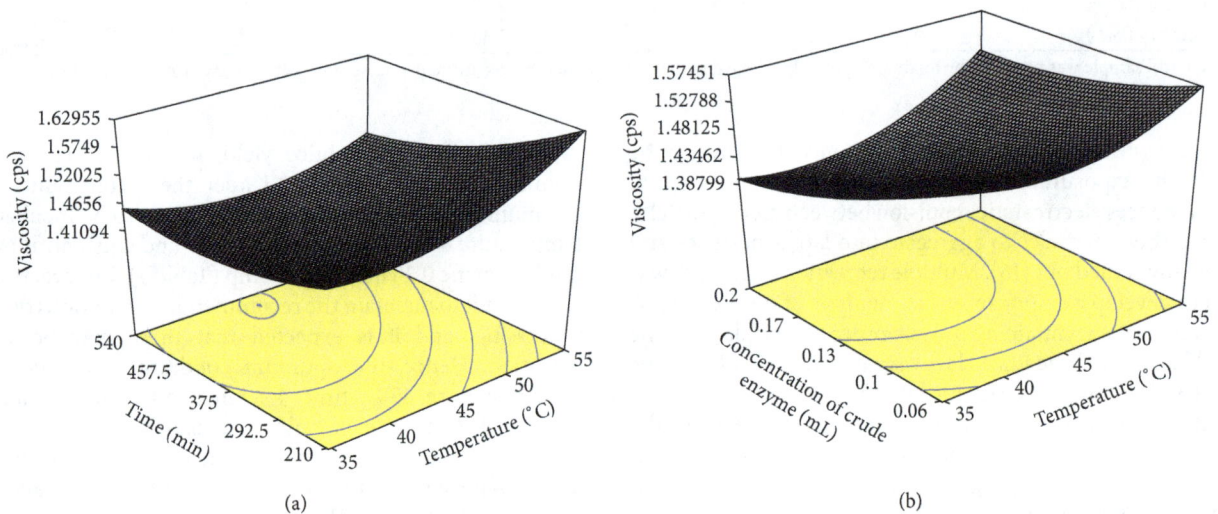

FIGURE 2: Response surface plots of viscosity of juice as a function of (a) time and temperature and (b) concentration of crude pectinase enzyme and temperature.

enzyme competitiveness may be a cumulative effect of other polysaccharases such as cellulases and hemicellulases along with the pectinases present in the crude enzyme.

Principal component analysis (PCA) was used to obtain a lesser number of uncorrelated variables from a large set of data. The objective of principal components analysis was to explain the maximum amount of variance with the least number of principal components. The results of the analysis are shown in Figures 4(a) and 4(b). The PCA plots provide an overview of the similarities and differences between the crude and commercial enzyme treated samples and of the interrelationships among the various measured properties. The distance between the locations of both the samples on the score plot is directly proportional to the degree of difference or similarity between them (Figure 4(b)).

The score plot (Figure 4(b)) clearly indicates that the commercial enzyme treated sample shows maximum variance in first principal component, whereas crude enzyme treated sample shows maximum variance in second principal component. Therefore, the variables of commercial enzyme treated samples can be considered as significant first principal components. Maximum covariance of 57% was obtained in first principal component, whereas in second principal component, variances of 68%, 73%, and 5.1% were found for yield, viscosity, and clarity, respectively. The maximum Eigen value in the correlation matrix was 3, which was the nearest to the variance derived from clarity; thereby, the variable clarity in second principal component can be considered as a significant variable to differentiate between crude and commercial enzymes treated juices. It is evident from

Comparative Effect of Crude and Commercial Enzyme on the Juice Recovery from Bael Fruit (Aegle marmelos Correa) Using Principal Component Analysis

39

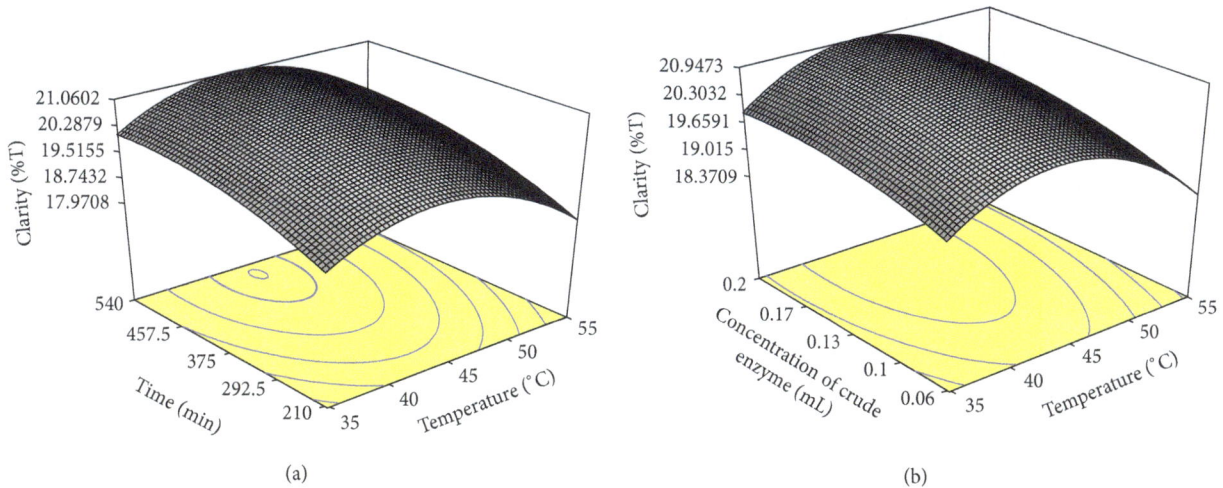

(a)

(b)

FIGURE 3: Response surface plots of clarity of juice as a function of (a) time and temperature and (b) concentration of crude pectinase enzyme and temperature.

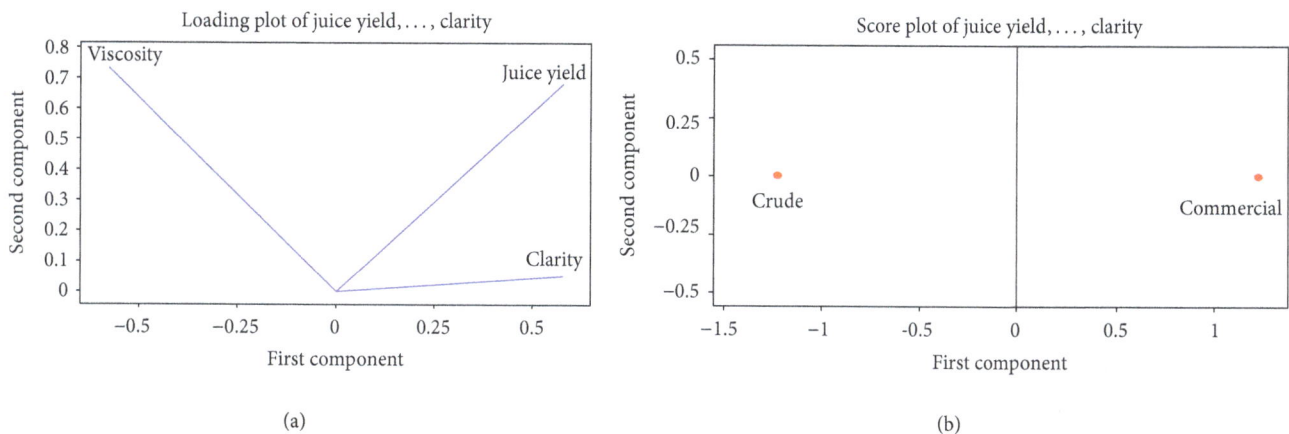

(a)

(b)

FIGURE 4: (a) Loading plot of yield, clarity, and viscosity of the juice under crude and commercial enzymes treatment. (b) Score plot of yield, clarity, and viscosity of the juice under crude and commercial enzymes treatment.

the loading plot of variables, that clarity and yield are correlated to higher degree as compared to the viscosity. It is suggested from this analysis that only two variables that is, clarity and yield, are to be studied for comparing samples of juice treated with either of enzymes.

4. Conclusions

The present study concluded that bael juice yield, viscosity, and clarity are functions of enzymatic hydrolysis conditions. Significant regression model describing the variation of juice yield, viscosity, and clarity with respect to the independent variables, temperature, time, and concentration of crude enzyme, was found adequately fit to predict the responses under study. The usage of either crude or commercial enzymes significantly enhanced juice yield and clarity as compared to the control. According to principal component analysis, it is suggested to study juice yield and clarity while comparing the different samples treated with either of the

enzymes. The study also indicates the equal effectiveness and competitiveness of crude enzyme as of commercial enzymes, and thus it may be one of the important considerations to reduce the processing cost.

References

[1] R. A. Kaushik, R. Yamdagni, and S. S. Dhawan, "Biochemical changes during storage of Bael preserve," *Haryana Journal of Horticulture Science*, vol. 31, no. 3-4, pp. 194–196, 2002.

[2] S. Charoensiddhi and P. Anprung, "Bioactive compounds and volatile compounds of Thai bael fruit (*Aegle marmelos* (L.) Correa) as a valuable source for functional food ingredients," *International Food Research Journal*, vol. 15, no. 3, pp. 287–295, 2008.

[3] J. Morton, "Bael fruit," in *Fruits of Warm Climates*, pp. 187–190, Julia F. Morton, Miami, Fla, USA, 1987.

[4] Rakesh, S. S. Dhawan, and S. S. Arya, "Processed products of Bael," *Process Food Industry*, vol. 8, no. 12, pp. 25–27, 2005.

[5] W. Pilnik and A. G. J. Voragen, "Pectic enzymes in juice manufacture," in *Enzymes in Food Processing*, T. Nagodawithana and G. Reed, Eds., Academic Press, New York, NY, USA, 1993.

[6] S. Kaur, B. C. Sarkar, H. K. Sharma, and C. Singh, "Optimization of enzymatic hydrolysis pretreatment conditions for enhanced juice recovery from guava fruit using response surface methodology," *Food and Bioprocess Technology*, vol. 2, no. 1, pp. 96–100, 2009.

[7] A. Singh, S. Kumar, and H. K. Sharma, "Effect of enzymatic hydrolysis on juice yield from Bael fruit (*Aegle marmelos correa*) pulp," *American Journal of Food Technology*, vol. 7, pp. 62–72, 2012.

[8] S. Kumar and H. K. Sharma, "Comparative effect of crude and commercial enzyme on the juice recovery from Pineapple (Ananas comosus) using Principal Component Analysis (PCA)," *Food Science and Biotechnology*, vol. 21, no. 4, pp. 959–967, 2012.

[9] S. Kumar, H. K. Sharma, and B. C. Sarkar, "Effect of substrate and fermentation conditions on pectinase and cellulase production by *Aspergillus niger* NCIM 548 in submerged (SmF) and solid state fermentation (SSF)," *Food Science and Biotechnology*, vol. 20, no. 5, pp. 1289–1298, 2011.

[10] D. C. Montgomery, *Design and Analysis of Experiments*, John Wiley & Sons, New York, NY, USA, 2001.

[11] J. J. P. Krop and W. Pilnik, "Effect of pectic acid and bivalent cation on cloud loss of citrus juice," *Lebensmittel-Wissenschaft & Technologie*, vol. 77, pp. 215–227, 1974.

[12] S. Ranganna, *Handbook of Analysis of Quality Control of Fruit and Vegetable Product*, Tata McGraw-Hill, New Delhi, India, 2nd edition, 1997.

[13] P. Kaushal, V. Kumar, and H. K. Sharma, "Comparative study of physicochemical, functional, antinutritional and pasting properties of taro (*Colocasia esculenta*), rice (*Oryza sativa*) flour, pigeonpea (*Cajanus cajan*) flour and their blends," *LWT Food Science and Technology*, vol. 48, no. 1, pp. 59–68, 2012.

[14] A. Singh and D. K. Das Gupta, "Effect of enzyme concentration, temperature and time of treatment on the quality of plum juice," *Processed Food Industry*, pp. 26–29, 2005.

[15] W. C. Lee, S. Yusof, N. S. A. Hamid, and B. S. Baharin, "Optimizing conditions for enzymatic clarification of banana juice using response surface methodology (RSM)," *Journal of Food Engineering*, vol. 73, no. 1, pp. 55–63, 2006.

[16] H. N. Sin, S. Yusof, N. S. A. Hamid, and R. A. Rahman, "Optimization of enzymatic clarification of sapodilla juice using response surface methodology," *Journal of Food Engineering*, vol. 73, no. 4, pp. 313–319, 2006.

[17] E. Karangwa, H. Khizar, L. Rao et al., "Optimization of processing parameters for clarification of blended carrot-orange juice and improvement of its carotene content," *Advance Journal of Food Science and Technology*, vol. 2, no. 5, pp. 268–278, 2010.

[18] N. Demir, J. Acar, K. Sarioglu, and M. Mutlu, "Use of commercial pectinase in fruit juice industry—part 3: immobilized pectinase for mash treatment," *Journal of Food Engineering*, vol. 47, no. 4, pp. 275–280, 2000.

6

Effect of Cassava Flour Characteristics on Properties of Cassava-Wheat-Maize Composite Bread Types

Maria Eduardo,[1,2,3] **Ulf Svanberg,**[2] **Jorge Oliveira,**[4] **and Lilia Ahrné**[2,3]

[1] *Departamento de Engenharia Química, Faculdade de Engenharia, Universidade Eduardo Mondlane, Maputo, Mozambique*
[2] *Department of Chemical and Biological Engineering/Food Science, Chalmers University of Technology, SE-41296 Gothenburg, Sweden*
[3] *SIK, The Swedish Institute for Food and Biotechnology, SE-402 29 Gothenburg, Sweden*
[4] *Department of Process and Chemical Engineering, University College Cork, Cork, Ireland*

Correspondence should be addressed to Ulf Svanberg; ulf.svanberg@chalmers.se

Academic Editor: Zoulikha Maache-Rezzoug

Replacement of wheat flour by other kinds of flour in bread making is economically important in South East Africa as wheat is mainly an imported commodity. Cassava is widely available in the region, but bread quality is impaired when large amounts of cassava are used in the bread formulation. Effect of differently processed cassavas (sun-dried, roasted and fermented) on composite cassava-wheat-maize bread quality containing cassava levels from 20 to 40% (w/w) was evaluated in combination with high-methylated pectin (HM-pectin) added at levels of 1 to 3% (w/w) according to a full factorial design. Addition of pectin to cassava flour made it possible to bake bread with acceptable bread quality even at concentration as high as 40%. In addition to cassava concentration, the type of cassava flour had the biggest effect on bread quality. With high level of cassava, bread with roasted cassava had a higher volume compared with sun-dried and fermented. The pectin level had a significant effect on improving the volume in high level roasted cassava bread. Crumb firmness similar to wheat bread could be obtained with sun-dried and roasted cassava flours. Roasted cassava bread was the only bread with crust colour similar to wheat bread.

1. Introduction

Bread is an important staple food in South East Africa, providing energy and many nutrients such as proteins, B-vitamins, vitamin E, and minerals. Traditionally in Mozambique, bread is produced by a mixture of wheat flour, yeast (*Saccharomyces cerevisiae*), salt, and water. As wheat is not grown in Mozambique, large quantities are therefore imported at high costs for the country. In order to cut the nation's expenses, the Government of Mozambique has thus mandated the use of composite flour in breadmaking including flour of cassava, maize, or millets that are locally available.

There is a growing interest in using composite flour for breadmaking owing to some economic, social, and health reasons. However, the partial substitution of wheat flour by other flour types presents considerable technological difficulties because their proteins lack the ability to form the necessary gluten network for holding the gas produced during the fermentation [1–3]. The dough formed is more difficult to handle, and the bread has poor loaf volume and crumb softness [4]. Composite flour with cassava has been evaluated in breadmaking, and general observations are reduced loaf volume, crust colour, and impaired sensory qualities as the level of substitution of wheat with other flours increased [5, 6]. Similar results were reported by Khalil et al. [7]; however, acceptable breadmaking potential could be obtained from partial substitution of wheat flour by cassava flour up to 20 and 30% with addition of 1% malt. Recent studies on bread quality from composite cassava-wheat flour have investigated the influence of the baking process, the cassava genotype, and effect of an added hydrocolloid, xanthan gum [8–10]. Very little attention has been paid to investigate the effect of differently processed cassava flour as ingredients in composite flour for breadmaking.

Cassava (*Manihot esculenta* Crantz) is the major food crop produced in Mozambique with maize (*Zea mays* L.) being the second one. A major constraint to cassava utilization is the rapid microbial degradation after harvest. Cassava roots have a shelf life of only 24–48 h after harvest [11]. One way to extend the shelf life of cassava is to prepare a dry product such as flour. In Mozambique three major cassava flour products are prepared for human consumption, sun-dried, fermented, and roasted. Traditionally, cassava flour can be produced from washed or peeled roots, that are grated, chipped, or sliced, then sun-dried on trays, and finally milled into flour [12, 13]. Grated roots may also be pressed to remove excess water and then toasted to produce roasted cassava flour with up to 95% of the starch gelatinized [14]. The grated cassava may also be fermented with lactic acid bacteria, dried, and subsequently roasted in a pan. The resulting flour is called gari and in Mozambique is known as "*rale.*" According to Numfor et al. [15] the fermentation stage improves the internal stability of the starch granules, reduces the swelling power, and decreases the amylose release during heat treatment.

Due to the unique ability of the wheat gluten proteins to form a viscoelastic dough that retains gas during dough leavening [16, 17], wheat flour cannot be substituted directly in a yeast-leavened product without formula modifications [16], for example, by adding bread improvers such as hydrocolloids, enzymes, and emulsifiers.

Hydrocolloids are used in baked goods primarily to increase moisture retention and to improve viscoelastic properties of the dough in order to improve bread volume. In gluten-free breads based on rice flour and corn starch, Lazaridou et al. [18] showed that addition of hydrocolloids improved dough strength and bread volume. A better specific volume, a softer crumb, and an improved moisture retention were obtained in wheat bread with addition of hydroxypropyl methylcellulose (HPMC) and κ-carrageenan [19, 20] and with HM-pectin [21]. Shittu et al. [10] demonstrated improved specific volume and crumb softness in bread made from composite cassava-wheat flour with added xanthan gum.

The method used to produce the cassava flour (sun-dried, roasted, or fermented) is expected to influence the functional properties of the flour and consequently the bread quality. Information is however scarce regarding the effect of cassava flour type combined with hydrocolloids on bread quality. Therefore, the objective of the present study was to evaluate the quality characteristics of bread baked on composite flour with sun-dried, roasted, and fermented cassava flour, respectively, and with HM-pectin as a baking additive. The proportion of cassava flour in the mixtures varied from 20 to 40% (w/w), and high-methylated pectin was added at levels of 1 to 3% (w/w).

2. Materials and Methods

2.1. Materials. The ingredients used were wheat flour at 65% extraction rate (Nord Mills), yellow maize flour (AB Risenta, Sweden), granulated white sugar (commercial), refined salt (commercial), instant dry yeast (KronJäst, Jästbolaget AB), margarine (Kondis UM UHF), L(+) ascorbic acid (GR,

E. Merck), soy lecithin (Lecico, GmbH), and HM-pectin (DANISCO). All these ingredients were purchased from commercial sources or directly from the suppliers, keeping the same specification in all experiments. About 300 kg of cassava roots (12 months old) was obtained from local producers in Mozambique and then were processed with different methods (sun-dried, roasted, and fermented). One hundred kg of roots were used for each processing method.

2.2. Methods

2.2.1. Processing of Cassava Flour. Three different traditional processing methods were used to produce flour from cassava roots at the Food Technology Laboratory at University Eduardo Mondlane. For sun-dried flour, the selected cassava roots were washed, peeled, and sliced in small pieces which were sun-dried 1-2 days. For roasted flour, the peeled cassava roots were chipped followed by pressing and screening in a mechanical machine and then toasted in a pan for 10 minutes. For fermented flour, the peeled cassava roots were immersed in water for 5 days that favours growth of lactic acid bacteria [12], and then the fermented roots were crumbled and sun-dried 1-2 days. The dried products per method yield approximately 36 kg, 20 kg, and 25.5 kg, respectively. All dried pieces were ground into flour with a laboratory mill, and excess fibre was removed by passing the ground material through a sieve DIN 4188 (0.125 mm aperture sieve).

2.2.2. Breadmaking Procedure. Bread dough samples were prepared according to the following formula: 300 g flour (containing different proportions of cassava flour and maize flour, and wheat flour at constant level, see Table 1), HM-pectin at 3, 6, or 9 g levels, 4.8 g yeast, 4.5 g salt, 6.0 g sugar, 9.0 g margarine, 0.3 g ascorbic acid, 1.2 g soy lecithin, and water according to the preceding baking tests. It should be noted that in order to obtain approximately equal consistency of the dough from each of the flour types, the amounts of water added varied as described in Table 1. These limits were set from preliminary experiments, by the quality of the dough being sufficiently good for working into proper bread. All the ingredients were mixed in a kitchen mixer KS90 (KitchenAid, USA) for 2 min at low speed (speed setting: 1) and for 8 min at medium speed (speed setting: 2), until the dough was well developed. The temperature of the dough was about 22°C. After mixing, the dough was covered with a kitchen cloth and left to ferment at room temperature for 45 min. After the first fermentation, the dough was divided into 50 g portions, rounded, placed into bread pans, and proofed for another 45 min in a fermentation cabinet (LabRum Klimat AB, Stockholm, Sweden) at 30°C and 80% relative humidity. The proofed dough was baked at 220°C for 7 min in an oven (Therma Grossküchen, *Le Chef*, Sweden) with air circulation. Ten miniloaves were thus produced in each batch. Then, the bread was cooled for 60 min at ambient temperature. A control wheat bread was prepared simultaneously in the same oven under identical conditions to those of the experimental design. All measurements were taken using one batch as one collective unit.

Table 1: Dough formulae of bread samples in the experimental design.

| Run order | Composition of the dough | | | | | Pretreat* |
| | Water (g) | Flour types(g) | | | Pectin (g) | |
		Wheat	Maize	Cassava		
13	230	150	90	60	3	S
5	234	150	30	120	3	S
14	233	150	90	60	9	S
11	236	150	30	120	9	S
6	236	150	90	60	3	R
10	254	150	30	120	3	R
4	236	150	90	60	9	R
8	265	150	30	120	9	R
7	230	150	90	60	3	F
15	233	150	30	120	3	F
1	232	150	90	60	9	F
3	235	150	30	120	9	F
12	232	150	60	90	6	S
2	232	150	60	90	6	S
9	232	150	60	90	6	S

*Pretreatment of the cassava flour. S: sun-dried; R: roasted; F: fermented.

2.2.3. Experimental Design.

The baking experiments were planned according to a full factorial design with 3 levels for types of cassava flour (sun-dried, roasted, and fermented) with 2 levels for amounts of cassava and HM-pectin, plus a center point (for sun-dried pretreated cassava flour only), replicated three times. This resulted in 15 experiments that were performed in random order. The design is shown in Table 1. It is noted that the amount of wheat flour was always 150 g, while that of maize + cassava flour was also always equal to 150 g, so the only control factor regarding flour composition is the ratio cassava-to-maize. The second control factor is the amount of HM-pectin (3, 6 or 9 g) while the third is the type of pretreatment. Water is actually a noise factor (source of variability and experimental error), which must vary as shown in Table 1 so that the consistency of doughs prepared would be similar.

Six miniloaves were analysed for each response in each batch, totalling $15 \times 6 = 90$ data points for each response.

2.2.4. Moisture Content.

The moisture content of the flour samples was measured by weight difference before and after drying of the samples in a vacuum oven [22]. All flour samples were analyzed in triplicate.

2.2.5. Water Absorption of Flour.

The water absorption of the flour samples was determined by the method modified by Anderson et al. [23]. Five gram of each sample was weighed into a centrifuged tube, and 30 mL of distilled water was added and vigorously mixed. The samples were allowed to stand for 30 min and centrifuged (Beckman GP, UK) at 3,000 rpm for 15 min. The top layer was decanted off and the sample weighed again. The amount of water in the sample was recorded as weight gain (g/g flour) and was taken as water absorbed.

2.2.6. Microstructure of Starch Granules.

The starch samples obtained from sun-dried, roasted, or fermented cassava roots were prepared using the smear method as reported by Hongsprabhas et al. [24] and examined with bright-field (BFM) and polarized light microscopy (PLM). The water slurries of cassava flour (soaked overnight at room temperature) were smeared onto an object glass. After drying, the samples were stained with Lugol's iodine solution and covered with a glass sealed with nail polish. The samples were thereafter examined with a Microphot FXA light microscope (Nikon, Tokyo, Japan) using a 10x and a 40x magnification. Images were taken with an Altra 20 Soft Imaging System camera (Olympus, Tokyo, Japan).

2.2.7. Bread Volume.

The loaf volume was measured one hour after the end of the baking process by the displacement method in which alfalfa seed was used instead of millet. The average of six measurements was recorded as the loaf volume (cm^3).

2.2.8. Crumb Structure.

The bread crumb structure was evaluated from images captured using a flatbed scanner (MiraScan, BenQ, Version 5.10). Images were scanned with full scale at 100 dots per inch.

2.2.9. Crust Colour.

The crust colour was measured on the bread surface using a Colour Reader CR-10 (Konica Minolta Sensing, Japan). $L^*a^*b^*$ values were recorded and the results reported as brownness index (BI), calculated according to Maskan [25]

$$BI = \frac{[100 \cdot (x - 0.31)]}{0.17}, \qquad (1)$$

where x is defined as

$$x = \frac{a + 1.75L}{5.645L + a - 3.01b}, \qquad (2)$$

where a^* is redness, b^* is yellowness, and L^* is lightness.

The average of four measurements was taken as the crust colour parameter.

2.2.10. Crumb Firmness.

Slices of each loaf (2.5 cm) were taken for analysis of the crumb firmness, measured 6 h after baking using an Instron Universal Testing Machine (UTM, model 5542). A modified AACC standard method 74-09 was used with a cylindrical probe (diameter 15 mm). The probe compressed the slices 40% at a test speed of 1.7 mm/s. The compression curves of the bread crumbs were recorded automatically by the BlueHill software (Merlin, version 5, Instron Corp., Canton, MA, USA). The Young modulus readings between 3 and 20% were taken as a measure of bread crumb firmness. The measurements were carried out on 15 mm thickness slice taken from the centre of the bread loaf. The measurements were carried out on four loaves from each batch.

TABLE 2: Direct and interaction effect of bread quality parameters.

Factor/interaction	%		
	Volume	Firmness	Brownness
Main effects			
Pretreatment (Pt)	6*	40*	85*
Cassava ratio (Cr)	27*	11*	3*
Pectin content (Pc)	5*	2*	1*
Interactions			
Pt × Cr	18*	20*	1
Pt × Pc	15*	17*	1
Cr × Pc	4*	5*	1*
Error	25	5	7
SS$_{total}$	12664	212.5	59418

*Effects are statistically significant at a 90% confidence level.

2.2.11. Statistical Analysis. A factorial ANOVA was applied to the full factorial design only (i.e., without using the data obtained for sun-dried cassava with middle settings for cassava level and pectin content), to quantify the relative significance of each of the control factors and all two-way interactions between factors, using Statistica software (v.8, StatSoft Inc.).

The data were also subjected to a least squares regression analysis with a multifactorial model using the type of cassava flour pretreatment as a dummy variable using Statistica (v8, StatSoft Inc.). The model is

$$Y = y_s + c_s C + p_s P + i_s CP + q_s \left(1 - D_f - D_r\right) C^2$$
$$+ y_f D_f + c_f D_f C + p_f D_f P + i_f D_f CP \qquad (3)$$
$$+ y_r D_r + c_r D_r C + p_r D_r P + i_r D_r CP,$$

where Y represents any of the responses measured (bread volume, firmness, and colour) and C and P are the coded values of the factors cassava level and pectin content, respectively (0, 0.5, or 1, as per Table 1). The model uses sun-dried bread with low cassava and pectin content as a reference, so y_s is the average model value for this bread. D_f is the dummy variable indicating whether the pretreatment used was fermentation (= 1 if yes, 0 otherwise), and D_r is the similar dummy variable indicating roasted pretreatment (= 1 if yes, 0 otherwise).

As the model is additive, it is straightforward to obtain the estimated value for any type of bread by simply adding the respective parameters. The results of the regression analysis are presented in modified Pareto charts.

3. Results and Discussion

3.1. Volume. Volume is an important quality characteristic of bread, and that is negatively affected when wheat is replaced by cassava. The results of the factorial ANOVA for bread volume are shown in Table 2. The cassava level was the factor that had the largest effect, 27%. The type of pretreatment and pectin content had less influence, while the interactive effect between cassava level and pectin content was also small.

However, the interactive effects of pretreatment with cassava level and with pectin content were larger.

The results of the regression model are shown in Figure 1. The results show that for roasted cassava a change in level or pectin content had no effect on bread volume. For the other types of cassava (fermented or sun-dried), increasing the cassava level clearly caused a volume decrease. All types of bread with cassava flour were smaller than the wheat flour control bread that had a volume of 135.0 cm^3; however the differences were considered acceptable. The largest bread volume of all cassava bread was obtained with fermented cassava (20%) and high pectin content, 118.3 cm^3 (115.0 − 15.67 + 19.00, see Figure 1). Increasing the fermented cassava level would negate the increase achieved with higher pectin (effects around +19 and −19, with negligible interactive effect).

In the baking experiments with the full factorial design the amount of water added in the dough preparations ranged from 230 to 236 g with sun-dried and fermented cassava flour and from 236 to 265 g with roasted cassava flour. The amount added was determined by taking in account the handling properties of the doughs. If the same amount of water was added to all the doughs, it would have been impossible to form bread from the dough. To evaluate the effect of water on bread volume, the amount of water was increased in doughs prepared with sun-dried and fermented cassava flour, to obtained similar amount of water as the one used for roasted. The addition of 255 g of water resulted in a very sticky dough but still workable, and slightly increased (not statistically significant) volume was observed.

Increasing the cassava ratio decreased the volume in the bread made with sun-dried and fermented flour which is quite evident from the results shown in Table 3. However, the cassava content had no effect on volume of roasted cassava bread, due to the interaction with pectin content, which once again was particularly dramatic for this pretreated flour. While increasing the cassava ratio with the lower pectin content had no effect on the bread volume, it resulted in a slightly increased bread volume with the higher pectin content.

The overall effect of a reduced bread volume with cassava flour in the flour mixture can be explained by reduced flour strength and a lower ability of the gluten network to enclose the carbon dioxide produced during fermentation. Similar findings on loaf volume of composite flours were reported by other researchers [1, 5–7]. In baked products, hydrocolloids influence the dough rheology and bread quality parameters. An increase in volume of composite bread formulations due to pectin addition may be attributed to improvement of dough development and gas retention by increasing dough viscosity [19, 26] and dough stability [27]. HM-pectin does contain hydrophobic groups that might induce interfacial activity with gluten and thereby forming gel networks during the breadmaking process. HM-pectin has also been shown to interact with wheat starch causing an increased paste viscosity during heating [28]. Both the increased viscosity and the gel network might strengthen the gas-holding properties of the expanding cells in the dough and consequently result in a higher loaf volume [29]. Lazaridou et al. [18] reported that pectin at 2% level was the only hydrocolloid of five tested

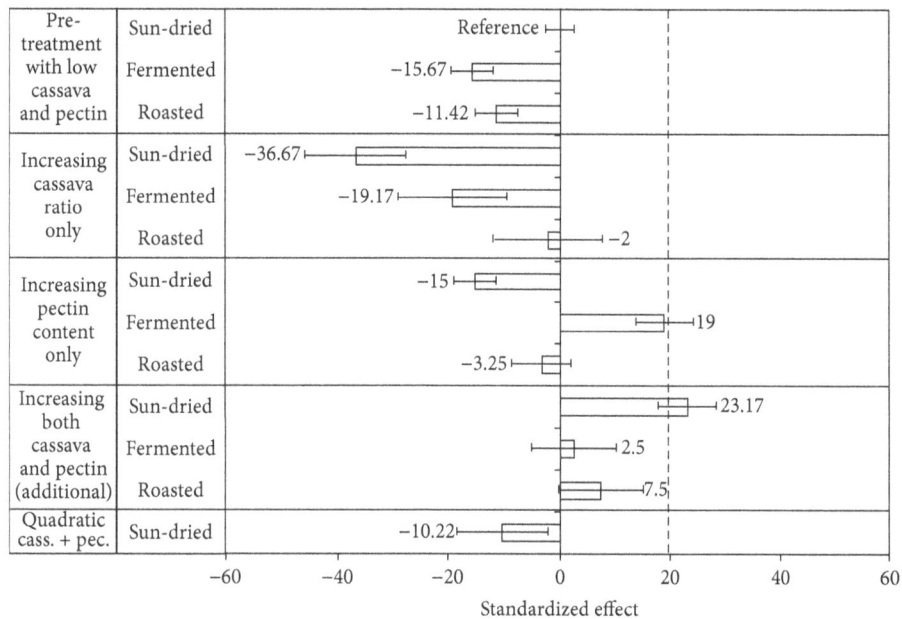

FIGURE 1: Modified Pareto chart showing the relative effects of pretreatments, flour composition factors, and the interaction of these on bread volume according to the regression model ($R^2 = 0.75$). Bread made with sun-dried flour with low cassava and low HM-pectin contents was set as reference bread. The volume of wheat bread is indicated by a dotted line. Effects with error bars not including point 0 are statistically significant at 95% confidence level.

(CMC, xanthan, Agarose, oat β-glucan) that increased the volume of gluten-free breads. High-methylated pectin has also shown to both increase dough volume and the specific volume of wheat bread [21]. The roasted cassava provides a soft agglomerate of gelatinized starch granules [14] that could provide an additional built-up of an interlinked network with the HM-pectin in the dough explaining the increased volume of the bread with roasted cassava compared with the other two cassava flour types.

The higher bread volume with fermented cassava and high pectin content could be related to the prereatment conditions given to the cassava flour. The fermentation process of cassava which took place during 5 days at ambient temperature and subsequent sun-drying has previously shown to result in better baking expansion during dough preparation [30]. Furthermore, these authors pointed out that the baking expansion will increase with starch disintegration and degradation during cooking. The influence of pectin on bread volume with fermented cassava addition might be explained by the increase in paste viscosity that slows down the rate of gas diffusion and allows its retention during the early stage of baking. Mestres et al. [30] found that the expansion ability of fermented cassava was negatively correlated with paste viscosity.

3.2. Firmness.

Bread volume has a direct influence on the firmness of bread; however firmness is also influenced by the strength of the crumb formed. The results of the factorial ANOVA for bread firmness are shown in Table 2, and the most influential factor on bread firmness is the type of pretreatment applied to cassava flour explaining 40% of the effect, followed by its interactive effects. In particular,

TABLE 3: Mean values of firmness (N) and volume (cm³) for bread types made with sun-dried, roasted, and fermented cassava. The means for the different cassava ratios with the standard error being due to white noise and to the influence of pectin content.

Cassava type	Firmness (N)	Volume (cm³)
Sun-dried		
20%	8.6 ± 0.4	107.5 ± 4.1
30%	11.0 ± 0.3	90.5 ± 2.9
40%	11.7 ± 0.2	82.4 ± 3.4
Roasted		
20%	7.8 ± 0.3	102.0 ± 2.2
40%	7.3 ± 0.2	103.7 ± 2.4
Fermented		
20%	8.8 ± 0.5	108.8 ± 4.3
40%	9.7 ± 0.6	90.9 ± 5.4

interactive effects of pretreatment with cassava ratio and pectin content vary for the differently pretreated cassava flour types.

Figure 2 shows more detailed results in a modified Pareto chart. The effects are related to the reference bread which was made with low level (20%) of sun-dried cassava and low pectin content, with a firmness of 7.66 ± 0.17 N. Bread with fermented cassava flour results in firmer bread crumb when cassava level and pectin content were both low, and softer bread types were obtained with roasted flour ($P < 0.05$). However, the bread types made with roasted cassava and sun-dried cassava both had a firmness that was similar to that of the control wheat bread, 7.33 ± 0.50 N (dotted

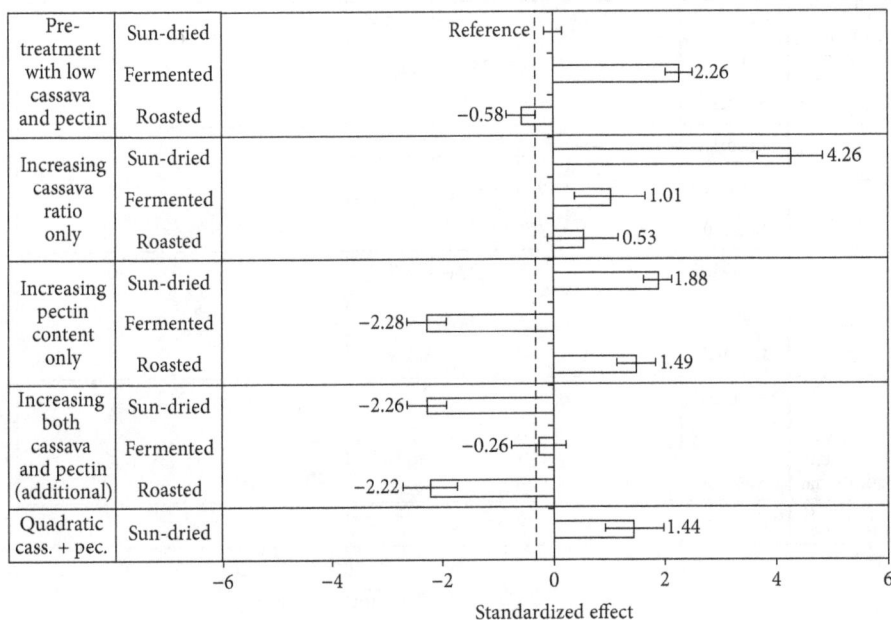

FIGURE 2: Modified Pareto chart showing the relative effects of pretreatments, flour composition factors, and the interaction of these on bread firmness (N) according to the regression model ($R^2 = 0.95$). Bread made with sun-dried flour with low cassava and low HM-pectin contents was set as reference bread. The firmness of wheat bread is indicated by a dotted line. Effects with error bars not including point 0 are statistically significant at 95% confidence level.

TABLE 4: Mean values of firmness (N), brownness index (BI), and volume (cm^3) of bread types made with roasted cassava and containing 1 or 3% pectin content.

| Cassava type | Firmness | | Brownness | | Volume | |
| | | | Pectin content | | | |
	1%	3%	1%	3%	1%	3%
Roasted						
20%	7.1 ± 0.1	8.6 ± 0.1	91.8 ± 0.2	126.4 ± 2.2	103.6 ± 5.0	100.3 ± 4.4
40%	7.6 ± 0.2	6.9 ± 0.4	109.0 ± 5.3	103.3 ± 4.7	101.6 ± 5.5	105.8 ± 3.8
Sun-dried	9.8 ± 0.9	10.5 ± 0.4				
Roasted	7.4 ± 0.1	7.7 ± 0.4				
Fermented	10.4 ± 0.3	8.0 ± 0.2				

line). Increasing the cassava level caused substantially higher firmness of the bread made with sun-dried cassava with an increase in firmness of 4.26 N. Increasing the pectin content had a softening effect in breads made with fermented flour compared with bread types of the other two cassava flour types. The effect of increasing both cassava ratio and pectin content, however, showed a significant interactive effect when using sun-dried or roasted cassava but not fermented.

Table 3 shows that the effect of the pretreatment is significant at higher cassava levels, with sun-dried flour giving firmer bread and roasted flour softer bread. The nature of the quadratic effect is also visible in a slight curvature of the effect of increasing the ratio of sun-dried cassava flour. Table 4 shows that the pectin content had very little effect in flours made with sun-dried or roasted cassava, but increasing its content with fermented cassava resulted in bread that were softer, becoming similar in firmness to

those made with roasted flour at the higher pectin content. Finally, the interactive effect between cassava level and pectin content showed that increasing the cassava level increased the firmness in all bread, except in the ones with roasted cassava and high pectin content (Table 4).

These results show that substitution of wheat flour with increasing amounts of cassava flour will affect the bread crumb firmness differently depending on the pretreatment of the cassava flour. A necessary property of the starch component is to support the elastic strength of the diluted gluten network responsible for the gas-holding capacity of the dough. Evidently, the starch properties of the three different cassava flour types interact differently with the protein-gluten network during the baking process resulting in different effects on the crumb structure and thus the firmness.

The softening effect of the roasted cassava flour might be attributed to pregelatinization of the cassava starch during

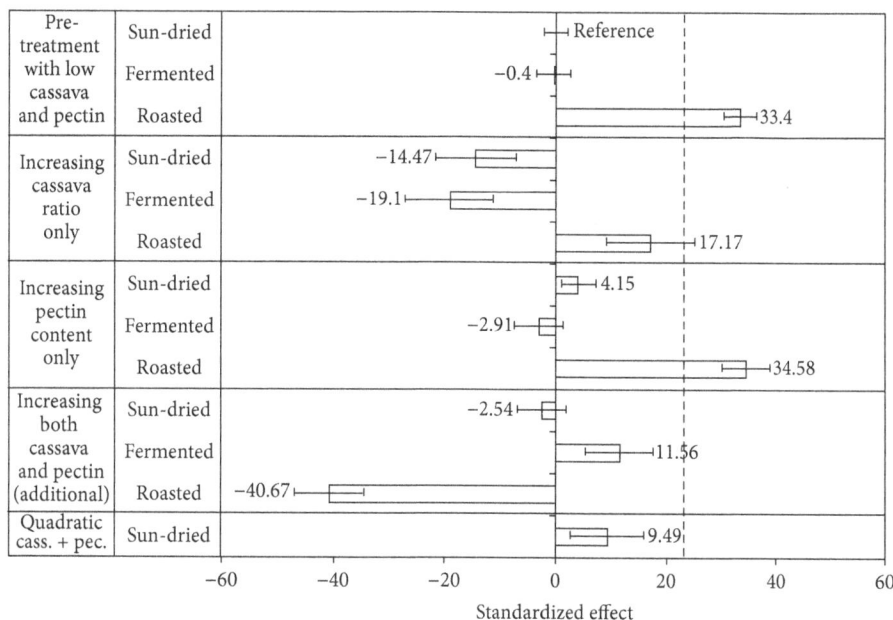

FIGURE 3: Modified Pareto chart showing the relative effects of pretreatments, flour composition factors, and the interaction of these on brownness according to the regression model ($R^2 = 0.93$). Bread made with sun-dried flour with low cassava and low HM-pectin contents was set as reference bread. The brownness index of wheat bread is indicated by a dotted line. Effects with error bars not including point 0 are statistically significant at 95% confidence level.

roasting (heat-moisture treatment), providing a high swelling capacity at the dough phase. Furthermore, pregelatinized cassava starch granules remain resistant against disintegration at higher temperatures, up to 90°C, and form soft agglomerates [14] that might contribute to a softer bread crumb structure during baking off.

Numfor et al. [31] observed that the average starch granule diameter, solubility, and swelling power were found to be depressed by fermentation. Formation of amylose-like fragments was suggested to interact with the starch granules thus resulting in greater internal granule stability. A more stable granule would account for the observed reduced solubility and swelling power of the fermented starches. This, in turn, would account for a reduced ability to support an elastic network during baking off and result in firmer bread crumb. However, adding pectin to this structure compensates for this effect, resulting in bread similar to the wheat bread control when using the higher cassava ratio.

The softening effect on bread crumb at high levels of HM-pectin (3%) that was obtained with fermented or roasted cassava might be explained by the interaction of the hydrocolloid with the amylose gel formation. Lower viscoelastic properties have been demonstrated in gels of amylose/hydrocolloid mixtures [32]. Furthermore, hydrocolloids have been shown to retard the retrogradation of cassava starch gels [33], and HM-pectin at 1% level resulted in a reduced crumb firmness of wheat bread [21].

3.3. Brownness. The colour of the bread is important for consumers, and it is related with the extension of the Maillard reactions. The results of the factorial ANOVA for brownness are shown in Table 2. Brownness is significantly influenced by the type of pretreatment which explains 85% of the effect, with pectin content and cassava level having a small influence.

It is evident from the model results shown in Figure 3 that roasting causes a significant increase in brownness, and this was the only treatment able to provide bread with a brown colour similar to that of the control wheat bread. The other pretreatments resulted in breads with a brownness index value of about 58 that was significantly lower than the value measured in wheat bread, 91.8 ± 2.3. Increasing the pectin content makes the bread made from roasted cassava even browner, although this effect has a strong interaction with increasing the cassava level. Increasing the pectin content seems to have little effect on the other bread types, while increasing the cassava ratio makes them less brown. This effect was also demonstrated by increasing the roasted cassava ratio in bread that became less brown with high pectin content, but with low pectin content the brownness increased substantially ($P < 0.05$), Table 4.

The effect of browning of the composite bread types can be attributed to Maillard reactions between wheat proteins and reducing sugar [34]. The Maillard reactions are also related to temperature, time, and the presence of water (moisture), and in bread crust the temperature and water activity might be optimal for browning reactions [35].

In bread with roasted cassava the pregelatinized starch released some amylose units that expose reducing glucose ends that can participate in browning reactions [14]. With high levels of pectin the amount of available water in the dough might be reduced resulting in a lower rate of browning.

3.4. Crumb Structure. Pictures of breads baked with 40% cassava and 3% HM-pectin (Figure 4) show the crust colour

(a)

| Wheat bread | Sun-dried cassava bread | Roasted cassava bread | Fermented cassava bread |

(b)

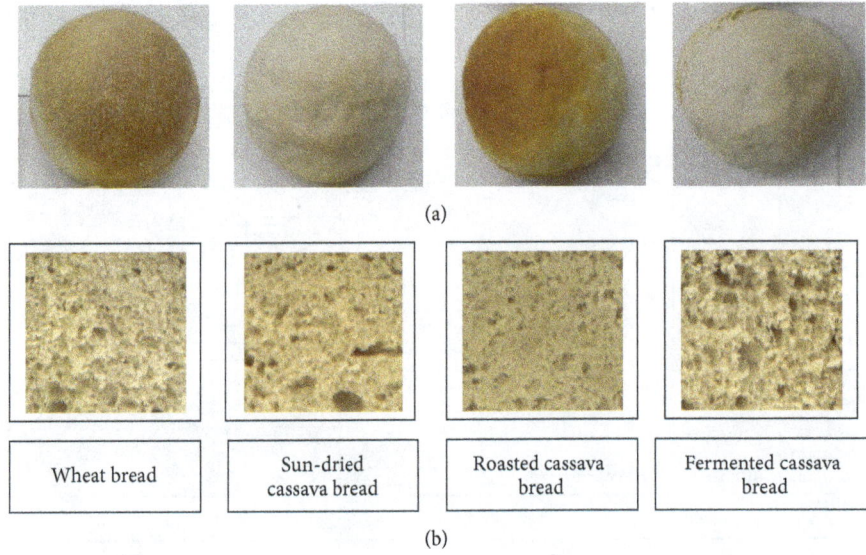

FIGURE 4: Composite bread made with 40% cassava flour and 3% of HM-pectin. Crust colour (a) and crumb structure (b).

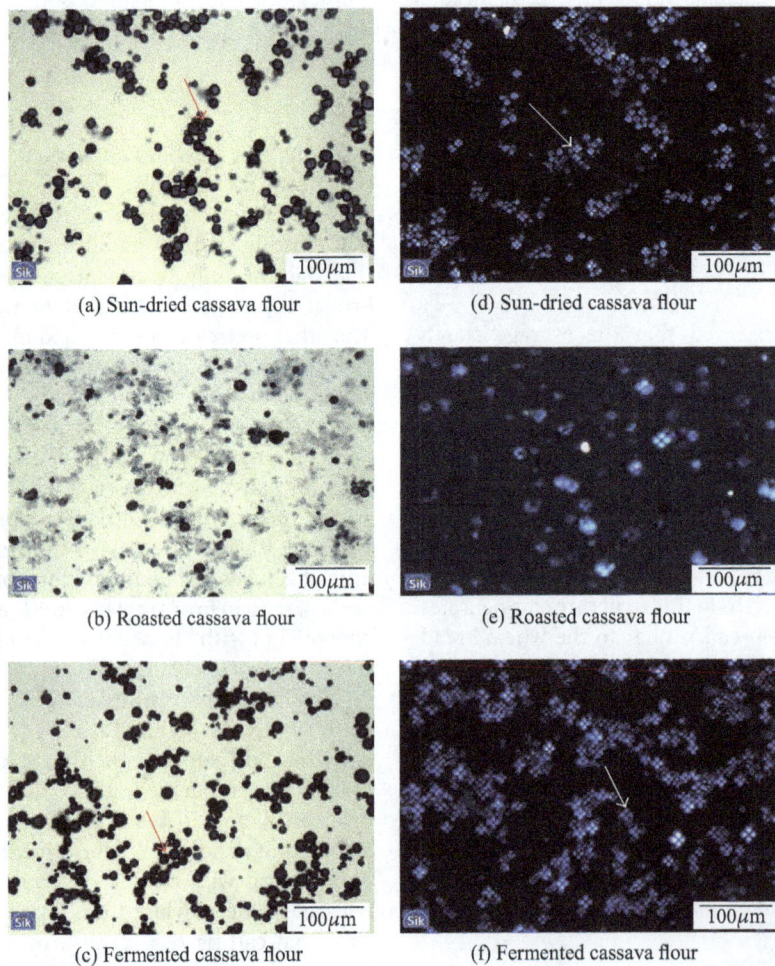

(a) Sun-dried cassava flour

(d) Sun-dried cassava flour

(b) Roasted cassava flour

(e) Roasted cassava flour

(c) Fermented cassava flour

(f) Fermented cassava flour

FIGURE 5: Bright-field (a, b, and c) and polarized Light (d, e, and f) micrographs of starch granules. Sun-dried and fermented cassava flour with intact granules ((a) and (c), red arrows) and roasted cassava flour with partly swollen granules, few Maltese crosses (e) compared with (d) and (f).

and crumb structure. The crumb structure of the breads prepared with roasted or sun-dried cassava flour was characterized by rather small and uniformly distributed pores with some few big pores of irregular shape. The crumb structures of these breads were similar to the wheat bread. However, an expandable crumb structure was noticed with fermented cassava flour in the bread. Compared to the other two cassava types, the bread baked with roasted flour had significantly better crust colour and similar to the wheat bread.

3.5. Cassava Starch Granules Microstructure. In order to understand the different bread characteristics obtained with the different cassava flour types, the microstructure of the cassava starch granules was analysed by bright-field (BF) and polarized light microscopy (PLM).

Micrographs of the cassava granules suggest that there is a structural difference among the three types of cassava. In roasted cassava flour, partially swollen starch granules are observed, whereas in others two intact granules are a dominating feature (Figures 5(a), 5(b), and 5(c)). The partially swollen starch granules resulted in significantly higher water absorption, 236% in the roasted cassava flour compared to 102% in the fermented cassava flour and 136% in the sun-dried cassava flour. It thus appears that starch characteristics are important factors that will influence water absorption in the three flour types of cassava. The water absorption data reflects that a larger amount of water is needed in the preparation of doughs with roasted cassava flour to obtain similar dough characteristics as with the other cassava flours. Similar findings have been reported by Defloor et al. [6].

Polarized light microscopy of sun-dried and fermented cassava starch granules showed the typical Maltese cross phenomenon (Figures 5(d), 5(e), and 5(f)) indicating a large number of nongelatinized starch granules. Mestres and Rouau [36] have earlier shown that fermentation and drying of cassava do not change the gelatinization properties of the starch granules.

4. Conclusion

The type of processed cassava flour has large effect on the quality parameters of cassava-wheat-maize composite breads. With high level of cassava, bread with roasted cassava flour had a significantly higher volume compared with bread made of sun-dried or fermented cassava flour. The pectin level had a significant effect on improving the volume in high level roasted cassava bread. Crumb firmness as well as crumb structure of sun-dried bread and of roasted cassava bread was similar to wheat bread. Composite bread types with roasted cassava were the only bread types with crust colour similar to wheat bread.

In relation to the important objective of achieving bread similar to that made with wheat flour, in terms of volume and firmness, roasted cassava flour is the most promising pretreatment.

In terms of sensory properties, a preliminary consumer acceptance test performed in Maputo, Mozambique, indicated that composite wheat-maize-cassava flour (sun-dried/roasted) bread had an overall acceptability similar to

wheat bread. A more comprehensive consumer acceptance study is underway to be performed.

Acknowledgments

Financial support from the Swedish International Development Agency (SIDA) programme under the Project "Technology processing of natural resources for sustainable development" is gratefully acknowledged. The Authors are also grateful to Camilla Öhgren for conducting the microscopy examinations and Professor Dr. José Francisco da Cruz for assisting with cassava roots processing.

References

[1] L. L. Navickis, "Corn flour addition to wheat flour doughs effect on rheological properties," *Cereal Chemistry*, vol. 64, pp. 307–310, 1987.

[2] E. K. Arendt, T. J. O'Brien, T. J. Schober, E. Gallagher, and T. R. Gormley, "Development of gluten-free cereal products," *Farm and Food*, vol. 12, pp. 21–27, 2002.

[3] E. Gallagher, T. R. Gormley, and E. K. Arendt, "Crust and crumb characteristics of gluten free breads," *Journal of Food Engineering*, vol. 56, no. 2-3, pp. 153–161, 2003.

[4] M. Mariotti, M. Lucisano, and M. Ambrogina Pagani, "Development of a baking procedure for the production of oat-supplemented wheat bread," *International Journal of Food Science and Technology*, vol. 41, no. 2, pp. 151–157, 2006.

[5] A. M. Almazan, "Effect of cassava flour variety and concentration on bread loaf quality," *Cereal Chemistry*, vol. 67, pp. 97–99, 1990.

[6] I. Defloor, M. Nys, and J. A. Delcour, "Wheat starch, cassava starch, and cassava flour impairment of the breadmaking potential of wheat flour," *Cereal Chemistry*, vol. 70, pp. 526–530, 1993.

[7] A. H. Khalil, E. H. Mansour, and F. M. Dawoud, "Influence of malt on rheological and baking properties of wheat-cassava composite flours," *LWT Food Science and Technology*, vol. 33, no. 3, pp. 159–164, 2000.

[8] T. A. Shittu, A. O. Raji, and L. O. Sanni, "Bread from composite cassava-wheat flour—I. Effect of baking time and temperature on some physical properties of bread loaf," *Food Research International*, vol. 40, no. 2, pp. 280–290, 2007.

[9] T. A. Shittu, A. Dixon, S. O. Awonorin, L. O. Sanni, and B. Maziya-Dixon, "Bread from composite cassava-wheat flour—II. Effect of cassava genotype and nitrogen fertilizer on bread quality," *Food Research International*, vol. 41, no. 6, pp. 569–578, 2008.

[10] T. A. Shittu, R. A. Aminu, and E. O. Abulude, "Functional effects of xanthan gum on composite cassava-wheat dough and bread," *Food Hydrocolloids*, vol. 23, no. 8, pp. 2254–2260, 2009.

[11] J. E. Wenham, "Post-harvest deterioration of cassava—a biotechnology perspective," FAO Plant Production and Protection Paper 130. FAO, Rome, Italy, 1995.

[12] A. Westby, "Cassava utilization, storage and small-scale processing," in *Cassava*, R. J. Hillocks, J. M. Thresh, and A. C. Belloti, Eds., pp. 281–300, CAB International, 2002.

[13] C. C. Wheatley, G. Chuzel, and N. Zakhia, "Cassava—uses as a raw material," in *Encyclopedia of Food Sciences and Nutrition*, B. Caballero, L. C. Trugo, and P. M. Finglas, Eds., pp. 969–974, Academic Press: Elsevier Science, London, 2nd edition, 2003.

[14] L. D. Tivana, P. Dejmek, and B. Bergenståhl, "Characterization of the agglomeration of roasted shredded cassava (Manihot esculenta crantz) roots," *Starch/Staerke*, vol. 62, no. 12, pp. 637–646, 2010.

[15] F. A. Numfor, W. M. Walter Jr., and S. J. Schwartz, "Emulsifiers affect the texture of pastes made from fermented and non-fermented cassava flours," *International Journal of Food Science and Technology*, vol. 33, no. 5, pp. 455–460, 1998.

[16] H. He and R. C. Hoseney, "Gas retention of different cereal flours," *Cereal Chemistry*, vol. 68, pp. 334–336, 1991.

[17] M. Collado-Fernández, "Dough fermentation," in *Encyclopedia of Food Sciences and Nutrition*, B. Caballero, L. C. Trugo, and P. M. Finglas, Eds., pp. 647–654, Academic Press, Elsevier Science, 2nd edition, 2003.

[18] A. Lazaridou, D. Duta, M. Papageorgiou, N. Belc, and C. G. Biliaderis, "Effects of hydrocolloids on dough rheology and bread quality parameters in gluten-free formulations," *Journal of Food Engineering*, vol. 79, no. 3, pp. 1033–1047, 2007.

[19] C. M. Rosell, J. A. Rojas, and C. Benedito de Barber, "Influence of hydrocolloids on dough rheology and bread quality," *Food Hydrocolloids*, vol. 15, no. 1, pp. 75–81, 2001.

[20] A. Guarda, C. M. Rosell, C. Benedito, and M. J. Galotto, "Different hydrocolloids as bread improvers and antistaling agents," *Food Hydrocolloids*, vol. 18, no. 2, pp. 241–247, 2004.

[21] C. M. Rosell and E. Santos, "Impact of fibers on physical characteristics of fresh and staled bake off bread," *Journal of Food Engineering*, vol. 98, no. 2, pp. 273–281, 2010.

[22] AACC, *Approved Methods of the American Association of Cereal Chemists*, AACC, St. Paul, Minn, USA, 9th edition, 1995.

[23] R. A. Anderson, H. F. Conway, V. F. Pfeifer, and E. L. Griffin, "Gelatinization of corn grits by roll and extrusion cooking," *Cereal Science Today*, vol. 14, pp. 4–6, 1969.

[24] P. Hongsprabhas, K. Israkarn, and C. Rattanawattanaprakit, "Architectural changes of heated mungbean, rice and cassava starch granules: effects of hydrocolloids and protein-containing envelope," *Carbohydrate Polymers*, vol. 67, no. 4, pp. 614–622, 2007.

[25] M. Maskan, "Kinetics of colour change of kiwifruits during hot air and microwave drying," *Journal of Food Engineering*, vol. 48, no. 2, pp. 169–175, 2001.

[26] J. A. Delcour, S. Vanhamel, and R. C. Hoseney, "Physicochemical and functional properties of rye nonstarch polysaccharides—II. Impact of a fraction containing water-soluble pentosans and proteins on gluten-starch loaf volumes," *Cereal Chemistry*, vol. 68, pp. 72–76, 1991.

[27] J. A. Rojas, C. M. Rosell, and C. Benedito De Barber, "Pasting properties of different wheat flour-hydrocolloid systems," *Food Hydrocolloids*, vol. 13, no. 1, pp. 27–33, 1999.

[28] M. E. Bárcenas, J. D. L. O-Keller, and C. M. Rosell, "Influence of different hydrocolloids on major wheat dough components (gluten and starch)," *Journal of Food Engineering*, vol. 94, no. 3-4, pp. 241–247, 2009.

[29] D. A. Bell, "Methylcellulose as a structure enhancer in bread baking," *Cereal Food World*, vol. 35, pp. 1001–1006, 1990.

[30] C. Mestres, O. Boungou, N. Akissoe, and N. Zakhia, "Comparison of the expansion ability of fermented maize flour and cassava starch during baking," *Journal of the Science of Food and Agriculture*, vol. 80, pp. 665–672, 2000.

[31] F. A. Numfor, W. M. Walter Jr., and S. J. Schwartz, "Physicochemical changes in cassava starch and flour associated with fermentation: effect on textural properties," *Starch*, vol. 47, pp. 86–91, 1995.

[32] M. Alloncle and J. L. Doublier, "Viscoelastic properties of maize starch/hydrocolloid pastes and gels," *Food Hydrocolloids*, vol. 5, pp. 455–467, 1991.

[33] J. Babić, D. Šubarić, Đ. Ačkar, V. Piližota, M. Kopjar, and N. Nedić Tiban, "Effects of pectin and carrageenan on thermophysical and rheological properties of tapioca starch," *Czech Journal of Food Science*, vol. 24, pp. 275–282, 2006.

[34] J. E. Hodge, "Origin of flavor in foods. Nonenzymatic browning reactions," in *Proceedings of the Symposium on Food Chemistry and Physiology of Flavors*, H. W. Schults, E. A. Day, and L. M. Libbey, Eds., pp. 465–491, The AVI Pub. Co., Westport, Conn, USA, 1967.

[35] F. Gögüs, C. Düzdemir, and S. Eren, "Effects of some hydrocolloids and water activity on nonenzymic browning of concentrated orange juice," *Nahrung*, vol. 44, no. 6, pp. 438–442, 2000.

[36] C. Mestres and X. Rouau, "Influence of natural fermentation and drying conditions on the physicochemical characteristics of cassava starch," *Journal of the Science of Food and Agriculture*, vol. 74, pp. 147–155, 1997.

7

Comparison of Dried Plum Puree, Rosemary Extract, and BHA/BHT as Antioxidants in Irradiated Ground Beef Patties

Iulia Movileanu,[1] Máryuri T. Núñez de González,[2] Brian Hafley,[3] Rhonda K. Miller,[4] and Jimmy T. Keeton[5]

[1] Department of Neurology, Upstate Medical University, 812 Jacobsen Hall, Syracuse, NY 13210, USA
[2] Department of Food Technology, Universidad de Oriente, Núcleo Nueva Esparta, Escuela de Ciencias Aplicadas del Mar, Isla de Margarita 6301, Venezuela
[3] Tyson Foods, 1825 Ford Avenue, Springdale, AR 72764, USA
[4] Department of Animal Science, Texas A&M University, 338 Kleberg Center, College Station, TX 77843-2471, USA
[5] Department of Nutrition and Food Science, Texas A&M University, 122 Kleberg Center, College Station, TX 77843-2253, USA

Correspondence should be addressed to Jimmy T. Keeton; jkeeton@tamu.edu

Academic Editor: Alejandro Castillo

Fresh ground beef patties with (1) no antioxidant (control), (2) 0.02% butylated hydroxyanisole/butylated hydroxytoluene (BHA/BHT), (3) 3% dried plum puree, or (4) 0.25% rosemary extract were aerobically packaged, irradiated at target doses of 0, 1.5, or 2.0 kGy (1.7 and 2.3 kGy actual doses), and stored at 4°C. The samples were evaluated for lipid oxidation on 0, 3, 7, 14, 21, and 28 days of storage after irradiation. When compared to the control, all antioxidant treatments were effective in retarding ($P < 0.05$) irradiation-induced lipid oxidation during storage as determined by 2-thiobarbituric acid reactive substances (TBARs) values. Rosemary extracts had the same antioxidant effect ($P > 0.05$) as BHA/BHT in irradiated and nonirradiated beef patties, followed by the dried plum puree treatment. Irradiation increased TBARs values, but no differences were noted in oxidation between irradiation dose levels.

1. Introduction

Irradiation of refrigerated and frozen red meat and poultry has been approved by the Food and Drug Administration (FDA) [1] and the United States Department of Agriculture-Food Safety and Inspection Service (USDA-FSIS) at levels up to 4.5–7.0 kGy and 3.0 kGy, respectively, to effectively reduce or eliminate pathogenic bacteria [2]. Recently, the FDA approved a maximum absorbed dose of ionizing radiation of 4.5 kGy to treat unrefrigerated (as well as refrigerated) uncooked meat, meat by-products, and certain meat food products to reduce levels of foodborne pathogens and extend shelf-life. Also, the regulation indicates that irradiation is a safe treatment for fresh (refrigerated and unrefrigerated) poultry food products and frozen poultry products at absorbed doses that do not exceed 4.5 kGy and 7.0 kGy, respectively [3]. This technology can eliminate foodborne pathogenic microorganisms in meats and help provide a safer food supply for human consumption [4]. Populations of the most common enteric pathogens such as *Escherichia coli* O157:H7, *Staphylococcus aureus*, *Salmonella*, *Listeria monocytogenes*, and *Aeromonas hydrophila* can be significantly decreased or eliminated by low dose (<3.0 kGy) treatments with ionizing radiation [5].

Use of irradiation on raw meat, however, results in the formation of radiolytic products [6] that can cause oxidation of myoglobin and lipids leading to discoloration, lipid oxidation, and formation of off-odor and off-flavor compounds [7, 8]. Free radicals and hydroperoxides generated by irradiation can also destroy endogenous antioxidants, reduce storage stability, and increase off-flavors from lipid and protein degradation [9, 10]. Although anaerobic packaging is an effective means of reducing oxidation, incorporation of antioxidants, synthetic or natural, is one of the major strategies for preventing lipid oxidation [11] and improving color stability of meat products. To be effective, antioxidants must compete with reactive meat components for free radicals generated by lipid oxidation and irradiation or inhibit the formation of free radicals induced by prooxidative metals [12, 13].

Rosemary (*Rosmarinus officinalis*) extracts have been found to be effective in inhibiting lipid oxidation in meat systems primarily through the presence of phenolic compounds [14, 15] which break free radical chain reactions by hydrogen atom donation [16]. Rosemary and rosemary extracts at concentrations ranging from 0.02% to 1% have been perhaps the most investigated natural antioxidants in meat products [17–21]. Various authors have reported the effectiveness of rosemary in ground beef [22], beef patties [23–26], pork patties [20, 27], pork sausage [28], and beef steaks [29].

Previous research has shown that dried plums have one of the highest antioxidant activities of both fruits and vegetables [30]. The principal phytochemicals in dried plums are phenolic compounds. These appear to be the primary contributors to antioxidant properties, are responsible for inhibition of low-density lipoprotein oxidation [31–33], and may play a role in suppressing cellular inflammation, a precursor to cancer development. The native antioxidant properties of dried plums may offer alternatives to the meat industry based on their potential to reduce microbial growth, retain moisture, and prevent off-flavor formation in meat [34].

The addition of 3% plum extract has been shown to improve mouthfeel and the antioxidant properties in irradiated turkey breast rolls [35]. Núñez de González et al. [36] demonstrated that dried plum puree at levels of 3% to 6% in precooked pork sausage was as effective as a butylated hydroxyanisole/butylated hydroxytoluene (BHA/BHT) combination in suppressing lipid oxidation. Moreover, Núñez de González et al. [37] reported that 2.5% fresh juice concentrate or dried plum juice concentrate could be incorporated into beef roasts to reduce lipid oxidation and potentially warmed-over flavor (WOF). Thus, the use of naturally occurring antioxidants (or extractives) with high antioxidant capacity may prove useful in retarding or reducing off-odors and off-flavors in irradiated meat products and particularly ground beef patties which are more prone to lipid oxidation.

Synthetic antioxidants such as butylated hydroxyanisole (BHA) and butylated hydroxytoluene (BHT) have been widely used in meat products [25, 28, 36, 38]. USDA regulations permit up to 0.01% (based on fat content) each of BHA and BHT in fresh sausage and up to 0.003% (based on total weight) each in dry sausage [39]. Addition of 0.02% BHA/BHT (0.01% BHA + 0.01% BHT) in ground beef has been shown to be more effective in retarding lipid oxidation, as determined by lower levels of 2-thiobarbituric acid reactive substances (TBARs) [19], when compared to natural plant extracts. TBARs values for the treated samples did not change significantly during storage.

The objective of this study was to compare the antioxidant effectiveness of dried plum puree (3%), BHA/BHT (0.02%), and rosemary extract (0.25%) in irradiated (0, 1.5, or 2.0 kGy) ground beef patties stored refrigerated over a 28-day period to assess their capacity in retarding lipid oxidation as measured by TBARs.

2. Materials and Methods

2.1. Beef Patty Preparation. Approximately 130 kg of fresh, coarse ground (1.27-cm plate) beef lean and fat trimmings were purchased from Ruffino Meats, Inc. (Bryan, TX), 3 to 4 d postmortem. Four 27.3 Kg batches were formulated to contain 20% fat and assigned to treatments of no antioxidant (control), BHA/BHT (0.02% combination based on the fat content, crystalline TENOX, Eastman Chemical Products, Kingsport, TN), rosemary extract (0.25% based on batch weight, liquid oil based HERBALOX, Type HT 25, Kalsec, Inc., Kalamazoo, MI), and dried plum puree (3% based on batch weight, Sunsweet Growers Inc., Yuba City, CA). No additional ingredients were included in the control treatment so that the true irradiation effect on 100% ground beef could be measured. The level of dried plum puree incorporated into ground beef patties was determined from a previous study in which 3% was found to be effective in retarding lipid oxidation in precooked pork sausage [36]. Moreover, Lee and Ahn [35] indicated that the addition of 3% plum extract was recommended to improve mouthfeel and antioxidant effect in irradiated turkey breast rolls.

Beef patties were manufactured in a state-inspected (Texas Department of Health) commercial-scale pilot plant located in the Rosenthal Meat Science and Technology Center at Texas A&M University. Each coarse-ground treatment was blended in a paddle mixer (Model 150 767, Butcher Boy, Lasar, MFG Inc., Los Angeles, CA) for 3 min and reground through a 0.635-cm plate (Biro Model 1056 22772, The Biro Mfg Co., Marblehead, OH). Dried plum puree, rosemary extract, and BHA/BHT were incorporated directly into the mixer in their original commercial form. Patties (113 g, 10.5 cm diameter, 1 cm thick) were made (Hollymatic JET FLOW Super Food Portioning Machine, 1033, Hollymatic Corp., Park Forest, IL) and stored at −40°C for 2 h. Approximately 220 patties per treatment were produced in this study. The crust frozen patties were then placed in Cryovac bags (Cryovac, Sealed Air Corporation, Duncan, SC) and aerobically packaged (a vacuum setting 3.5 on a 10 point scale where 10 = maximum vacuum) to a constant level of air using an Ultravac packaging machine (Model 2100 D, KOCH Inc, Kansas City, MO). An aerobic package environment was chosen to exacerbate lipid oxidation and challenge the antioxidant treatments when exposed to a constant concentration of air and two electron-beam irradiation levels. Pulling a full vacuum (setting of 10) would have reduced irradiation mediated oxidation of each treatment and would have reduced the oxidative environment to stress each antioxidant's capacity for suppressing oxidation. After packaging, the patties were rapidly frozen (−40°C) prior to shipping to an electron beam irradiation facility.

2.2. Irradiation of Samples and Storage. Frozen patties were placed in insulated containers with ice packs and shipped overnight to the Surebeam Corporation irradiation facility

TABLE 1: Experimental design[b] of patty allocation by treatment and storage time.

Treatment		Storage time (day)						
Actual irradiation dose (kGy)	0	3	7	14	21	28	Total	
Control[a]	0	12	12	12	12	12	12	72
	1.7	12	12	12	12	12	12	72
	2.3	12	12	12	12	12	12	72
BHA/BHT[a]	0	12	12	12	12	12	12	72
	1.7	12	12	12	12	12	12	72
	2.3	12	12	12	12	12	12	72
Dried plum puree[a]	0	12	12	12	12	12	12	72
	1.7	12	12	12	12	12	12	72
	2.3	12	12	12	12	12	12	72
Rosemary extract[a]	0	12	12	12	12	12	12	72
	1.7	12	12	12	12	12	12	72
	2.3	12	12	12	12	12	12	72
Total								864

[a]Control = no antioxidant; BHA/BHT = butylated hydroxyanisole/butylated hydroxytoluene (0.02% of fat content); dried plum puree = 3.0% based on batch weight; rosemary extract = 0.25% based on batch weight.

[b]A total of three replications were performed in this experiment.

(Glendale Heights, IL). Samples were tempered and irradiated at $-4°C$ at targeted doses of 0 (control), 1.5, and 2 kGy while passing through a linear accelerator. The targeted doses were determined based on previous studies [40, 41]. For each batch, a dosimetry profile was established for uniform packages of 12 patties stacked three high with an actual mean dose level of 1.7 and 2.3 kGy, respectively, for the irradiation treatments. To confirm the target dose, 2 alanine dosimeters per cart were attached to the top and bottom surface of a sample. All samples were hard frozen $(-20°C)$ after irradiation and returned via overnight carrier to the Department of Animal Science at Texas A&M University. The patties remained in their original package as described above and were then stored and refrigerated at $4°C$ and evaluated on storage days 0, 3, 7, 14, 21, and 28. Day 0 was designated the next day after the return of the irradiated samples.

2.3. 2-Thiobarbituric Acid Reactive Substances (TBARs). Lipid oxidation of the ground beef patties from each treatment, irradiation level, and storage day was determined using the distillation 2-thiobarbituric acid (TBA) procedure of Tarladgis et al. [42], with the addition of propyl gallate and ethylenediaminetetraacetic acid (EDTA) at the sample blending step to prevent further lipid oxidation [43]. TBARs determinations were performed in duplicate, and results were expressed as mg of malonaldehyde per kg of meat sample. The percent reduction in TBARs by different treatments in comparison to control was also determined.

2.4. Experimental Design and Statistical Analysis. The experiment was designed as a factorial arrangement with a split-plot design. Batch treatments included (1) control (no antioxidant); (2) BHA/BHT; (3) dried plum puree, and (4) rosemary extract. The patties were aerobically packaged, irradiated at actual levels of 0, 1.7, and 2.3 kGy, and stored at $4°C$ for 0,

3, 7, 14, 21, and 28 days. Seventy-two patties were randomly assigned to the batch treatment, dose level and storage condition (Table 1). One trial replication was processed to establish shipping requirements and analytical logistics followed by three complete experimental replications that were conducted individually approximately 6-7 weeks apart.

Analysis of variance was conducted using the General Linear Model (GLM) procedure of statistical analysis system [44]. Data was analyzed as a factorial arrangement with a split-plot design. The whole-plot included the main effects of batch and treatment. The error term for the whole plot was the batch by treatment interaction. The split-plot included the main effects of radiation dose and storage days and their interactions. The residual error was used to test split-plot effects at a significance value of $P < 0.05$. Least square means were separated using the PDIFF statement of GLM [44]. Least square means were generated and reported for the main effects and for significant interactions.

3. Results and Discussion

TBARs values of beef patties containing different antioxidants and irradiated at two dose levels over 28 days at refrigerated conditions are presented in Tables 2 and 3. No differences in TBARs values $(P < 0.05)$ were noted between doses 1.7 and 2.3 kGy, regardless of antioxidant treatment (Table 2). The inclusion of BHA/BHT or rosemary extract retarded the oxidative process regardless of irradiation dose level, and TBARs values were lower than control patties not receiving irradiation. Lee et al. [38] found that thiobarbituric acid values were not very different during storage regardless of the irradiation dose or the type of antioxidant (rosemary extracts and BHA). In the present study, irradiated samples without antioxidants (control) had significantly higher TBARs values $(P < 0.05)$ than the 1.7 and 2.3 kGy irradiation levels with

TABLE 2: Least square means for TBARs values of ground beef patties with/without antioxidants and irradiated at two dose levels.

Treatment[d]	Dose[e] (kGy)	TBARs (mg malonaldehyde/kg sample)
Control	0	1.56[b]
	1.7	2.70[c]
	2.3	2.68[c]
BHA/BHT	0	0.42[a]
	1.7	0.37[a]
	2.3	0.52[a]
Dried plum puree	0	0.43[a]
	1.7	1.60[b]
	2.3	1.61[b]
Rosemary extract	0	0.57[a]
	1.7	0.47[a]
	2.3	0.48[a]
RMS[f]		0.55

[a–c]Least square means in the same column with superscripts that differ are significantly different ($P < 0.05$).
[d]Control = no antioxidant; BHA/BHT = butylated hydroxyanisole/butylated hydroxytoluene (0.02% of fat content); dried plum puree = 3.0% based on batch weight; rosemary extract = 0.25% based on batch weight.
[e]Actual dosimetry levels were 1.7 and 2.3 kGy.
[f]RMS = root mean square error.

TABLE 3: Least square means for TBARS values of irradiated ground beef patties treated with/without antioxidants and stored at 4°C for 28 days.

Treatment[h]	Storage day	TBARS (mg malonaldehyde/kg sample)
Control	0	0.97[bcd]
	3	1.97[e]
	7	2.52[f]
	14	2.79[fg]
	21	2.58[fg]
	28	3.04[g]
BHA/BHT	0	0.30[a]
	3	0.32[a]
	7	0.36[a]
	14	0.44[a]
	21	0.74[abc]
	28	0.46[ab]
Dried plum puree	0	0.51[ab]
	3	0.83[abc]
	7	1.02[cd]
	14	1.56[e]
	21	1.89[e]
	28	1.46[de]
Rosemary extract	0	0.32[a]
	3	0.39[a]
	7	0.44[a]
	14	0.54[abc]
	21	0.53[abc]
	28	0.83[abc]
RMS[i]		0.55

[a–g]Least square means in the same column with superscripts that differ are significantly different ($P < 0.05$).
[h]Control = no antioxidant; BHA/BHT = butylated hydroxyanisole/butylated hydroxytoluene (0.02% of fat content); dried plum puree = 3.0% based on batch weight; rosemary extract = 0.25% based on batch weight.
[i]RMS = root mean square error.

antioxidants (Table 2). Previous studies have indicated that irradiation accelerated lipid oxidation of raw meat under aerobic conditions [13, 40, 45]. Nawar [46] reported that irradiation in the presence of oxygen accelerated the autoxidation of fats by one of three possible reactions: (1) formation of free radicals which combine with oxygen to form hydroperoxides; (2) breakdown of hydroperoxides; and/or (3) destruction of antioxidants.

TBARs values showed that dried plum puree lessened the effect of irradiation-induced oxidation but also showed that it was less effective than BHA/BHT or rosemary treatments (Table 2). Overall, TBARs values were higher for dried plum puree than BHA/BHT or rosemary extract, even though dried plum puree partially reduced the oxidative effects of irradiation. It is postulated that the antioxidant component(s) in 3% dried plum puree might not be as concentrated as that found in the rosemary extract or BHA/BHT. Thus, equivalent levels of the antioxidant components in dried plum puree need to be identified and characterized prior to additional comparisons. The protective antioxidant effect of BHA/BHT and rosemary extract was not altered by irradiation dose. However, it cannot be ruled out that oxidation might occur in beef patties treated with higher levels of irradiation. In conclusion, rosemary extract and BHA/BHT treatments maintained lower and equivalent TBARs values regardless of dose level (1.7 or 2.3 kGy), while dried plum puree patties had slightly higher TBARs values. These values, however, were still significantly lower ($P < 0.05$) than the irradiated control.

All antioxidant treatments reduced TBARs values of all irradiated samples during the storage period when compared to the control (Table 3). Lipid oxidation, as measured by TBARs, of the control treatment increased significantly ($P < 0.05$) through day 7, remained static to day 21, and increased slightly from day 21 to day 28. Among the antioxidant treatments, samples containing BHA/BHT and rosemary extract showed the greatest antioxidant effect. There was an increase in TBARs values over time for the dried plum puree, but not to the same degree as the control. This increase was likely due to the inability of the dried plum puree's antioxidant component to completely suppress the oxidative effect of irradiation. Although, BHA/BHT and rosemary extract TBARs values increased slightly numerically, they were not statistically ($P > 0.05$) significant within a treatment over the storage period. When compared to the control, BHA/BHT and rosemary reduced TBARs by 69 and 67%, respectively,

TABLE 4: Percent decrease in TBARs values for irradiated ground beef patties treated with/without antioxidants and stored at 4°C for 28 days.

Treatment[b]	Percent decrease in TBARs as compared to control[a]					
	Storage day					
	0	3	7	14	21	28
BHA/BHT	69.1	83.8	85.7	84.2	71.3	84.9
Rosemary extract	67.0	80.2	82.5	80.6	79.4	72.7
Dried plum puree	47.4	57.9	59.5	44.0	26.7	52.0

[a] Data not statistically analyzed, but percentages were computed from the statistical data in Table 3.
[b] BHA/BHT = butylated hydroxyanisole/butylated hydroxytoluene (0.02% of fat content); dried plum puree = 3.0% based on batch weight; rosemary extract = 0.25% based on batch weight.

at day 0 (Table 4) and continued to have consistently high antioxidant activity throughout the 28-day storage period. Nam and Ahn [13] compared the TBARs values of irradiated pork homogenates containing sesamol (200 μM) and Trolox (200 μM) and found them to reduce TBARs values by as much as 72% when compared to other antioxidants after 5 days of storage. Nam et al. [12] indicated that irradiated pork loins treated with rosemary-tocopherol/double packaging had lower TBARs values than a vacuum-packaged control after 10 days of refrigerated storage. Rosemary oleoresin (200 ppm based on fat content) in pork patties, irradiated at 4.5 kGy and stored at 4°C for 7 days, was shown to be effective only for the first 3 days of the storage period [9], while BHT (0.01%) was effective across all storage days.

In this study, BHA/BHT and rosemary TBARs values did not differ ($P > 0.05$) from one another during storage of irradiated beef patties. Similarly, rosemary extract was noted to be equally effective as BHA/BHT in maintaining low TBARs values of nonirradiated refrigerated, fresh pork sausage and cooked-frozen sausage [28]. McBride et al. [25] indicated that rosemary extract was more effective in controlling lipid oxidation in aerobically stored beef samples as compared with a control or a BHA/BHT treatment. Additional antioxidant effects of rosemary in meats and meat products have been reported extensively in the scientific literature [20, 22–24, 26, 29].

Dried plum puree was not as effective as rosemary extract or BHA/BHT (Tables 3 and 4) in inhibiting oxidation over the storage period but did reduce ($P < 0.05$) TBARs values by approximately 50% when compared to the control. At day 0 (Table 4), dried plum puree decreased TBARs values by 47.4% but was less effective as an antioxidant from day 14 to 21. Lee and Ahn [35] likewise reported that the addition of 2% or 3% of plum extract puree was effective in controlling lipid oxidation in irradiated turkey breast rolls.

Dried plum puree contains numerous components such as dextrose, sorbitol, and other compounds that could dilute and lessen its antioxidant effectiveness. Based on this observation, further work is needed to extract, identify, and standardize the principle antioxidant components of dried plum puree and then test their efficacy. Thus, a dried plum puree extract could potentially have the same or possibly greater antioxidant capacity as that of rosemary extract or a BHA/BHT combination in irradiated ground beef patties.

4. Conclusions

All antioxidant treatments decreased TBARs values in ground beef patties irradiated at 1.7 and 2.3 kGy. Rosemary (0.25%) and BHA/BHT (0.02%) were the most effective antioxidants, while dried plum puree (3%) also reduced lipid oxidation but to a lesser degree. Irradiation doses of 1.7 and 2.3 kGy increased TBARs values, but no differences in oxidation were noted between irradiation dose levels.

Acknowledgment

This project was funded by Texas A&M AgriLife Research, an agency of the Texas A&M University System.

References

[1] CFR Code of Federal Regulations, *Ionizing Radiation for the Treatment of Food. Title 21, Part 179.26*, Government Printing Office, Washington, DC, USA, 2001.

[2] D. G. Olson, "Meat irradiation and meat safety," in *Proceedings of the 51st Reciprocal Meat Conference*, pp. 149–152, Storrs, CT, USA, June-July 1998.

[3] FDA, "Irradiation in the production, processing and handling of food," *Federal Register*, vol. 77, no. 231, pp. 71312–71321, 2012.

[4] Y. Liu, X. Fan, Y. R. Chen, and D. W. Thayer, "Changes in structure and color characteristics of irradiated chicken breasts as a function of dosage and storage time," *Meat Science*, vol. 63, no. 3, pp. 301–307, 2003.

[5] D. W. Thayer, "Use of irradiation to kill enteric pathogens on meat and poultry," *Journal of Food Safety*, vol. 15, pp. 181–192, 1995.

[6] R. J. Woods and A. K. Pikaev, "Interaction of radiation with matter," in *Applied Radiation Chemistry: Radiation Processing*, pp. 59–89, John Wiley & Sons, New York, NY, USA, 1994.

[7] E. A. Murano, "Irradiation of fresh meats," *Food Technology*, vol. 49, no. 12, pp. 52–54, 1995.

[8] K. E. Nanke, J. G. Sebranek, and D. G. Olson, "Color characteristics of irradiated vacuum-packaged pork, beef, Turkey," *Journal of Food Science*, vol. 63, no. 6, pp. 1001–1006, 1998.

[9] X. Chen, C. Jo, J. I. Lee, and D. U. Ahn, "Lipid oxidation, volatiles and color changes of irradiated pork patties as affected by antioxidants," *Journal of Food Science*, vol. 64, no. 1, pp. 16–19, 1999.

[10] C. Jo and D. U. Ahn, "Volatiles and oxidative changes in irradiated pork sausage with different fatty acid composition and tocopherol content," *Journal of Food Science*, vol. 65, no. 2, pp. 270–275, 2000.

[11] M. C. Rojas and M. S. Brewer, "Effect of natural antioxidants on oxidative stability of cooked, refrigerated beef and pork," *Journal of Food Science*, vol. 72, no. 4, pp. S282–S288, 2007.

[12] K. C. Nam, K. Y. Ko, B. R. Min et al., "Influence of rosemary-tocopherol/packaging combination on meat quality and the survival of pathogens in restructured irradiated pork loins," *Meat Science*, vol. 74, no. 2, pp. 380–387, 2006.

[13] K. C. Nam and D. U. Ahn, "Use of antioxidants to reduce lipid oxidation and off-odor volatiles of irradiated pork homogenates and patties," *Meat Science*, vol. 63, no. 1, pp. 1–8, 2003.

[14] C. M. Houlihan, C. T. Ho, and S. S. Chang, "Elucidation of the chemical structure of a novel antioxidant, rosmaridiphenol, isolated from rosemary," *Journal of the American Oil Chemists' Society*, vol. 61, no. 6, pp. 1036–1039, 1984.

[15] W. Zheng and S. Y. Wang, "Antioxidant activity and phenolic compounds in selected herbs," *Journal of Agricultural and Food Chemistry*, vol. 49, no. 11, pp. 5165–5170, 2001.

[16] H. Basaga, C. Tekkaya, and F. Acikel, "Antioxidative and free radical scavenging properties of rosemary extract," *LWT— Food Science and Technology*, vol. 30, no. 1, pp. 105–108, 1997.

[17] A. J. St. Angelo, K. L. Crippin, H. P. Dupuy, and C. James Jr., "Chemical and sensory studies of antioxidant-treated beef," *Journal of Food Science*, vol. 55, no. 6, pp. 1501–1505, 1990.

[18] T. L. McCarthy, J. P. Kerry, J. F. Kerry, P. B. Lynch, and D. J. Buckley, "Evaluation of the antioxidant potential of natural food/plant extracts as compared with synthetic antioxidants and vitamin e in raw and cooked pork patties," *Meat Science*, vol. 58, no. 1, pp. 45–52, 2001.

[19] J. Ahn, I. U. Grün, and L. N. Fernando, "Antioxidant properties of natural plant extracts containing polyphenolic compounds in cooked ground beef," *Journal of Food Science*, vol. 67, no. 4, pp. 1364–1369, 2002.

[20] L. R. Nissen, D. V. Byrne, G. Bertelsen, and L. H. Skibsted, "The antioxidative activity of plant extracts in cooked pork patties as evaluated by descriptive sensory profiling and chemical analysis," *Meat Science*, vol. 68, no. 3, pp. 485–495, 2004.

[21] J. Han and K. S. Rhee, "Antioxidant properties of selected Oriental non-culinary/nutraceutical herb extracts as evaluated in raw and cooked meat," *Meat Science*, vol. 70, no. 1, pp. 25–33, 2005.

[22] C. W. Balentine, P. G. Crandall, C. A. O'Bryan, D. Q. Duong, and F. W. Pohlman, "The pre- and post-grinding application of rosemary and its effects on lipid oxidation and color during storage of ground beef," *Meat Science*, vol. 73, no. 3, pp. 413–421, 2006.

[23] A. Sánchez-Escalante, D. Djenane, G. Torescano, J. A. Beltrán, and P. Roncalés, "Antioxidant action of borage, rosemary, oregano, and ascorbic acid in beef patties packaged in modified atmosphere," *Journal of Food Science*, vol. 68, no. 1, pp. 330–344, 2003.

[24] A. Sánchez-Escalante, D. Djenane, G. Torrescano, J. A. Beltrán, and P. Roncalés, "The effects of ascorbic acid, taurine, carnosine and rosemary powder on colour and lipid stability of beef patties packaged in modified atmosphere," *Meat Science*, vol. 58, no. 4, pp. 421–429, 2001.

[25] N. T. M. McBride, S. A. Hogan, and J. P. Kerry, "Comparative addition of rosemary extract and additives on sensory and antioxidant properties of retail packaged beef," *International Journal of Food Science and Technology*, vol. 42, no. 10, pp. 1201–1207, 2007.

[26] M. N. Lund, M. S. Hviid, and L. H. Skibsted, "The combined effect of antioxidants and modified atmosphere packaging on protein and lipid oxidation in beef patties during chill storage," *Meat Science*, vol. 76, no. 2, pp. 226–233, 2007.

[27] K. C. Nam, K. Y. Ko, B. R. Min et al., "Effects of oleoresin-tocopherol combinations on lipid oxidation, off-odor, and color of irradiated raw and cooked pork patties," *Meat Science*, vol. 75, no. 1, pp. 61–70, 2007.

[28] J. G. Sebranek, V. J. H. Sewalt, K. L. Robbins, and T. A. Houser, "Comparison of a natural rosemary extract and BHA/BHT for relative antioxidant effectiveness in pork sausage," *Meat Science*, vol. 69, no. 2, pp. 289–296, 2005.

[29] D. Djenane, A. Sánchez-Escalante, J. A. Beltrán, and P. Roncalés, "Ability of a-tocopherol, taurine and rosemary, in combination with vitamin C, to increase the oxidative stability of beef steaks packaged in modified atmosphere," *Food Chemistry*, vol. 76, no. 4, pp. 407–415, 2002.

[30] H. Wang, G. Cao, and R. L. Prior, "Total antioxidant capacity of fruits," *Journal of Agricultural and Food Chemistry*, vol. 44, no. 3, pp. 701–705, 1996.

[31] M. Stacewicz-Sapuntzakis, P. E. Bowen, E. A. Hussain, B. I. Damayanti-Wood, and N. R. Farnsworth, "Chemical composition and potential health effects of prunes: a functional food?" *Critical Reviews in Food Science and Nutrition*, vol. 41, no. 4, pp. 251–286, 2001.

[32] J. L. Donovan, A. S. Meyer, and A. L. Waterhouse, "Phenolic composition and antioxidant activity of prunes and prune juice (*Prunus domestica*)," *Journal of Agricultural and Food Chemistry*, vol. 46, no. 4, pp. 1247–1252, 1998.

[33] S. E. Kasim-Karakas, R. U. Almario, L. Gregory, H. Todd, R. Wong, and B. L. Lasley, "Effects of prune consumption on the ratio of 2-hydroxyestrone to 16α-hydroxyestrone," *American Journal of Clinical Nutrition*, vol. 76, no. 6, pp. 1422–1427, 2002.

[34] H. Kreuzer, *Dried Plums Solve Meaty Issues. Food Product Design*, Weeks Publishing, Northbrook, Ill, USA, 2001.

[35] E. J. Lee and D. U. Ahn, "Quality characteristics of irradiated turkey breast rolls formulated with plum extract," *Meat Science*, vol. 71, no. 2, pp. 300–305, 2005.

[36] M. T. Núñez de González, R. M. Boleman, R. K. Miller, J. T. Keeton, and K. S. Rhee, "Antioxidant properties of dried plum ingredients in raw and precooked pork sausage," *Journal of Food Science*, vol. 73, no. 5, pp. H63–H71, 2008.

[37] M. T. Núñez de González, B. S. Hafley, R. M. Boleman, R. K. Miller, K. S. Rhee, and J. T. Keeton, "Antioxidant properties of plum concentrates and powder in precooked roast beef to reduce lipid oxidation," *Meat Science*, vol. 80, no. 4, pp. 997–1004, 2008.

[38] J. W. Lee, K. S. Park, J. G. Kim et al., "Combined effects of gamma irradiation and rosemary extract on the shelf-life of a ready-to-eat hamburger steak," *Radiation Physics and Chemistry*, vol. 72, no. 1, pp. 49–56, 2005.

[39] CFR Code of Federal Regulations, *Food Ingredients and Sources of Irradiation. Title 9, Part 424, Subpart C*, Government Printing Office, Washington, DC, USA, 2009, http://www.ecfr.gov/cgi-bin/text-idx?c=ecfr&sid=ff85ab30d9f 72231cf408cf2293e83da&rgn=div8&view=text&node=9:2.0.2.4 .41.2.70.1&idno=9.

[40] S. E. Luchsinger, D. H. Kropf, C. M. García Zepeda et al., "Color and oxidative properties of irradiated ground beef patties," *Journal of Muscle Foods*, vol. 8, no. 4, pp. 445–464, 1997.

[41] K. E. Nanke, J. G. Sebranek, and D. G. Olson, "Color characteristics of irradiated aerobically packaged pork, beef, and turkey," *Journal of Food Science*, vol. 64, no. 2, pp. 272–278, 1999.

[42] B. G. Tarladgis, B. M. Watts, M. T. Younathan, and L. Dugan, "A distillation method for the quantitative determination of malonaldehyde in rancid foods," *Journal of the American Oil Chemists' Society*, vol. 37, no. 1, pp. 44–48, 1960.

[43] K. S. Rhee, "Minimization of further lipid peroxidation in the distillation 2-thibarbituric acid test of fish and meat," *Journal of Food Science*, vol. 43, no. 6, pp. 1776–1778, 1978.

[44] SAS, *SAS USer's Guide*, Statistical Analysis System Institute, Cary, NC, USA, 1995.

[45] Y. H. Kim, K. C. Nam, and D. U. Ahn, "Volatile profiles, lipid oxidation and sensory characteristics of irradiated meat from different animal species," *Meat Science*, vol. 61, no. 3, pp. 257–265, 2002.

[46] W. W. Nawar, "Lipids," in *Food Chemistry*, O. Fennema, Ed., pp. 139–244, Marcel Dekker, Inc., New York, NY, USA, 1985.

8

Identification of Imitation Cheese and Imitation Ice Cream Based on Vegetable Fat Using NMR Spectroscopy and Chemometrics

Yulia B. Monakhova,[1,2,3] **Rolf Godelmann,**[1] **Claudia Andlauer,**[1]
Thomas Kuballa,[1] **and Dirk W. Lachenmeier**[1,4]

[1] *Chemisches und Veterinäruntersuchungsamt (CVUA) Karlsruhe, Weissenburger Strasse 3, 76187 Karlsruhe, Germany*
[2] *Department of Chemistry, Saratov State University, Astrakhanskaya Street 83, 410012 Saratov, Russia*
[3] *Bruker Biospin GmbH, Silbersteifen, 76287 Rheinstetten, Germany*
[4] *Ministry of Rural Affairs and Consumer Protection, Kernerplatz 10, 70182 Stuttgart, Germany*

Correspondence should be addressed to Dirk W. Lachenmeier; lachenmeier@web.de

Academic Editor: Carl J. Schaschke

Vegetable oils and fats may be used as cheap substitutes for milk fat to manufacture imitation cheese or imitation ice cream. In this study, 400 MHz nuclear magnetic resonance (NMR) spectroscopy of the fat fraction of the products was used in the context of food surveillance to validate the labeling of milk-based products. For sample preparation, the fat was extracted using an automated Weibull-Stoldt methodology. Using principal component analysis (PCA), imitation products can be easily detected. In both cheese and ice cream, a differentiation according to the type of raw material (milk fat and vegetable fat) was possible. The loadings plot shows that imitation products were distinguishable by differences in their fatty acid ratios. Furthermore, a differentiation of several types of cheese (Edamer, Gouda, Emmentaler, and Feta) was possible. Quantitative data regarding the composition of the investigated products can also be predicted from the same spectra using partial least squares (PLS) regression. The models obtained for 13 compounds in cheese (R^2 0.75–0.95) and 17 compounds in ice cream (R^2 0.83–0.99) (e.g., fatty acids and esters) were suitable for a screening analysis. NMR spectroscopy was judged as suitable for the routine analysis of dairy products based on milk or on vegetable fat substitutes.

1. Introduction

Due to industry efforts to provide low-cost foods or due to general ethical considerations against cow's milk consumption [1], imitation dairy products have recently appeared on the market [2–5]. Cheese analogues or imitation cheese are cheese-like products in which milk fat, milk protein, or both are partially or completely replaced with nonmilk-based components such as soy [2], starch [6], or vegetable replacers [3]. Other alternative products for consumers with cow milk intolerance [7] based on goat [8, 9] or sheep milk [9] can also be found on the market. Vegetable oils and fats are most commonly used as cheap substitutes for milk fat to manufacture imitation cheese or imitation ice cream. While not being harmful to health, the imitation products may be of lesser nutritional quality (e.g., by lower calcium content) and contain several artificial flavors and food colors [10]. Unfortunately, such imitation products may be offered without the necessary labeling, which is a deception of the consumer. Pizza topping is a good example of such a possibility [10]. It has therefore become necessary to develop a reliable technique able to detect such products in the market.

Chromatographic methods are the most popular choice for analysis of organic substances in cheese. For example, gas chromatography (GC) with mass-spectrometric detection [11–13] or flame ionization detection [11] and high-performance liquid chromatography [9, 13, 14] were previously applied. With these methods, precise and diverse information about volatile profiles of the particular type of cheese could be obtained. This has been done, for example, for Reggianito Argentino cheese [11], Italian mountain cheese (Bitto) [15], Majorcan cheese [16], Kuflu Turkish cheese [13], and different varieties of goat and sheep cheese [9, 12]. However, due to the matrix complexity of dairy products chromatographic analysis usually involves pretreatment steps such as solid-phase extraction (SPE) [17] or headspace sorptive extraction (HSSE) combined with thermal desorption (TD) [15]. Therefore, it can be concluded that chromatographic techniques are accurate but laborious, expensive, and time consuming.

Other methods based on spectroscopic techniques are also available. These include Fourier transform infrared (FTIR) spectroscopy [18–20], visible-near infrared reflectance spectroscopy [21], near infrared (NIR) spectroscopy [20, 22], atomic absorption spectroscopy [23], inductively coupled plasma optical emission spectrometry [24, 25], and fluorescence spectroscopy [20]. Fluorescence spectroscopy is the most sensitive method but only few compounds give rise to fluorescence signals. In FTIR and NIR spectra, strong and broad signals of water prevent the informative characterization of dairy products.

Among spectroscopic techniques in the area of food analysis, NMR is currently on the rise [26]. Previous application areas include beer [27], juice [28], grapes [29], infant formulas, [30] or pine nuts [31]. The application of NMR spectroscopy to cheese and ice cream analysis has been also presented in several studies. ^{1}H NMR was used to investigate the influence of packaging on the degradation of soft cheese [32]. Full ^{1}H NMR assignments of signals of the water fractions of different types of cheese were recently provided [25, 33, 34]. ^{1}H NMR is also able to provide reliable qualitative and quantitative analysis of amino acids [35] and biogenic amines [36] in cheese. ^{31}P and ^{23}Na NMR were used for the investigation of both phosphate and sodium ion distribution in semihard cheese [37]. A time-domain nuclear magnetic resonance (TD-NMR) was applied to the quick determination of moisture profiles during cheese drying [38]. Regarding NMR analysis of ice cream, TD-NMR was previously used for the investigation of the aggregation state (liquid or solid) of water and fat [39–41]. Another article utilized site-specific natural isotope fractionation NMR to detect adulteration of vanillin in ice cream [42].

Despite the mentioned diverse studies about composition of dairy products, there are only few articles dealing with the detection of their adulteration. For example, it was demonstrated that it is possible to identify the presence of cow milk in buffalo mozzarella by the use of electrophoretic mobility of cow and buffalo casein [43]. A method based on triacylglycerol composition obtained with GC-FID of cheese was also proposed to detect the levels of foreign fat [44].

Other techniques based on the determination of particular markers were also reviewed [4]. All of them are based on time-consuming chromatographic measurements and, what is more important, are able to detect only specific types of adulteration.

In the view of these facts, NMR seems to be promising to provide accurate classification of dairy products according to the raw material origin. Therefore, the main objective of this research was to investigate the ability of NMR spectroscopy to differentiate milk fat products from vegetable fat substitutes. Cheese and ice cream were chosen as examples.

2. Experimental Section

2.1. Samples. A total of 109 cheese samples and 112 ice cream samples based on milk fat were analyzed. The products were either purchased at local stores in Karlsruhe, Germany, or submitted to the CVUA Karlsruhe for official food control purposes in Baden-Württemberg, Germany. Samples were selected to cover all possible imitation products available on the German market and a wide composition variability of milk fat products. Furthermore, imitation products based on vegetable fat (or vegetable fat/milk fat mixture) were analyzed ($n = 11$ cheese and $n = 11$ ice cream). All samples were subjected to the standard GC/MS analysis that confirmed the labeling information in every case.

2.2. Sample Preparation and Validation. Sample preparation of cheese and ice cream was conducted by the German reference Weibull-Stoldt methodology for fat hydrolysis and extraction. The hydrolysis of the sample was conducted using the automated hydrolysis system HYDROTHERM (Gerhardt Analytical Systems, Königswinter, Germany) as shown in Figure 1. Briefly, a representative average sample (at least 200 g) is minced and homogenized. Then 10 g of the homogenized sample is weighed and put into the digestion beaker for automated hydrochloric acid hydrolysis. After the addition of hydrochloric acid (4 mol/L, 150 mL), the liquid is then quickly brought to boil and simmered for about 1 hour. At the end of hydrolysis the digestion mixture is diluted with hot water (100 mL) to the double amount and then is immediately filtered through pleated filter, which has been moistened automatically by the system with water (number of moisture cycles = 3 and water amount per cycle = 40 mL). After the program has finished, the filter is placed on a watch glass and dried for up to 1.5 h at $103 \pm 2°C$ in a drying oven. After cooling off, the fat is extracted using Soxhlet extraction with petroleum ether.

After finishing, the extraction flasks are dried in the drying oven for 60 minutes at $103 \pm 2°C$. Then, they are placed in a desiccator, left to cool down to room temperature. After a constant weight was achieved, the fat phase was ready for NMR analysis, for which 200 mg of the fat fraction is mixed with 0.80 mL of $CDCl_3$ containing 0.1% tetramethylsilane (TMS). 0.6 mL of the mixture is poured into an NMR tube and directly measured.

To investigate the reproducibility of the sample preparation, two different imitation cheese samples were prepared

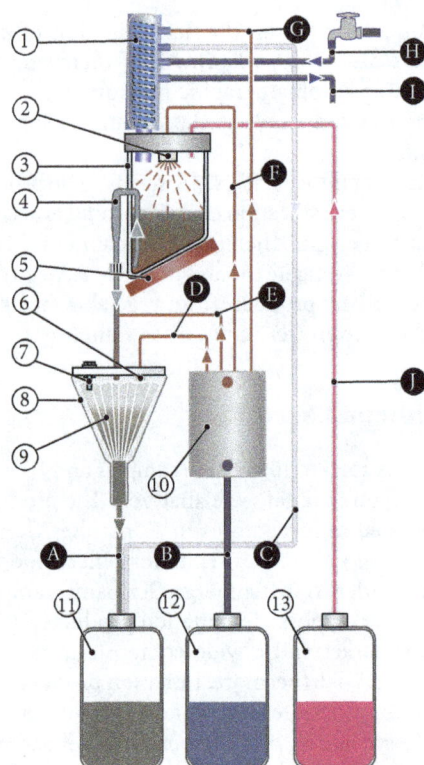

FIGURE 1: Schematic illustration of the automatic sample hydrolysis process necessary for Weibull-Stoldt fat extraction (reproduced with permission from Gerhardt Analytical Systems). 1 Condenser, 2 shower, 3 hydrolysis beaker, 4 sample transfer device, 5 hotplate, 6 shower for filter, 7 level sensor funnel, 8 funnel, 9 folded filter, 10 hot water generator, 11 tank for sample waste, 12 tank for H_2O, and 13 tank for HCl, A sample drainage, B distilled water addition, C air ventilation for condenser, D hot water addition-filter moisture, E hot water addition-rinsing sample transfer, F hot water addition-rinsing hydrolysis beaker, G hot water addition-rinsing condenser, H cooling water inlet, I cooling water outlet, J hydrochloric acid addition.

twice and several resonances were integrated: 9.76–9.74 ppm (triplet), 4.33–4.30 ppm (doublet), and 2.80–2.72 ppm (triplet). The reproducibility was then calculated as relative standard deviation (RSD) between replicates.

2.3. NMR Measurements at 400 MHz.
All NMR measurements were performed on a Bruker Avance 400 Ultrashield spectrometer (Bruker Biospin, Rheinstetten, Germany) equipped with a 5 mm SEI probe with Z-gradient coils, using a Bruker Automatic Sample Changer (B-ACS 120). All spectra were acquired at 300.0 K. The data were acquired automatically under the control of ICON-NMR (Bruker Biospin, Rheinstetten, Germany), requiring about 12 min (^{1}H NMR) and 30 min (^{13}C NMR) per sample. All NMR spectra were phased and baseline corrected.

2.4. NMR Spectra Acquisition.
^{1}H NMR spectra were acquired using the Bruker 1D zg pulse sequence with 128 scans (NS) and 2 prior dummy scans (DS). The sweep width (SW) was

20.5503 ppm and the time domain (TD) of the free induction decays (FIDs) was 131 k, acquisition time (AQ) was 7.97 s, and the repetition time (D1) was 1.0 s. Receiver gain (RG) value was set to 8.0. For acquisition of ^{13}C NMR spectra, the Bruker pulse sequence zgpg was used. After the application of 4 DS, 8 FIDs (NS = 1024) were collected into a TD of 131072 (131 k) complex data points using a 238.8728 ppm SW and a RG of 2050 (AQ = 1.38 s and D1 = 2.00 s).

2.5. Nontargeted Analysis and Chemometrics.
The resulting spectra were analyzed using the software Unscrambler X version 10.0.1 (Camo Software AS, Oslo, Norway). We tested several spectral regions for calculation: aliphatic (0.25–3 ppm), midfield (3–6 ppm), aromatic (6–10 ppm) as well as the 0.25–6 ppm region with 0.01 ppm bucket width. Details on the bucketing process of NMR spectra for multivariate data analysis were previously described [27].

The technique of cross-validation was applied to determine the number of principal components (PCs) needed. For cheese spectra differentiation, we used 7 PCs (explained variance 97%) for ^{1}H NMR and 8 PCs (explained variance 98%) for ^{13}C NMR. The PCA model for ice cream spectra required 8 PCs (explained variance 97%). Using PLS regression, the NMR spectra were correlated with reference GC analysis data. PCA and PLS models were validated via full cross-validation. Furthermore, the PLS models were evaluated via test set validation (n = 10), and results are compared with those obtained from a standard GC method.

3. Results and Discussion

3.1. Sample Preparation and Spectra Analysis.
Cheese and ice cream cannot be directly measured with liquid-state NMR, so that a sample preparation has to occur aiming to provide a measurable liquid solution. According to the literature, the most meaningful information about discrimination between types of cheese is contained in the fat fraction [45–47]. For this reason, we decided to apply the fat obtained with the Weibull-Stoldt methodology for our NMR analysis. According to Weibull-Stoldt, the sample is first hydrolyzed to free the fat, and then the fat fraction was extracted from the rest of the sample using a solvent. The use of the Weibull-Stoldt fat had also the advantage that this methodology is already conducted for nearly all samples of cheese and ice cream that reach our laboratory, as the labeled fat content on the package has to be controlled. The Weibull-Stoldt fat was also used for standard GC analysis. To simplify the Weibull-Stoldt protocol, we applied an automated device for the hydrolysis, which is the first system worldwide that was recently commercialized for this purpose. Prior to its application, we conducted this procedure using manual hydrolysis according to the Weibull-Stoldt method. The NMR spectra obtained with both methods showed the same fatty acid profile; however, the automated device was considerably more efficient as it is possible to prepare 6 samples at once without human intervention.

To demonstrate the reproducibility of this method, replicate measurements of different samples were performed.

The relative standard deviations (RSD) between the two measurements were found to range between 0.1% and 2.1% (9.76–9.74 ppm), 1.0% and 1.2% (4.33–4.30 ppm), and 0.7 and 3.1% (2.80–2.72 ppm) for imitation cheese samples. The data indicated that the sample preparation procedure is adequately reproducible to facilitate a comparison between different cheese and ice cream samples.

Figure 2(a) showed the ^{1}H NMR spectrum of a representative sample of Gouda cheese. The signal of triglycerides and fatty acids dominated the spectrum [46]. Imitation cheese displayed a similar fatty acid profile (Figure 2(b)). By inspection of these spectra, we found the differences in the intensity of resonances relative to methyl (1.00–0.90 ppm) and bis-allylic protons (5.00–4.90 ppm) between the two groups of the products. In the ^{13}C NMR spectra, differences in the 173–170 ppm region (butyric acid) can also be observed. The same findings were valid for ice cream samples. Nevertheless, it can be concluded that NMR spectra of cheese and ice cream are very complex and a strong overlap of the resonances occurs. In the following, dairy products properties were uncovered from the NMR spectra using multivariate data analysis.

3.2. Nontargeted Analysis.

The spectroscopic data were visualized either through PCA scatter plots, in which each point represents an individual sample, or through loadings plots, which permit the identification of the most important spectral regions to separate the clusters and, therefore, reveal markers (compounds that are responsible for differentiation).

At first, PCA was performed on NMR spectra of cheese samples. The best grouping of similar samples was observed in the PCA scores plots of PC1-PC2 (^{1}H NMR spectra, Figure 3(a)) and PC3-PC6 (^{13}C NMR spectra, Figure 3(b)). On both plots, the imitation cheese samples were clearly separated from all of the remaining ones and were clustered in the range of negative values of PC1 (^{1}H NMR) or positive values of PC3 (^{13}C NMR). Furthermore, on both plots two especially conspicuous imitation samples were observed (marked with stars on Figures 3(a) and 3(b)). These two products represented tzatziki (a traditional Greek appetizer), which consists of both milk fat with vegetable fat and olive oil addition. Additionally, one outlier was located in the positive values of PC1 (Figure 3(a)). In addition to vegetable fat, this sample contained also about 3% of milk fat (as proven by GC analysis). Cheese made from milk (cow, goat, and sheep) was clustered around 0 in both PCA scores plots. Overall, we think that PCA in the aliphatic ^{1}H NMR region (Figure 3(a)) provided better differentiation of cheese samples. In this case, the samples in the imitation cluster were located closer to each other and the distance between milk fat/vegetable fat clusters was larger than that obtained with ^{13}C NMR data. Furthermore, vegetable fat/milk fat mixtures can also be recognized with ^{1}H NMR spectra.

The PCA scatter plot of ^{1}H NMR spectra (3–0.25 ppm) of ice cream samples was shown in Figure 4. In this case, seven PCs were found sufficient for differentiation. Unlike cheese samples, for which a good discrimination was observed between the first two PCs (Figure 3(a)), for ice cream the best model was constructed between PC4 and PC7. For ice cream samples, therefore, the variability in minor compound concentrations (such as alcohols and long-chain fatty acids) influenced the discrimination.

The chemical shifts and the associated functional groups that were responsible for the differentiation of the dairy products can be identified using the loadings plots. In the loadings plot, each chemical shift was plotted against its importance in discriminating the samples. In our case, the spectral regions 1.00–0.90 and 5.00–4.90 ppm were found to be important in the milk fat/vegetable fat product differentiation (for both cheese and ice cream). These regions consisted of the signals from methyl groups of different compounds and olefinic protons of all unsaturated chains [46]. The buckets at 2.32 and 2.30 ppm (most probably methylenic protons bonded to C2 of all fatty acid chains) [46] were also important for ice cream products.

Due to the low number of different imitation products currently available on the German market (probably because of a recent media campaign against these products [10]), we were not able to analyze our data with classification methods such soft independent modeling of class analogy (SIMCA) or linear discriminant analysis (LDA), which would require a larger dataset for training and validation. However, new samples can be distinguished by adding them to the developed PCA model.

While the differences of fat material used for cheese manufacture can be seen within the first several PCs (Figure 3(a)), higher PCs could uncover further clustering. To do this, we removed the imitation products and repeated the PCA. It can seen from Figure 5 that grouping in respect to the cheese types (Edamer, Gouda, Feta, and Emmentaler) is observed. It is not surprising because every cheese type was produced differently and, therefore, had a unique fatty acid profile. The two separate Gouda clusters were separated probably due to different ripening times similar to what was previously found for Italian Parmigiano Reggiano cheese [34]. It should be noted that ^{13}C NMR cannot provide such clear differentiation (only Emmentaler, Gouda, and Feta can be classified).

A recent review discussed the potential of different techniques coupled with chemometric analysis for the determination of the quality and the authenticity of dairy products, from which NMR plays an important role [47]. In the study of Rodrigues et al. [33] it was shown that metabolic profiling obtained by NMR combined with multivariate analysis allows to distinguish cheese samples in terms of maturation time, as well as added probiotic and prebiotic substances. PCA analysis was also performed on ^{1}H NMR spectra of Italian Parmigiano Reggiano cheese to control time of ripening [34]. In the same research, the authors were able to provide geographical differentiation of cheese samples with Partial Least Squares-Discriminant Analysis (PLS-DA). ^{1}H and ^{13}C NMR coupled with PCA was used to differentiate PDO Asiago cheese produced in different areas [46]. High-resolution magic angle spinning (HR MAS) NMR together with PCA was able to distinguish Emmental cheese samples according to geographical region [48]. It should be mentioned, however, that data sets in all of these studies involved not more than 30 samples and focused on one specific group or origin.

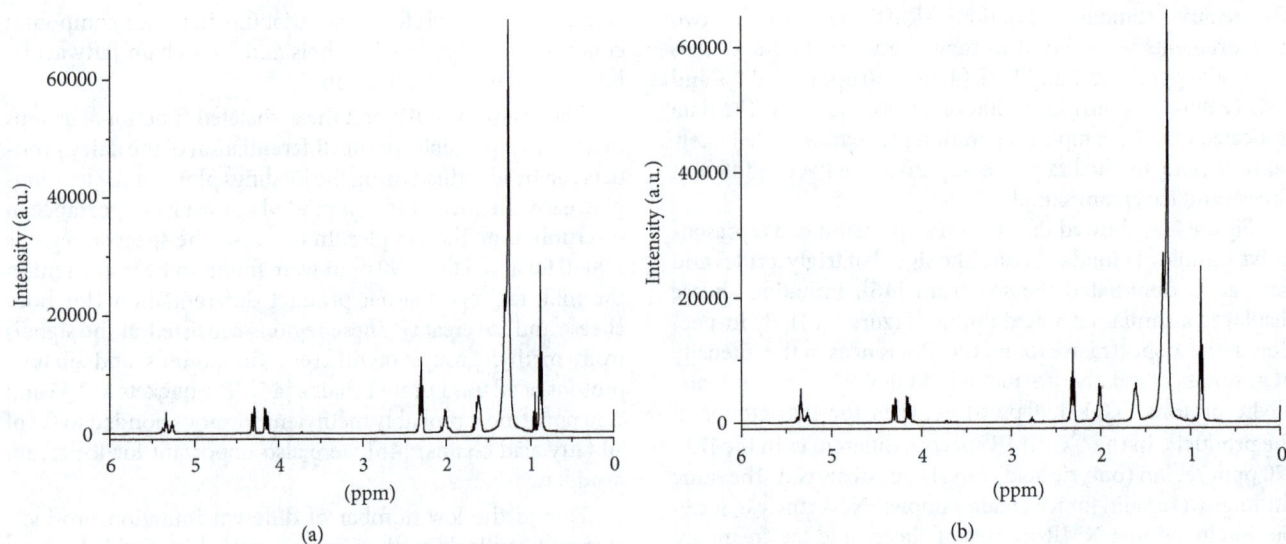

FIGURE 2: ^{1}H NMR spectra of Gouda cheese (a) compared to an imitation cheese based on vegetable fat (b).

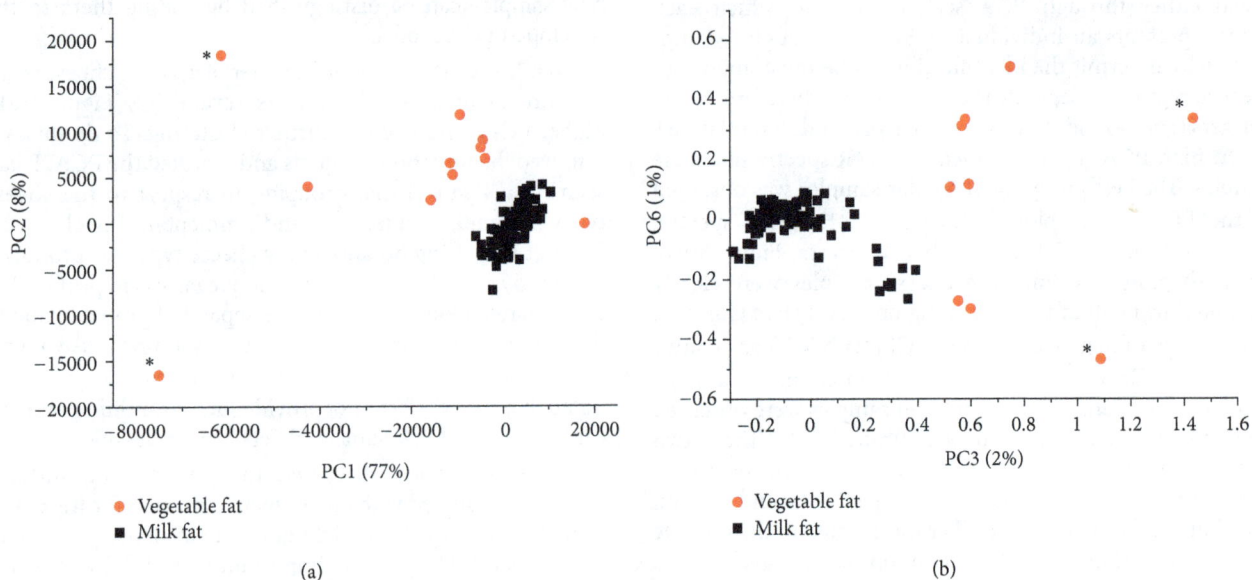

FIGURE 3: Scatter plot of the PCA scores for ^{1}H NMR ((a), 3.0–0.25 ppm) and ^{13}C NMR ((b), 200–0.25 ppm) for cheese samples (stars denote tzatziki samples).

Therefore, our investigation is the first to apply ^{1}H NMR spectroscopy with multivariate methods to characterize a large number of commercial cheese and ice cream samples. Furthermore, to the best of our knowledge, no previous studies evaluated the performance of NMR spectroscopy to reveal vegetable fat adulteration of cheese and ice cream so far.

3.3. Quantitative Prediction of Dairy Product Composition.
Besides the qualitative classification of our samples, in order to perform quality control of dairy products, it is also

necessary to quantify certain compounds (e.g., saturated and unsaturated fatty acids and their esters). GC is among the most common methods for determining the fatty acid composition of cheese and ice cream [9, 11, 12, 15, 17]. ^{1}H NMR spectroscopic methods based on direct integration were also proposed for this purpose [45]. However, only a limited number of fatty acids can be quantified because they display similar and overlapping signals in the NMR spectra (Figure 2), which make simple quantification by integration of a distinct peak not possible. Moreover, in some cases the results can strongly deviate from the reference values [45].

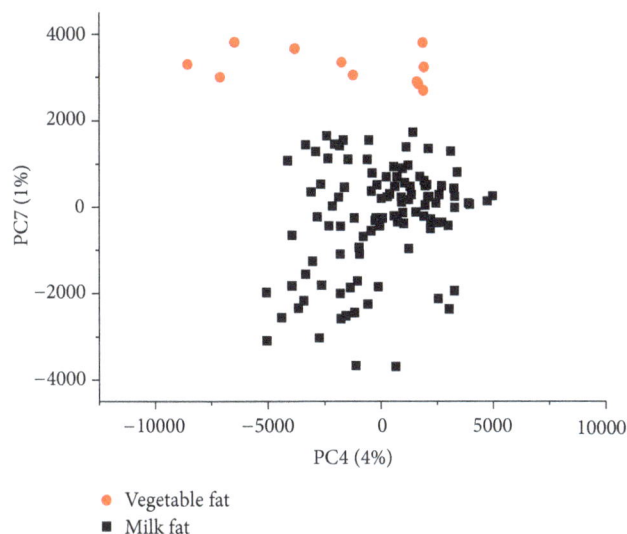

FIGURE 4: Scatter plot of the PCA ^{1}H NMR scores in the 3.0–0.25 ppm region for ice cream samples.

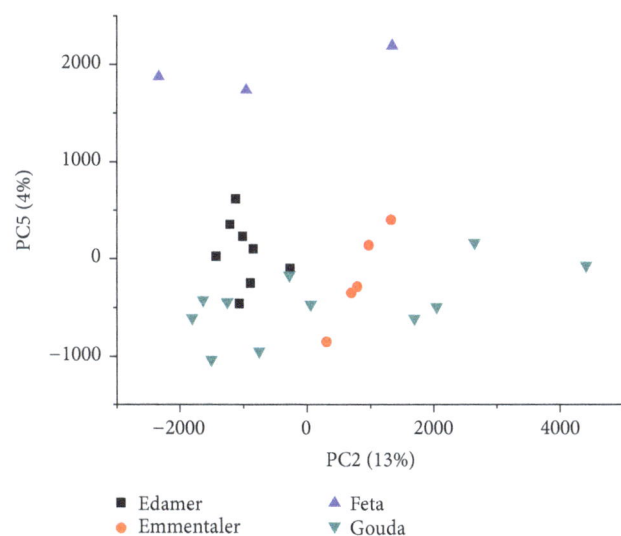

FIGURE 5: PCA of cheese types in the aliphatic region (3.0–0.25 ppm).

To overcome these problems, we used Partial Least Squares regression (PLS) to correlate NMR spectra (in the 6–0.25 ppm range) to the data of reference GC analysis. Results (i.e., root mean square error (RMSE), correlation coefficient (R^2), the number of PLS factors as well as NMR range used) of the best-fitting PLS models for ice creams were listed in Table 1. Fourteen of seventeen models exhibited correlation coefficients greater than 0.90. The correlation coefficients for butyric acid ($R = 0.89$), octadecanoic acid ($R = 0.83$), and nonadecanoic acid ($R = 0.89$) also appeared to be adequate for a screening procedure. The separate PLS models constructed for cheese showed slightly lower but comparable performance (R^2 values were in the range of 0.75–0.95). Inadequate PLS models ($R^2 < 0.50$) were only obtained for

pentadecanoic acid, margaric acid, octadecanoic acid, and nonadecanoic acid for cheese samples (in comparison to ice cream, Table 1) due to the small concentrations of these compounds.

The PLS prediction models described previously were validated with an independent set of ten randomly chosen samples (only the acceptable models with $R^2 > 0.90$ were considered). The average relative deviations of the predicted values from the GC ones were in the range of 2–10% and 5–15% for ice cream and cheese samples. These data clearly demonstrated the reliability of the models and the potential of this technique for simultaneous quantitative analysis of dairy product characteristics along with the nontargeted control.

Therefore, we have extended quantitative NMR spectroscopy to 17 compounds that have to be analyzed during food control. In principle, NMR spectra contain the same information as GC but can be gathered much faster and more efficiently. The results emphasized the capability of NMR to rapidly and reliably predict cheese and ice cream characteristics.

4. Conclusions

Traditional analytical strategies to uncover adulteration of food rely on targeted analysis (in the course of which only certain marker compounds are analyzed). This approach has obstacles on many points, in particular, starting from time of analysis, use of sophisticated analytical equipment, or usage of expensive and/or toxic reagents. What is more important is that new forms of adulteration could not be uncovered as in the case of melamine adulteration of milk by applying the unspecific Kjeldahl assay [30]. A demand exists for rapid, accurate, and cheaper methods for direct quality measurement of food and food ingredients, which should include a nontargeted approach.

Spectroscopic techniques combined with chemometric methods are a possible solution of this problem. From the wide range of spectroscopic methods, the high amount of spectral information of NMR is ideally suited for nontargeted analysis. This approach has only recently entered routine analysis in food control institutions. Examples are the investigation of infant formulas [30], pine nuts [31], or milk and milk substitutes [49]. This study has further shown that NMR is an efficient tool to detect fraud in the dairy product industry, protecting consumers against improper practices, and guaranteeing fair trade.

The method of automated sample preparation (Weibull Stoldt hydrolysis) used in this paper appears to be extremely efficient, especially when dealing with a high number of samples to be analyzed using multivariate methods. This procedure allowed the collection of high-quality spectra with distinct spectral features that were consistent within each sample. The developed technique could be further applied to solve other classification problems (differentiation between milk fat from different animals besides cows, or geographic discrimination) and shows great promise as a rapid tool for cheese analysis.

TABLE 1: PLS correlation between data of reference GC analysis and NMR spectra (6.0–0.25 ppm) for ice cream ($n = 99$).

Analytes	Reference range	PLS factors	Calibration		Validation	
			RMSE[a]	R^2	RMSE	R^2
Butyric acid (C4:0) (%)	0–25	7	0.21	0.89	0.24	0.85
Caproic acid (C6:0) (%)	0–1.9	7	0.15	0.87	0.17	0.84
Octanoic acid (C8:0) (%)	0–8.4	4	0.33	0.93	0.37	0.92
n-Capric acid (C10:0) (%)	0–6.7	4	0.28	0.91	0.33	0.87
Dodecanoic acid (C12:0) (%)	0–43.5	6	1.2	0.97	1.6	0.96
Tetradecanoic acid (C14:0) (%)	0–20.3	5	0.70	0.94	0.84	0.92
Myristoleic acid (C14:1) (%)	0–1.1	6	0.072	0.91	0.089	0.87
Hexadecanoic acid (C16:0) (%)	9.7–40.6	6	1.1	0.96	1.39	0.94
Palmitoleic acid (C16:1) (%)	0–1.7	6	0.09	0.94	0.11	0.91
Oleic acid (C18:1) (%)	5.2–60.7	4	1.0	0.99	1.5	0.97
Pentadecanoic acid (C15:0) (%)	0–1.6	7	0.099	0.93	0.12	0.90
Margaric acid (C17:0) (%)	0–0.8	7	0.04	0.92	0.05	0.89
Octadecanoic acid (C18:0) (%)	3.9–36	5	1.0	0.83	1.2	0.76
Nonadecanoic acid (C19:0) (%)	0–1.1	7	0.02	0.89	0.04	0.78
cis. cis-9.12-Octadecadienoic acid (C18:2) (%)	0.3–22.8	7	0.25	0.99	0.39	0.97
Methyl butanoate (g/100 g fat)	0–3.8	7	0.11	0.99	0.18	0.97
Hexanoic acid methyl ester (g/100 g fat)	0.04–2.24	7	0.07	0.98	0.09	0.96

[a]Root mean-squared error (RMSE) values are expressed in the same units as the analytes.

Acknowledgments

The authors warmly thank H. Heger, M. Böhm, B. Siebler, and J. Geisser for excellent technical assistance. G. Raiber and M. Kranz (Gerhardt Analytical Systems) are thanked for their support in introducing the automatic Weibull-Stoldt hydrolysis to the authors' routine analysis. The views expressed in this paper do not necessarily reflect those of the Ministry of Rural Affairs and Consumer Protection.

References

[1] A. Beardsworth and T. Keil, "The vegetarian option: varieties, conversions, motives and careers," *Sociological Review*, vol. 40, pp. 253–293, 1992.

[2] N. Sutar, P. P. Sutar, and G. Singh, "Evaluation of different soybean varieties for manufacture of soy ice cream," *International Journal of Dairy Technology*, vol. 63, no. 1, pp. 136–142, 2010.

[3] A. I. W. Rosnani, I. N. Aini, A. M. M. Yazid, and M. H. Dzulkifly, "Flow properties of ice cream mix prepared from palm oil: anhydrous milk fat blends," *Pakistan Journal of Biological Sciences*, vol. 10, no. 10, pp. 1691–1696, 2007.

[4] H. Bachmann, "Cheese analogues: a review," *International Dairy Journal*, vol. 11, no. 4–7, pp. 505–515, 2001.

[5] C. R. Cunha, A. I. Dias, and W. H. Viotto, "Microstructure, texture, colour and sensory evaluation of a spreadable processed cheese analogue made with vegetable fat," *Food Research International*, vol. 43, no. 3, pp. 723–729, 2010.

[6] N. Noronha, E. D. O'Riordan, and M. O'Sullivan, "Replacement of fat with functional fibre in imitation cheese," *International Dairy Journal*, vol. 17, no. 9, pp. 1073–1082, 2007.

[7] M. Montalto, V. Curigliano, L. Santoro et al., "Management and treatment of lactose malabsorption," *World Journal of Gastroenterology*, vol. 12, no. 2, pp. 187–191, 2006.

[8] D. Sanchez-Macias, M. Fresno, I. Moreno-Indias et al., "Physicochemical analysis of full-fat, reduced-fat, and low-fat artisanstyle goat cheese," *Journal of Dairy Science*, vol. 93, no. 9, pp. 3950–3956, 2010.

[9] K. Raynal-Ljutovac, G. Lagriffoul, P. Paccard, I. Guillet, and Y. Chilliard, "Composition of goat and sheep milk products: an update," *Small Ruminant Research*, vol. 79, no. 1, pp. 57–72, 2008.

[10] A. Rehm, "Käse ist nicht gleich Käse," *Ernährung im Fokus*, vol. 11, pp. 538–541, 2011.

[11] I. V. Wolf, M. C. Perotti, S. M. Bernal, and C. A. Zalazar, "Study of the chemical composition, proteolysis, lipolysis and volatile compounds profile of commercial Reggianito Argentino cheese: characterization of Reggianito Argentino cheese," *Food Research International*, vol. 43, no. 4, pp. 1204–1211, 2010.

[12] J. M. Poveda, E. Sánchez-Palomo, M. S. Pérez-Coello, and L. Cabezas, "Volatile composition, olfactometry profile and sensory evaluation of semi-hard Spanish goat cheeses," *Dairy Science and Technology*, vol. 88, no. 3, pp. 355–367, 2008.

[13] A. A. Hayaloglu, E. Y. Brechany, K. C. Deegan, and P. L. H. McSweeney, "Characterization of the chemistry, biochemistry and volatile profile of Kuflu cheese, a mould-ripened variety," *Lebensmittel-Wissenschaft & Technologie*, vol. 41, no. 7, pp. 1323–1334, 2008.

[14] S. Bonetta, J. D. Coïsson, D. Barile et al., "Microbiological and chemical characterization of a typical Italian cheese: Robiola di Roccaverano," *Journal of Agricultural and Food Chemistry*, vol. 56, no. 16, pp. 7223–7230, 2008.

[15] S. Panseri, I. Giani, T. Mentasti, F. Bellagamba, F. Caprino, and V. M. Moretti, "Determination of flavour compounds in a mountain cheese by headspace sorptive extraction-thermal desorption-capillary gas chromatography-mass spectrometry," *Lebensmittel-Wissenschaft & Technologie*, vol. 41, no. 2, pp. 185–192, 2008.

[16] A. Castell-Palou, C. Rosselló, A. Femenia, and S. Simal, "Application of multivariate statistical analysis to chemical, physical and sensory characteristics of Majorcan cheese," *International Journal of Food Engineering*, vol. 6, no. 2, article 9, 2010.

[17] S. Hauff and W. Vetter, "Quantification of branched chain fatty acids in polar and neutral lipids of cheese and fish samples," *Journal of Agricultural and Food Chemistry*, vol. 58, no. 2, pp. 707–712, 2010.

[18] A. Subramanian, W. J. Harper, and L. E. Rodriguez-Saona, "Rapid prediction of composition and flavor quality of cheddar cheese using ATR-FTIR spectroscopy," *Journal of Food Science*, vol. 74, no. 3, pp. C292–C297, 2009.

[19] N. A. Kocaoglu-Vurma, A. Eliardi, M. A. Drake, L. E. Rodriguez-Saona, and W. J. Harper, "Rapid profiling of swiss cheese by attenuated total reflectance (ATR) infrared spectroscopy and descriptive sensory analysis," *Journal of Food Science*, vol. 74, no. 6, pp. S232–S239, 2009.

[20] C. M. Andersen, M. B. Frøst, and N. Viereck, "Spectroscopic characterization of low- and non-fat cream cheeses," *International Dairy Journal*, vol. 20, no. 1, pp. 32–39, 2010.

[21] A. Lucas, D. Andueza, E. Rock, and B. Martin, "Prediction of dry matter, fat, pH, vitamins, minerals, carotenoids, total antioxidant capacity, and color in fresh and freeze-dried cheeses by visible-near-infrared reflectance spectroscopy," *Journal of Agricultural and Food Chemistry*, vol. 56, no. 16, pp. 6801–6808, 2008.

[22] G. Cozzi, J. Ferlito, G. Pasini, B. Contiero, and F. Gottardo, "Application of near-infrared spectroscopy as an alternative to chemical and color analysis to discriminate the production chains of Asiago d'Allevo cheese," *Journal of Agricultural and Food Chemistry*, vol. 57, no. 24, pp. 11449–11454, 2009.

[23] R. Moreno-Rojas, P. J. Sánchez-Segarra, F. Cámara-Martos, and M. A. Amaro-López, "Multivariate analysis techniques as tools for categorization of Southern Spanish cheeses: nutritional composition and mineral content," *European Food Research and Technology*, vol. 231, no. 6, pp. 841–851, 2010.

[24] A. Lante, G. Lomolino, M. Cagnin, and P. Spettoli, "Content and characterisation of minerals in milk and in Crescenza and Squacquerone Italian fresh cheeses by ICP-OES," *Food Control*, vol. 17, no. 3, pp. 229–233, 2006.

[25] M. A. Brescia, M. Monfreda, A. Buccolieri, and C. Carrino, "Characterisation of the geographical origin of buffalo milk and mozzarella cheese by means of analytical and spectroscopic determinations," *Food Chemistry*, vol. 89, no. 1, pp. 139–147, 2005.

[26] G. Le Gall and I. J. Colquhoun, *Food Authenticity and Traceability*, M. Lees eds, Woodhead Publishing, Cambridge, UK, 2003.

[27] D. W. Lachenmeier, W. Frank, E. Humpfer et al., "Quality control of beer using high-resolution nuclear magnetic resonance spectroscopy and multivariate analysis," *European Food Research and Technology*, vol. 220, no. 2, pp. 215–221, 2005.

[28] P. S. Belton, I. J. Colquhoun, E. K. Kemsley et al., "Application of chemometrics to the ^{1}H NMR spectra of apple juices: discrimination between apple varieties," *Food Chemistry*, vol. 61, no. 1-2, pp. 207–213, 1998.

[29] L. Forveille, J. Vercauteren, and D. N. Rutledge, "Multivariate statistical analysis of two-dimensional NMR data to differentiate grapevine cultivars and clones," *Food Chemistry*, vol. 57, no. 3, pp. 441–450, 1996.

[30] D. W. Lachenmeier, H. Eberhard, F. Fang et al., "NMR-spectroscopy for nontargeted screening and simultaneous quantification of health-relevant compounds in foods: the example of melamine," *Journal of Agricultural and Food Chemistry*, vol. 57, no. 16, pp. 7194–7199, 2009.

[31] H. Köbler, Y. B. Monakhova, T. Kuballa et al., "Nuclear magnetic resonance spectroscopy and chemometrics to identify pine nuts that cause taste disturbance," *Journal of Agricultural and Food Chemistry*, vol. 59, no. 13, pp. 6877–6881, 2011.

[32] R. Lamanna, I. Piscioneri, V. Romanelli, and N. Sharma, "A preliminary study of soft cheese degradation in different packaging conditions by ^{1}H-NMR," *Magnetic Resonance in Chemistry*, vol. 46, no. 9, pp. 828–831, 2008.

[33] D. Rodrigues, C. H. Santos, T. A. P. Rocha-Santos, A. M. Gomes, B. J. Goodfellow, and A. C. Freitas, "Metabolic profiling of potential probiotic or synbiotic cheeses by nuclear magnetic resonance (NMR) spectroscopy," *Journal of Agricultural and Food Chemistry*, vol. 59, no. 9, pp. 4955–4961, 2011.

[34] R. Consonni and L. R. Cagliani, "Ripening and geographical characterization of Parmigiano Reggiano cheese by ^{1}H NMR spectroscopy," *Talanta*, vol. 76, no. 1, pp. 200–205, 2008.

[35] S. De Angelis Curtis, R. Curini, M. Delfini, E. Brosio, F. D'Ascenzo, and B. Bocca, "Amino acid profile in the ripening of Grana Padano cheese: a NMR study," *Food Chemistry*, vol. 71, no. 4, pp. 495–502, 2000.

[36] E. Schievano, K. Guardini, and S. Mammi, "Fast determination of histamine in cheese by nuclear magnetic resonance (NMR)," *Journal of Agricultural and Food Chemistry*, vol. 57, no. 7, pp. 2647–2652, 2009.

[37] M. Gobet, C. Rondeau-Mouro, S. Buchin et al., "Distribution and mobility of phosphates and sodium ions in cheese by solid-state ^{31}P and double-quantum filtered ^{23}Na NMR spectroscopy," *Magnetic Resonance in Chemistry*, vol. 48, no. 4, pp. 297–303, 2010.

[38] A. Castell-Palou, C. Rosselló, A. Femenia, J. Bon, and S. Simal, "Moisture profiles in cheese drying determined by TD-NMR: mathematical modeling of mass transfer," *Journal of Food Engineering*, vol. 104, no. 4, pp. 525–531, 2011.

[39] T. Lucas, D. Le Ray, P. Barey, and F. Mariette, "NMR assessment of ice cream: effect of formulation on liquid and solid fat," *International Dairy Journal*, vol. 15, no. 12, pp. 1225–1233, 2005.

[40] T. Lucas, M. Wagener, P. Barey, and F. Mariette, "NMR assessment of mix and ice cream. Effect of formulation on liquid water and ice," *International Dairy Journal*, vol. 15, no. 10, pp. 1064–1073, 2005.

[41] F. Mariette and T. Lucas, "NMR signal analysis to attribute the components to the solid/liquid phases present in mixes and ice creams," *Journal of Agricultural and Food Chemistry*, vol. 53, no. 5, pp. 1317–1327, 2005.

[42] G. S. Remaud, Y. Martin, G. G. Martin, and G. J. Martin, "Detection of sophisticated adulterations of natural vanilla flavors and extracts: application of the SNIF-NMR to method vanillin and *p*-hydroxybenzaldehyde," *Journal of Agricultural and Food Chemistry*, vol. 45, no. 3, pp. 859–866, 1997.

[43] F. Locci, R. Ghiglietti, S. Francolino et al., "Detection of cow milk in cooked buffalo Mozzarella used as Pizza topping," *Food Chemistry*, vol. 107, no. 3, pp. 1337–1341, 2008.

[44] J. Fontecha, I. Mayo, G. Toledano, and M. Juárez, "Triacylglycerol composition of protected designation of origin cheeses during ripening. Authenticity of milk fat," *Journal of Dairy Science*, vol. 89, no. 3, pp. 882–887, 2006.

[45] G. Knothe and J. A. Kenar, "Determination of the fatty acid profile by ^{1}H-NMR spectroscopy," *European Journal of Lipid Science and Technology*, vol. 106, no. 2, pp. 88–96, 2004.

[46] E. Schievano, G. Pasini, G. Cozzi, and S. Mammi, "Identification of the production chain of Asiago d'Allevo cheese by nuclear magnetic resonance spectroscopy and principal component analysis," *Journal of Agricultural and Food Chemistry*, vol. 56, no. 16, pp. 7208–7214, 2008.

[47] R. Karoui and J. De Baerdemaeker, "A review of the analytical methods coupled with chemometric tools for the determination of the quality and identity of dairy products," *Food Chemistry*, vol. 102, no. 3, pp. 621–640, 2007.

[48] L. Shintu and S. Caldarelli, "Toward the determination of the geographical origin of emmental(er) cheese via high resolution MAS NMR: a preliminary investigation," *Journal of Agricultural and Food Chemistry*, vol. 54, no. 12, pp. 4148–4154, 2006.

[49] Y. B. Monakhova, T. Kuballa, J. Leitz et al., "NMR spectroscopy as a screening tool to validate nutrition labeling of milk, lactose-free milk, and milk substitutes based on soy and grains," *Dairy Science and Technology*, vol. 92, no. 2, pp. 109–120, 2012.

9

Quality Parameters of Six Cultivars of Blueberry Using Computer Vision

Silvia Matiacevich,[1] **Daniela Celis Cofré,**[1] **Patricia Silva,**[1]
Javier Enrione,[2] **and Fernando Osorio**[1]

[1] *Departamento de Ciencia y Tecnología de los Alimentos, Facultad Tecnológica, Universidad de Santiago de Chile,*
 Avenida Libertador Bernardo O'Higgins No. 3363, Estación Central, 9170022 Santiago, Chile
[2] *Departamento de Nutrición y Dietética, Facultad de Medicina, Universidad de los Andes, San Carlos de Apoquindo 2200,*
 Las Condes, 7620001 Santiago, Chile

Correspondence should be addressed to Silvia Matiacevich; silvia.matiacevich@usach.cl

Academic Editor: Carl J. Schaschke

Background. Blueberries are considered an important source of health benefits. This work studied six blueberry cultivars: "Duke," "Brigitta", "Elliott", "Centurion", "Star," and "Jewel", measuring quality parameters such as °Brix, pH, moisture content using standard techniques and shape, color, and fungal presence obtained by computer vision. The storage conditions were time (0–21 days), temperature (4 and 15°C), and relative humidity (75 and 90%). *Results*. Significant differences ($P < 0.05$) were detected between fresh cultivars in pH, °Brix, shape, and color. However, the main parameters which changed depending on storage conditions, increasing at higher temperature, were color (from blue to red) and fungal presence (from 0 to 15%), both detected using computer vision, which is important to determine a shelf life of 14 days for all cultivars. Similar behavior during storage was obtained for all cultivars. *Conclusion*. Computer vision proved to be a reliable and simple method to objectively determine blueberry decay during storage that can be used as an alternative approach to currently used subjective measurements.

1. Introduction

Blueberries have an increasing demand for popular consumption because of their nutraceutical properties [1, 2], including their high content of phenolic compounds with a wide spectrum of biochemical activities such as antioxidant, antimutagenic, cardiovascular protection, antidiabetic, vision improvement properties, and carcinogenesis inhibition [3].

Blueberries are little blue fruits of the genus *Vaccinium* that have short shelf life. It has been stated that under refrigeration temperatures (0°C), the shelf life of blueberries is about 14–20 days [4, 5]. The main quality indicators of the fruit are appearance (color, size, and shape), firmness or texture, flavor (soluble solids and pH), and nutritive value [6]. The color ranges from light blue to deep black blue depending on the cultivar and the presence of an epicuticular wax, which gives its attractive appearance [4]. Color changes during

storage may have a profound effect on consumer acceptability [7].

Consumers demand high quality fruits which are dependent on harvest methods, cultivar characteristics, postharvest handling, and storage temperatures [1]. Computer vision (CV) is a nondestructive technology used for acquiring and analyzing digital images to obtain information of heterogeneous products. It has been regarded as a valuable tool which helps to improve the automatic assessment of food quality [8, 9]. CV has been recently used in the food industry for quality and color evaluation, detection of defects, grading and sorting of fruits and vegetables, among other applications [7–11].

The objectives of this work were to study important quality factors of six different blueberry cultivars harvested in Chile under different storage conditions and to compare these cultivars as part of a comprehensive study of blueberry conservation, including the innovative technology of computer

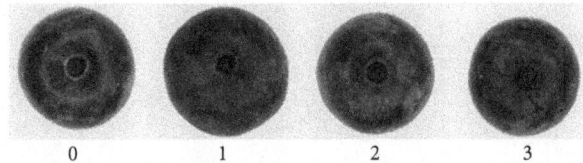

FIGURE 1: Fruit dehydration degree visually observed following the norms for quality of fresh blueberries from Chilean Blueberry Committee (CBBC, 2011).

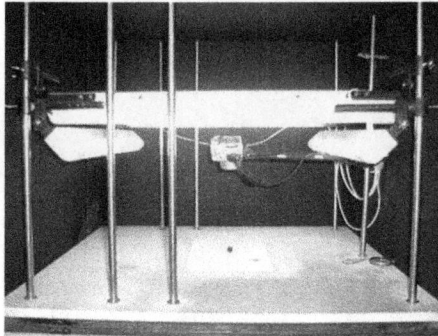

FIGURE 2: Computer vision system. Elements distribution in the digital image acquisition.

vision, which was applied in this research as a preliminary study to determine blueberry decay objectively instead of measuring it subjectively as is done nowadays.

2. Materials and Methods

2.1. Plant Material. This study was conducted during the 2009-2010 harvest season. All cultivars were donated by the Chilean Association of Exporters (ASOEX-Chile). Three blueberry cultivars ("Brigitta", "Elliott," and "Duke") from Southern Highbush variety (*Vaccinium darrowii*), two cultivars ("Jewel" and "Star") from Northern Highbush variety (*Vaccinium corymbosum*), and the cultivar "Centurion" from Rabbiteye variety (*Vaccinium virgatum*) were used. All cultivars were hand-harvested at full maturity from commercial plantations located in the central valley of the Metropolitan Region in Chile (Curacaví, Hortifrut S.A) during mid-December ("Duke" and "Jewel"), January ("Brigitta"), mid-February ("Elliott" and "Centurion"), and mid-March ("Star") and transported to the laboratory on the same day. High quality blueberries ($n = 50$) at each storage condition were obtained by random ($n = 10$ of each clamshell) from 12 clamshells of approximately 125 mg each, presorted by hand; discarding the excessively small, soft, visually damaged, nonblue fruits, and those with the presence of pedicel and floral remains as sample set for all experiments.

2.2. Storage Conditions. Blueberries ($n = 50$ at each condition) were stored at 4 and 15°C and equilibrated under different relative humidities (RH) using saturated solutions of NaCl (75% RH) and KCl (90% RH) [12] during different storage times (0, 7, 14, and 21 days).

2.3. Fruit Quality Indicators

2.3.1. pH and Total Soluble Solids Content. Blueberry juice was preparedfrom 5 blueberries randomly selected (in triplicate) at each storage time. The pH was measured with a pH meter (Jenway, UK) using a liquid electrode (Jenway, 924-001 model 3505) calibrated according to OMA, 1975. Total soluble solids concentration was determined by placing a drop of this blueberry juice (1 mL) on a calibrated portable refractometer (0–32 °Brix, RHB-32ATC). The mean and standard deviation of three replicates were recorded and expressed as °Brix.

2.3.2. Fruit Size and Form. The equatorial and polar diameter of each blueberry was measured with a digital caliper (Bull Tools, USA), and the roundness index (RI) was determined from (1). Mean and standard deviation of all blueberries ($n = 50$) measured at each storage condition were reported. These parameters were also obtained from image analysis, correlating linearly with experimental data ($R^2 = 0.998$).

$$RI = \frac{\text{polar diameter}}{\text{equatorial diameter}}. \qquad (1)$$

2.3.3. Water Content and Fruit-Dehydrated Percentage. Water content was gravimetrically determined using an analytical balance (Mettler Toledo, Switzerland). Twelve blueberries (in triplicate) were dried for 24 h in an oven (Wiseven, Korea) at 105°C until constant weight. Water content was expressed as wet basis percentage (g water/100 g wet sample).

Dehydrated fruit percentage was evaluated visually using the photographies obtained by computer vision, taking into account the different degrees of fruit dehydration (Figure 1) following the norms for quality of fresh blueberries from Chilean Blueberry Committee [13]. Fruits with dehydration degree of 2 or 3 were counted as dehydrated fruit.

2.4. Image Analysis

2.4.1. Color. Digital images of each blueberry (of two opposite sides) were taken at each storage time in order to obtain the surface color of the fruit using a computer vision system (Figure 2), which consisted of a black box with four 18 W natural light tubes (D65, Philips) and a digital camera (Canon 10 MP, PowerShot G4) placed in a vertical position 22.5 cm from the samples (camera lens angle and lights at 45°) [9]. All images were obtained under the same conditions; the camera was remotely controlled by ZoomBrowser software (v.6.0, Canon). Surface color data were measured in the CIEL

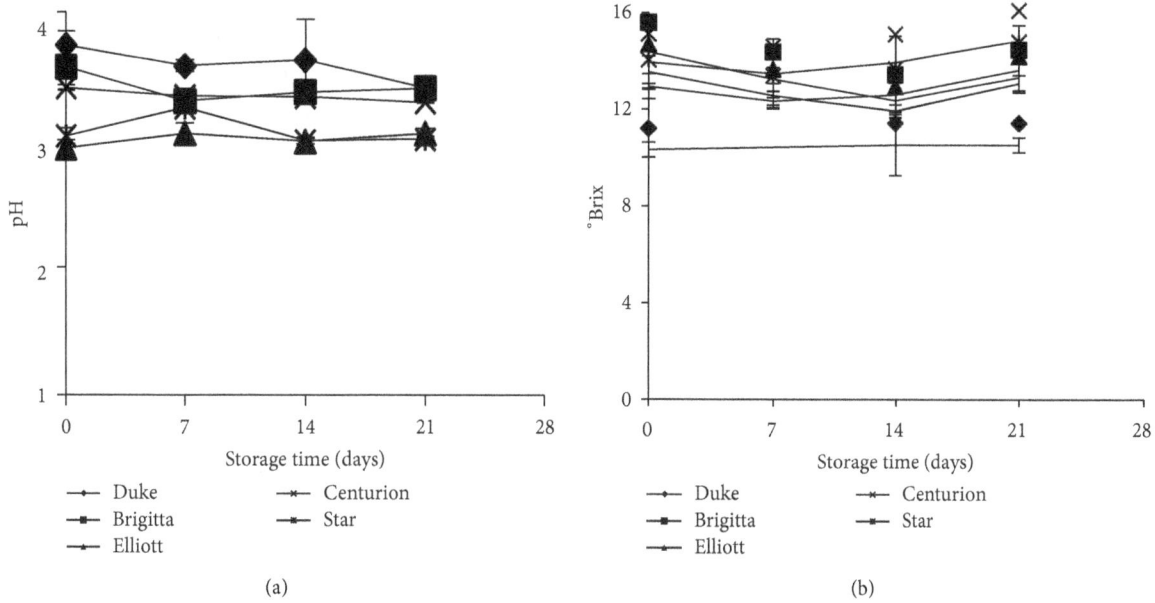

(a)   (b)

FIGURE 3: pH and °Brix variations of different cultivars for different storage times at 4°C and 90% RH.

TABLE 1: Quality parameter differences between fresh cultivars.

Cultivar	pH	°Brix	RI[1]	Water content (% wb)[2]	L^{*}[3]
"Duke"	$3.74 \pm 0.11^{(a)}$	$10.33 \pm 0.31^{(a)}$	$0.73 \pm 0.01^{(a)}$	$86.50 \pm 0.52^{(a)}$	$64.96 \pm 2.82^{(a)}$
"Brigitta"	$3.57 \pm 0.17^{(a,c)}$	$12.70 \pm 0.53^{(a,b)}$	$0.74 \pm 0.01^{(a)}$	$87.59 \pm 0.72^{(a)}$	$71.49 \pm 1.97^{(a)}$
"Elliott"	$2.94 \pm 0.06^{(b)}$	$13.20 \pm 0.10^{(b)}$	$0.72 \pm 0.01^{(a)}$	$83.46 \pm 0.79^{(b)}$	$72.85 \pm 5.20^{(a)}$
"Centurion"	$3.41 \pm 0.02^{(c)}$	$13.93 \pm 1.50^{(b)}$	$0.95 \pm 0.03^{(b)}$	$82.36 \pm 0.83^{(b)}$	$71.85 \pm 1.94^{(a)}$
"Star"	$3.03 \pm 0.06^{(b)}$	$12.93 \pm 0.11^{(b)}$	$0.78 \pm 0.01^{(c)}$	$83.97 \pm 0.51^{(b)}$	$71.78 \pm 3.00^{(a)}$
"Jewel"	$3.50 \pm 0.10^{(c)}$	$11.90 \pm 0.38^{(a)}$	$0.82 \pm 0.02^{(d)}$	$79.86 \pm 0.85^{(c)}$	$72.20 \pm 2.50^{(a)}$

Different superscript letters for the same column indicate values to be significantly different ($P < 0.05$).
[1]RI: roundness index (1).
[2]% wb means wet basis percentage.
[3]L^{*}: color parameter of lightness.

TABLE 2: Percentages (%) of dehydrated fruits take into account visually surface dehydration at different storage conditions of temperature (°C) and relative humidity (RH) after 21 days of storage.

Cultivar	4°C 75% RH	4°C 90% RH	15°C 75% RH	15°C 90% RH
"Duke"	12	10.7	21	19
"Brigitta"	5	2.7	35.1	33.3
"Elliott"	20.1	15.5	35.7	35.4
"Centurion"	20	18.7	40.2	39.6
"Star"	37	36	38.1	38
"Jewel"	15.5	12.5	38.1	37.5

$a^{*}b^{*}$ space and image analysis was performed with the Balu Toolbox in Matlab software (v7) [8]. The camera parameters and Balu software were calibrated using 30-color charts with a Minolta colorimeter. Therefore, $L^{*}a^{*}b^{*}$ values obtained from image analysis were equal to the values from the colorimeter.

Color variation, CIEDE2000, or ΔE_{00}, is regarded as the best uniform color difference model coinciding with subjective visual perception, which can reflect the color difference between two images. The color variation (ΔE_{00}, (2)) during storage time was determined using the formulas which include the concepts of chroma (C^{*}, (3)) and hue (h', (4)) for $L^{*}a^{*}b^{*}$ values [14]. The color grade differences (ΔE_{00}) between two samples (1, 2) were determined as follows: the perception of color differences was taken as imperceptible if $\Delta E_{00} < 1.5$, noticeable if $\Delta E_{00} < 3$, and appreciable if $\Delta E_{00} < 6$ [15].

$$\Delta E_{00} = \sqrt{\left(\frac{\Delta L'}{K_L S_L}\right)^2 + \left(\frac{\Delta C'}{K_C S_C}\right)^2 + \left(\frac{\Delta H'}{K_H S_H}\right)^2 + R_T \left(\frac{\Delta C'}{K_C S_C}\right)\left(\frac{\Delta H'}{K_H S_H}\right)}, \qquad (2)$$

$$C_{i,ab}^* = \sqrt{(a_i^*)^2 + (b_i^*)^2}, \quad i = 1, 2, \tag{3}$$

$$h_i' = \tan^{-1}\left(\frac{b_i^*}{a_i'}\right), \quad i = 1, 2, \tag{4}$$

where

$$a_i' = (1 + G)\, a_i^*, \quad i = 1, 2,$$

$$G = 0.5\left(1 - \sqrt{\frac{\overline{C_{ab}^*}^2}{\overline{C_{ab}^*}^7 + 25^7}}\right),$$

$$\overline{C_{ab}^*} = \frac{\left(C_{1,ab}^* + C_{2,ab}^*\right)}{2},$$

$$C_i' = \sqrt{(a_i^*)^2 + (b_i^*)^2}, \quad i = 1, 2,$$

$$\Delta L' = L_2^* - L_1^*,$$

$$\Delta C' = C_2^* - C_1^*,$$

$$\Delta h' = h_2' - h_1',$$

$$\Delta H' = 2\sqrt{C_1' C_2'}\, \sin\left(\frac{\Delta h'}{2}\right),$$

$$\overline{L'} = \frac{(L_1^* + L_2^*)}{2},$$

$$\overline{C'} = \frac{(C_1' + C_2')}{2},$$

$$\overline{h'} = \frac{(h_1' + h_2')}{2},$$

$$S_L = 1 + \frac{0.015\left(\overline{L'} - 50\right)^2}{\sqrt{20 + \left(\overline{L'} - 50\right)^2}},$$

$$S_C = 1 + 0.045\overline{C'},$$

$$S_H = 1 + 0.015\overline{C'}T,$$

$$T = 1 - 0.17\cos\left(\overline{h'} - 30\right) + 0.24\cos\left(2\overline{h'}\right)$$
$$+ 0.32\cos\left(3\overline{h'} + 6\right) - 0.20\cos\left(4\overline{h'} - 63\right),$$

$$R_T = \frac{-\sin(2\Delta\theta)}{R_C},$$

$$\Delta\theta = 30\exp\left\{-\left[\frac{\left(\overline{h'} - 275°\right)}{25}\right]^2\right\},$$

$$R_C = 2\sqrt{\frac{\overline{C'}^7}{\left(\overline{C'}^7 + 25^7\right)}},$$

$$K_L = K_h = K_C = 1 \tag{5}$$

(see [16]).

2.4.2. Fungal Presence. Fungal presence percentage was obtained by image analysis by computer vision, taking as positive blueberry when fungal filaments were visually observed according to (6):

$$\%\text{Fungal presence} = \frac{\text{No. of positive blueberries}}{\text{No. of total blueberries}} \times 100. \tag{6}$$

In order to validate the fungal presence percentage visually observed by image analysis, fungal filaments from fruits were extracted by immersion of fruits in 5 mL of distilled water for 1 min with manual agitation. Turbidity of extracted aqueous samples was measured by absorbance at 720 nm. A linear correlation of fungal presence on fruits that measured both turbidity and image analysis ($R^2 = 0.995$) was obtained. Pearson correlation coefficient was 0.99, indicating a good positive correlation between the values reported by absorbance and image analysis using (6) [7].

2.5. Statistical Analysis. Statistical analysis was made by analysis of variance (ANOVA) and Tukey's posthoctest, considering significant differences if $P \leq 0.05$. Pearson's correlation coefficient (P) was also calculated.

3. Results and Discussions

In first place, a characterization of quality indicators of six cultivars was determined at initial time (fresh blueberries) to obtain possible differences among cultivars. Then, a characterization of each quality indicator was made during storage time under different storage conditions in order to obtain the shelf life and the behaviour of blueberries under the studied storage conditions.

3.1. pH and °Brix. All evaluated cultivars were from the same field conditions (Summer 2009-2010) in order to avoid the influence of growing conditions such as soil, pH, and water that could affect the pH of the fruits [17]. The pH and °Brix obtained from each fresh cultivar are shown in Table 1.

The initial pH (fresh fruits) founded in "Duke" (pH = 3.74 ± 0.11) was higher than other blueberry cultivars, while the lowest value was obtained for "Elliott" (pH = 2.9 ± 0.06). The reason for these differences would be the citric acid concentration present in each cultivar, which also depends on genetic differences [17]. The pH values obtained for all the cultivars were in the range of 2.75–3.81, in agreement with previous literature reports for other different blueberry cultivars [18].

The highest °Brix values were found in "Centurion" (°Brix = 13.9 ± 0.5), and "Elliott" (°Brix = 13.5 ± 0.5) and the lowest value was found for "Jewel" (°Brix = 11.9 ± 0.4). These results are in agreement with the expected range of 11.2–14.3 °Brix reported for other blueberry cultivars [18, 19].

Although pH and °Brix values were different among cultivars of fresh fruits, these values remained constant with respect to the initial values ($P > 0.05$) during the storage time, regardless of storage conditions (Figure 3). It has been reported that the increasing pH in the fruit is due to maturing time on the plant and also to dehydration during postharvest storage [1, 20]. Therefore, the constant values of pH and °Brix during postharvest storage obtained in this study could be an indicative that no significant dehydration occurred during evaluated storage conditions.

3.2. Fruit Size and Shape.

The equatorial diameter and roundness index (RI, (1)) of each fresh cultivar are shown in Table 1. All the cultivars presented an equatorial diameter greater than 1 cm, which is required to satisfy Chilean export specifications. Therefore, these blueberries were not considered as low calibre [13]. However, the percentage variation of the equatorial diameter of all the cultivars (50–60%) was higher than 3%, which is established for export standards.

The roundness index (RI) indicated that "Centurion" had a significant RI difference ($P < 0.05$) compared to the other cultivars, presenting a more spherical shape (0.95 ± 0.03), while "Elliott" presented the smallest RI (0.72 ± 0.01).

After harvest, fruit size could be altered by both water content, which is kept within in the cell by osmotic forces, and degradation of peptic substances, which weakens the cell walls. Consequently, the fruits cannot retain their shape and integrity [21]. However, under the two storage conditions studied (75% and 95% RH; 4 and 15°C) no significant differences ($P > 0.05$) in both round index (RI) and equatorial diameter were found among the different cultivars (Figure 4). These results indicated that neither significant dehydration nor pectin degradation occurred during storage time under both storage conditions studied.

3.3. Water Content and Dehydrated Fruits.

The water content of each fresh cultivar is shown in Table 1. The cultivar with the highest water content value was "Brigitta" (87.5 ± 0.7% wb), and the lowest water content value was obtained for "Jewel" (79.9 ± 0.8% wb).

At the storage conditions, no significant differences ($P > 0.05$) were found in water content in each cultivar during the whole storage period, which indicates that the blueberries did not undergo significant dehydration ($P > 0.05$) during storage under the controlled temperature and humidity.

The percentage of fruits with presence of surface dehydration at final storage time (21 days), calculated using image analysis following dehydration degree of Chilean Blueberry Committee [13], is presented in Table 2. The results showed

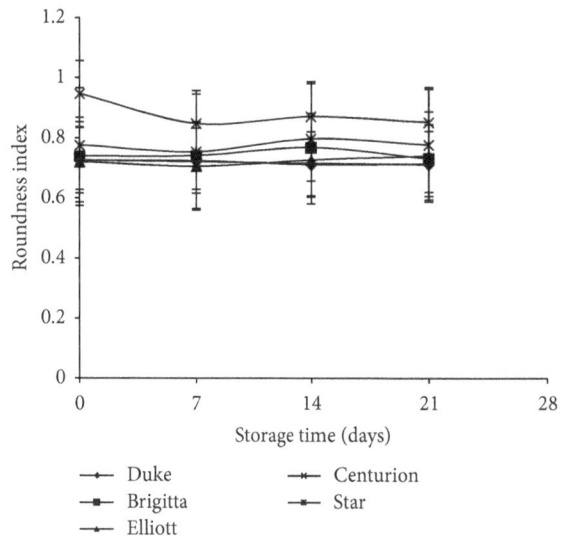

FIGURE 4: Roundness index (RI) of different cultivars over time at 4°C and 90% RH.

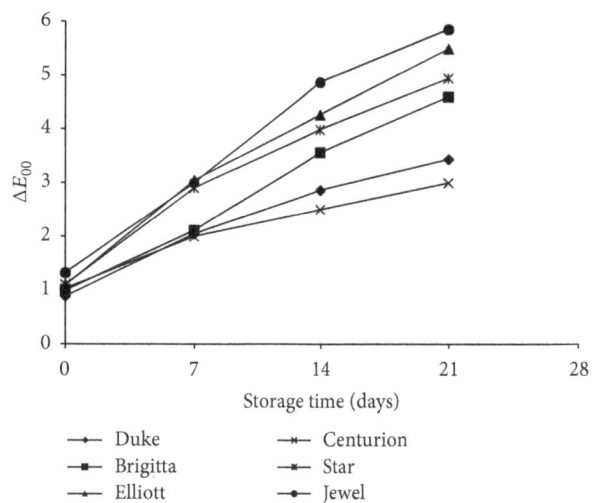

FIGURE 5: Color change (ΔE_{00}) of different cultivars over storage time at 4°C and 90% RH.

that although water content values are constant among storage time for all cultivars, the percentage of fruits with surface dehydration degree higher than 2 on all evaluated cultivars increased with high temperature and low relative humidity, as expected.

Therefore, the results showed that the quality parameters of pH, °Brix, shape, roundness index, and water content of fresh blueberries are different among the evaluated six cultivars hand-harvested in Chile. However, the storage behaviour was similar between them independently of temperature and humidity conditions, indicating that in the selected storage conditions no significant differences ($P > 0.05$) were obtained during storage time. Regarding these important quality parameters, which remained constant during storage time, could not indicate deteriorative changes on the evaluated storage conditions. However, a surface dehydration of

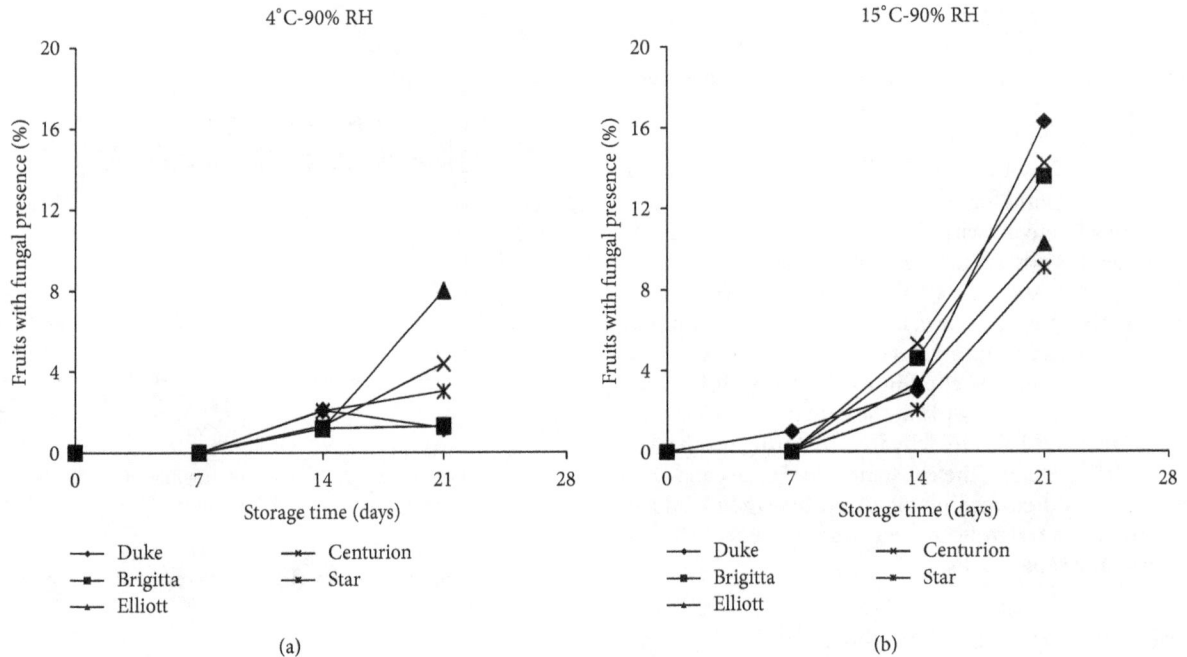

FIGURE 6: Percentage of fruits with fungal presence (6) in six blueberry cultivars stored under different storage conditions (4 and 15°C and 90% RH).

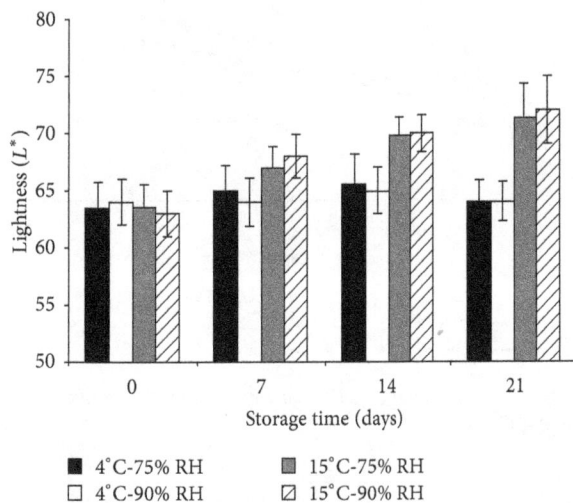

FIGURE 7: Total lightness (L^*) of "Brigitta" at different temperatures (°C) and relative humidities (RH) of storage conditions. Similar behavior was observed for other cultivars.

cuticle of fruits was detected visually using image analysis. This is an important approach to define a damage pattern with different quality degree levels, which can be designed by automatic classification algorithms to be implemented in the industry, reducing overall batch rejections for the market.

3.4. Color. The use of image analysis using computer vision allowed differentiating the blueberry color of different cultivars at various storage times.

The lightness (L^* value) of the blueberry surface showed significant differences ($P < 0.05$) in the initial color of fresh "Duke" cultivar compared to the other cultivars, among which no significant differences ($P > 0.05$) were found (Table 1). This L^* value for "Duke" could not be associated with a lower presence of epicuticular wax on this fruit's surface because the percentage of epicuticular wax ($30 \pm 5\%$) was similar among the cultivars, as expected [7]. Similar results ($28 \pm 8\%$) were found by Matiacevich et al. [7] comparing "Duke", "Jewel," and "Elliott" cultivars.

Color change during storage, observed as ΔE_{00} (2), showed that the behaviour of each cultivar does not differ significantly ($P > 0.05$) as a function of storage time at 4°C and 90% RH (Figure 5). At 7 days of storage, the color changes are imperceptible for all cultivars. However, appreciable color differences ($\Delta E_{00} > 3$) were obtained for "Jewel", "Elliott", "Star," and "Brigitta" at final storage time of 21 days. Only "Duke" retained a noticeable range of color change value at the end of the storage period.

The observed behaviour in Figure 5 was similar to those fruits under the other storage conditions, indicating that the color variation was mainly due to differences among cultivars and not to storage temperature and relative humidity conditions. The color change was appreciable from the initial blue to a red color, which was observed by changes in both a^* and b^* values, indicating senescence of the fruit. The a^* values for all fresh blueberries analysed in this study were obtained in the range from −5 to 5 and b^* values from −10 to 5, as in agreement with those found for the same cultivars [7]. However, these values increased during storage time showing a range for a^* of 0–12 and for b^* from −2 to 6. As a function of storage time, the color change occurred in more than 90% of the blueberries for all cultivars, except for "Centurion", where the color change occurred in around of 75% of the fruits.

These results showed the importance of color fruit as an indicative of deterioration of blueberry quality, which was determined using image analysis.

3.5. Fungal Presence. As expected, fungal presence obtained through image analysis by computer vision using (6) was affected significantly ($P < 0.05$) by storage conditions: temperature and time. Figure 5 shows that fruits with fungal presence increased with increasing storage temperature and time for all cultivars analyzed, showing a lower growth kinetic at 4°C (2%) than at 15°C (up to 14%) at both RHs after 21 days.

It is important to note that the behavior under each storage condition was different for all cultivars, where "Jewel" was more susceptible to fungal growth, emphasizing that the development of *Botrytis* in blueberries could be due to the presence of fungi in the fruit in the initial phase of the study as natural inoculums, and it was not exposed intentionally or inoculated during its storage. Therefore, the differences between cultivars may mainly be due to differences of initial inoculums obtained in the field and not to different genomic susceptibility between cultivars.

The color parameter, lightness (L^*) increased during storage time (Figure 7). This increase was related to fungal presence in all cultivars, which was attributed to *Botrytis cinerea* identified taxonomically [22]. Pearson correlation coefficient between lightness and fungal presence was higher than 0.9 for all cultivars, indicating that total lightness increased principally due, the characteristic white-gray color of *Botrytis* filaments [22].

Therefore, the results showed that color changes outside the initial color range of each cultivar are an important quality factor to define another damage pattern, which could be measured by image analysis and, therefore, is possible for designed automatic algorithms using computer vision.

3.6. Shelf Life. The shelf life of blueberries is based mainly on visual choice by consumers during its consumption, where the conditions that may change during storage time are the most important ones to be taken into account. Therefore, the factors considered on determination of shelf-life in this study were fungal presence and color change.

Since the determination of shelf-life is a subjective parameter, its determination was based on the occurrence of some of the following conditions: (i) color change from blue to red in more than 45% of the samples and/or (ii) fungal presence higher than 2% due to the only presence of fungal filaments is unacceptable by the consumers [7].

According to information delivered by blueberry producers and used in this research, the shelf-life of different cultivars at 90% RH and 0–4°C is 20 days. However, experimental data indicated that shelf life was 14 days for all cultivars, principally due to fungal presence higher than 2%, as shown in Figure 6.

4. Conclusions

Quality parameters (pH, °Brix, shape, water content, and color) were different among fresh cultivars, as expected, due principally to genetic differences between them and these values did not changed during the evaluated storage conditions. Other parameters depended on storage conditions, as expected such as color changes from blue to red by time in all conditions, surface dehydration, and fungal presence, which both increased principally with temperature and time. However, fungal presence was considered the most important quality parameter to determine a shelf-life due to its unacceptability by consumers.

Despite the differences on temperature and humidity of the storage conditions, the shelf life (taking into account more than 2% of fruits with presence of fungal filaments) was calculated as 14 days for all cultivars independently of the storage conditions.

Moreover, the innovative technology of computer vision applied in this research was a useful tool to determine blueberry decay such as color, surface dehydration, and fungal presence, in an objective manner instead of measuring it subjectively as is done nowadays.

Practical applications of the results obtained in this study are related to the knowledge of the important quality factors from six different blueberry cultivars harvested in Chile under different storage conditions, as part of a comprehensive study of blueberry conservation. Computer vision could be used as an approach to obtain damage patterns that define different quality levels of blueberries. This technology allows designing automatic classification algorithms to be implemented in the industry based on its simplicity, allowing also the analysis of heterogeneous materials such as fresh fruits.

Disclosure

This paper has not been published elsewhere and has not been submitted for publication elsewhere.

Acknowledgments

The authors gratefully acknowledge the financial support of Project Innova Chile-Corfo CT11 PUT-20 and CONICYT by Project PBCT-PSD-62 and FONDECYT Project Grants 11100209 and 1110607.

References

[1] V. Chiabrando, G. Giacalone, and L. Rolle, "Mechanical behaviour and quality traits of highbush blueberry during postharvest storage," *Journal of the Science of Food and Agriculture*, vol. 89, no. 6, pp. 989–992, 2009.

[2] N. Sinelli, A. Spinardi, V. Di Egidio, I. Mignani, and E. Casiraghi, "Evaluation of quality and nutraceutical content of blueberries (*Vaccinium corymbosum L.*) by near and mid-infrared spectroscopy," *Postharvest Biology and Technology*, vol. 50, no. 1, pp. 31–36, 2008.

[3] G. Antonio, F. Faria, C. Takeiti, and K. Park, "Rheological behovior of blueberry," *Ciencia y Tecnología de Alimentos*, vol. 29, pp. 732–737, 2009.

[4] M. C. Nunes, J. P. Emon, and J. K. Brecht, "Quality curves for Highbush blueberries as a function of the storage temperature," in *Proceedings of the 9th North American Blueberry Research and Extension Workers Conference; and In Small Fruits Review*, pp. 423–438, Food Product press. Haworth Press, 2004.

[5] A. Yommi and C. Godoy, "Arándanos, fisiología y tecnologías de postcosecha," 2002, http://anterior.inta.gov.ar/f/?url=http://anterior.inta.gob.ar/balcarce/info/documentos/agric/posco/fruyhort/arandano.htm.

[6] C. Duarte, M. Guerra, P. Daniel, A. L. Camelo, and A. Yommi, "Quality changes of highbush blueberries fruit stored in CA with different CO_2 levels," *Journal of Food Science*, vol. 74, no. 4, pp. S154–S159, 2009.

[7] S. Matiacevich, P. Silva, F. Osorio, and J. Enrione, "Evaluation of blueberry color during storage using image analysis," in *Color in Food: Technological and Psychophysical Aspects*, J. L. Caivano and M. P. Buera, Eds., pp. 211–218, CRC Publisher, 2011.

[8] D. Mery, J. J. Chanona-Peréz, A. Soto et al., "Quality classification of corn tortillas using computer vision," *Journal of Food Engineering*, vol. 101, pp. 357–364, 2010.

[9] F. Pedreschi, J. León, D. Mery, and P. Moyano, "Development of a computer vision system to measure the color of potato chips," *Food Research International*, vol. 39, no. 10, pp. 1092–1098, 2006.

[10] S. Gunasekaran and K. Ding, "Using computer vision for food quality evaluation," *Food Technology*, vol. 48, no. 6, pp. 151–154, 1994.

[11] V. Leemans, H. Magein, and M. F. Destain, "Defects segmentation on "Golden Delicious" apples by using colour machine vision," *Computers and Electronics in Agriculture*, vol. 20, no. 2, pp. 117–130, 1998.

[12] L. Greenspan, "Humidity fixed points of binary saturated aqueous solutions," *Journal of Research of the National Bureau of Standards*, vol. 81, no. 1, pp. 89–96, 1977.

[13] Chilean Blueberry Comittee, "Norma de calidad arándano fresco de exportación," pp. 1–9, 2011.

[14] G. Sharma, W. Wu, and E. N. Dalal, "The CIEDE2000 color-difference formula: implementation notes, supplementary test data, and mathematical observations," *Color Research and Application*, vol. 30, no. 1, pp. 21–30, 2005.

[15] Y. Yang, J. Ming, and N. Yu, "Color image quality assessment based on CIEDE2000," *Advances in Multimedia*, vol. 2012, Article ID 273723, 6 pages, 2012.

[16] M. R. Luo, G. Cui, and B. Rigg, "The development of the CIE 2000 colour-difference formula: CIEDE2000," *Color Research and Application*, vol. 26, no. 5, pp. 340–350, 2001.

[17] J. M. Molina, D. Calvo, J. J. Medina, C. Barrau, and F. Romero, "Fruit quality parameters of some southern highbush blueberries (*Vaccinium xcorymbosum* L.) grown in Andalusia (Spain)," *Spanish Journal of Agricultural Research*, vol. 6, no. 4, pp. 671–676, 2008.

[18] J. Duan, R. Wu, B. C. Strik, and Y. Zhao, "Effect of edible coatings on the quality of fresh blueberries (Duke and Elliott) under commercial storage conditions," *Postharvest Biology and Technology*, vol. 59, no. 1, pp. 71–79, 2011.

[19] W. Kalt and J. E. McDonald, "Chemical composition of lowbush blueberry cultivars," *Journal of the American Society for Horticultural Science*, vol. 121, no. 1, pp. 142–146, 1996.

[20] C. Godoy, "Conservación de dos variedades de arándano alto en condiciones de frío convencional," *Revista Facultad de Ciencias Agrarias de UNCuyo*, vol. 36, pp. 53–61, 2004.

[21] V. Graciela Echeverría, V. Juan Cañumir, and G. Humberto Serri, "Postharvest behavior of highbush blueberry fruits cv. O'Neal cultivated with different organic fertilization treatments," *Chilean Journal of Agricultural Research*, vol. 69, no. 3, pp. 391–399, 2009.

[22] S. Mirzaei, E. M. Goltapeh, and M. Shams-bakhsh, "Taxonomical studies on the genus Botrytis in Iran," *Journal of Agricultural Technology*, vol. 3, pp. 65–76, 2007.

10

Polydextrose Enhances Calcium Absorption and Bone Retention in Ovariectomized Rats

Adriana R. Weisstaub,[1] **Victoria Abdala,**[1] **Macarena Gonzales Chaves,**[2]
Patricia Mandalunis,[3] **Ángela Zuleta,**[1] **and Susana Zeni**[2]

[1] *Food Science and Nutritional Department, School of Pharmacy and Biochemistry,*
  Buenos Aires University (UBA), 1114 Buenos Aires, Argentina
[2] *Metabolic Bone Diseases Laboratory, Clinical Hospital, Immunology, Genetic and Metabolism Institute (INIGEM),*
  National Council for Scientific and Technologic Research (CONICET), UBA, 1114 Buenos Aires, Argentina
[3] *Histologycal and Embryology Department, School of Dentistry, UBA, 1114 Buenos Aires, Argentina*

Correspondence should be addressed to Adriana R. Weisstaub; arweiss@ffyb.uba.ar

Academic Editor: Elad Tako

Purpose. To evaluate the effect of polydextrose (PDX) on Ca bioavailability and prevention of loss of bone mass. *Methods.* Twenty-four two-month-old ovariectomized rats were fed three isocaloric diets only varied in fiber source and content up to 60 days (FOS group, a commercial mixture of short- and long-chain fructooligosaccharide, OVX group fed AIN 93 diet, and PDX). A SHAM group was included as control. Apparent Ca absorption percentage (%ABS), changes in total skeleton bone mineral content (tsBMC) and bone mineral density (BMD) and femur BMD, % Bone Volume, Ca and organic femur content, caecal weight, and pH were evaluated. *Results.* %ABS and caecum weight of PDX and FOS were higher, and caecum pH was lower compared to OVX and SHAM. PDX reached a higher pH and lower caecum weight than FOS possibly because PDX is not completely fermented in the colon. Changes in tsBMC and femur BMD in FOS and PDX were significant lower than SHAM but significantly higher than OVX. % Bone Volume and femur % of Ca in PDX were significantly higher than OVX and FOS but lower than SHAM. *Conclusions.* PDX increased Ca absorption and prevented bone loss in OVX rats.

1. Introduction

Although an optimal calcium intake (CaI) is vital throughout the life cycle, a great percentage of population consumes levels that are far below the recommended amounts. Ca nutrition adequacy is necessary for both bone accretion during growth to achieve an optimum peak bone mass and maintenance in the adult life to suppress bone turnover and, therefore, bone loss [1, 2]. Bone health requires an adequate consumption of Ca to maintain its homeostasis. When Ca levels decrease, parathyroid hormone (PTH) is released to increase bone resorption and to indirectly absorb the available Ca to elevate levels of serum Ca to normal ranges.

Although the increment in CaI would be the most effective strategy for avoiding Ca deficit, if the intake remains inadequate, the ability to improve absorption becomes an important tool to optimize bone health. Ca absorption average only 30% in the adult and, regardless of the CaI, losses through endogenous secretions approximate 120 mg/d [3]. Most of the Ca absorption occurs in the small intestine; however, if the mineral remains ionized and in solution, about 5% occurs in the colon. Several dietary factors could enhance Ca absorption. In this regard, the solubilisation of Ca salts by the acids generated through microbial fermentation in the large intestine has been proposed as one of the mechanisms responsible for the increase in Ca absorption observed following ingestion of highly fermentable, indigestible materials [4].

Nondigestible oligosaccharides can modify the colonic microbiota by increasing the proliferation and activity of beneficial flora [5] which induces changes in the enzymatic activity and produces compounds that enhance paracellular and transcellular absorption of Ca [6]. Several type fructans compounds including inulin, oligofructose, or a mixture of short- and long-chain products (Synergy) [7] have been investigated related to their effect on Ca absorption and bone health [8, 9].

Fiber is an essential nutrient in a healthy diet, contributing to health maintenance and preventing the occurrence of different diseases. Polydextrose (PDX), a compound widely used as an additive for more than 15 years, is a soluble fiber, that is, not digested in the upper gastrointestinal tract [10, 11]. The beneficial effect on gastrointestinal function of PDX has been largely studied; however, few reports were carried out regarding the effect on Ca absorption and bone retention, and several of them were done in normal or gastrectomized rats [4].

The adult ovariectomized (OVX) rat is a suitable animal model to mimic the estrogen withdrawal of postmenopausal women [12, 13]. Bone loss by estrogen withdrawal is a well-known consequence of OVX. As the increment of CaI decreases bone turnover, a recommended manner to prevent bone loss would be increasing the Ca absorption.

On these bases, the aim of the present report was to evaluate the effect of feeding an adequate Ca diet containing 5 g/kg of PDX on Ca absorption and retention into bone in young-OVX rats. In addition, the effect of this diet in preventing the loss of bone mass in OVX rats was compared with the same Ca diet containing a commercial mixture of short- and long-chain fructo-oligosaccharide (FOS) products (Synergy) whose effects on these studied parameters are well documented.

2. Methods and Materials

2.1. Rats and Diets.
A total of 36 two-month-old female virgin Wistar rats (195.0 ± 9.0 g) were obtained from the Laboratory Animal Service of the School of Biochemistry and Pharmacy, Buenos Aires University (Argentina). Throughout the experiment, animals were allowed free access to deionized water and food and were housed in individual stainless steel cages in a temperature- ($21 \pm 1°C$) and humidity- ($60 \pm 10\%$) controlled room with a 12 h light/dark cycle. They were fed a commercial stock diet for laboratory rats during 7 days for acclimatization (Gavave SA, Argentina). After that, ovariectomy was performed by a dorsal approach under light anesthesia (0.1 mg/100 g body weight (BW) of ketamine hydrochloride + 0.1 mg/100 g BW of acepromazine maleate) (Holliday-Scott S A, Buenos Aires, Argentina) and randomly placed on one of the 3 following diets ($n = 8$/group) during a 60-day period (Table 1):

(i) SHAM groups: rats were fed a semisynthetic diet prepared according to the American Institute of Nutrition Diet (AIN 93 M) [14];

(ii) OVX group: rats were fed a semisynthetic diet prepared according to the American Institute of Nutrition Diet (AIN 93 M);

(iii) FOS group (FOS): rats were fed the AIN 93 M diet containing 10 g/100 g diet of Synergy (Orafti BENEO, Tienen, Belgium) replacing equal amounts of cornstarch and sucrose;

(iv) PDX group (PDX): rats were fed the AIN 93 M containing 5 g/100 g diet of polydextrose (Litesse^R, Danisco) replacing equal amounts of cornstarch and sucrose.

Diets were isocaloric and supplied a similar amount of Ca (0.5%) and phosphorus (P) (0.3%).

BW was recorded once a week throughout the study. Food intakes were recorded every three days throughout the experiment, and the total intake, daily intake, and daily CaI were then calculated. This study was carried out in accordance with the National Institute of Health Guide for the Care and Use of Laboratory Animals and approved by the Committee of Health Guide for the Care and Use of Laboratory Animals of the Facultad de Farmacia y Bioquímica, Universidad de Buenos Aires.

2.2. Apparent Calcium Absorption.
Food intake and feces recorded during the last three days of the experiment were used to calculate the apparent Ca absorption (%ABS) as follows: [(CaI – Fecal Ca)/CaI] × 100.

2.3. Bone Measurements.
At the beginning ($t = 0$) and at the end of the experiment ($t = 60$), total skeleton bone mineral content (tsBMC) and bone mineral density (tsBMD) were determined "in vivo" under light anesthesia with a total body scanner by dual energy X-ray absorptiometry (DXA) provided with a specifically designed software for small animals (DPX Alpha, Small Animal Software, Lunar Radiation Corp., Madison, WI, USA) as previously described [15]. In brief, all rats were scanned using an identical scan procedure. Precision was assessed by measuring one rat five times with repositioning between scans on the same and on different days. The coefficient of variation (CV) was 0.9% for total skeleton BMD and 3.0% for BMC. The analysis of the different subareas was carried out on the image of the animal on the screen using a ROI for each segment. The BMD CV was 2.2% for the femur. To minimize the interobserver variations, all analyses were carried out by the same technician. To avoid possible differences in BMC for a different BW, the data of BMC were expressed as percentage of BW.

At the end of the experience, rats were placed under anesthesia (0.1 mg/100 g body weight of ketamine hydrochloride + 0.1 mg/100 g BW of acepromazine maleate), and the right femur was excised at sacrifice for biochemical analysis. The right tibia was resected and fixed by immersion in buffered formalin for 48 h, decalcified in 10% ethylenediamine tetraacetic acid (EDTA) (pH 7) during 25 days, and embedded in paraffin. An 8- to 10-μm-thick longitudinally oriented section of subchondral bone was obtained at the level of the middle third, including primary and secondary spongiosa. It was stained with haematoxylin-eosin and microphotographed (AXIOSKOP, Carl Zeiss) to perform bone volume % (BV%) on the central area of the metaphyseal bone displayed on the digitalized image [16].

2.4. Caecal pH.
After the rats were killed, the caecum from each rat was excised, weighed, and split open, and the pH recorded.

TABLE 1: Composition of the experimental diets.

Ingredient (g/100 g diet)	AIN 93 M diet	FOS diet	PDX diet
Protein[*]	12.0	12.0	12.0
Lipids[**]	4.0	4.0	4.0
Mineral mix[***]	3.5	3.5	3.5
Vitamin mix[****]	1.0	1.0	1.0
L-Cystine	0.18	0.18	0.18
Choline	0.15	0.15	0.15
Cellulose	5.0		
Synergy[#]		10.0	
Polydextrose[##]			5.0
Dextrin	To complete 100 g	To complete 100 g	To complete 100 g

[*]Potassum Caseinate, Nestlé Argentina S.A, containing 85.1% of protein and 0.095 g% Ca.
[**]Commercial soy oil.
[***]Composition according AIN 93 M-MX.
[****]Composition according AIN 93-VX.
[#]Orafti BENEO, Tienen, Belgium.
[##]Litesse[R], Danisco.
Dextrin was added as carbohydrate source to achieve 100 g of diet.

2.5. Analytical Procedures.

Feces were dried under infrared light and pounded. Diets and feces were wet ashes with nitric acid using Parr bombs [17].

Femurs were cleaned of any adhering soft tissue and dried at 100°C for 72 hours, and fat was extracted by immersion for 15 days in chloroform-methanol (3 : 1) mixture which was removed and replaced every three days. Finally, it was dried for 48 hours at 100°C. The fat-free and dried femurs were weighed and ashes were obtained at 700°C until white and crystalline. Thereafter, they were dissolved in HCl and diluted for Ca and P analysis. The amounts of Ca and P were calculated as total content and percentage content of dried fat-free tissue, and the femur Ca/P ratio was also calculated.

Ca concentration in diets, feces, and femur was determined using an atomic absorption spectrophotometer [18]. Lanthanum chloride (6500 mg/L in the final solution) was added to avoid interferences. P concentrations were measured according to the Gomori method [19]. NIST reference material RM 8435 (whole milk powder) was also subjected to identical treatment to verify accuracy of the analytical procedures and treated with each bath of samples to ensure accuracy and reproducibility of mineral analysis.

2.6. Statistical Analysis.

Results were expressed as mean ± standard deviation (SD). Differences were tested by one-way analysis of variance (ANOVA) using the INSTAT package. When ANOVA presented statistical differences ($P < 0.05$), intragroup comparisons were tested by Tukey test.

3. Results

3.1. Food Consumption, Calcium Intake, and Body Weight Gain.

As expected, Table 2 shows a lower food consumption (g/60 days), daily diet intake (g/day), daily CaI (mg/day), and BW gain (g/60 days) in SHAM group as compared to the 3 OVX groups which did not present differences among them.

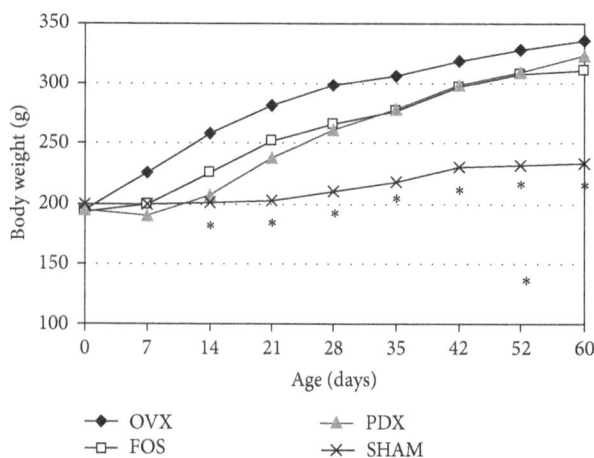

FIGURE 1: Body weight (g) throughout the study. $^{*}P < 0.01$ versus the other three studied groups. Differences were analyzed by Bonferroni test after ANOVA.

Figure 1 showed the BW throughout the experiment. BW was significantly lower in SHAM group as compared to OVX, FOS, and PDX groups ($P < 0.01$) throughout the study; BW was also significantly lower in OVX versus FOS and PDX groups from the 1st to the 5th week of the study ($P < 0.01$) without differences between them. Thereafter, no significant differences among the 3 OVX groups were found until the end of the experiment (8th week). During the studied period, the BW gain was higher in the 3 OVX groups versus SHAM group ($P < 0.05$); no differences between PDX and FOS groups were observed which showed a tendency to reach lower values as compared to OVX groups (Table 2).

3.2. Caecum pH and Weight and Calcium Absorption.

The Ca fecal excretion was significantly lower, and the %ABS was significantly higher in PDX and FOS as compared to OVX and SHAM groups ($P < 0.01$) without differences between

TABLE 2: Food consumption throughout the study, daily intake, daily Ca intake (Ca I) and body weight (BW) gain.

	Intake (g/60 days)	Daily intake (g/day)	Ca I (mg/day)	BW gain (g/60 days)
SHAM	**967.6 ± 45.3**	**14.8 ± 0.1**	**77.0 ± 1.1**	**95.2 ± 22.4**
OVX	1073.8 ± 23.4*	16.6 ± 0.3*	83.1 ± 1.3*	143.0 ± 31.2*
FOS	987.0 ± 96.3*	15.7 ± 1.4*	78.3 ± 6.9*	129.4 ± 27.3*
PDX	1022.9 ± 109.4*	16.4 ± 1.7*	81.9 ± 8.8*	137.1 ± 33.4*

Data are expressed as mean ± SD. *$P < 0.05$ versus SHAM. Differences among groups were analyzed by Bonferroni test after ANOVA.

TABLE 3: Daily fecal Ca excretion, percentage of apparent Ca absorption (% ABS) during the last three days of the experiment, and caecum weight and caecum pH at the end of the study.

	Fecal Ca (mg/day)	% ABS	Caecum weight (g)	Caecum pH
SHAM	**39.9 ± 17.2**	**52.3 ± 6.1**	**2.51 ± 0.37**	**7.06 ± 0.16**
OVX	45.6 ± 14.0*	43.2 ± 5.5*	2.40 ± 0.45	7.06 ± 0.19
FOS	28.0 ± 8.7*,**	61.6 ± 4.9*,**	6.05 ± 0.59*,**	5.77 ± 0.21*,**
PDX	28.1 ± 10.3*,**	58.0 ± 4.1*,**	4.13 ± 0.84*,**,#	6.61 ± 0.18*,**,#

% Abs was determined during the last three days of the experience. Data are expressed as mean ± SD. *$P < 0.01$ versus SHAM; **$P < 0.01$ versus OVX and #$P < 0.05$ versus FOS. Differences among groups were analyzed by Bonferroni test after ANOVA.

them; in addition, no differences were observed between OVX and SHAM groups while caecum pH was higher and caecum weight was lower in PDX versus FOS group ($P < 0.01$) (Table 3).

3.3. Bone Mineral Content and Density, Femur Composition, and Bone Volume.

Total skeleton BMC (tsBMC) and BMD (tsBMD) and their changes between the end and the beginning of the study are shown in Table 4. No significant differences were observed at $t = 0$ among groups; however, tsBMC at $t = 60$ was lower in the 3 OVX groups as compared to SHAM group ($P < 0.05$) without differences between them. The changes in BMC between $t = 60$ and $t = 0$ of FOS and PDX groups were significantly lower than SHAM group ($P < 0.05$) but significantly higher than OVX group, expressed as total content ($P < 0.01$) or normalized per BW ($P < 0.01$) without significant differences between them. The tsBMD and its changes were higher in SHAM than in OVX groups ($P < 0.05$) and in FOS and PDX than in OVX group ($P < 0.01$) without differences between them. Femur BMD (fBMD) changes were significantly lower in OVX groups as compared to SHAM group ($P < 0.05$) and significantly higher in FOS and PDX than in OVX group ($P < 0.05$), without differences between them.

Table 5 shows right femur ashes and organic content and femoral Ca and P content at the end of the experiment. All these parameters were significantly lower in OVX than in SHAM group ($P < 0.05$), and only ashes content reached significance as compared to FOS and PDX groups ($P < 0.05$) which did not present significant differences between them, but they were also lower than SHAM group ($P < 0.05$). The ashes/organic content ratio was higher in SHAM than in the 3 OVX groups ($P < 0.05$) which did not present differences among them while no significant differences were observed in femoral Ca/P ratio.

In agreement with BMD and femoral Ca content, the highest bone volume % was observed in SHAM group ($P < 0.01$); in addition, PDX and FOS presented a significantly

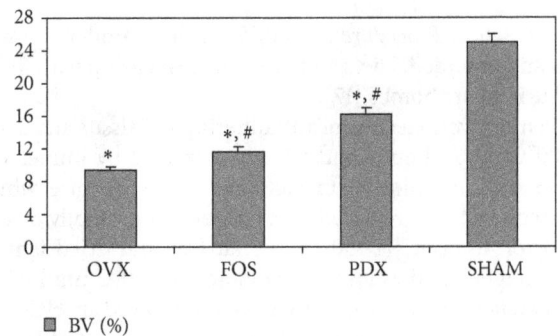

FIGURE 2: Femoral bone volume percentage (BV%) at the end of the study. *$P < 0.01$ versus SHAM; #$P < 0.05$ versus OVX. Differences were analyzed by Bonferroni test after ANOVA.

higher value than OVX group ($P < 0.05$), and there was a tendency to be higher in PDX than in FOS group without reaching significance (Figure 2).

4. Discussion

The results of the present report showed that OVX rats feeding a diet according to AIN 93 M that contains a 5% of PDX during 60 days increased Ca absorption and bone Ca content.

When food intake and BW were compared, the data showed that no differences among OVX groups were observed confirming that PDX diet did not impair diet consumption or BW gain. In addition, as the diet had the same Ca content, no differences in CaI were observed; however, faecal Ca was lower in both DPX and FOS groups. As a result, PDX Ca absorption was as higher as that observed in the FOS-containing diet. Moreover, bone Ca bioavailability was also enhanced in the PDX diet as observed by the increment in BMC, femoral Ca content, and bone volume.

Ca absorption and bioavailability depend not only on luminal Ca concentration, but also on age. Ca absorption

TABLE 4: Total skeleton bone mineral content (tsBMC) (expressed as mg and as mg/g BW); total skeleton bone mineral density (tsBMD) and femur BMD (fBMD) at the beginning ($t = 0$) and at the end of the study ($t = 60$) and their respective changes between $t = 60$ and $t = 0$.

	SHAM	OVX	FOS	PDX
tsBMC $t = 0$ (mg)	**2936 ± 431**	3125 ± 630	2643 ± 312	2759 ± 265
tsBMC $t = 60$ (mg)	**6228 ± 304**	5410 ± 416[*]	5517 ± 659[*]	5947 ± 592[*]
tsBMC $t = 60$–tBMC $t = 0$ (mg)	**4587 ± 216**	2285 ± 214[*]	3096 ± 611[*,**]	3189 ± 472[*,**]
tBMC $t = 0$ (mg/g BW)	**13.8 ± 1.6**	14.2 ± 2.6	13.9 ± 1.6	14.2 ± 1.5
tBMC $t = 60$ (mg/g)	**19.7 ± 2.1**	16.0 ± 2.6[*]	18.2 ± 3.1[*,#]	18.1 ± 2.6[*,#]
tsBMC $t = 60$–tBMC $t = 0$ (mg/g)	**6.5 ± 1.4**	1.33 ± 0.04[*]	5.23 ± 3.55[*,**]	4.48 ± 2.41[*,**]
tsBMD $t = 0$ (mg/cm^2)	**248.2 ± 2.1**	251.0 ± 5.5	247.7 ± 2.6	247.6 ± 8.0
tsBMD $t = 60$ (mg/cm^2)	**281.3 ± 1.9**	278.5 ± 2.7	274.0 ± 7.5	278.1 ± 5.0
tBMD $t = 60$–tBMD $t = 0$ (mg/cm^2)	**34.2 ± 2.2**	27.5 ± 2.7[*]	29.4 ± 1.9[*,#]	30.7 ± 8.8[*,#]
fBMD $t = 0$ (mg/cm^2)	**235.3 ± 25.8**	232.0 ± 35.4	231.5 ± 17.0	227.7 ± 9.6
fBMD $t = 60$ (mg/cm^2)	**298.3 ± 10.9**	253.5 ± 13.4[*]	278.7 ± 19.7[*]	272.0 ± 9.3[*]
fBMD $t = 60$–fBMD $t = 0$ (mg/cm^2)	**62.1 ± 10.8**	21.5 ± 17.0[*]	47.2 ± 11.1[*,**]	44.3 ± 14.6[*,**]

Data are expressed as mean ± SD. [*]$P < 0.05$ versus SHAM group; [#]$P < 0.05$, [**]$P < 0.01$ versus OVX group. Differences were analyzed by Bonferroni test after ANOVA.

TABLE 5: Ashes and organic femur content; ashes/organic ratio, Ca and P femur content and femur Ca/P ratio.

	SHAM	OVX	FOS	PDX
Ashes content (mg/100 g BW)	**72.3 ± 2.4**	44.1 ± 2.5[*]	51.2 ± 11.5[*,#]	54.8 ± 1.9[*,#]
Organic content (mg/100 g BW)	**74.4 ± 1.2**	55.9 ± 2.4[*]	57.2 ± 1.5[*]	55.2 ± 1.9[*]
Ashes/Organic ratio (mg/mg)	**0.98 ± 0.22**	0.79 ± 0.08[*]	0.89 ± 0.04[#]	0.93 ± 0.16[#]
Femoral Ca content (mg/100 g BW)	**29.9 ± 3.6**	15.4 ± 4.1[*]	18.7 ± 2.2[*,#]	20.4 ± 1.2[*,#]
Femoral P content (mg/100 g BW)	**15.1 ± 3.3**	7.6 ± 1.6[*]	9.5 ± 1.1[*,#]	10.2 ± 1.2[*,#]
Femoral Ca/P ratio (mg/mg)	**2.01 ± 0.11**	2.00 ± 0.12	2.07 ± 0.20	2.11 ± 0.25

Femur parameters were determined at the end of the experience. Data are expressed as mean ± SD. [*]$P < 0.05$ versus SHAM group; [#]$P < 0.05$ versus SHAM. Differences were analyzed by Bonferroni test after ANOVA.

in normal rats feeding the recommended dietary Ca levels reached the highest levels at weaning and decreased thereafter to reach the lowest values in adult life [20]. All the OVX studied rats herein were of similar age and fed diets containing the same level of Ca; then, Ca absorption may have been affected by the fiber source and content. The mechanism of Ca absorption involves active, saturable, transcellular movement that takes place largely in the duodenum and passive, nonsaturable, paracellular movement that takes place throughout the small intestine. The sojourn time that a soluble ion remains in the small intestine segments determines how much it is absorbed in each segment. Ca duodenum sojourn time is a matter of minutes, whereas in the lower part of the small intestine is about 2 hr in people and 3 hr in rats [21, 22]. The Ca absorption percentage is about 30%, depending on nutritional status, type, and content of dietary Ca (dairy products versus others), bioavailability, physiological status, and so forth. [21, 22]. Then, a significant more Ca could be potentially absorbed if the insoluble, unabsorbed Ca coming from the small intestine is maintained to an anionic form in the colon.

Soluble fibers and oligosaccharides are being investigated for their potential to improve bone health largely through their influence on mineral absorption and retention. The beneficial effect of inulin, a long-chain fructo-oligosaccharide (FOS) often obtained from chicory root, and other FOSs has

been extensively studied in several experimental models [23]. In this regard, the effect on Ca metabolism and bone health of different substances, such as inulin, oligofructose, or a mixture of short- and long-chain products (Synergy) was tested [7]. Although FOSs enhance Ca absorption, only inulin-FOS increases Ca retention in rats [24, 25]. PDX is a soluble fiber that could be fermented in the gut with production of short-chain fatty acids (SCFA) and reduction of pH. Several studies showed that PDX increases Ca absorption and retention [26, 27]. Mineo et al. [26] "in vitro" found that PDX enhances net Ca transport from the small and large intestine epithelium of rats. Hara et al. and dos Santos et al. [4, 27] demonstrated that the ingestion of PDX increases Ca absorption in normal and total gastrectomized rats suggesting that the small intestine rather than the large one is the responsible for Ca absorption increase. However, no studies were done to demonstrate PDX effect on OVX rats in which an extra Ca supplied may be relevant for avoiding bone mass diminution.

In the present paper the source of fiber in the 3 diets was different. While cellulose was the fiber source of the diet fed by the OVX and SHAM groups (AIN 93 M diet), FOS and PDX diets contained a 10% and 5% of different soluble fibers, a mixture of short- and long-chain products (Synergy) and PDX, respectively. It is important to point out that we performed a preliminary test with a diet containing 10% PDX; however, such diet was not well tolerated by the rats. One possible explanation for this effect would be that PDX is not fully fermented in the colon which induced intolerance and diarrhea. Synergy and PDX induce changes in the intestinal microflora which produces SCFA by fermentation that decrease caecal pH [28]. The low pH maintains Ca and other minerals in solution which improves their absorptions [23, 29–31]. Furthermore, it is also known that the rate and degree of fermentation and, consequently, the acidity could vary with the type and concentration of polysaccharide. According to the literature, the colonic fermentation range for various fibers is broad, from approximately 5% for cellulose to nearly complete for pectin [32]. In the present report, a reduction in the caecal pH was observed in the diets containing PDX or FOS; however, PDX groups showed a low pH reduction as compared to FOS. This effect may be partially explained by the fact that PDX is not completly fermented in the colon; as a consequence, it may also induce a lower intestinal wall thickening leading to the lower caecum weight than was also observed in PDX.

SCFA directly or indirectly participated in Ca absorption. In this regard, Ca acetate and propionate pass across cell membranes even more readily than ionized Ca alone [33]. In addition, butyrate improves the active and passive absorption of Ca. Indeed, butyrate is a potent stimulator of CaBP-9 kDa expression involved in the active Ca absorption [34]. In addition, the passive transepithelial absorption via the activation of tight junctions can be also promoted by FOS and PDX [26, 35]. Therefore, the greatest impact on mineral utilization would depend on the greatest SCFA production and consequently pH lowering. Nevertheless, the mechanism of enhanced Ca absorption with PDX supplementation has been debated. Some authors found a decreased of pH and an increment of SCFA production but others not. Oliveira et al. [36] observed that the addition of PDX to fermented milk produced the highest postacidification when compared to maltodextrin and oligofructose. Mäkeläinen et al. [37], working with a 4-stage colon simulator, founded an increment of the concentrations of all SCFA, especially acetate and propionate. In humans, Hengst et al. [38] showed a decrease on pH value of fecal content but did not find changes on SCFA concentrations. Conversely, other authors [39–41] "in vitro" found that PDX alone or mixed with other FOS lowered gas production rates, butyrate amount, and total SCFA production. In the present report, SCFA production was not measured; however, data showed a significant reduction in caecal pH and a caecum content twice higher than that observed in OVX group confirming the good fermentability of PDX diet [42]. In addition, despite the differences observed in these parameters when PDX and FOS groups were compared, no differences in Ca absorption were found between them. These results are in agreement with a previous report showing that Ca absorption was increased in rats feeding an inulin-FOS diet despite an increment in the total fecal mass [29]. Several papers in OVX rats had observed an increment in Ca absorption by feeding a diet containing 5% FOS and the double of Ca (1%) used in present study [43] or with the same Ca content (0.5%) but a higher (10%) FOS level [7, 32, 43]. In the present study, a concentration of 50 g PDX/kg diet was used because, as previously mentioned, a

higher concentration was no well tolerated by the animals. Nevertheless, Ca absorption was higher than OVX groups and not different from FOS diet indicating a stimulating effect on Ca absorption by PDX.

The amount of Ca absorption and bone retention depends on several experimental conditions such as age, sex, hormonal status, duration of experiment, and diet composition including level of Ca and fructooligosaccharides. Although the absorption measurements could only evidence changes on Ca metabolism for a short time, determinations such as Ca in bone, bone density, and/or histology results showed a long-term impact. In this regard, it has been shown that inulin-type fructans, under certain conditions including OVX, stimulate Ca absorption and enhance bone Ca content [8, 9, 28, 44], although results were conflicting. In this regard, Scholz-Ahrens and Schrezenmeir [8] found that 5% of oligofructose given with the recommended amount of Ca was too low to stimulate Ca absorption and retention in adult OVX rats. The authors explained their findings suggesting a decrease in intestinal and renal functions associated with the rats aging. Conversely, Zafar et al. [25] found that 5.5% of Synergy in a normal Ca diet had positive effects on Ca absorption and retention in OVX rats. Moreover, Taguchi et al. [44] demonstrated an increment in trabecular bone, although Ca absorption was unaffected by feeding a diet containing 5% FOS and 0.5% Ca. The results of the present report showed that Ca absorption was improved in both AIN 93 M diets containing 5% PDX or 10% FOS supplying an extraamount of Ca to maintain homeostasis reducing Ca bone resorption and bone loss. Indeed, under our experimental conditions, bone Ca loss induced by OVX was partially prevented by feeding the PDX and FOS diets evidenced by the higher bone volume, femur ashes content, and tsBMC. Moreover, although no changes in tsBMD were observed, the BMD showed an increase at the femur site. Such differences can be explained taking into account that the total skeleton has a great percentage of cortical bone, while femur and proximal tibia sites have a higher percentage of trabecular bone, metabolic more active and consequently more susceptible to changes in bone remodeling.

Although further studies are required, the results of the present report demonstrated that PDX has a prebiotic effect. Moreover, this ingredient given jointly with an adequate amount of Ca optimizes its absorption and bone retention in OVX rats improving bone health. This effect may have important implications in preventing osteoporosis.

Conflict of Interests

The authors declare no conflict of interests.

Acknowledgments

The authors thank technicians Ricardo Orzuza and Cecilia Mambrin for their technical assistance. This study was supported by the Buenos Aires University and CONICET. UBACyT 20020090200037 funding.

References

[1] V. Coxam and M. N. Horcajada, *Prevention Nutritionnelle de l'Osteoporose*, EM Inter, Cachon, France, Lavoisier edition, 2004.

[2] K. F. Michaelsen, A. V. Astrup, L. Mosekilde, B. Richelsen, M. Schroll, and O. H. Sørensen, "The importance of nutrition for the prevention of osteoporosis," *Ugeskrift for Laeger*, vol. 156, no. 7, pp. 958–963, 1994.

[3] C. M. Weaver and M. Liebman, "Biomarkers of bone health appropriate for evaluating functional foods designed to reduce risk of osteoporosis," *British Journal of Nutrition*, vol. 88, no. 2, pp. S225–S232, 2002.

[4] H. Hara, T. Suzuki, and Y. Aoyama, "Ingestion of the soluble dietary fibre, polydextrose, increases calcium absorption and bone mineralization in normal and total-gastrectomized rats," *British Journal of Nutrition*, vol. 84, no. 5, pp. 655–661, 2000.

[5] M. B. Roberfroid, "Functional foods: concepts and application to inulin and oligofructose," *British Journal of Nutrition*, vol. 87, no. 2, pp. S139–S143, 2002.

[6] J. H. Cummings and G. T. MacFarlane, "Gastrointestinal effects of prebiotics," *British Journal of Nutrition*, vol. 87, no. 2, pp. S145–S151, 2002.

[7] K. E. Scholz-Ahrens and J. Schrezenmeir, "Inulin and oligofructose and mineral metabolism: the evidence from animal trials," *The Journal of Nutrition*, vol. 137, no. 11, pp. 2513S–2523S, 2007.

[8] K. E. Scholz-Ahrens and J. Schrezenmeir, "Inulin, oligofructose and mineral metabolism—experimental data and mechanism," *British Journal of Nutrition*, vol. 87, no. 2, pp. S179–S186, 2002.

[9] K. E. Scholz-Ahrens, G. Schaafsma, E. G. H. M. Van den Heuvel, and J. Schrezenmeir, "Effects of prebiotics on mineral metabolism," *American Journal of Clinical Nutrition*, vol. 73, no. 2, 2001.

[10] S. A. S. Craig, J. F. Holden, J. P. Troup, M. H. Auerbach, and H. I. Frier, "Polydextrose as soluble fiber: physiological and analytical aspects," *Cereal Foods World*, vol. 43, no. 5, pp. 370–376, 1998.

[11] P. García Peris and C. Velasco Gimeno, "Evolution in the knowledge on fiber," *Nutricion Hospitalaria*, vol. 22, no. 2, pp. 20–25, 2007.

[12] T. J. Wronski and C. F. Yen, "The ovariectomized rat as an animal model for postmenopausal bone loss," *Cells and Materials*, supplement 1, pp. 69–74, 1991.

[13] D. N. Kalu, "The ovariectomized rat model of postmenopausal bone loss," *Bone and Mineral*, vol. 15, no. 3, pp. 175–191, 1991.

[14] P. G. Reeves, F. H. Nielsen, and G. C. Fahey, "AIN-93 purified diets for laboratory rodents: final report of the American Institute of Nutrition ad hoc writing committee on the reformulation of the AIN-76A rodent diet," *The Journal of Nutrition*, vol. 123, no. 11, pp. 1939–1951, 1993.

[15] S. N. Zeni, S. Di Gregorio, C. Gomez Acotto, and C. Mautalen, "Olpadronate prevents the bone loss induced by cyclosporine in the rat," *Calcified Tissue International*, vol. 70, pp. 48–53, 2002.

[16] A. M. Parfitt, M. K. Drezner, F. H. Glorieux et al., "Bone histomorphometry: standardization of nomenclature, symbols, and units. Report of the ASBMR Histomorphometry Nomenclature Committee," *Journal of Bone and Mineral Research*, vol. 2, no. 6, pp. 595–610, 1987.

[17] R. E. Sapp and S. D. Davidson, "Microwave digestion of multi-component foods for sodium analysis by atomic absorption spectrometry," *Journal of Food Science*, vol. 56, p. 1412, 1991.

[18] Perkin Elmer Corp, *Analytical Method for Atomic Absorption Spectrophotometry*, Perkin Elmer Corp., Norwalk, Conn, USA, 1971.

[19] G. A. Gomori, "A modification of the colorimeter phosphorus determination for use with the photoelectric colorimeter," *Journal of Laboratory and Clinical Medicine*, vol. 27, pp. 955–960, 1942.

[20] D. Pansu, C. Bellaton, and F. Bronner, "Developmental changes in the mechanisms of duodenal calcium transport in the rat," *The American Journal of Physiology*, vol. 244, no. 1, pp. G20–26, 1983.

[21] F. Bronner and D. Pansu, "Nutritional aspects of calcium absorption," *The Journal of Nutrition*, vol. 129, no. 1, pp. 9–12, 1999.

[22] F. Bronner, "Recent developments in intestinal calcium absorption," *Nutrition Reviews*, vol. 67, no. 2, pp. 109–113, 2009.

[23] C. M. Weaver, "Inulin, oligofructose and bone health: experimental approaches and mechanisms," *British Journal of Nutrition*, vol. 93, pp. S99–S103, 2005.

[24] C. Coudray, J. C. Tressol, E. Gueux, and Y. Rayssiguier, "Effects of inulin-type fructans of different chain length and type of branching on intestinal absorption and balance of calcium and magnesium in rats," *European Journal of Nutrition*, vol. 42, no. 2, pp. 91–98, 2003.

[25] T. A. Zafar, C. M. Weaver, Y. Zhao, B. R. Martin, and M. E. Wastney, "Nondigestible oligosaccharides increase calcium absorption and suppress bone resorption in ovariectomized rats," *The Journal of Nutrition*, vol. 134, no. 2, pp. 399–402, 2004.

[26] H. Mineo, H. Hara, H. Kikuchi, H. Sakurai, and F. Tomita, "Various indigestible saccharides enhance net calcium transport from the epithelium of the small and large intestine of rats in vitro," *The Journal of Nutrition*, vol. 131, no. 12, pp. 3243–3246, 2001.

[27] E. F. dos Santos, K. H. Tsuboi, M. R. Araújo, A. C. Ouwehand, N. A. Andreollo, and C. K. Miyasaka, "Dietary polydextrose increases calcium absorption in normal rats," *Arquivos Brasileiros de Cirurgia Digestiva*, vol. 22, no. 4, pp. 201–205, 2009.

[28] C. Coudray, C. Feillet-Coudray, J. C. Tressol et al., "Stimulatory effect of inulin on intestinal absorption of calcium and magnesium in rats is modulated by dietary calcium intakes: short- and long-term balance studies," *European Journal of Nutrition*, vol. 44, no. 5, pp. 293–302, 2005.

[29] A. Ohta, M. Ohtsuki, S. Baba, T. Adachi, T. Sakata, and E. Sakaguchi, "Calcium and magnesium absorption from the colon and rectum are increased in rats fed fructooligosaccharides," *The Journal of Nutrition*, vol. 125, no. 9, pp. 2417–2424, 1995.

[30] M. Rossi, C. Corradini, A. Amaretti et al., "Fermentation of fructooligosaccharides and inulin by bifidobacteria: a comparative study of pure and fecal cultures," *Applied and Environmental Microbiology*, vol. 71, no. 10, pp. 6150–6158, 2005.

[31] D. L. Topping and P. M. Clifton, "Short-chain fatty acids and human colonic function: roles of resistant starch and nonstarch polysaccharides," *Physiological Reviews*, vol. 81, no. 3, pp. 1031–1064, 2001.

[32] S. I. Cook and J. H. Sellin, "Review article: short chain fatty acids in health and disease," *Alimentary Pharmacology and Therapeutics*, vol. 12, no. 6, pp. 499–507, 1998.

[33] T. P. Trinidad, T. M. S. Wolever, and L. U. Thompson, "Effect of acetate and propionate on calcium absorption from the rectum and distal colon of humans," *American Journal of Clinical Nutrition*, vol. 63, no. 4, pp. 574–578, 1996.

[34] A. C. Maiyar and A. W. Norman, "Effects of sodium butyrate on 1,25-dihydroxyvitamin D3 receptor activity in primary chick kidney cells," *Molecular and Cellular Endocrinology*, vol. 84, no. 1-2, pp. 99–107, 1992.

[35] T. Shimazaki, M. Tomita, S. Sadahiro, M. Hayashi, and S. Awazu, "Absorption-enhancing effects of sodium caprate and palmitoyl carnitine in rat and human colons," *Digestive Diseases and Sciences*, vol. 43, no. 3, pp. 641–645, 1998.

[36] R. P. S. Oliveira, A. C. R. Florence, R. C. Silva et al., "Effect of different prebiotics on the fermentation kinetics, probiotic survival and fatty acids profiles in nonfat symbiotic fermented milk," *International Journal of Food Microbiology*, vol. 128, no. 3, pp. 467–472, 2009.

[37] H. S. Mäkeläinen, H. A. Mäkivuokko, S. J. Salminen, N. E. Rautonen, and A. C. Ouwehand, "The effects of polydextrose and xylitol on microbial community and activity in a 4-stage colon simulator," *Journal of Food Science*, vol. 72, no. 5, pp. M153–M159, 2007.

[38] C. Hengst, S. Ptok, A. Roessler, A. Fechner, and G. Jahreis, "Effects of polydextrose supplementation on different faecal parameters in healthy volunteers," *International Journal of Food Sciences and Nutrition*, vol. 60, no. 5, pp. 96–105, 2009.

[39] B. M. Vester Boler, D. C. Hernot, T. W. Boileau et al., "Carbohydrates blended with polydextrose lower gas production and short-chain fatty acid production in an in vitro system," *Nutrition Research*, vol. 29, no. 9, pp. 631–639, 2009.

[40] D. C. Hernot, T. W. Boileau, L. L. Bauer et al., "In vitro fermentation profiles, gas production rates, and microbiota modulation as affected by certain fructans, galactooligosaccharides, and polydextrose," *Journal of Agricultural and Food Chemistry*, vol. 57, no. 4, pp. 1354–1361, 2009.

[41] H. B. Ghoddusi, M. A. Grandison, A. S. Grandison, and K. M. Tuohy, "In vitro study on gas generation and prebiotic effects of some carbohydrates and their mixtures," *Anaerobe*, vol. 13, no. 5-6, pp. 193–199, 2007.

[42] C. M. Weaver, B. R. Martin, J. A. Story, I. Hutchinson, and L. Sanders, "Novel fibers increase bone calcium content and strength beyond efficiency of large intestine fermentation," *Journal of Agricultural and Food Chemistry*, vol. 58, no. 16, pp. 8952–8957, 2010.

[43] K. E. Scholz-Ahrens, Y. Açil, and J. Schrezenmeir, "Effect of oligofructose or dietary calcium on repeated calcium and phosphorus balances, bone mineralization and trabecular structure in ovariectomized rats," *British Journal of Nutrition*, vol. 88, no. 4, pp. 365–377, 2002.

[44] A. Taguchi, A. Ohta, M. Abe et al., "The influence of fructooligosaccharides on the bone of model rats with ovariectomized osteoporosis," *Scientific Reports of Meiji Seika Kaisha*, vol. 33, pp. 37–44, 1994.

11

Enhanced Production of Xylitol from Corncob by *Pachysolen tannophilus* Using Response Surface Methodology

S. Ramesh, R. Muthuvelayudham, R. Rajesh Kannan, and T. Viruthagiri

Department of Chemical Engineering, Annamalai University, Annamalainagar 608002, Tamil Nadu, India

Correspondence should be addressed to S. Ramesh; ramesh_lecturer@yahoo.co.in

Academic Editor: Bruce A. Welt

Optimization of the culture medium and process variables for xylitol production using corncob hemicellulose hydrolysate by *Pachysolen tannophilus* (MTTC 1077) was performed with statistical methodology based on experimental designs. The screening of nine nutrients for their influence on xylitol production was achieved using a Plackett-Burman design. Peptone, xylose, $MgSO_4 \cdot 7H_2O$, and yeast extract were selected based on their positive influence on xylitol production. The selected components were optimized with Box-Behnken design using response surface methodology (RSM). The optimum levels (g/L) were peptone: 6.03, xylose: 10.62, $MgSO_4 \cdot 7H_2O$: 1.39, yeast extract: 4.66. The influence of various process variables on the xylitol production was evaluated. The optimal levels of these variables were quantified by the central composite design using RSM, for establishment of a significant mathematical model with a coefficient determination of $R^2 = 0.91$. The validation experimental was consistent with the prediction model. The optimum levels of process variables were temperature (36.56°C), pH (7.27), substrate concentration (3.55 g/L), inoculum size (3.69 mL), and agitation speed (194.44 rpm). These conditions were validated experimentally which revealed an enhanced xylitol yield of 0.80 g/g.

1. Introduction

Lignocellulosic materials represent an abundant and inexpensive source of sugars and can be microbiologically converted to industrial products. Xylitol ($C_5H_{12}O_5$), a sugar alcohol obtained from xylose, is generated during the metabolism of carbohydrates in animals and humans. Its concentration in human blood varies from 0.03 to 0.06 mg/100 mL [1]. Xylitol was present in fruits and vegetables [2], at low concentration, which makes its production from these sources economically unfeasible [3]. As a sweetener, xylitol is a substitute for conventional sugars [4]. Its sweetening power was comparable to that of sucrose and is higher than that of sorbitol and mannitol [5]. Furthermore, xylitol has anticariogenic properties. Because it is not consumed by *streptococcus mutans*, xylitol prevents the formation of acids that attack tooth enamel [6]. In addition to reducing dental caries, xylitol also promotes tooth enamel remineralization by reversing small lesions. This happens because, when in contact with xylitol, the saliva seems to be favorably influenced; the chemical composition of xylitol induces the calcium ions and phosphate [7]. For these characteristics, xylitol was a feed stock of great interest to food, odontological, and pharmaceutical industries [1].

Currently, xylitol is produced by chemical hydrogenation using nickel as a catalyst [8]. However it was expensive and it requires several steps of xylose purification before the chemical reaction [4, 9, 10]. Xylitol production through bioconversion has been proposed as for utilizing microorganism such as yeast, bacteria, and fungi [11, 12]. Among these, yeast has shown to possess desirable properties for xylitol production [13, 14]. Therefore, for the present study, yeast strain *Pachysolen tannophilus* was selected for xylitol production. Furthermore studies have shown that nutritional factors including sources of carbon and nitrogen can influence xylitol production [15].

Corncob is a large volume solid waste for using sweet corn processing industry in India. They are currently used as animal feed or returned to the harvested field for land application [16]. Corncob contains approximately over 40% of the dry matter in residues [17] and thus has value has a raw material for production of xylose, xylitol, arabinose, xylobiose, and xylo oligosaccharides. The hemicelluloses fraction in corncob can be easily hydrolysed to constituent carbohydrates. These carbohydrates mainly consist of the xylose and other minor pentose [18–20]. Among various agricultural wastes, corncob was regarded as promising agricultural resources for microbial xylitol production.

In microbial production of xylitol from corncob, the cobs were first hydrolysed to produce from hemicelluloses by acid hydrolysis and the corncob hydrolysate is then used as the medium for xylitol production. The bioconversion of xylitol is influenced by the concentration of various ingredients in culture medium. So their optimization study was very important. This study also investigates the effect of process variables such as pH, temperature, substrate concentration, inoculum size, and agitation speed on xylitol yield. Response surface methodology (RSM) is a mathematical and statistical analysis, which is useful for the modeling and analysis problems that the response of interest is influenced by several variables [21]. RSM was utilized extensively for optimizing different biotechnological process [22, 23].

In the present study, the screening and optimization of medium composition and process variables for xylitol production by *Pachysolen tannophilus* using Plackett-Burman and RSM were reported. The Plackett-Burman screening design was applied for knowing the most significant nutrients enhancing xylitol production. Then, Box-Behnken design and central composite design (CCD) were applied to determine the optimum level of significant nutrients and process variables, respectively.

2. Materials and Methods

2.1. Microorganisms and Maintenance. The yeast strain *Pachysolen tannophilus* (MTCC 1077) was collected from Microbial Type Culture Collection and Gene bank, Chandigarh. The lyophilized stock cultures were maintained at 4°C on culture medium supplemented with 20 g of agar. The medium composition (g/L) was compressed of the following: malt extract: 3.0; yeast extract: 3.0; peptone: 5.0; glucose: 10.0 at pH: 7. It was subcultured every thirty days to maintain viability.

2.2. Size Reduction. Corncob was collected from perambalur farms, Tamil Nadu, India, and was dried in sunlight for 2 days, crushed, and sieved for different mesh size ranging from 0.45 mm to 0.9 mm (20–40 mesh) and used for further studies. The composition of the corncob (g/L): xylose: 28.7, glucose: 5.4, arabinose: 3.7, cellobiose: 0.5, galactose: 0.7, mannose: 0.4, acetic acid: 2, furfural: 0.8, hydroxymethyl furfural: 0.2 was used for xylitol production.

2.3. Acid Hydrolysis. The pretreatment was carried out in 500 mL glass flasks. 2 g of corncobs at a solid loading of 10% (w/w) was mixed with dilute sulfuric acid (0.1% (w/v)) and pretreated in an autoclave at 120°C with residence time of 1 hour. The liquid fraction was separated by filtration and the unhydrolysed solid residue was washed with warm water (60°C). The filtrate and wash liquid were pooled together.

2.4. Detoxification. Hemicellulose acid hydrolysate was heated at 100°C for 15 min to reduce the volatile components. The hydrolysate was overlined with solid $Ca(OH)_2$ up to pH 10, in combination with 0.1% sodium sulfite, and filtered to remove the insoluble materials. The filtrate was adjusted to pH 7 with H_2SO_4. The water phase was treated with activated charcoal.

2.5. Activated Charcoal Treatment. Activated charcoal treatment was an efficient and economic method of reduction in the amount of phenolic compounds, acetic acid, aromatic compounds, furfural, and hydroxymethylfurfural normally found in hemicellulosic hydrolysates. After centrifugation, the solutions were mixed with powdered charcoal at 5% (w/v) for 30 and stirred (100 rpm) at 30°C. The liquor was recovered by filtration, chemically characterized, and used for culture media.

2.6. Fermentation. Fermentation was carried out in 250 mL Erlenmeyer flasks with 100 mL of pretreated corncob hemicelluloses hydrolysate is adjusted to pH 7 with 2 M H_2SO_4 or 3 M NaOH and supplemented with different nutrients concentration for tests according to the selected factorial design, were used for fermentation medium and sterilized at 120°C for 20 mins. After cooling the flasks were inoculated with 1 mL of grown culture broth. The flasks were maintained at 30°C for agitation at 200 rpm for 48 hours. After the optimization of medium composition, the fermentation was carried out with different parameter levels (Table 5) with the optimized media for tests according to the selected factorial design. During the preliminary screening, the experiments were carried out for 5 days and the maximum production was obtained in 48 hours. Hence experiments were carried out for 48 hours.

2.7. Analytical Methods. Sugar and sugar alcohol in the culture broth were measured by high-performance liquid chromatography (HPLC), model LC-10-AD (Shimadzu, Tokyo, Japan) equipped with a refractive index (RI) detector. The chromatography column used was a Aminex HPX-87H (300 × 7.8 mm) column at 80°C with 5 mm H_2SO_4 as mobile phase at a flow rate of 0.4 mL/min, and the injected sample volume was 20 μL.

2.8. Optimization of Xylitol Production: Design of Experiment (DOE). The RSM has several classes of designs, with its own properties and characteristics. Central composite design (CCD), Box-Behnken design, and three-level factorial design are the most popular designs applied by the researchers. A prior knowledge with understanding of the related bioprocesses is necessary for a realistic modeling approach.

2.9. Plackett-Burman Experimental Design. It assumes that there are no interactions between the different variables in the range under consideration. A linear approach is considered to be sufficient for screening. Plackett-Burman experimental design is a fractional factorial design and the main effects of such a design may be simply calculated as the difference between the average of measurements made at the high level

(+1) of the factor and the average of measurements at the low level (−1). To determine which variables significantly affect xylitol production, Plackett-Burman design is used. Nine variables were screened in 12 experimental runs (Table 1), and insignificant ones are eliminated in order to obtain a smaller, manageable set of factors. The low level (−1) and high level (+1) of each factor (−1, +1) were listed as follows (g/L): K_2HPO_4 (6.6, 7), yeast extract (1.5, 5), peptone (2, 5), KH_2PO_4 (1.2, 3.6), xylose (9.8, 10.2), $(NH_4)_2SO_4$ (1, 4), $MgSO_4 \cdot 7H_2O$ (0.7, 1.3), malt (2.8, 3.2), and glucose (9.8, 10.2), and they were coded with $A, B, C, D, E, F, G, H, I$, respectively. The statistical software package "Minitab 16" is used for analyzing the experimental data. Once the critical factors are identified through the screening, the Box-Behnken design was used to obtain a quadratic model after the central composite design (CCD) was used to optimize the process variables and obtain a quadratic model.

The Box-Behnken design and CCD was used to study the effects of the variables towards their responses and subsequently in the optimization studies. This method was suitable for fitting a quadratic surface, and it helps to optimize the effective parameters with a minimum number of experiments, as well as to analyze the interaction between the parameters. In order to determine the relationship between the factors and response variables, the data collected were analyzed in statistical manner. A regression design was employed to model a response as a mathematical function (either known or empirical) for few continuous factors, and good model parameter estimates are desired [21].

The coded values of the process parameters are determined by the following equation:

$$x_i = \frac{X_i - X_0}{\Delta x}, \tag{1}$$

where x_i is coded value of the ith variable, X_i is uncoded value of the ith test variable, and X_0 is uncoded value of the ith test variable at center point. The regression analysis is performed to estimate the response function as a second-order polynomial:

$$Y = \beta_0 + \sum_{i=1}^{K} \beta_i X_i + \sum_{i=1}^{K} \beta_{ii} X_i^2 \sum_{i=1, i<j}^{K-1} \sum_{j=2}^{K} \beta_{ij} X_i X_j, \tag{2}$$

where Y is the predicted response, β_0 constant, and β_i, β_{ii}, β_{ij} are coefficients estimated from regression. They represent the linear, quadratic, and cross-products of X_i and X_j on response.

2.10. Model Fitting and Statistical Analysis. The regression and graphical analysis are carried out using Design-Expert software (version 7.1.5, Stat-Ease, Inc., Minneapolis, USA). The optimum values of the process variables were obtained from the regression equation. The adequacy of the models is further justified through analysis of variance (ANOVA). Lack-of-fit is a special diagnostic test for adequacy of a model that compares the pure error, based on the replicate measurements to the other lack of fit, based on the model performance [24]. F value, calculated as the ratio between

TABLE 1: Plackett-Burman experimental design for nine variables.

Run order	A	B	C	D	E	F	G	H	I	Xylitol yield (g/g)
1	−1	1	−1	−1	−1	1	1	1	−1	0.47
2	1	−1	−1	−1	1	1	1	−1	1	0.34
3	1	1	−1	1	1	−1	1	−1	−1	0.44
4	−1	−1	−1	−1	−1	−1	−1	−1	−1	0.35
5	−1	−1	−1	1	1	1	−1	1	1	0.26
6	−1	−1	1	1	1	−1	1	1	−1	0.50
7	1	1	1	−1	1	1	−1	1	−1	0.49
8	−1	1	1	−1	1	−1	−1	−1	1	0.48
9	1	−1	1	−1	−1	−1	1	1	1	0.59
10	1	1	−1	1	−1	−1	−1	1	1	0.45
11	−1	1	1	1	−1	1	1	−1	1	0.69
12	1	−1	1	1	−1	1	−1	−1	−1	0.65

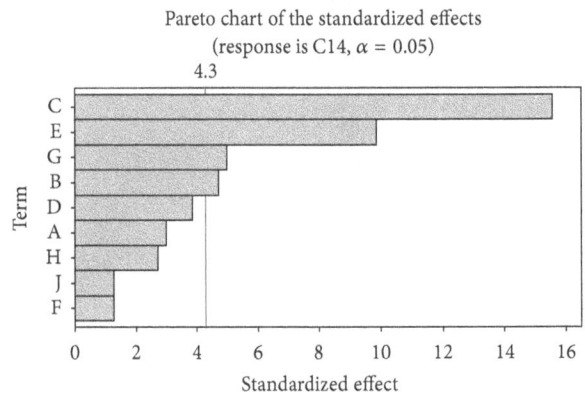

FIGURE 1: Pareto chart showing the effect of media components on xylitol production.

the lack-of-fit mean square and the pure error mean square, is the statistic parameter used to determine whether the lack-of-fit is significant or not, at a significance level. The statistical model was validated with respect to xylitol production under the conditions predicted by the model in shake-flask level. Samples were drawn at the desired intervals and xylitol production was determined as described above.

3. Results and Discussions

Plackett-Burman experiments (Table 1) showed a wide variation in xylitol production. This variation reflected the importance of optimization to attain higher productivity. From the Pareto chart shown in Figure 1 the variables, namely, peptone, xylose, $MgSO_4 \cdot 7H_2O$, and yeast extract, were selected for further optimization to attain a maximum response.

The levels of factors and the effect of their interactions on xylitol production were determined by Box-Behnken design of RSM. The design matrix of experimental results by tests was planned according to the 29 full factorial designs. Twenty-nine experiments were performed at different combinations of the factors shown in Table 2, and the central point was repeated five times. The predicted and observed responses along with design matrix are presented in Table 3,

TABLE 2: Ranges of variables used in Box-Behnken design.

S. no	Variables	Code	Levels (g/L)		
			−1	0	1
1	Peptone	A	3	5	7
2	Xylose	B	8	10	12
3	$MgSO_4 \cdot 7H_2O$	C	1	2	3
4	Yeast extract	D	2	4	6

TABLE 3: Box-Behnken design in coded levels with xylitol yield as response.

Runs	A	B	C	D	Xylitol Yield (g/g)	
					Experimental	Predicted
1	0	−1	1	0	0.44	0.47
2	0	1	0	−1	0.30	0.32
3	0	0	1	1	0.59	0.59
4	1	0	1	0	0.62	0.64
5	0	0	0	0	0.70	0.69
6	0	1	0	1	0.55	0.58
7	−1	0	0	1	0.50	0.50
8	−1	1	0	0	0.34	0.33
9	0	0	1	−1	0.60	0.56
10	0	−1	−1	0	0.39	0.41
11	−1	0	0	−1	0.36	0.41
12	0	0	0	0	0.70	0.69
13	0	0	−1	−1	0.40	0.36
14	−1	−1	0	0	0.51	0.48
15	−1	0	1	0	0.55	0.54
16	−1	0	−1	0	0.43	0.42
17	0	0	−1	1	0.55	0.55
18	1	0	0	1	0.66	0.62
19	1	0	0	−1	0.48	0.49
20	1	1	0	0	0.63	0.62
21	0	0	0	0	0.69	0.69
22	1	−1	0	0	0.43	0.40
23	0	1	1	0	0.58	0.57
24	1	0	−1	0	0.51	0.53
25	0	−1	0	1	0.39	0.39
26	0	1	−1	0	0.42	0.40
27	0	−1	0	−1	0.45	0.44
28	0	0	0	0	0.70	0.69
29	0	0	0	0	0.70	0.69

and the results were analyzed by ANOVA. The second-order regression equation provided the levels of xylitol production as a function of peptone, xylose, $MgSO_4 \cdot 7H_2O$, and yeast extract, which can be presented in terms of coded factors as in the following equation:

$$Y = 0.70 + 0.053A + 0.018B + 0.057C + 0.054D$$
$$+ 0.092AB - (2.500E - 003)\,AC$$
$$+ (1.000E - 002)\,AD + 0.028BC + 0.077BD$$
$$- 0.040CD - 0.083A^2 - 0.16B^2 - 0.076C^2 - 0.11D^2, \tag{3}$$

where Y is the xylitol yield (g/g) and A, B, C, and D were peptone, xylose, $MgSO_4 \cdot 7H_2O$, and yeast extract, respectively. ANOVA for the response surface was shown in Table 4. The model F value of 26.29 implies that the model is significant. There is only a 0.01% chance that a "Model F-value" this large could occur due to noise. Values of "prob > F" less than 0.05 indicate that model terms are significant. Values greater than 0.1 indicate that model terms are not significant. In the present work, linear terms of A, C, and D and all the square effects of A, B, C, and D and the combination of $A * B$, $B * D$, and $C * D$ were significant for xylitol production. The coefficient of determination (R^2) for xylitol production was calculated as 0.9634, which is very close to 1 and can explain up to 96.00% variability of the response. The predicted R^2 value of 0.7898 was in reasonable agreement with the adjusted R^2 value of 0.9267. An adequate precision value greater than 4 is desirable. The adequate precision value of 16.010 indicates an adequate signal and suggests that the model can navigate the design space.

The above model can be used to predict the xylitol production within the limits of the experimental factors that the actual response values agree well with the predicted response values.

Experimental conditions for optimization of the process variables for xylitol yield were determined by CCD of RSM. Five process variables are assessed at 5 coded levels as shown in Table 5. The design matrix of experimental results by tests was planned according to the 50 full factorial designs, and the central point was repeated eight times. The predicted and observed responses along with design matrix are presented in Table 6 and the results were analyzed by ANOVA.

The second-order regression equation provided the levels of xylitol production as a function of temperature, substrate concentration, pH, agitation speed, and inoculums size,

which can be presented in terms of coded factors as in the following equation:

$$Y = 0.79 + 0.025A + 0.043B + 0.049C$$
$$+ 0.030D + 0.038E - 0.029AB - (4.063E - 003)\,AC$$
$$+ 0.018AD + (2.187E - 003)\,AE - (9.688E - 003)\,BC$$
$$+ 0.014BD + (5.312E - 003)\,BE + (1.562E - 003)\,CD$$
$$+ (5.312E - 003)\,CE - (3.125E - 004)\,DE - 0.040A^2$$
$$- 0.041B^2 - 0.046C^2 - 0.027D^2 - 0.034E^2, \tag{4}$$

where Y was the xylitol yield (g/g) and A, B, C, D, and E are temperature, substrate concentration, pH, agitation speed, and inoculums size, respectively. ANOVA for the response surface was shown in Table 7. The model F value of 15.58 implies that the model is significant. There is only a 0.01% chance that a "Model F-value" this large could occur due to

TABLE 4: Analyses of variance (ANOVA) for response surface quadratic model for the production of xylitol using Box-Behnker design.

Source	Sum of square	df	Mean square value	F value	P value
Model	0.40	14	0.028	26.29	<0.0001
A-peptone	0.034	1	0.034	31.67	<0.0001
B-xylose	$3.675E - 003$	1	$3.675E - 003$	3.41	0.0861
C-MgSO$_4 \cdot$7H$_2$O	0.039	1	0.039	35.75	<0.0001
D-yeast extract	0.035	1	0.035	32.67	<0.0001
AB	0.034	1	0.034	31.76	<0.0001
AC	$2.500E - 005$	1	$2.500E - 005$	0.023	0.8811
AD	$4.000E - 004$	1	$4.000E - 004$	0.37	0.5521
BC	$3.025E - 003$	1	$3.025E - 003$	2.81	0.1160
BD	0.024	1	0.024	22.29	0.0003
CD	$6.400E - 003$	1	$6.400E - 003$	5.94	0.0288
A^2	0.045	1	0.045	41.63	<0.0001
B^2	0.16	1	0.16	148.20	<0.0001
C^2	0.037	1	0.037	34.46	<0.0001
D^2	0.074	1	0.074	68.80	<0.0001
Residual	0.015	14	$1.078E - 003$		
Lack of fit	0.015	10	$1.501E - 003$	75.04	0.0004
Pure error	$8.000E - 005$	4	$2.000E - 005$		
Cor total	0.41	28			

TABLE 5: Ranges of variables used in CCD.

S. no	Variables	Code	Levels				
			−2.37	−1	0	1	2.37
1	Temperature (°C)	A	20	25	30	35	40
2	Substrate concentration (g/L)	B	1	2	3	4	5
3	pH	C	6	6.5	7	7.5	8
4	Agitation speed (rpm)	D	50	100	150	200	250
5	Inoculum size (mL)	E	1	2	3	4	5

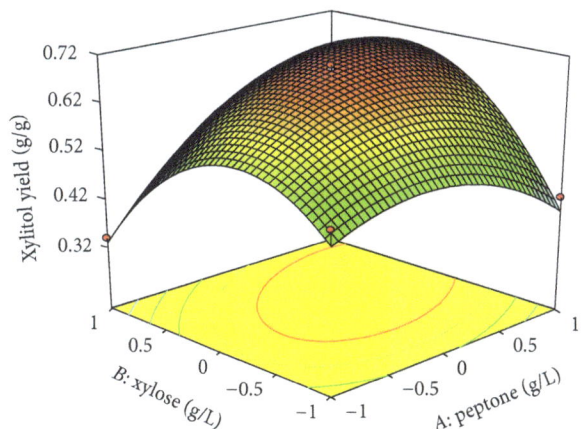

FIGURE 2: 3D plot showing the effect of peptone and xylose on xylitol yield.

noise. Values of "prob > F" less than 0.05 indicate that model terms are significant. Values greater than 0.1 indicate that model terms are not significant. In the present work, linear terms and all the square effects of A, B, C, D, and E and the combination of $A * B$ and $A * D$ were significant for xylitol production. The coefficient of determination (R^2) for xylitol production was calculated as 0.9148, which is very close to 1 and can explain up to 91.00% variability of the response. The predicted R^2 value of 0.6867 was in reasonable agreement with the adjusted R^2 value of 0.8561. An adequate precision value greater than 4 is desirable. The adequate precision value of 12.951 indicates an adequate signal and suggests that the model can navigate the design space.

In both designs the interaction effects of variables on xylitol production were studied by plotting 3D surface curves against any two independent variables, while keeping another variable at its central (0) level. The 3D curves of the calculated response (xylitol yield) and contour plots from the interactions between the variables were obtained. Figure 2 shows the dependency of xylitol on peptone and xylose. The xylitol production increased with increase in peptone to about 6 g/L, and thereafter xylitol production decreased with

further increase in peptone. The same trend was observed in Figures 3 and 4. This evidence from above figures shows the dependency of xylose, MgSO$_4 \cdot$7H$_2$O, yeast extract on xylitol production. The optimal operation conditions of peptone, xylose, MgSO$_4 \cdot$7H$_2$O, and yeast extract for maximum xylitol production were determined by response surface analysis and also estimated by regression equation. The predicted results were shown in Table 3. The predicted values from the regression equation closely agreed with that obtained from experimental values.

In CCD Figure 5 shows the dependency of xylitol on temperature and substrate concentration. The xylitol production increased with increase in temperature to about 36°C, and thereafter xylitol production decreased with further increase

TABLE 6: Central composite design (CCD) in coded levels with xylitol yield as response.

Runs	A	B	C	D	E	Xylitol yield (g/g)	
						Experiment	Predicted
1	−2.37	0	0	0	0	0.55	0.50
2	−1	1	1	1	1	0.74	0.76
3	−1	−1	1	1	−1	0.47	0.53
4	0	0	0	0	0	0.80	0.78
5	1	1	1	1	−1	0.62	0.68
6	−1	1	1	−1	−1	0.60	0.61
7	0	0	0	0	0	0.81	0.78
8	1	1	−1	−1	−1	0.43	0.50
9	0	0	0	−2.37	0	0.61	0.57
10	−1	1	−1	1	−1	0.58	0.59
11	1	−1	1	1	1	0.74	0.74
12	1	1	1	1	1	0.80	0.79
13	0	0	−2.37	0	0	0.44	0.41
14	0	−2.37	0	0	0	0.47	0.46
15	−1	−1	−1	−1	1	0.45	0.47
16	0	0	0	0	−2.37	0.59	0.51
17	−1	1	−1	1	1	0.64	0.66
18	2.37	0	0	0	0	0.65	0.62
19	0	0	0	0	0	0.81	0.78
20	−1	−1	1	1	1	0.65	0.60
21	1	1	−1	1	1	0.68	0.70
22	−1	1	1	−1	1	0.65	0.71
23	−1	−1	−1	1	1	0.43	0.46
24	−1	1	−1	−1	−1	0.50	0.54
25	0	0	0	0	0	0.81	0.78
26	0	2.37	0	0	0	0.72	0.66
27	0	0	2.37	0	0	0.69	0.65
28	0	0	0	0	0	0.81	0.78
29	1	−1	1	−1	−1	0.57	0.59
30	−1	1	−1	−1	1	0.67	0.62
31	−1	−1	−1	1	−1	0.42	0.41
32	0	0	0	0	0	0.78	0.78
33	1	1	1	−1	1	0.67	0.66
34	0	0	0	0	0	0.73	0.78
35	0	0	0	0	2.37	0.68	0.68
36	−1	1	1	1	−1	0.67	0.67
37	0	0	0	2.37	0	0.74	0.71
38	−1	−1	1	−1	−1	0.53	0.53
39	1	−1	−1	1	−1	0.53	0.56
40	−1	−1	−1	−1	−1	0.40	0.42
41	1	−1	1	1	−1	0.66	0.66
42	1	1	−1	1	−1	0.65	0.62
43	1	1	1	−1	−1	0.58	0.56
44	1	−1	−1	1	1	0.63	0.62
45	1	−1	−1	−1	1	0.57	0.55
46	1	−1	1	−1	1	0.63	0.67
47	1	1	−1	−1	1	0.57	0.58
48	−1	−1	1	−1	1	0.61	0.60
49	0	0	0	0	0	0.65	0.77
50	1	−1	−1	−1	−1	0.51	0.50

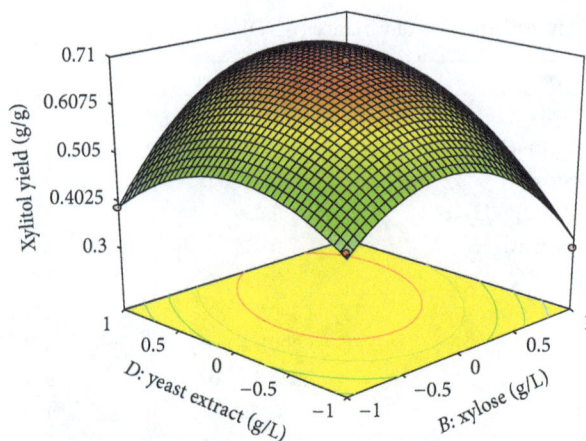

FIGURE 3: 3D plot showing the effect of xylose and yeast extract on xylitol yield.

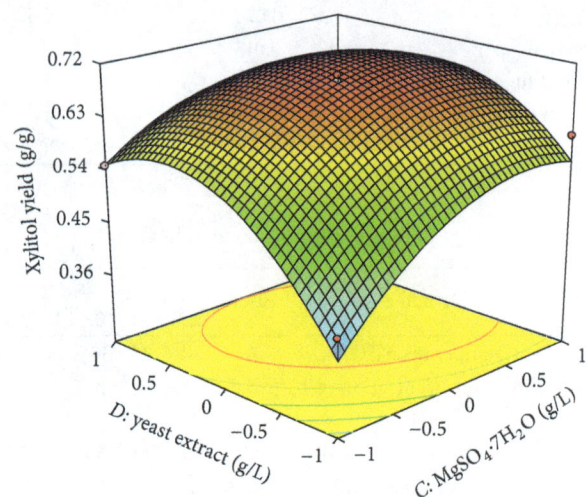

FIGURE 4: 3D plot showing the effect of $MgSO_4 \cdot 7H_2O$ and yeast extract on xylitol yield.

in temperature. The same trend was observed in Figure 6 and figures of other variables. This evidence from above figures shows the dependency of pH, substrate concentration, agitation speed, and inoculum size on xylitol production. The optimal operation conditions of temperature, substrate concentration, pH, agitation speed, and inoculum size for maximum xylitol production were determined by response surface analysis and also estimated by regression equation. The predicted results were shown in Table 6. The predicted values from the regression equation closely agreed with that obtained from experimental values.

3.1. Validation of the Experimental Model. Validation of the experimental model was tested by carrying out the batch experiment under optimal operation conditions which are (g/L): peptone: 6.03, xylose: 10.62, $MgSO_4 \cdot 7H_2O$: 1.39, yeast extract: 4.66 established by the regression model. Under optimal process variables levels are temperature (36.56°C), pH (7.27), substrate concentration (3.55 g/L), inoculum size

TABLE 7: Analyses of variance (ANOVA) for response surface quadratic model for the production of xylitol using CCD.

Source	Sum of square	df	Mean square value	F value	P value
Model	0.63	20	0.031	15.58	<0.0001
A temperature (°C)	0.026	1	0.026	13.07	<0.0001
B-substrate concentration (g/L)	0.079	1	0.079	38.99	<0.0001
C-pH	0.10	1	0.10	51.74	<0.0001
D-agitation speed (rpm)	0.038	1	0.038	18.76	0.0002
E-inoculum size (mL)	0.061	1	0.061	30.26	<0.0001
AB	0.027	1	0.027	13.42	0.0010
AC	$5.281E - 004$	1	$5.281E - 004$	0.26	0.6125
AD	0.011	1	0.011	5.40	0.0273
AE	$1.531E - 004$	1	$1.531E - 004$	0.076	0.7847
BC	$3.003E - 003$	1	$3.003E - 003$	1.49	0.2319
BD	$6.328E - 003$	1	$6.328E - 003$	3.14	0.0868
BE	$9.031E - 004$	1	$9.031E - 004$	0.45	0.5084
CD	$7.812E - 005$	1	$7.812E - 005$	0.039	0.8452
CE	$9.031E - 004$	1	$9.031E - 004$	0.45	0.5084
DE	$3.125E - 006$	1	$3.125E - 006$	$1.552E - 003$	0.9688
A^2	0.088	1	0.088	43.81	<0.0001
B^2	0.092	1	0.092	45.78	<0.0001
C^2	0.12	1	0.12	58.49	<0.0001
D^2	0.039	1	0.039	19.48	<0.0001
E^2	0.063	1	0.063	31.25	<0.0001
Residual	0.058	29	$2.014E - 003$		
Lack of fit	0.053	22	$2.400E - 003$	3	0.0699
Pure error	$5.600E - 003$	7	$8.000E - 004$		
Cor total	0.69	49			

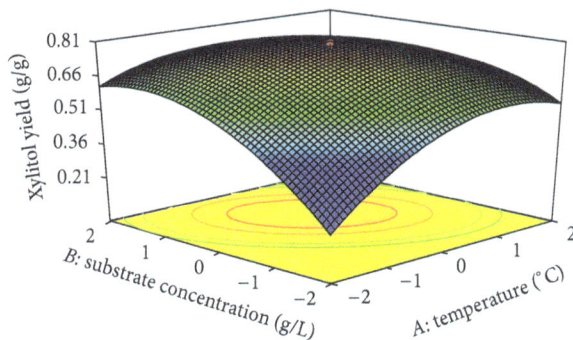

FIGURE 5: 3D plot showing the effect of temperature and substrate concentration on xylitol yield.

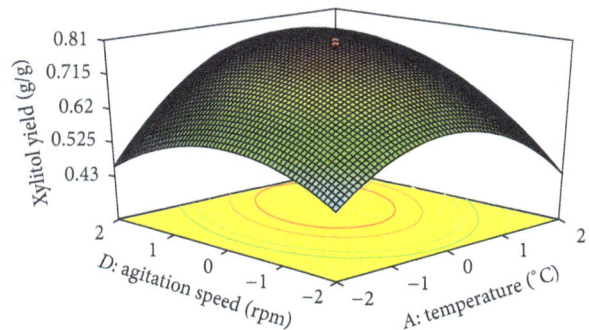

FIGURE 6: 3D plot showing the effect of temperature and agitation speed on xylitol yield.

(3.69 mL), and agitation speed (194.44 rpm). Four repeated experiments were performed and the results are compared. The xylitol production (0.80 g/g) obtained from experiments was very close to the actual response (0.78 g/g) predicted by the regression model, which proved the validity of the model.

4. Conclusion

In this work, Plackett-Burman design was used to test the relative importance of medium components on xylitol production. Among the variables, peptone, xylose, $MgSO_4 \cdot 7H_2O$,

and yeast extract were found to be the most significant variables. From further optimization studies the optimized values of the nutrients for xylitol production were as follows (g/L): peptone: 6.03, xylose: 10.62, $MgSO_4 \cdot 7H_2O$: 1.39, and yeast extract: 4.66. Then the influence of various process variables, namely, temperature, pH, substrate concentration, agitation speed, and inoculum size on the xylitol production was evaluated by CCD. The optimum levels of process variables are temperature (36.56°C), pH (7.27), substrate concentration (3.55 g/L), inoculum size (3.69 mL), and agitation speed (194.44 rpm). This study showed that the corncob is a good source for the production of xylitol. Using the optimized conditions, the xylitol yield reaches 0.80 g/g. The results show a close concordance between the expected and obtained production level.

Acknowledgments

The authors wish to express their gratitude for the support extended by the authorities of Annamalai University, Annamalai, Nagar, India, in carrying out the research work in Bioprocess Laboratory, Department of Chemical Engineering. They assure that this research work is not funded by any organizations.

References

[1] U. Manz, E. Vanninen, and F. Voirol, "Xylitol: its properties and use as a sugar," in *Proceedings of the Symposium on Sugar and Sugar Replacements*, R. A. Food, Ed., London, UK, 1973.

[2] Y. M. Wang and J. van Eys, "Nutritional significance of fructose and sugar alcohols," *Annual Review of Nutrition*, vol. 1, pp. 437–475, 1981.

[3] A. Emidi, "Xylitol its properties and food application," *Food Technology*, vol. 32, pp. 28–32, 1978.

[4] L. Hyvonen, P. Koivistoinen, and F. Voirol, "Food technological evaluation of xylitol," *Advances in Food Research*, vol. 28, pp. 373–403, 1982.

[5] A. Bar, "Xylitol," in *Alternative Sweeteners*, L. O. Nabors and L. Gelardi, Eds., pp. 185–216, Marcel Deeker, New York, NY, USA, 1986.

[6] O. Aguirre-Zero, D. T. Zero, and H. M. Proskin, "Effect of chewing xylitol chewing gum on salivary flow rate and the acidogenic potential of dental plaque," *Caries Research*, vol. 27, no. 1, pp. 55–59, 1993.

[7] K. K. Makinen, "The sugar that prevents that tooth decay," *The Futurist*, pp. 135–139, 1976.

[8] J.-P. Mikkola and T. Salmi, "Three-phase catalytic hydrogenation of xylose to xylitol—prolonging the catalyst activity by means of on-line ultrasonic treatment," *Catalysis Today*, vol. 64, no. 3-4, pp. 271–277, 2001.

[9] A. J. Melaji and L. Hamalainen, US patent no. 4. 008 285, 1997.

[10] E. Winkelhausen and S. Kuzmanova, "Microbial conversion of D-xylose to xylitol," *Journal of Fermentation and Bioengineering*, vol. 86, no. 1, pp. 1–14, 1998.

[11] A. Converti and J. M. Dominguez, "Influence of temperature and pH on xylitol production from xylose by *Debaromyces hansenii*," *Biotechnology and Bioengineering*, vol. 75, pp. 39–45, 2001.

[12] A. Converti, P. Perego, A. Sordi, and P. Torre, "Effect of starting xylose concentration on the microaerobic metabolism of *Debaryomyces hansenii*: the use of carbon material balances," *Applied Biochemistry and Biotechnology A*, vol. 101, no. 1, pp. 15–29, 2002.

[13] J. M. Dominguez, C. S. Gong, and G. Tsao, "Production of xylitol from D-xylose by *Debaryomyces hansenii*," *Applied Biochemistry and Biotechnology*, vol. 63, pp. 117–127, 1997.

[14] F. M. Girio, J. C. Roseiro, P. Sa-Machado, A. R. Duarte-Reis, and M. T. Amaral-Collaco, "Effect of oxygen transfer rate on levels of key enzymes of xylose metabolism in *Debaryomyces hansenii*," *Enzyme and Microbial Technology*, vol. 16, no. 12, pp. 1074–1078, 1994.

[15] H. Ling, K. Cheng, J. Ge, and W. Ping, "Statistical optimization of xylitol production from corncob hemicellulose hydrolysate by *Candida tropicalis* HDY-02," *New Biotechnology*, vol. 28, no. 6, pp. 673–678, 2011.

[16] G. E. Inglett, *Corn: Culture, Processing and Production*, AVI Publishing, Westport, Conn, USA, 1970.

[17] B. Barl, C. G. Biliaderis, E. D. Murray, and A. W. MacGregor, "Combined chemical and enzymic treatments of corn husk lignocellulosic," *Journal of Science of Food Agriculture*, vol. 56, pp. 195–214, 1991.

[18] L. Olsson and B. Hahn-Hägerdal, "Fermentation of lignocellulosic hydrolysates for ethanol production," *Enzyme and Microbial Technology*, vol. 18, no. 5, pp. 312–331, 1996.

[19] V. Balan, B. Bals, S. P. S. Chundawat, D. Marshall, and B. E. Dale, "Ligocellulosic biomass pretreatment using AFEX," *Methods in Molecular Biology*, vol. 581, pp. 61–77, 2009.

[20] W. Liaw, C. Chen, W. Chang, and K. Chen, "Xylitol production from rice straw hemicellulose hydrolyzate by polyacrylic hydrogel thin films with immobilized candida subtropicalis WF79," *Journal of Bioscience and Bioengineering*, vol. 105, no. 2, pp. 97–105, 2008.

[21] D. C. Montgomery, *Design and Analysis of Experiments*, John Wiley & Sons, New York, NY, USA, 2001.

[22] W. Li, W. Du, and D. Liu, "Optimization of whole cell-catalyzed methanolysis of soybean oil for biodiesel production using response surface methodology," *Journal of Molecular Catalysis B*, vol. 45, no. 3-4, pp. 122–127, 2007.

[23] B. J. Naveena, M. Altaf, K. Bhadrayya, and G. Reddy, "Direct fermentation of starch to L(+) lactic acid in SSF by Lactobacillus amylophilus GV6 using wheat bran as support and substrate: medium optimization using RSM," *Process Biochemistry*, vol. 40, no. 2, pp. 681–690, 2005.

[24] M. Y. Noordin, V. C. Venkatesh, S. Sharif, S. Elting, and A. Abdullah, "Application of response surface methodology in describing the performance of coated carbide tools when turning AISI 1045 steel," *Journal of Materials Processing Technology*, vol. 145, no. 1, pp. 46–58, 2004.

12

Retrogradation of Waxy Rice Starch Gel in the Vicinity of the Glass Transition Temperature

Sanguansri Charoenrein and Sunsanee Udomrati

Department of Food Science and Technology, Faculty of Agro-Industry, Kasetsart University, Bangkok 10900, Thailand

Correspondence should be addressed to Sanguansri Charoenrein; fagisscr@ku.ac.th

Academic Editor: Fernanda Fonseca

The retrogradation rate of waxy rice starch gel was investigated during storage at temperatures in the vicinity of the glass transition temperature of a maximally concentrated system (T'_g), as it was hypothesized that such temperatures might cause different effects on retrogradation. The T'_g value of fully gelatinized waxy rice starch gel with 50% water content and the enthalpy of melting retrograded amylopectin in the gels were investigated using differential scanning calorimetry. Starch gels were frozen to −30°C and stored at 4, 0, −3, −5, and −8°C for 5 days. The results indicated that the T'_g value of gelatinized starch gel annealed at −7°C for 15 min was −3.5°C. Waxy rice starch gels retrograded significantly when stored at 4°C with a decrease in the enthalpy of melting retrograded starch in samples stored for 5 days at −3, −5, and −8°C, respectively, perhaps due to the more rigid glass matrix and less molecular mobility facilitating starch chain recrystallization at temperatures below T'_g. This suggests that retardation of retrogradation of waxy rice starch gel can be achieved at temperature below T'_g.

1. Introduction

The retrogradation of starch has been defined as a process which occurs when the molecules comprising gelatinized starch begin to reassociate in an ordered structure [1]. Retrogradation can lead to an obvious increase in the firmness of stored baked goods [2] and frozen cooked rice [3], making them unattractive to consumers. However, in some products, retrogradation can provide a desirable quality such as in the manufacture of rice stick noodles [4], resistant starch type 3 [5] croutons, and bread crumb [6]. For this reason, numerous studies have been performed to examine the factors affecting the retrogradation of starch. Water content and storage temperatures play key roles in the extent of retrogradation. The maximum extent of retrogradation of most starches is attained in starch gels containing 50%–60% solids [7–10]. It was also found that starch gels retrograde faster when stored at 4–6°C [11, 12].

Glass transition is a second-order phase transition that occurs over the temperature range at which amorphous solid materials (glassy materials) are transformed to a metastable leathery state [13]. A special glass transition temperature, denoted as T'_g, has been defined as the glass transition of a maximally freeze-concentrated system [14]. T'_g plays a key role in the quality and storage stability of frozen products because the rate of deteriorative changes in frozen food is closely related to T'_g [13, 15]. Below T'_g, where the food matrix is in a glassy state, molecular mobility decreases and consequently reduces the rate of deteriorative changes involving molecular mobility. Due to the fact that retrogradation of gelatinized starch involves the movement and rearrangement of starch chains to form a junction zone [16], we hypothesized that the extent of retrogradation process of starch gels at temperatures below and above T'_g should be different. Although there have been some studies on the influence of storage temperatures on the extent of retrogradation, these studies used ambient temperature, refrigeration (4 or 5°C), and frozen storage (−20°C) [11, 17] but not temperatures in the vicinity of T'_g. Moreover, all of these experiments used samples which were gelatinized in differential scanning calorimeter (DSC) pans. This static gelatinization might not mimic the real heating process of starch suspension which usually includes stirring the suspension during application of heat. Heating with shear would completely paste the starch suspension while heating without shear might cause incomplete pasting. Moreover previous studies [10] have shown that maximum retrogradation occurred in rice starch gels with 40%–60% water content. In this paper, we present an alternative method to prepare completely gelatinized starch gels and then limit the water content to 50% to provide maximum retrogradation. This research will make a contribution toward an improvement in the acceleration or retardation of retrogradation of starch-based products.

2. Materials and Methods

2.1. Materials. Waxy rice starch was made from Thai waxy rice grains (RD 6), grown in the Kalasin province area and aged at least six months.

2.2. Flour and Starch Preparation. Rice kernels were soaked in water for 4 h and then ground with water using a double-disc stone mill. The slurry was centrifuged for 15 min and dried at 45°C for 15 h. The flour was ground in a hammer mill and passed through a 100-mesh sieve; then it was stored at room temperature in sealed plastic bags. For the isolation of waxy rice starch from waxy rice flour, the method of Hogan [18] was used. Waxy rice flour was mixed with sodium hydroxide solution. The slurry was stirred, filtered, and centrifuged. The sediment was washed with water, neutralized with hydrochloric acid solution, and dried. The rice starch was then ground in a hammer mill and passed through a 100-mesh sieve. The granule size of was 2.8–5.1 μm with mean diameter of 3.8 ± 0.7 μm. The protein, fat, and ash contents of waxy rice starch were 0.20 ± 0.00, 0.05 ± 0.00, and 0.03 ± 0.00 g/100 g dried solid by the AACC Method 46-12, 30-25, and 08-01, respectively [19]. The moisture content was 11.07 ± 0.10 g/100 g by the AACC Method 44-15A [19]. The amylose content of waxy rice starch was 6.49 ± 0.13 g/100 g dried solid by the method of Juliano [20].

2.3. Waxy Rice Starch Gel Preparation. A starch suspension (10% w/w) was prepared by mixing the waxy rice starch in distilled water and stirring continuously at 250 rpm for 1 h followed by 200 rpm at 95°C for 1 h. The gelatinized starch sample was then put in stainless steel DSC pans. Water in each sample was allowed to evaporate at room temperature until the final water content was 50% as determined by weighing. The final sample weight in each pan was 20–30 mg. The pans were hermetically sealed to prevent moisture loss. The sealed pans were separated into two sets with one set being used for glass transition determination and the other set being used in the retrogradation study.

2.4. Glass Transition Temperature (T_g') Determination. A Pyris-1 DSC (Perkin Elmer, Norwalk, CT, USA) equipped with an intracooler subambient accessory was used. Nitrogen gas was used as the purge gas at a flow rate of 20 mL/min during calibration and measurements. The instrument was calibrated using indium and ice. An empty pan was used as a reference sample. Each sample in a sealed DSC pan was cooled to −60°C and then heated to 25°C at 5°C/min to determine the glass transition temperature of a nonannealed state. For the isothermal annealing treatment, the samples were cooled to −60°C and held for 15 min, heated to three different annealing temperatures (−4, −7, and −10°C) in the vicinity of the T_g' of a nonannealed sample and held at this temperature for 15 min, and cooled back to −60°C at 10°C/min and reheated to 25°C at 5°C/min to locate the T_g'. The T_g' was indicated by an onset temperature of the heat capacity change, which was determined using the computer software program associated with the Perkin Elmer instrument. All measurements were run in triplicates.

2.5. Amylopectin Retrogradation Analysis. The sealed DSC pans with gelatinized starch were frozen to −30°C in a cooling bath and held for 1 h and then immediately stored at −8, −5, −3, 0, and 4°C for 5 days. Storage at −8, −5, and −3°C was done in a cooling bath while the samples at 0 and 4°C were kept in a low temperature incubator (Model IPP 400, Memmert, Germany). After storage, the pans were left to stand for 30 min at room temperature to equilibrate and then heated from 25 to 120°C in the DSC at 10°C/min to observe the melting peak of the retrograded starch gels. All measurements were performed in triplicate.

3. Results and Discussion

3.1. Glass Transition. In a system that is allowed to form the maximum amount of ice, the glass transition of this maximally freeze-concentrated matrix occurs at T_g' and is independent of the initial solids fraction (before freezing) [21]. However, if the maximum amount of ice is not formed in the system, the resulting unfrozen matrix will be more dilute. Annealing is a way to form a maximally freeze-concentrated phase. The DSC thermograms showing the T_g' values of the gelatinized waxy rice starch isothermally annealed at different temperatures are presented in Figure 1. The T_g' value of the gelatinized waxy rice starch was about −5°C in the nonannealed state. At the three different annealing temperatures of −4, −7, and −10°C, the T_g' value of gelatinized waxy rice starch was −4, −3.5, and −4.2°C, respectively. The annealing temperature of −7°C resulted in the highest T_g' value and the most clearly detectable among the three annealing temperatures. This might have occurred because this temperature, which was slightly below the T_g' of the gelatinized waxy rice starch (−5°C), was high enough to have sufficient molecular mobility for ice formation and yet also low enough to maintain the amorphous glass matrix as discussed by Lim et al. [22]. Our T_g' result of −3.5°C agreed relatively well with Slade and Levine [23], Roos and Karel [24], and Israkarn and Charoenrein [25] who found that the T_g' values of gelatinized wheat starch, gelatinized waxy corn starch, and cooked rice stick noodles were −5, −6, and −4°C, respectively.

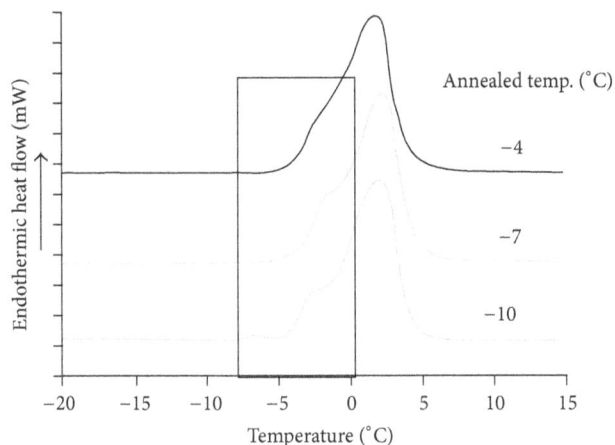

FIGURE 1: Differential scanning calorimetry thermograms showing glass transition temperature (T_g') of waxy rice starch gel samples which were isothermally annealed at −4, −7, and −10°C for 15 min.

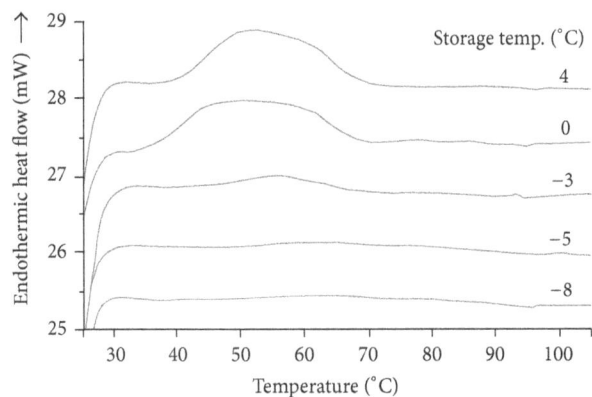

FIGURE 2: Differential scanning calorimetry melting endotherms obtained from waxy rice starch gel samples containing 50% water content stored at 4, 0, −3, −5, and −8°C for 5 days.

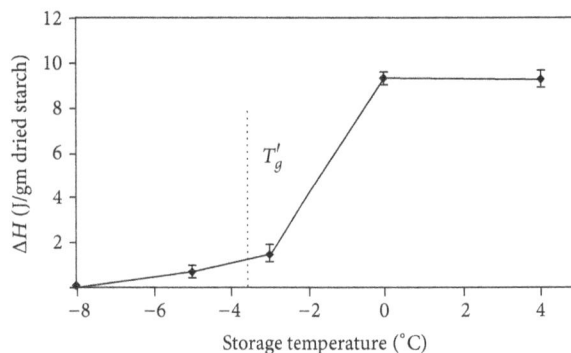

FIGURE 3: Enthalpy of melting retrograded waxy rice starch gel samples containing 50% water content as a function of storage temperature after 5-day storage.

degree of recrystallization of starch than those stored at 4°C. However, their intervals of storage temperature were greater than those in our studies.

4. Conclusions

The onset of T_g' of the gelatinized waxy rice starch gel annealed at −7°C for 15 min was −3.5°C. The results showed that rice starch gels retrograded substantially when stored at 4°C for 5 days. Decreases in the enthalpy of melting of retrograded starch gel were observed in samples stored at −3, −5, and −8°C. This was due to the more rigid glass matrix and less molecular mobility at temperatures below T_g'. These results suggested that the retardation of retrogradation of waxy rice starch gel can be manipulated by a temperature below T_g' and acceleration could be carried out at temperature above T_g'.

References

[1] W. A. Atwell, L. F. Hood, E. Varriano-Marston, and H. F. Zobel, "The terminology and methodology associated with basic starch phenomena," *Cereal Foods World*, vol. 33, pp. 306–311, 1988.

[2] N. Sozer, R. Bruins, C. Dietzel, W. Franke, and J. L. Kokini, "Improvement of shelf life stability of cakes," *Journal of Food Quality*, vol. 34, no. 3, pp. 151–162, 2011.

[3] S. Yu, Y. Ma, and D.-W. Sun, "Effects of freezing rates on starch retrogradation and textural properties of cooked rice during storage," *LWT-Food Science and Technology*, vol. 43, no. 7, pp. 1138–1143, 2010.

[4] P. Satmalee and S. Charoenrein, "Acceleration of ageing in rice stick noodle sheets using low temperature," *International Journal of Food Science and Technology*, vol. 44, no. 7, pp. 1367–1372, 2009.

[5] R. C. Eerlingen, H. Jacobs, and J. A. Delcour, "Enzyme-resistant starch. 5: effect of retrogradation of waxy maize starch on enzyme susceptibility," *Cereal Chemistry*, vol. 71, pp. 351–355, 1994.

[6] M.-A. Ottenhof and I. A. Farhat, "Starch retrogradation," *Biotechnology and Genetic Engineering Reviews*, vol. 21, pp. 215–228, 2004.

3.2. Amylopectin Retrogradation. In this study, waxy rice starch was selected because it showed a well-defined melting peak of retrograded amylopectin at temperature range of 40–75°C. The method of gel preparation used in this study included (1) shearing during heating, which provided a fully gelatinized starch sample, and (2) control of the water content at 50% to obtain the maximum retrogradation extent. Gelatinized waxy rice starch samples stored at various temperatures in the vicinity of the T_g' for 5 days showed differences in enthalpy of melting of the retrograded starch gels (Figures 2 and 3). The peak of melting of retrograded starch was large in samples stored at 4 and 0°C. Samples stored at −3°C showed a small peak while a very small peak and no peak were shown in samples stored at −5 and −8°C, respectively, which were temperatures below T_g' (at −3.5°C). This indicated that at temperature below the glass transition temperature, the movement of starch chains to form junction zone of retrogradation was hindered. The results also agreed with Baik et al. [26] who reported that waxy rice starch gels stored at subzero temperature (−12°C) showed a lower

[7] J. K. Zeleznak and R. C. Hoseney, "The role of water in the retrogradation of wheat starch gels and bread crumb," *Cereal Chemistry*, vol. 63, pp. 407–411, 1986.

[8] L. Slade and H. Levine, "Recent advances in starch retrogradation," in *Recent Development in Industrial Polysaccharides*, S. S. Stivala, V. Crescenzi, and M. C. I. Dea, Eds., pp. 387–430, 1987.

[9] Q. Liu and D. B. Thompson, "Effects of moisture content and different gelatinization heating temperatures on retrogradation of waxy-type maize starches," *Carbohydrate Research*, vol. 314, no. 3-4, pp. 221–235, 1998.

[10] S. Charoenrein and S. Udomrati, "The role of water in the retrogradation of rice starch gels," in *Starch: From Polysaccharides To Granules, Simple and Mixture Gels*, V. P. Yuryev, P. Tomasik, and H. Ruck, Eds., pp. 195–201, 2004.

[11] J. K. Zeleznek and R. C. Hoseney, "Characterization of starch from bread aged at different temperature," *Starch/ Stärke*, vol. 39, pp. 231–233, 1987.

[12] R. D. L. Marsh and J. M. V. Blanshard, "The application of polymer crystal growth theory to the kinetics of formation of the B-amylose polymorph in a 50% wheat-starch gel," *Carbohydrate Polymers*, vol. 9, no. 4, pp. 301–317, 1988.

[13] Y. H. Roos, *Phase Transitions in Foods*, Academic Press, San Diego, Calif, USA, 1995.

[14] H. Levine and L. Slade, "Principles of cryostabilization technology from structure/property relationships of carbohydrate/water systems: a review," *Cryo-Letters*, vol. 9, pp. 21–63, 1988.

[15] B. Bhandari and T. Howes, "Glass transition in processing and stability of food," *Food Australia*, vol. 52, no. 12, pp. 579–585, 2000.

[16] V. J. Morris, "Starch gelation and retrogradation," *Trends in Food Science and Technology*, vol. 1, pp. 2–6, 1990.

[17] Y. J. Wang and J. Jane, "Correlation between glass transition temperature and starch retrogradation in the presence of sugars and maltodextrins," *Cereal Chemistry*, vol. 71, pp. 527–531, 1994.

[18] T. J. Hogan, "The manufacture of rice starch," in *Starch: Chemistry and Technology*, L. R. Whistler and F. E. Paschall FE, Eds., vol. 2, pp. 65–76, Academic Press, New York, NY, USA, 1967.

[19] AACC, *Approved Methods of the American Association of Cereal Chemists*, The American Association of Cereal Chemists, Inc., 10th edition, 2000.

[20] O. B. Juliano, "A simplified assay for milled-rice amylose," *Cereal Science Today*, vol. 6, pp. 335–337, 1971.

[21] M. P. Buera, Y. Roos, H. Levine et al., "State diagrams for improving processing and storage of foods, biological materials, and pharmaceuticals (IUPAC Technical Report)," *Pure and Applied Chemistry*, vol. 83, no. 8, pp. 1567–1617, 2011.

[22] M. H. Lim, H. Wu, and D. S. Reid, "The effect of starch gelatinization and solute concentrations on T_g' of starch model system," *Journal of the Science of Food and Agriculture*, vol. 80, pp. 1757–1762, 2000.

[23] L. Slade and H. Levine, "Non-equilibrium melting of native granular starch. Part I. Temperature location of the glass transition associated with gelatinization of A-type cereal starches," *Carbohydrate Polymers*, vol. 8, no. 3, pp. 183–208, 1988.

[24] Y. H. Roos and M. Karel, "Water and molecular weight effects on glass transitions in amorphous carbohydrates and carbohydrate solutions," *Journal of Food Science*, vol. 56, pp. 1676–1681, 1991.

[25] K. Israkarn and S. Charoenrein, "Influence of annealing temperature on T_g' of cooked rice stick noodles," *International Journal of Food Properties*, vol. 9, no. 4, pp. 759–766, 2006.

[26] M.-Y. Baik, K.-J. Kim, K.-C. Cheon, Y.-C. Ha, and W.-S. Kim, "Recrystallization kinetics and glass transition of rice starch gel system," *Journal of Agricultural and Food Chemistry*, vol. 45, no. 11, pp. 4242–4248, 1997.

13

Survivability of *Vibrio cholerae* O1 in Cooked Rice, Coffee, and Tea

John Yew Huat Tang,[1] **Bariah Ibrahim Izenty,**[1] **Ahmad Juanda Nur' Izzati,**[1]
Siti Rahmah Masran,[1] **Chew Chieng Yeo,**[2] **Arshad Roslan,**[1] **and Che Abdullah Abu Bakar**[1]

[1] *Faculty of Food Technology, Universiti Sultan Zainal Abidin, Gong Badak Campus, 21300 Kuala Terengganu, Terengganu, Malaysia*
[2] *Faculty of Agriculture and Biotechnology, Universiti Sultan Zainal Abidin, Gong Badak Campus, 21300 Kuala Terengganu, Terengganu, Malaysia*

Correspondence should be addressed to John Yew Huat Tang; jyhtang@unisza.edu.my

Academic Editor: Alejandro Castillo

This study aimed to investigate the survival of *Vibrio cholerae* O1 in 3 types of preparation for cooked rice, *Oryza sativa* L., (plain rice, rice with coconut milk, and rice with ginger); coffee, *Coffea canephora*, (plain coffee, coffee with sugar, and coffee with sweetened condensed milk); and tea, *Camellia sinensis*, (plain tea, tea with sugar, and tea with sweetened condensed milk) held at room temperature ($27°C$). The survival of *V. cholerae* O1 was determined by spread plate method on TCBS agar. Initial cultures of $8.00 \log$ CFU/mL were inoculated into each food sample. After 6 h incubation, significant growth was only detected in rice with coconut milk ($9.67 \log$ CFU/mL; $P < 0.05$). However, all 3 types of rice preparation showed significant growth of *V. cholerae* after 24 h ($P < 0.05$). For coffee and tea preparations, *V. cholerae* survived up to 6 h in tea with condensed milk ($4.72 \log$ CFU/mL) but not in similar preparation of coffee. This study showed evidence for the survivability of *V. cholerae* in rice, coffee, and tea. Thus, holding these food and beverages for an extended period of time at room temperature should be avoided.

1. Introduction

The World Health Organization [1] reported the reemergence of cholera as a major infectious disease which posed a global threat to public health, especially in developing countries. In 2009, cholera recorded a fatality rate of 2.24% from a total of 221,226 reported cases [2]. Sporadic outbreaks of cholera can happen in any areas with inadequate supply of water, sanitation, food safety, and hygiene practices [3]. Outbreaks of cholera have been reported since the dawn of ancient civilization in the Ganges delta, India [4].

Although *Vibrio cholerae* is the causal agent for cholera, it is noteworthy that only a small number of *V. cholerae* are capable of producing cholera toxin which causes the clinical symptom [5] presented as acute diarrhea, commonly known as "rice water" stool.

Water is recognised as the main mode of cholera transmission in areas where clean and adequate water supply is limited. In countries where potable water is available, food would be the main vehicle for cholera outbreaks [6]. *V. cholerae* have been reported to cause several outbreaks which implicated food contaminated with *V. cholerae* O1 [7], but person-to-person transmission is uncommon [7]. Outbreaks have been reported to be associated with the consumption of raw or undercooked seafood, and this has been recognized to be a very important factor contributing to foodborne cholera [8]. However, other food such as contaminated cooked rice and fresh produce had also been reported [8, 9]. Ready-to-eat (RTE) foods can be contaminated with *Vibrio* following poor hygiene practices by food handlers [10–12], and the organism can multiply rapidly at ambient temperatures.

Rice is a staple food in the Southeast Asian countries and is thus a main source for *V. cholerae* contamination. There is currently very little known regarding the survivability of *V. cholerae* in various types of cooked rice and beverages that are commonly found in Malaysia. Our previous study

[13] had reported on the survival pattern of *V. cholerae* O1 on three types of rice kept in an opened container, and the current study aims to investigate the survivability of *V. cholerae* O1 in different preparations of cooked rice, as well as the common beverages tea and coffee, along with other ingredients in an enclosed container. This study will provide useful information with regards to *V. cholerae* behaviour in food and beverages that are common in Malaysia. As such, the intervention and proper practices can be determined to reduce the risk of cholera outbreak by food handlers, especially street foods.

2. Materials and Methods

2.1. Vibrio cholerae. *V. cholerae* O1 serotype Inaba was obtained from Kyoto University, Japan. The bacterium was kept in glycerol stock ($-20°C$) and was used throughout this study.

2.1.1. Preparation of V. cholerae for Experiments. *V. cholerae* was prepared as described in our previous study [13]. *V. cholerae* O1 from a stock culture was streaked onto TCBS agar and incubated at 37°C for 24 h. Isolated *V. cholerae* colony was inoculated into alkaline peptone water and incubated in shaker incubator (150 rpm) (Infors HT Ecotron, Basel, Switzerland) at 37°C for 22 h. The culture was centrifuged, and the bacterial pellet was resuspended in phosphate-buffered saline (PBS). Absorbance of the bacterial suspension at 625-nm wavelength was adjusted to a reading of 1.89, which corresponded to about 5×10^9 CFU/mL.

2.1.2. Enumeration of Spiked V. cholerae. The spread plate method was performed as described in the previous study [13]. Each dilution was plated in triplicates, and the plates were incubated at 37°C for 24 h. Yellow colonies on replicate plates were counted and expressed as mean *V. cholerae* CFU per (g/mL).

2.1.3. Preparation of Cooked Rice for Experiments. Nonglutinous white rice (*Oryza sativa* L.) was used throughout this study. Plain rice was prepared using 5 g of white rice washed twice with 10 mL of sterile distilled water. The washed rice was transferred to a large test tube, then 8 mL of sterile distilled water was added, and the mixture was cooked in a beaker with boiling water. The cooked rice with coconut milk was prepared using 5 mL of distilled water together with 3 mL of coconut milk while the cooked rice with ginger used 8 mL of distilled water together with 0.1 g of ginger. The time taken to cook the rice in boiling water was about 15 min for the white rice. All cooked rice preparations were left in an incubator (Infors HT Ecotron) to cool down to 27°C. A new batch was freshly prepared on the day of each experiment.

2.1.4. Preparation of Coffee for Experiments. Instant coffee powder (*Coffea canephora*) was used throughout this study. Plain coffee was prepared by adding 1 g of instant coffee powder into 100 mL of boiled distilled water. Coffee with sucrose (hereafter referred to as sugar) was prepared using 1 g of instant coffee powder with 7 g of sugar in 100 mL of boiled distilled water, while coffee with condensed milk was prepared using 1 g of instant coffee powder with 15 g of condensed milk in 100 mL of boiled distilled water. All the coffee preparations were left in the incubator (Infors HT Ecotron) to cool down to 27°C. A new batch was freshly prepared on the day of each experiment.

2.1.5. Preparation of Tea for Experiments. Loose black tea (*Camelia sinensis*) was used throughout this study. Plain tea was prepared by adding 1 g of loose black tea to 100 mL of boiled distilled water. Tea with sugar was prepared using 1 g of loose black tea together with 5 g of sugar in 100 mL of boiled distilled water, while tea with condensed milk was prepared using 1 g of loose black tea together with 15 g of condensed milk in 100 mL of boiled distilled water. All the tea preparations were left in the incubator (Infors HT Ecotron) to cool down to 27°C. A new batch was freshly prepared on the day of each experiment.

2.1.6. Survival of V. cholerae Inoculated onto Cooked Rice. Three grams of the cooked rice was transferred to each of the five universal bottles [13]. An estimated 1×10^8 CFU of *V. cholerae* in 20 µL of PBS was dispensed randomly as small droplets on each of the rice clumps for incubation time of 0, 1, 3, 6, and 24 h at 27°C in the incubator (Infors HT Ecotron). The bottles were loosely capped. Immediately after the last sample was inoculated (0 h exposure), 5 mL of PBS was added to the bottle and the rice grains in it were mixed by vortex for 1 min. Determination of viable counts of *V. cholerae* was performed with 100 µL of serial diluted rice suspension on TCBS agar. The same procedure was performed with the remaining inoculated samples after 1, 3, 6, and 24 h of incubation time. *V. cholerae* inoculated into empty bottles were kept under similar conditions to serve as control. Four replicated experiments were performed on each rice preparation.

2.1.7. Survival of V. cholerae Inoculated into Coffee and Tea. Ten millilitres of the prepared coffee (or tea) was transferred to each of the five universal bottles. An estimated 1×10^8 CFU of *V. cholerae* in 20 µL of PBS was inoculated into the coffee (or tea) for incubation time of 0, 1, 3, and 6 h at 27°C in incubator (Infors HT Ecotron). The bottles were loosely capped. Immediately after the last sample was inoculated (0 h exposure), the sample was mixed by vortex for 1 min. Determination of viable counts of *V. cholerae* was performed with 100 µL of serial diluted coffee (or tea) suspension on TCBS agar. The same procedure was performed with the remaining inoculated samples after 1, 3, and 6 h of incubation time. *V. cholerae* inoculated into 10 mL of sterile distilled water was kept under similar conditions and served as a control. Four replicated experiments were performed on each coffee (or tea) preparation.

2.1.8. Determination of pH and a_w of Cooked Rice, Coffee, and Tea. The pH and a_w of cooked rice, coffee, and tea were taken using pH 211 Microprocessor pH meter (Hanna Instrument,

India) and HygroLab 3 Bench (Rotronic Instrument Corp., NY, USA), respectively.

2.1.9. Statistical Analysis. Data collected during the experiment was analyzed using SPSS 17.0 software. The data were analyzed using Kruskal-Wallis one-way analysis of variance and Mann-Whitney U test. The significance level was set at $P < 0.05$.

3. Results

Figure 1 shows the survival of *V. cholerae* inoculated onto three types of cooked rice. *V. cholerae* survived well in the enclosed empty bottle which was used as the control where its survivability was stable up to 6 h before declining significantly ($P < 0.05$) after 24 h incubation. There was no statistically significant difference up to 6 h incubation for all test samples (control, plain rice, rice with coconut milk, and rice with ginger). There was significant growth ($P < 0.05$) of *V. cholerae* on plain rice, rice with coconut milk, and rice with ginger after 24 h incubation. However, there was no significant difference of *V. cholerae* survivability with regards to the ingredients added. The mean pH recorded for the plain rice, rice with coconut milk, and rice with ginger was 5.34 ± 0.02, 5.63 ± 0.02 and 5.28 ± 0.02, respectively. Mean a_w recorded for the plain rice, rice with coconut milk, and rice with ginger were all 0.99 ± 0.00.

The survival of *V. cholerae* inoculated in three types of coffee preparation is shown in Figure 2. *V. cholerae* survived poorly in the control sterile distilled water with no detectable *V. cholerae* after 1 h incubation. An immediate inactivation of *V. cholerae* was observed in plain coffee with no detection of the organism at 0 h. *V. cholerae* was only detected at 0 h in coffee with sugar while longer survival (up to 3 h) was observed in coffee with condensed milk. There was no *V. cholerae* detected after 6 h incubation in all preparations of coffee. The mean pH recorded for the plain coffee, coffee with sugar, and coffee with condensed milk was 4.68 ± 0.02, 4.81 ± 0.01, and 6.01 ± 0.02, respectively. The mean a_w recorded for all three preparations of coffee were 0.99 ± 0.00.

Similar results were obtained for preparations of tea (Figure 3). An immediate inactivation of *V. cholerae* was observed in plain tea with no detection of the organism at 0 h. *V. cholerae* was only detected at 0 h in tea with sugar while longer survival (up to 6 h) was observed in tea with condensed milk. The mean pH recorded for the plain tea, tea with sugar, and tea with condensed milk was 5.18 ± 0.01, 5.29 ± 0.01, and 6.10 ± 0.02, respectively, whereas the mean a_w recorded for the three tea preparations were 0.99 ± 0.00.

4. Discussion

This study revealed that *V. cholerae* O1 was able to survive in street food and beverages commonly found in Malaysia. Studies have shown that poor food handling had resulted in numerous cases of food contamination with pathogens which eventually resulted in outbreaks [8]. Our previous study had shown that rice was able to support the growth of *V. cholerae*

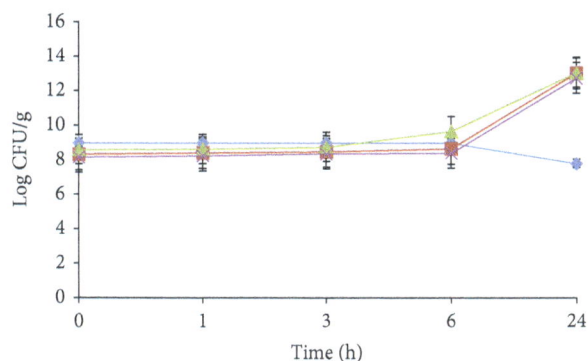

FIGURE 1: Survival of *V. cholerae* on empty capped bottle as control (♦), clumps of cooked plain rice (■), rice with coconut milk (▲), and rice with ginger (×). Each point is the mean of four replicate experiments.

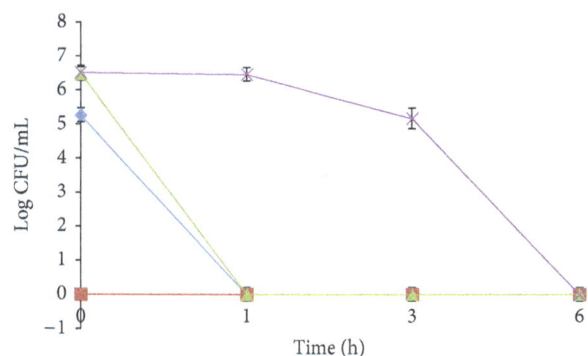

FIGURE 2: Survival of *V. cholerae* in sterile distilled water as control (♦), plain coffee (■), coffee with sugar (▲), and coffee with condensed milk (×). Each point is the mean of four replicate experiments.

O1 [13]. Several reports also showed the relationship between rice and cholera outbreaks [9, 14–16]. *Nasi lemak* is a popular local food that uses coconut milk and ginger as the ingredients for cooking the rice, and it has been implicated in several cholera outbreaks in Malaysia [6]. The phytoconstituents of ginger have been known for their antibacterial properties. Studies have reported that zingerone and shogaols in ginger have antimicrobial properties against *Staphylococcus aureus*, *Streptococcus pyogenes*, *Salmonella enterica* serovar Typhi, and *V. cholerae* [17–19]. However, the amount of ginger that has been added to the rice sample may be too small to show any significant inhibition of *V. cholerae* in the present study.

In the present study, plain coffee showed instant inactivation of *V. cholerae* O1, and this may be due to the acidic nature of coffee [20]. According to the Public Health Agency of Canada [21], *V. cholerae* can grow very fast under optimum conditions which include pH from 5.00 to 9.60 and water activity of approximately 0.97 to 0.99. Other factors that could inhibit the growth of *V. cholerae* O1 in plain coffee may be the presence of antimicrobial components such as caffeine, volatile and nonvolatile organic acid, phenols, and other aromatic compounds [22–24]. Studies have shown that

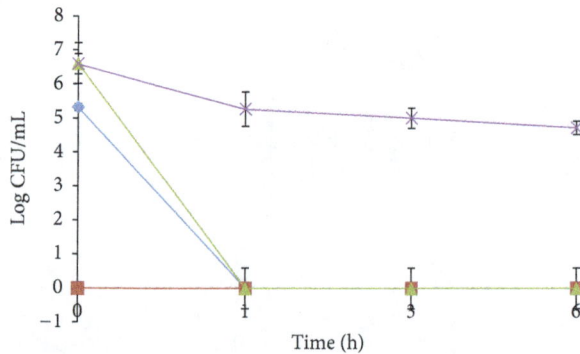

FIGURE 3: Survival of *V. cholerae* in sterile distilled water as control (♦), plain tea (■), tea with sugar (▲), and tea with condensed milk (×). Each point is the mean of four replicate experiments.

the coffee extracts were capable of inhibiting bacteria such as *Escherichia coli*, *S. aureus*, and *Streptococcus mutans* [22–24]. Daglia et al. [23] reported that instant coffee possessed high antimicrobial compound in which the acid extract of coffee during roasting process showed significant antibacterial properties. The relative survival of *V. cholerae* O1 was affected by the addition of sugar in the coffee. *V. cholerae* O1 survived up to 1 h in the coffee added with sugar that contains 99.5 to 99.9% pure sucrose. Felsenfeld [25] reported that the presence of sucrose is favourable for the survival of the Vibrios. However, the survival of *V. cholerae* O1 in coffee with sugar did not exceed one hour. In contrast, *V. cholerae* O1 survived up to 3 hours period in the coffee added with condensed milk. An increase in pH to 6.01 of coffee with condensed milk might contribute to the better survival of *V. cholerae* as studies discovered that the near-neutral pH facilitates the survival and multiplication of *V. cholerae* with increasing pH [26].

Tea has been well known for its antioxidant and antimicrobial properties [20, 27, 28]. The antimicrobial properties of tea are found in its polyphenol fractions, in particular the compounds epicatechin gallate (ECG) and epigallocatechin gallate (EGCG). *V. cholerae* was not detected after inoculation into the plain tea probably because of the antimicrobial properties of the tea itself [20, 27, 28]. Rabbani and Greenough III [29] discussed the physicochemical characteristics of foods that support the survival and growth of *V. cholerae* O1 which include neutral or an alkaline pH, low temperature, and high-organic content. This was proven in this study as *V. cholerae* was not detectable in the acidic condition (pH 5.18) of the tea samples. Nevertheless, the addition of sugar and condensed milk in tea had improved the survival of *V. cholerae* in the tea as observed in this study.

As a conclusion, this study clearly showed that *V. cholerae* is able to survive for an extended period of time in rice and coffee or tea that has been prepared with condensed milk, which are popular street food and beverages in Malaysia. Poor hygiene in food and beverage preparations by food handlers may result in cross-contamination that is made worse as the contaminated food (especially rice) is usually held at ambient temperatures for long periods of time, thereby allowing the pathogens to multiply rapidly before the detectable spoilage sign is observed.

Acknowledgment

This research was supported by Universiti Sultan Zainal Abidin Research Grants UniSZA/11/GU(016) and UniSZA/11/GU(037).

References

[1] World Health Organization, "Cholera," Fact Sheets 107, WHO, Geneva, Switzerland, 2008, http://www.who.int/mediacentre/factsheets/fs107/en/.

[2] WHO, "Cholera," Global Health Observatory (GHO), 2009, http://www.who.int/gho/epidemic_diseases/cholera/en/.

[3] WHO, "Water sanitation and health," 2011, http://www.who.int/water_sanitation_health/diseases/cholera/en/index.html.

[4] M. G. Prouty and K. E. Klose, "*Vibrio cholera*: the genetics of pathogenesis and environmental persistence," in *The Biology of Vibrios*, F. L. Thompson, B. Austin, and J. Swings, Eds., pp. 311–339, American Society for Microbiology, 2006.

[5] M. Nishibuchi, "Molecular identification," in *The Biology of Vibrios*, F. L. Thompson, B. Austin, and J. Swings, Eds., chapter 4, pp. 44–64, American Society for Microbiology, 2006.

[6] V. K. Lim, "Cholera: a re-emerging infection," *Medical Journal of Malaysia*, vol. 56, no. 1, pp. 1–3, 2001.

[7] S. Mandal, M. D. Mandal, and N. K. Pal, "Cholera: a great global concern," *Asian Pacific Journal of Tropical Medicine*, vol. 4, no. 7, pp. 573–580, 2011.

[8] M. Maheshwari, K. Nelapati, and B. Kiranmayi, "*Vibrio cholerae*—a review," *Veterinary World*, vol. 4, no. 9, pp. 423–428, 2011.

[9] M. E. St. Louis, J. D. Porter, A. Helal et al., "Epidemic cholera in West Africa: the role of food handling and high-risk foods," *American Journal of Epidemiology*, vol. 131, no. 4, pp. 719–728, 1990.

[10] M. Ackers, R. Pagaduan, G. Hart et al., "Cholera and sliced fruit: probable secondary transmission from an asymptomatic carrier in the United States," *International Journal of Infectious Diseases*, vol. 1, no. 4, pp. 212–214, 1997.

[11] R. Tunung, S. P. Margaret, P. Jeyaletchumi et al., "Prevalence and quantification of *Vibrio parahaemolyticus* in raw salad vegetables at retail level," *Journal of Microbiology and Biotechnology*, vol. 20, no. 2, pp. 391–396, 2010.

[12] A. Ubong, R. Tunung, A. Noorlis et al., "Prevalence and detection of *Vibrio* spp. and *Vibrio cholerae* in fruit juices and flavored drinks," *International Food Research Journal*, vol. 18, no. 3, pp. 1163–1169, 2011.

[13] J. Y. H. Tang, K. L. Yap, and H. L. Hing, "Laboratory study of *Vibrio cholerae* O1 survival on three types of boiled rice (*Oryza sativa* L.) held at room temperature," *Journal of Food Protection*, vol. 71, no. 12, pp. 2453–2459, 2008.

[14] T. G. Boyce, E. D. Mintz, K. D. Greene et al., "*Vibrio cholerae* O139 Bengal infections among tourists to Southeast Asia: an intercontinental foodborne outbreak," *Journal of Infectious Diseases*, vol. 172, no. 5, pp. 1401–1404, 1995.

[15] J. M. Johnston, D. L. Martin, and J. Perdue, "Cholera on a Gulf Coast oil rig," *New England Journal of Medicine*, vol. 309, no. 9, pp. 523–526, 1983.

[16] A. A. Ries, D. J. Vugia, L. Beingolea et al., "Cholera in Piura, Peru: a modern urban epidemic," *Journal of Infectious Diseases*, vol. 166, no. 6, pp. 1429–1433, 1992.

[17] R. Khan, M. Zakir, S. H. Afaq, A. Latif, and A. U. Khan, "Activity of solvent extracts of *Prosopis spicigera, Zingiber officinale and Trachyspermum ammi* against multidrug resistant bacterial and fungal strains," *Journal of Infection in Developing Countries*, vol. 4, no. 5, pp. 292–300, 2010.

[18] A. Sebiomo, A. D. Awofodu, A. O. Awosanya, F. E. Awotona, and A. J. Ajayi, "Comparative studies of antibacterial effect of some antibiotics and ginger (*Zingiber officinale*) on two pathogenic bacteria," *Journal of Microbiology and Antimicrobials*, vol. 3, pp. 18–22, 2011.

[19] J. A. J. Sunilson, R. Suraj, G. Rejitha, K. Anandarajagopal, A. V. Anita Gnana Kumari, and P. Promwichit, "In vitro antimicrobial evaluation of *Zingiber officinale, Curcuma longa* and *Alpinia galanga* extracts as natural food preservatives," *American Journal of Food Technology*, vol. 4, no. 5, pp. 192–200, 2009.

[20] M. Shetty, K. Subbannayya, and P. G. Shivananda, "Antibacterial activity of tea (*Camellia sinensis*) and coffee (*Coffee arabica*) with special reference to *Salmonella typhimurium*," *Journal of Communicable Diseases*, vol. 26, no. 3, pp. 147–150, 1994.

[21] Public Health Agency of Canada, "*Vibrio cholerae*," 2001, http://www.phac-aspc.gc.ca/lab-bio/res/psds-ftss/vibrio-cholerae-eng.php.

[22] A. A. P. Almeida, A. Farah, D. A. M. Silva, E. A. Nunan, and M. B. A. Glória, "Antibacterial activity of coffee extracts and selected coffee chemical compounds against enterobacteria," *Journal of Agricultural and Food Chemistry*, vol. 54, no. 23, pp. 8738–8743, 2006.

[23] M. Daglia, A. Papetti, C. Dacarro, and G. Gazzani, "Isolation of an antibacterial component from roasted coffee," *Journal of Pharmaceutical and Biomedical Analysis*, vol. 18, no. 1-2, pp. 219–225, 1998.

[24] J. A. Rufián-Henares and S. P. de la Cueva, "Antimicrobial activity of coffee melanoidins—a study of their metal-chelating properties," *Journal of Agricultural and Food Chemistry*, vol. 57, no. 2, pp. 432–438, 2009.

[25] O. Felsenfeld, "Notes on food, beverages and fomites contaminated with *Vibrio cholerae*," *Bulletin of World Health Organization*, vol. 33, no. 5, pp. 725–734, 1965.

[26] M. Patel, M. Isaäcson, and E. Gouws, "Effect of iron and pH on the survival of *Vibrio cholerae* in water," *Transactions of the Royal Society of Tropical Medicine and Hygiene*, vol. 89, no. 2, pp. 175–177, 1995.

[27] D. Bandyopadhyay, T. K. Chatterjee, A. Dasgupta, J. Lourduraja, and S. G. Dastidar, "In vitro and in vivo antimicrobial action of tea: the commonest beverage of asia," *Biological and Pharmaceutical Bulletin*, vol. 28, no. 11, pp. 2125–2127, 2005.

[28] S. Mandal, M. DebMandal, N. K. Pal, and K. Saha, "Inhibitory and killing activities of black tea (*Camellia sinensis*) extract against *Salmonella enterica* serovar Typhi and *Vibrio cholerae* O1 biotype El Tor serotype Ogawa isolates," *Jundishapur Journal of Microbiology*, vol. 4, no. 2, pp. 115–121, 2011.

[29] G. H. Rabbani and W. B. Greenough III, "Food as a vehicle of transmission of cholera," *Journal of Diarrhoeal Diseases Research*, vol. 17, no. 1, pp. 1–9, 1999.

14

Extrusion Conditions and Amylose Content Affect Physicochemical Properties of Extrudates Obtained from Brown Rice Grains

Rolando José González,[1] Elena Pastor Cavada,[2] Javier Vioque Peña,[2] Roberto Luis Torres,[1] Dardo Mario De Greef,[1] and Silvina Rosa Drago[1,3]

[1] *Instituto de Tecnología de Alimentos, Universidad Nacional del Litoral, 1 de Mayo 3250, 3000 Santa Fe, Argentina*
[2] *Instituto de la Grasa (CSIC), Avenida Padre García Tejero 4, 41012 Seville, Spain*
[3] *Consejo Nacional de Investigaciones Científicas y Técnicas (CONICET), Avenida Rivadavia 1917,*
 1033 Ciudad Autónoma de Buenos Aires, Argentina

Correspondence should be addressed to Rolando José González; rolgonza@fiq.unl.edu.ar and
Silvina Rosa Drago; sdrago@fiq.unl.edu.ar

Academic Editor: Marie Walsh

The utilization of whole grains in food formulations is nowadays recommended. Extrusion cooking allows obtaining precooked cereal products and a wide range of ready-to-eat foods. Two rice varieties having different amylose content (Fortuna 16% and Paso 144, 27%) were extruded using a Brabender single screw extruder. Factorial experimental design was used to study the effects of extrusion temperature (160, 175, and 190°C) and grits moisture content (14%, 16.5%, and 19%) on extrudate properties. Specific mechanical energy consumption (SMEC), radial expansion (E), specific volume (SV), water absorption (WA), and solubility (S) were determined on each extrudate sample. In general, Fortuna variety showed higher values of SMEC and S (703–409 versus 637–407 J/g; 33.0–21.0 versus 20.1–11.0%, resp.) than those of Paso 144; on the contrary SV (8.64–3.47 versus 8.27–4.53 mL/g) and WA tended to be lower (7.7–5.1 versus 8.4–6.6 mL/g). Both varieties showed similar values of expansion rate (3.60–2.18). Physical characteristics depended on extrusion conditions and rice variety used. The degree of cooking reached by Paso rice samples was lower than that obtained for Fortuna. It is suggested that the presence of germ and bran interfered with the cooking process, decreasing friction level and broadening residence time distribution.

1. Introduction

The utilization of whole grains in food formulations is nowadays much recommended. The beneficial effects of including whole grains in the diet have been demonstrated by several authors [1]. Whole grains are rich in nutritive, functional, and phytochemical compounds [2].

Cereal food manufactures have responded to these advantages with the development of new whole grains and rich fiber products, most of them being in the form of flakes or blends such as muslin. Beside that, the demand of precooked cereal products, in the form of snacks, breakfast cereals, and flours, has increased during the last decades. Market globalization and changes on food consumption model of developing countries have been the main factors for this increasing demand [3].

The most used processes for the production of precooked cereal foods are flaking, puffing, and extrusion [4]. Extrusion cooking is considered not only an HTST process, but also a versatile one to obtain a wide range of ready-to-eat cereal foods. HTST process is recommended when keeping the nutritional value is important.

Among cereals, rice is one of the three most important. It is considered a staple food for more than 3,000 million people and provides them with about 60% of their daily energy intake [5]. The unique properties of rice, such as its hypoallergenicity, bland taste, white color, and easy digestion, make it a very desirable grain for new food development [6–9].

Rice is most consumed as grain in its three traditional forms: white, parboiled, and brown rice.

It is well known that much of the vitamins and minerals are lost during the polishing step of white rice processing. Even though protein content of rice is lower than that in maize and wheat, it is richer in lysine [10]. On the other hand, an increase of demand of nongluten products based on rice has been verified in the last decades [11], and there is a need of novel formulas in order to diversify the use of rice other than the traditional one.

Amylose content affects quality and texture of cooked rice [12, 13]. Rice varieties containing higher amylose content are normally more resistant to hydration [14] and give more viscous flour dispersion during water cooking [15]. Amylose content affecting the texture of gluten-free bread is controversial. It has been reported that flour from rice containing less than 20% amylose is desirable to use in bread without gluten formulas since it produces much less retrogradation [16], but other authors [7] have shown that the use of rice varieties, having medium and high amylose content (18–27%), gave good results for gluten-free bread. Moreover, Juliano [17] pointed out that waxy rice is preferred for cakes and pudding, because its flour is easy to hydrate and gives to the products better stability. On the other hand, high amylose rice variety is recommended for the production of noodles and is preferred for the production of expanded snacks products [15, 18, 19].

The advantages of extrusion as a cooking process are well known and have been discussed by several authors [20, 21]. The wide range of degree of cooking of starchy materials that can be obtained by extrusion is remarkable, since it is possible to produce samples with low degree of cooking (including loss of crystallinity and granule destruction) to fully cooked samples (highly destructed granules and total loss of native crystallinity) [18, 22, 23]. Water dispersion prepared with precooked flour obtained by extrusion under condition of high degree of cooking has much less viscosity than dispersion prepared with flour precooked by other processes, which can be considered a nutritional advantage [3].

The effects of extrusion variables on structural changes and product properties of starchy materials have been extensively studied by several authors [23–27] and particularly for rice [6–8, 19, 28–32]. However, studies regarding the processing effect and product properties from brown rice having different amylose content are scarce.

In the present work, two commercial rice varieties with different amylose content were selected to analyze the effect of extrusion variables on energy consumption and whole grains extrudate properties, using surface response methodology.

2. Materials and Methods

Two commercial long rice varieties, Paso 144 (28% amylose) and Fortuna (16% amylose), with length/width ratio of 3.5 and 2.3, respectively, were provided by Molino Trimacer (Santa Fe-Argentina) in the form of dehulled (or brown) grain. Brown rice was milled to obtain grits with a particle size between 1.190 and 0.420 mm, using a Buhler Miag, roll mill, according to a milling diagram [24, 33]. Moisture content, crude protein, fat content (expressed as petroleum ether extract), and ash content were determined by AOAC methods [34].

The extrusion process was carried out with a Brabender 20 DN single screw extruder, using a $4:1$ compression ratio screw, a 3/20 mm (diameter/length) die, and a screw speed of 150 rpm. The effects of grits moisture (M) and extrusion temperature (T) were analyzed by surface response methodology using a factorial design 3^2 with triplicate of the central point, resulting in 11 runs for each rice variety. The levels of each variable were as follows: T: 160–175–190°C and M: 14–16.5–19%. Rice grits samples were conditioned by adding water to reach the moisture level corresponding to each experimental sample, 1 h before each run.

Each extruded sample was obtained as soon as the stationary condition was reached, with torque and mass output simultaneously measured. These values were used to determine the specific mechanical energy consumption (SMEC) [23, 24] using the following formula: SMEC ($J g^{-1}$) = $k \cdot T \cdot N \cdot Qa^{-1}$, where k is: $61.6 \ 10^{-3}$, T is torque in Brabender units (BU), N is screw rpm, and Qa (g/min) is the mass output, referring to feeding moisture level. The k value takes into account unit conversion and constants. All extruded samples were air-dried in an oven at 50°C until a moisture content of 6% was reached, this moisture level being considered adequate for texture evaluation. Each dried sample was divided into several portions and kept in plastic bags hermetically sealed until their evaluation. Diameters were measured with a Vernier caliper on ten pieces of sample, and axial expansion (E) was determined as the ratio $E = D \cdot d^{-1}$, where D is the extrudate diameter (average of ten determinations) and d is the die diameter. Extrudate specific volume (SV) was obtained by calculating the volume/weight (d·b) ratio (as cm^3/g), corresponding to an extrudate piece of about 15 cm long. This procedure was applied to ten pieces and the average is reported.

Water solubility and water absorption were determined according to González et al. [23]. An amount (150 g) representative of each extrudate sample was first ground with a laboratory hammer mill (Retsch-Muhle-Germany) using a 2 mm sieve and then with a Cyclotec mill (UD Corp Boulder Colorado-USA) using a 1 mm sieve.

Water solubility was done by dispersing 2.5 g of flour in 50 mL water, agitating during 30 min. and centrifuging at 2000 g; soluble solids were obtained after evaporation in an oven at 105°C and calculated as soluble solids 100 g of flour (d·b). Water absorption was determined using Bauman method and expressed as mL of water/g of sample.

One extruded sample from each rice variety was selected to evaluate flour dispersion viscosity, and rheograms were carried out for two solids concentrations (8 and 11%, W/W), at 60°C, using an RV3 Haake Rotovisco viscometer (Germany). In every case, power law parameters k and n ($\tau = kG^n$) were estimated by regression.

2.1. Statistical Analysis. The average of duplicate of each determination is reported. Analysis of Variance was carried

TABLE 1: Composition of raw rice[**].

Sample	Proteins[*] ($g\,kg^{-1}$)	Ash[*] ($g\,kg^{-1}$)	Moisture ($g\,kg^{-1}$)	Lipids (ether extract)[*] ($g\,kg^{-1}$)	Total Dietary Fibre[*] ($g\,kg^{-1}$)
Fortuna	71.5 ± 2.0	11.8 ± 0.4	118.6 ± 1.2	19.8 ± 0.3	52.7 ± 1.7
Paso	75.9 ± 0.9	15.0 ± 0.1	115.7 ± 0.7	23.6 ± 0.4	55.3 ± 1.7

[*] Dry base; [**] average $\pm$ SD.

TABLE 2: Specific mechanical energy consumption (SMEC), expansion (E), specific volume (SV), water absorption (WA), and solubility ($S\%$) corresponding to the eleven experimental extruded samples and for the two rice varieties.

Extrusion conditions		SMEC (J/g)		E		SV (mL/g)		WA (mL/g)		$S\%$	
M (%)	T (°C)	F	P	F	P	F	P	F	P	F	P
14	160	703	637	3.36	3.46	5.39	7.26	5.1	6.6	33.0	20.1
14	175	671	590	3.38	3.50	6.52	8.27	5.9	8.4	30.8	17.0
14	190	620	493	3.36	3.42	8.64	7.89	6,5	7.9	31.9	17.5
16.5	160	609	585	3.60	3.40	5.43	6.12	5.7	6.9	24.9	19.5
16.5	175	547	575	3.14	3.01	5.22	6.65	7.5	7.5	23.4	13.5
16.5	175	534	543	3.03	3.16	5.65	6.98	7.3	7.3	23.3	12.1
16.5	175	541	566	3.12	3.17	5.29	6.90	7.2	7.22	25.8	12.9
16.5	190	541	483	2.89	2.95	7.23	7.83	7.0	8.3	27.5	17.2
19	160	483	511	2.55	2.92	3.47	4.53	6.0	7.6	21.5	14.8
19	175	462	501	2.77	2.76	4.29	5.39	7.2	7.0	21.0	11.1
19	190	409	407	2.46	2.18	5.94	5.84	7.7	8.1	26.1	15.5

M: moisture; T: temperature; F: Fortuna rice variety; P: Paso 144 rice variety.

out using the software Statgraphics Plus 3.0, and the statistical differences among samples were determined using the LSD test.

3. Results and Discussion

Table 1 shows the composition of the raw materials. The values are similar to those reported in the literature for brown rice [35].

Table 2 shows results of specific mechanical energy consumption (SMEC), expansion, specific volume (SV), water absorption (WA), and solubility ($S\%$) corresponding to the eleven experimental extruded samples and for the two rice varieties.

Table 3 shows the degree of significance (P value), corresponding to each term of the regression model obtained for each response. The lack of fit was not significant ($P > 0.5$), except for WA corresponding to Paso 144.

It is observed that, for all responses, one or more terms of the regression model were significant ($P < 0.05$). Linear terms M and T were significant in all cases, except for M for WA corresponding to Paso 144 ($P < 0.6779$) and T for solubility corresponding to both rice varieties ($P < 0.2544$ for F and $P < 0.1337$ for P). Quadratic terms (M^2 and T^2) were significant only in the following cases: (a) M^2 in expansion and WA for Fortuna variety and in expansion for Paso 144 variety (b) T^2 in SV for Fortuna and also in $S\%$ for Paso 144.

3.1. Specific Mechanical Energy Consumption. SMEC surface response corresponding to each rice variety is observed in Figures 1(a) and 1(b).

The surface corresponding to Fortuna is planar, but that of Paso 144 shows a curvature in the T direction. According to the ANOVA results (Table 2), only the linear terms (H and T) were significant ($P < 0.05$), but, in the case of Paso 144, the degree of significance of the term T^2 is not negligible ($P < 0.097$) and that explains the curvature in T direction. In both cases, SMEC was inversely related to both H and T, since as H and T increased, friction level in the extruder decreased and consequently SMEC decreased [23, 36]. Several authors working with others materials, such as wheat, quinoa, or barley, have found similar effect [32, 37–39].

SMEC values of both varieties were in similar range, although those corresponding to Fortuna tend to be slightly higher, particularly at low M level, which is in agreement with the results of water solubility (Table 2) which is discussed later on this paper. Since starch fraction of Fortuna rice variety reached higher degree of cooking during extrusion (higher S), energy dissipation (SMEC) would be also higher than that in the case of higher amylose content. Moreover, the presence of fiber and amylose content would reduce SMEC, since when low and high amylose polished rice varieties are extruded, higher level of solubility was obtained [22].

3.2. Radial Expansion. Radial expansion values of both varieties were in a similar range. Expansion response surfaces corresponding to each rice variety are observed in Figures 2(a) and 2(b). In the case of Fortuna, the curvature in M direction is explained by the significance of the term M^2 ($P < 0.0448$). While in the case of Paso 144, the expected plane is modified by the influence of the term $M \times T$, whose significance is not negligible (0.0614).

Extrusion Conditions and Amylose Content Affect Physicochemical Properties of Extrudates Obtained from
Brown Rice Grains

103

TABLE 3: Degrees of significance (P value) of each term of the regression model (source of variation), corresponding to each response and for the two rice varieties.

Source of variation	SMEC		E		SV		WA		S%	
	f	p	f	p	f	f	f	p	f	p
M	**0.0006**	**0.0174**	**0.0043**	**0.0043**	**0.0067**	**0.0030**	**0.0043**	0.6779	**0.0204**	**0.0169**
T:	**0.0048**	**0.0129**	**0.0351**	**0.0351**	**0.0055**	**0.0129**	**0.0351**	**0.0128**	0.2554	0.1337
M^2	0.7563	0.2007	**0.0448**	**0.0448**	0.2361	0.0968	**0.0448**	0.2160	0.3121	0.7205
$M \times T$	0.5381	0.3411	0.5534	0.5534	0.2331	0.1886	0.5534	0.1270	0.2171	0.1510
T^2	0.1403	0.0978	0.9086	0.9086	**0.0415**	0.2467	0.9086	0.8320	0.2432	**0.0112**
Lack of Fit	0.0794	0.7997	0.0515	0.0515	0.2398	0.1593	0.0515	**0.0396**	0.8966	0.1951

M: moisture; T: temperature; F: Fortuna rice variety; P: Paso 144 rice variety; specific mechanical energy consumption (SMEC); expansion (E); specific volume (SV); water absorption (WA); and solubility (S%).

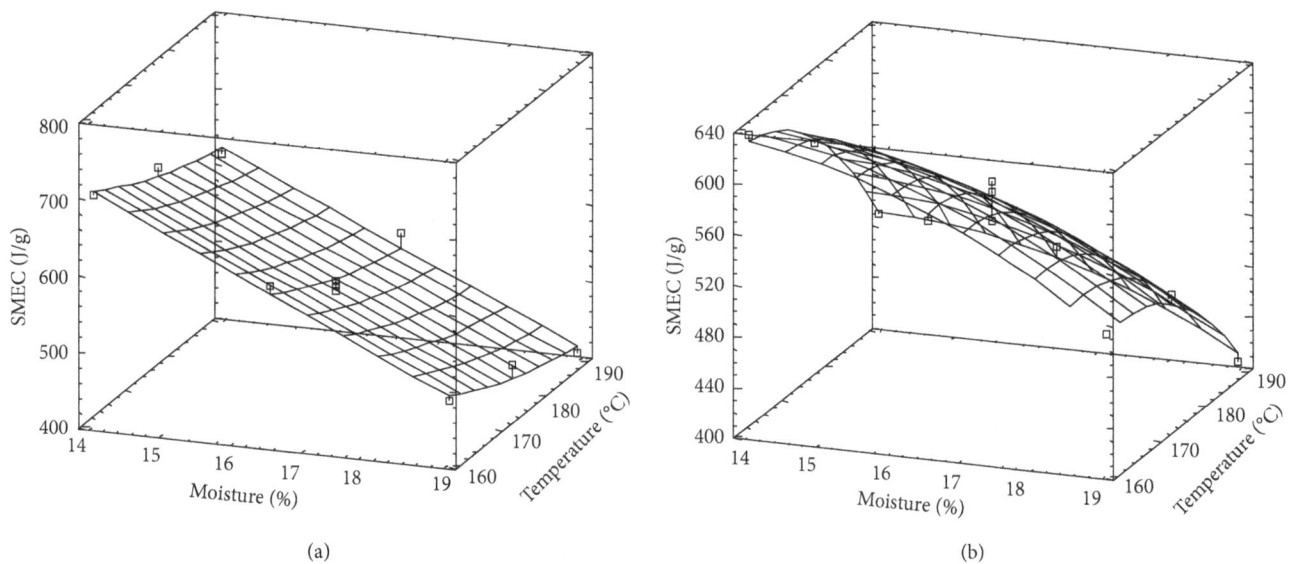

(a)

(b)

FIGURE 1: Surface response for SMEC corresponding to Fortuna rice variety (a) and Paso 144 rice variety (b).

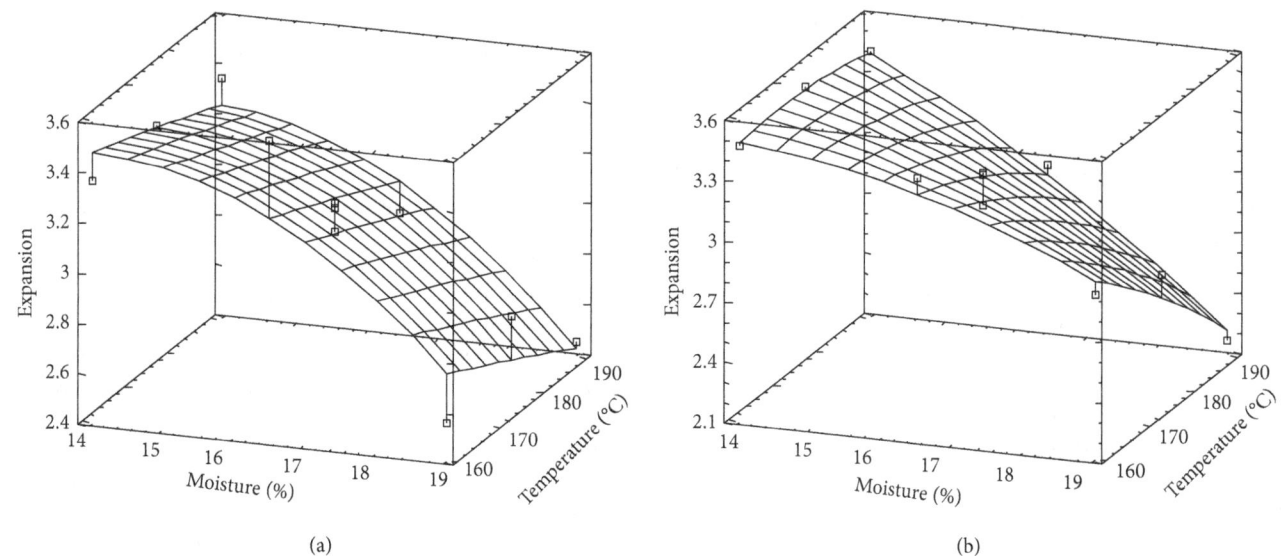

(a)

(b)

FIGURE 2: Surface response for expansion corresponding to Fortuna rice variety (a) and Paso 144 rice variety (b).

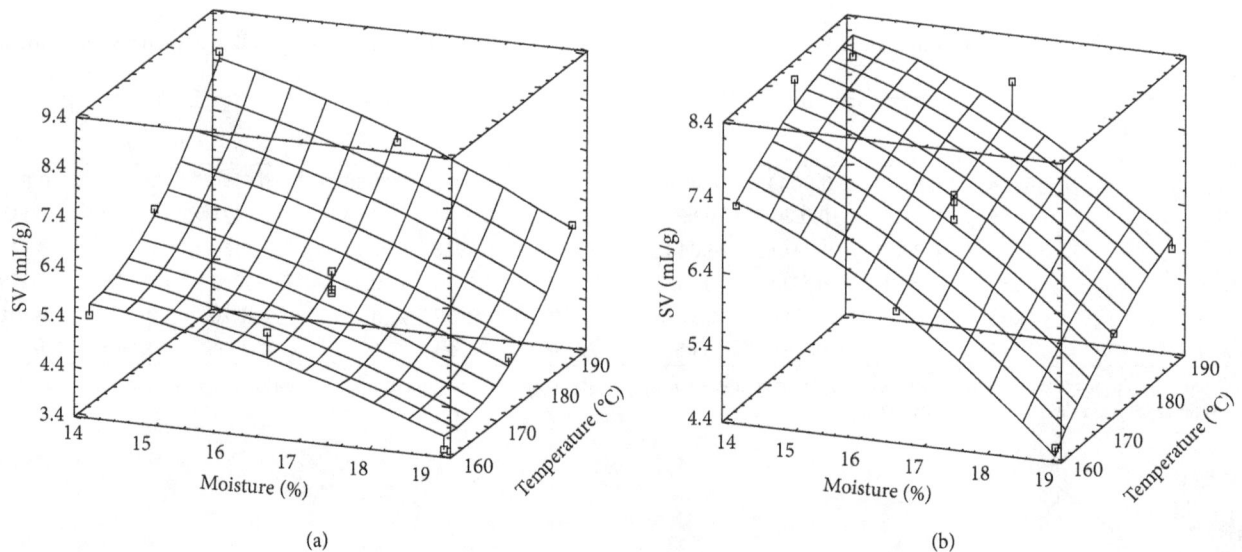

FIGURE 3: Surface response for specific volume (SV) corresponding to Fortuna rice variety (a) and Paso 144 rice variety (b).

Both varieties show similar tendency as that of SMEC, supporting the idea that expansion rate is directly related to the degree of friction inside the extruder [23, 40]. As it is expected, expansion values corresponding to both varieties are higher at the lower level of M, since this extrudate property is positively correlated with the elastic component of the melt coming out from the die, which would decrease as T or M increases [3]. This inverse relationship between E and M was also found by Hagenimana et al. [6] and Pedrosa Silva Clérici and El-Dash [7]. Regarding T effect, Hagenimana et al. [6] and Singh et al. [32] have shown that a maximum of E exists at intermediate T (130–150°C). However, this temperature range is lower than that we used.

3.3. Specific Volume.
Specific Volume has been proposed as a good indicator of degree of cooking (DC) for cereal extruded products, because DC is directly related to granule structure destruction [23].

In most of the cases, SV values corresponding to Paso were slightly higher than those of Fortuna. Response surfaces corresponding to each rice variety are observed in Figures 3(a) and 3(b). In the case of Fortuna, the curvature in T direction is explained by the significance of the term T^2 ($P < 0.0415$), which also can explain the high dependence of this response on temperature. In the case of Paso 144, the expected planar surface is modified by the term M^2 whose significance is not too low ($P < 0.0968$).

The effects of variables followed similar tendency for both varieties. SV decreases as M increases, while an inverse effect is observed for T. This tendency could be explained taking into account that as M increases and T decreases, DC decreases and consequently SV decreases. Similar tendency was also found for other materials such as barley and corn-bean mixtures [6, 23, 36]. Moreover, González et al. [41] have found that as the extrusion temperature increases, extrudate structure became lighter, as a consequence of the reduction of wall thickness of the pores.

It is interesting to point out that the effects of T on SV is higher for Fortuna than for Paso 144. This is in agreement with the results obtained by González et al. [15] which have shown that rice variety, containing lower amylose content, are more sensible to thermal effects than those of higher amylose content.

3.4. Water Absorption.
Water absorption values corresponding to Paso are higher than those of Fortuna. According to the ANOVA results, the lack of fit corresponding to Paso 144 was significant ($P < 0.0396$); thus the regression model does not explain adequately the response variability, suggesting that some other factor, not controlled during the experiment, played some role.

For Fortuna rice, WA was directly related to both variables and the highest value corresponded to the sample obtained at 190°C and 19% (Figure 4). There is an important difference in the tendency observed for the moisture effect, because other responses evaluated, decreased as M increased. This would suggest that WA is more affected by thermal effect than that of friction. It is known that as extrusion conditions minimize friction effects, destruction of granular structure is also minimized and consequently WA is improved [18].

3.5. Solubility.
S values corresponding to Paso were significantly lower than those of Fortuna, which is in agreement with the results obtained by other authors, who have found an inverse relation between amylose content and solubility [6, 8, 19, 29, 32]. Moreover, it is also important to point out that as DC increases, S increases as a consequence of starch granule destruction, implying a reduction of WA [23].

Solubility (S) response surfaces corresponding to each rice variety are observed in Figures 5(a) and 5(b).

In the case of Fortuna, the surface is almost a plane and S is only affected by M, suggesting that friction effects are the most important ones. As it was observed by other authors [23, 42], S is directly related with SMEC and SV. Again, the

Extrusion Conditions and Amylose Content Affect Physicochemical Properties of Extrudates Obtained from
Brown Rice Grains

105

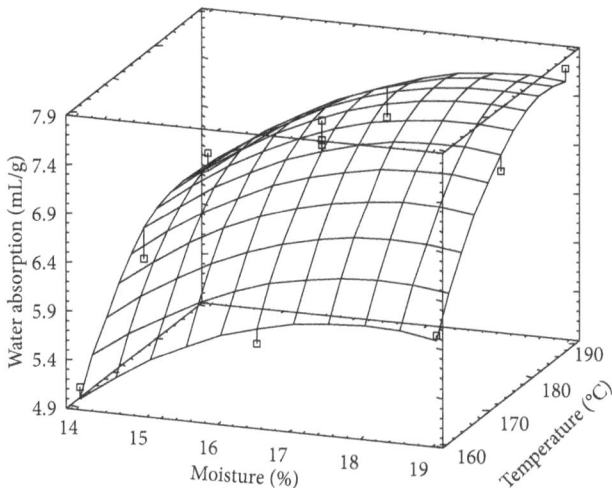

FIGURE 4: Surface response for water absorption corresponding to Fortuna rice variety.

lower amylose content of this variety could explain the high sensitivity to friction.

In the case of Paso 144, the surface shows a curvature in T direction, as it is expected, since T^2 term was highly significant ($P < 0.00112$). As it was observed for WA, S decreases as M increases and similar explanation could be done regarding friction effects. The highest S values corresponded to samples obtained at low M level (14%) and at 160°C and 190°C.

The existence (at any M level in the range of this study) of minimum values of S, at intermediate T level (175°C), is difficult to explain. S results suggest that the behavior of both varieties inside the extruder would be different. On one hand, Fortuna shows much higher S values than Paso 144, indicating that a higher degree of cooking was reached by Fortuna rice. This is in agreement with that expected for the variety with lower amylose content. Moreover, in the extrusion of rice grits obtained by milling white rice (without the germ and the bran) and for similar extrusion conditions [15], S values were much higher than those obtained in this study, suggesting that lower degree of cooking is obtained when grits from whole grains is used in comparison with those coming from degermed and dehulled grains. Regarding that, Pan et al. [19] reported that the addition of oil and bran to rice grits caused a reduction of energy consumption, indicating a reduction of degree of cooking.

Beside that, DC is, in general, directly related to T; in our case, it is true only at T higher than 175°C and for any moisture level. It seems that an increase of T from 160°C to 175°C produces a reduction of DC (lower S), so the effect of the reduction of friction caused by this increase of T would be higher than the thermal effect on DC. Then, above 175°C the relative importance of these two effects is inversed and the expected effect of T is verified.

These results would indicate that the presence of germ (oil) and bran (fiber) could reduce the DC reached by the extrudate. This was confirmed by sample observation under microscope, which revealed the presence of native starch in all Paso samples. Even though no quantitative determination

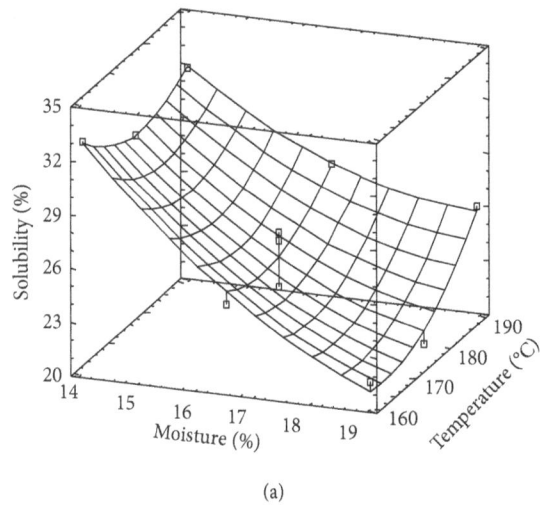

(a)

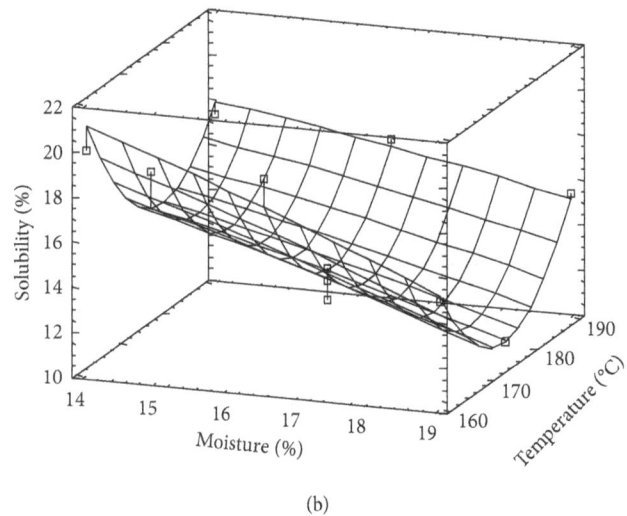

(b)

FIGURE 5: Surface response for solubility corresponding to Fortuna rice (a) and Paso 144 rice variety (b).

was done, the proportion of native starch present in samples corresponding to Paso 144 was much higher than that corresponding to Fortuna; moreover, no native starch was observed in Fortuna samples obtained at low-level moisture (14%). This incomplete cooking process is attributed to per- turbation of the particles transport inside the extruder that not only could retard the cooking process of starchy particles caused by friction, but also could broaden the residence time distribution of particles inside the extruder. So, some particles would reach the die very fast, without suffering much change (remaining in native state). This interference on the cooking process of starch caused by the presence of oil and fiber could be overlapped by changing the extrusion condition in order to increase DC, for example, reduce die diameter, increase screw compression ratio, reduce moisture content, and reduce particle size.

3.6. Dispersion Viscosity Evaluation. Since the milled extru- date could be used as a base for gruel and cream soup formulations, dispersion viscosity becomes an important factor to be taken into account.

TABLE 4: Power law parameters (k and n), regression coefficient (r), and viscosity at 135 s^{-1} (Pa·s^{-1}), corresponding to flour dispersion at 8% and 11% solid concentration and for each rice varieties.

Sample	W/W %	τ_0 (Pa)	k	n	r	η 135 s^{-1}(Pa·s^{-1})
Fortuna	8	0	0,436	0,497	0,9948	0,037
	11,0	4,5	1,053	0,636	0,9999	0,210
Paso	8	0	0,686	0,433	0,9994	0,042
	11,0	7,2	3,670	0,550	0,9988	0,451

The evaluation of dispersion viscosity was done on the samples obtained at 14% and 160°C, since these extrusion conditions permitted the highest DC. Viscosity was obtained from rheograms at 135 s^{-1} shear rate according to Pérez et al. [3] since this level of shear rate would be in range of that exerted inside the mouth when a fluid food is ingested. Table 4 shows the results of extrudate flour viscosity evaluation: power law parameters (k and n) and regression coefficient obtained from the rheograms made for each solid concentration (8% and 11%) and for each rice varieties.

It is observed that power law model adjusted adequately the rheograms data ($r > 0.994$). Dispersion prepared at 11% solids concentration showed plastic behavior (yield stress). Dispersion corresponding to Fortuna variety showed lower viscosity than Paso, indicating that samples having lower degree of cooking give thicker dispersion. Beside that, a direct correlation was observed between water absorption and viscosity ($r > 0.945$), as it was expected, since water absorption is an indicator of dispersion thickness. These results are in agreement with those obtained in other works [3, 43].

4. Conclusions

Physical characteristics of extrudates depend on extrusion conditions and rice variety used. Low amylose variety showed values of SMEC, SV, and S higher than those of higher content; on the contrary, WA was lower, indicating that the last one reached lower degree of cooking. Both varieties showed similar values of expansion rate.

Regarding the extrusion conditions, 14% M and 175°C could be selected to obtain a good expanded product, with high expansion and intermediate degree of cooking.

Since degrees of cooking reached by both rice samples (under the same extrusion condition) were lower than those obtained using grits from polish rice, it is suggested that the presence of germ and bran interferes the cooking process by both decreasing friction level and by broadening residence time distribution.

The higher protein, fiber, and mineral content of brown rice in comparison to polish rice implies nutritional advantages, so the use of brown rice to produce snacks products is highly recommended.

Acknowledgments

This work supported by PICT 1105 and CAI+D 2009 PI 54-259. The authors thank Oliva D. A. and Bonaldo A. G. for the technical support.

References

[1] D. P. Richardson, "Wholegrain health claims in Europe," *Proceedings of the Nutrition Society*, vol. 62, no. 1, pp. 161–169, 2003.

[2] J. L. Slavin, "Whole grains and human health," *Nutrition Research Reviews*, vol. 17, pp. 99–110, 2004.

[3] A. A. Pérez, S. R. Drago, C. R. Carrara, D. M. De Greef, R. L. Torres, and R. J. González, "Extrusion cooking of a maize/soybean mixture: factors affecting expanded product characteristics and flour dispersion viscosity," *Journal of Food Engineering*, vol. 87, no. 3, pp. 333–340, 2008.

[4] M. Mariotti, C. Alamprese, M. A. Pagani, and M. Lucisano, "Effect of puffing on ultrastructure and physical characteristics of cereal grains and flours," *Journal of Cereal Science*, vol. 43, no. 1, pp. 47–56, 2006.

[5] K. F. Nwanze, "International fund for agricultural development," in *Proceedings of the 3rd International Rice Congress*, Hanoi, Vietnam, 2010.

[6] A. Hagenimana, X. Ding, and T. Fang, "Evaluation of rice flour modified by extrusion cooking," *Journal of Cereal Science*, vol. 43, no. 1, pp. 38–46, 2006.

[7] M. T. Pedrosa Silva Clérici and A. A. El-Dash, "Características tecnológicas de farinhas de arroz pregelatinizadas obtidas por extrusão termoplástica," *Ciência e Agrotecnologia*, vol. 32, no. 5, pp. 1543–1550, 2008.

[8] S. H. Yang, J. Peng, W. B. Lui, and J. Lin, "Effects of adlay species and rice flour ratio on the physicochemical properties and texture characteristic of adlay-based extrudates," *Journal of Food Engineering*, vol. 84, no. 3, pp. 489–494, 2008.

[9] H. Yano, "Improvements in the bread-making quality of gluten-free rice batter by glutathione," *Journal of Agricultural and Food Chemistry*, vol. 58, no. 13, pp. 7949–7954, 2010.

[10] K. F. Kiple, *Cambridge World Encyclopaedia of Food, Volume I, Animal, Marine and Vegetable Oils*, Cambridge University Press, Cambridge, UK, 2000.

[11] H. D. Sanchez, C. A. Osella, and M. A. De la Torre, "Optimization of gluten-free bread prepared from cornstarch, rice flour, and cassava starch," *Journal of Food Science*, vol. 67, no. 1, pp. 416–419, 2002.

[12] B. O. Juliano, "Varietal impact on rice quality," *Cereal Foods World*, vol. 43, no. 4, pp. 207–222, 1998.

[13] C. M. Sowbhagya, B. S. Ramesh, and S. Z. Alí, "Hydration, Swelling and solubility behaviour of rice in relation to other physicochemical properties," *Journal of the Science of Food and Agriculture*, vol. 64, pp. 1–7, 1994.

[14] B. K. Yadav and V. K. Jindal, "Water uptake and solid loss during cooking of milled rice (Oryza sativa L.) in relation to its physicochemical properties," *Journal of Food Engineering*, vol. 80, no. 1, pp. 46–54, 2007.

[15] R. J. González, R. L. Torres, and D. M. De Greef, "El arroz como Alimento: el grano y la harina, parámetros de caracterización

y de calidad," in *El Arroz: Su Cultivo y Sustentabilidad en Entre Ríos*, R. Benavides, Ed., chapter 1, pp. 19–52, 2006.

[16] M. M. Ben and K. D. Nishita, "Rice flours for baking," in *Rice Chemistry and Technology*, B. O. Juliano, Ed., pp. 539–556, American Association for Clinical Chemistry, St. Paul, Minn, USA, 1985.

[17] B. O. Juliano, "Sakurai, Miscellaneous rice products," in *Rice Chemistry and Technology*, B. O. Juliano, Ed., pp. 569–618, American Association for Clinical Chemistry, St. Paul, Minn, USA, 1985.

[18] R. J. González, R. L. Torres, and D. M. De Greef, "Comportamiento a la cocción de variedades de arroz y maíz utilizando el amilógrafo y dos diseños de extrusores," *Información Tecnológica*, vol. 9, pp. 35–43, 1998.

[19] B. S. Pan, M. S. Kong, and H. H. Chen, "Twin-Screw extrusion for expanded rice products: processing parameters and formulation of extrudate properties," in *Food Extrusion Science and Technology*, J. L. Kokini, C. T. Ho, and M. V. Karwe, Eds., pp. 693–709, Marcel Dekker, New York, NY, USA, 2002.

[20] J. M. Harper, "Food extruders and their applications," in *Extrusion Cooking*, C. Mercier, P. Linko, and J. M. Harper, Eds., pp. 1–16, American Association of Cereal Chemists, St. Paul, Minn, USA, 1989.

[21] M. Bhattacharya, S. Y. Zee, and H. Corke, "Physicochemical properties related to quality of rice noodles," *Cereal Chemistry*, vol. 76, no. 6, pp. 861–867, 1999.

[22] R. J. González, R. L. Torres, and M. C. Añón, "Comparison of rice and corn cooking characteristics before and after extrusion," *Polish Journal of Food And Nutrition Sciences*, vol. 9, no. 50, pp. 29–34, 2000.

[23] R. J. González, R. L. Torres, and D. M. De Greef, "Extrusión-cocción de cereales," *Sociedade Brasileira de Ciência e Tecnologia de Alimentos*, vol. 36, no. 2, pp. 104–115, 2002.

[24] R. J. Gonzalez, R. L. Torres, and D. M. De Greef, "El arroz como Alimento: el grano y la harina, parámetros de caracterización y de calidad," in *El Arroz: Su Cultivo y Sustentabilidad en Entre Ríos*, R. Benavides, Ed., chapter 1, pp. 19–52, 2006.

[25] R. J. Bryant, R. S. Kadan, E. T. Champagne, B. T. Vinyard, and D. Boykin, "Functional and digestive characteristics of extruded rice flour," *Cereal Chemistry*, vol. 78, no. 2, pp. 131–137, 2001.

[26] R. S. Kadan, R. J. Bryant, and A. B. Pepperman, "Functional properties of extruded rice flours," *Journal of Food Science*, vol. 68, no. 5, pp. 1669–1672, 2003.

[27] G. Sacchetti, G. G. Pinnavaia, E. Guidolin, and M. D. Rosa, "Effects of extrusion temperature and feed composition on the functional, physical and sensory properties of chestnut and rice flour-based snack-like products," *Food Research International*, vol. 37, no. 5, pp. 527–534, 2004.

[28] R. S. Kadan and A. B. Pepperman, "Physicochemical properties of starch in extruded rice flours," *Cereal Chemistry*, vol. 79, no. 4, pp. 476–480, 2002.

[29] M. T. Pedrosa Silva Clerici and A. El-Dash, "Farinha extrusada de arroz comosubstituto de glúten na produção de pão de arroz," *Alan*, vol. 56, no. 3, pp. 288–298, 2006.

[30] H. D. Sánchez, R. J. Gonzalez, C. A. Osella, R. L. Torres, and M. A. G. De la Torre, "Elaboración de pan sin gluten con harinas de arroz extrudidas," *Ciencia y Tecnología Alimentaria*, vol. 6, no. 2, pp. 109–116, 2008.

[31] Q. B. Ding, P. Ainsworth, G. Tucker, and H. Marson, "The effect of extrusion conditions on the physicochemical properties and sensory characteristics of rice-based expanded snacks," *Journal of Food Engineering*, vol. 66, no. 3, pp. 283–289, 2005.

[32] B. Singh, K. S. Sekhon, and N. Singh, "Effects of moisture, temperature and level of pea grits on extrusion behaviour and product characteristics of rice," *Food Chemistry*, vol. 100, no. 1, pp. 198–202, 2007.

[33] J. Robutti, F. Borrás, R. González, R. Torres, and D. De Greef, "Endosperm properties and extrusion cooking behavior of maize cultivars," *LWT—Food Science and Technology*, vol. 35, no. 8, pp. 663–669, 2002.

[34] AOAC, *Official Methods of Analysis*, Association of Official Analytical Chemists, Arlington, Tex, USA, 16 edition, 1999.

[35] F. Senser and H. Scherz, *El Pequeño "Souci-Fachmann-Kraut"*. *Tablas de Composición de Alimentos*, Acribia, Zaragoza, España, 1999.

[36] J. Ruiz-Ruiz, A. Martínez-Ayala, S. Drago, R. González, D. Betancur-Ancona, and L. Chel-Guerrero, "Extrusion of a hard-to-cook bean (Phaseolus vulgaris L.) and quality protein maize (Zea mays L.) flour blend," *LWT—Food Science and Technology*, vol. 41, no. 10, pp. 1799–1807, 2008.

[37] G. H. Ryu and P. K. Ng, "Effects of selected process parameters on expansion and mechanical properties of wheat flour and whole cornmeal extrudates," *Starch*, vol. 53, pp. 147–154, 2001.

[38] H. Köksel, G. H. Ryu, A. Başman, H. Demiralp, and P. K. W. Ng, "Effects of extrusion variables on the properties of waxy hulless barley extrudates," *Nahrung—Food*, vol. 48, no. 1, pp. 19–24, 2004.

[39] H. Doğan and M. V. Karwe, "Physicochemical properties of quinoa extrudates," *Food Science and Technology International*, vol. 9, no. 2, pp. 101–114, 2003.

[40] S. J. Mulvaney, C. Onwulata, Q. Lu, F. Hsieh, and J. J. Brent, "Effect of screw speed on the process dynamics and product properties of cormeal for a twin extruder," in *Food Extrusion Science and Technology*, J. L. Kokini, C. T. Ho, and M. V. Karwe, Eds., chapter 34, pp. 527–537, Marcel Dekker, New York, NY, USA, 1992.

[41] R. J. González, D. M. De Greef, R. L. Torres, and N. A. Gordo, "Efectos de algunas variables de extrusión sobre la harina de maíz," *Alan*, vol. 37, no. 3, pp. 578–591, 1987.

[42] A. Smith, "Studies on the physical structure of starch based materials in the extrusion cooking process," in *Food Extrusion Science and Technology*, J. L. Kokini, C. T. Ho, and M. V. Karwe, Eds., chapter 36, pp. 573–618, Marcel Dekker, New York, NY, USA, 1992.

[43] R. J. González, R. L. Torres, D. M. De Greef, N. A. Gordo, and M. E. Velocci, "Influencia de las condiciones de extrusión en las características de la harina de maíz para elaborar sopas instantâneas," *Revista Agroquímica y Tecnología. Alimentaria*, vol. 31, no. 1, pp. 87–96, 1991.

15

Ethoxyquin: An Antioxidant Used in Animal Feed

Alina Błaszczyk,[1] Aleksandra Augustyniak,[1] and Janusz Skolimowski[2]

[1] Department of General Genetics, Molecular Biology and Plant Biotechnology, Faculty of Biology and Environmental Protection, University of Łódź, Banacha 12/16, 90-237 Łódź, Poland
[2] Department of Organic Chemistry, Faculty of Chemistry, University of Łódź, Tamka 12, 91-403 Łódź, Poland

Correspondence should be addressed to Alina Błaszczyk; ablasz@biol.uni.lodz.pl

Academic Editor: Ángel Medina-Vayá

Ethoxyquin (EQ, 6-ethoxy-1,2-dihydro-2,2,4-trimethylquinoline) is widely used in animal feed in order to protect it against lipid peroxidation. EQ cannot be used in any food for human consumption (except spices, e.g., chili), but it can pass from feed to farmed fish, poultry, and eggs, so human beings can be exposed to this antioxidant. The manufacturer Monsanto Company (USA) performed a series of tests on ethoxyquin which showed its safety. Nevertheless, some harmful effects in animals and people occupationally exposed to it were observed in 1980's which resulted in the new studies undertaken to reevaluate its toxicity. Here, we present the characteristics of the compound and results of the research, concerning, for example, products of its metabolism and oxidation or searching for new antioxidants on the EQ backbone.

1. Introduction

During storage of animal feed many different processes may occur which alter their initial natural proprieties. First of all, lipids undergo peroxidation, the process during which they are deteriorated in a free radical autocatalytic oxidation chain reaction with atmospheric oxygen. Lipid autooxidation is a cascade phenomenon ensuring continuous delivery of free radicals, which initiate continuous peroxidation. This process results in food rancidity which manifests itself as the change of taste, scent, and color as well as decrease in shelf life of the product. Natural or synthetic antioxidants are usually used to slow down or stop lipid peroxidation and in consequence to preserve freshness of the product. Many natural antioxidants, such as tocopherols, vitamin C, flavonoids, for a short period, may be effective in food preserving, but in many cases such protection is not sufficient. Therefore synthetic antioxidants are widely used, among which BHT (butylated hydroxytoluene), BHA (butylated hydroxyanisole), and EQ (ethoxyquin) are the most frequent. However, some effects of synthetic antioxidants are not always beneficial for our health. Antioxidants such as BHA or BHT have been widely used for many years to preserve freshness, flavor, and colour of foods and animal feeds as well as to improve the stability of

pharmaceuticals and cosmetics. There are many controversies about the use of these two antioxidants in foods. Some experimental studies have reported that both BHT and BHA have tumour-promoting activity [1, 2]. On the other hand, there were reports on anticarcinogenic properties of these antioxidants when they are used at low concentrations [3]. Human exposures are at least 1000-fold below those associated with any neoplastic actions in laboratory animals thus it is assumed that they are not harmful for human beings [3, 4].

The third compound, EQ, is one of the best known feed antioxidants for domestic animal and fish. Its unquestionable advantage is its high antioxidant capacity and low production costs. However, some of the authors have suggested that it is responsible for a wide range of health-related problems in dogs as well as in humans [5–9]. Due to the increased use of this antioxidant it was nominated by FDA (US Food and Drug Administration) for carcinogenicity testing [10]. The tests were carried out by Monsanto Company (USA), EQ producer, and after that in 1977 FDA requested for optional lowering of the maximum level of EQ in complete dog foods from allowed 150 ppm (0.015%) to 75 ppm (0.0075%). At the same time new studies were started by the Pet Food Institute to determine whether even lower EQ levels (between 30 and

60 ppm) would provide antioxidant protection for dog food [11].

Ethoxyquin is also known as Santoquin, Santoflex, Quinol. It was originally developed in rubber industry to prevent rubber from cracking due to oxidation of isoprene [12]. The Monsanto Company (USA) taking into account its high antioxidant efficiency and stability as well as low costs of synthesis refined it later for use as a preservative in animal feeds because it protects against lipid peroxidation and stabilizes fat soluble vitamins (A, E). Presently, ethoxyquin is used primarily as an antioxidant in canned pet food and in feed intended for farmed fish or poultry.

The use of ethoxyquin is not permitted in foods intended for human, except preserving powdered paprika and chili colour and using it as an antiscald agent in pears and apples (inhibition of "brown spots" development). However, because EQ is used as a feed antioxidant it can be also found in other products intended for human consumption like fish meal, fish oils, and other oils, fats, and meat (Table 1). An acceptable daily intake (ADI) of EQ for human (0–0.005 mg kg^{-1} bw) based on the results obtained from studies on dogs was established in 1998 [13, 14].

This paper presents characteristics of ethoxyquin with regard to its properties, metabolism, toxicity, possible carcinogenicity, and antioxidant activity.

2. Physical and Chemical Properties of Ethoxyquin (EQ)

For the first time EQ was synthesized in 1921 by Knoevenagel [15]. The synthesis was based on condensation of aniline with molecules of acetone or its analogues [15, 16]. Błaszczyk et al. [16] synthesized EQ from p-phenetidine (4-ethoxyaniline) and diacetone alcohol in the presence of p-toluenesulfonic acid or iodine. Pure ethoxyquin (EQ; 6-ethoxy-1,2-dihydro-2,2,4-trimethylquinoline; CAS number 91-53-2; Figure 1) is a light yellow liquid, but it changes color to brown if it is exposed to oxygen [12]. It also tends to polymerize on exposure to light and air. The scent of EQ is described as mercaptan like. As a nonpolar substance EQ is soluble only in organic solvents. Some of the additional properties of EQ are presented in Table 2.

3. Biotransformation of EQ

3.1. Animals. Ethoxyquin is rapidly absorbed from gastrointestinal tract of laboratory animals like rats and mice. Peak blood concentration of the compound is observed within 1 h. Distribution of EQ in animal body is similar when it is administered orally and intravenously. Small amounts of parent EQ were detected in liver, kidney, and adipose tissue and fish muscles [20–24]. It is excreted predominantly as metabolites via urine. Metabolism of EQ was studied in rats, mice, dogs, chickens, and fish, as well as in plants [13, 21, 25]. It is not fully described but some metabolites were identified (Table 3). The most important EQ metabolites observed in rat urine and bile result from O-deethylation at position 6-C, and then conjugation with sulphate or glucuronide residues.

TABLE 1: Permitted amounts of EQ in different products approved by FDA*.

Product	Dose [ppm]
Animal feed	150
Spices	150
Uncooked fat of meat from animals (except poultry)	5
Uncooked liver and fat of poultry	3
Uncooked muscle meat of animals	0.5
Poultry eggs	0.5
Milk	0
Pear	3

* According to the Code of Federal Regulations (CFR), Title 21, Parts 573.380, 573.400, 172.140.

The other metabolic pathways include hydroxylation and glucuronidation at position 8-C, deethylation at 6-C and epoxidation between positions 3-C and 4-C [21]. The main metabolites observed may be different depending on animal species. In mice mainly glucuronide metabolites were detected while in rats those result from conjugation of EQ with sulfate.

In the studies of Bohne et al. [23, 26] parent EQ, demethylated EQ (DEQ), quinone imine (QI), and EQ dimer (EQDM) were observed in salmonid fish after long-term dietary exposure to EQ. It was in agreement with the results obtained earlier by Skaare and Roald [27]. EQ is considered as a model inducer of phase II enzymes involved in the metabolism of xenobiotics, but influence of EQ on phase I enzyme gene transcript levels was also observed [28, 29]. The key role in mediating phase I reactions (e.g., oxidation or reduction) producing more hydrophilic compounds is played by the CYP (cytochrome P450) enzyme family. Bohne et al. [28, 29] observed the alteration of CYP3A gene expression; an increase in the amount of CYP3A transcripts was detected in salmon after feeding them with the diet containing EQ at the highest dose used (1800 mg kg^{-1}). The authors speculate that EQ may regulate CYP3 gene expression by interaction, for example, with pregnane X nuclear receptor (PXR) whose function is to sense the presence of toxic xenobiotics and in response enhance the expression of proteins involved in their detoxification. On the other hand, CYP1A1 gene expression, which was described as an exposure biomarker to both endogenous and exogenous compounds [30], was not increased after dietary exposure of salmonid fish to EQ and during the depuration period a trend toward downregulation was noted [28, 29]. Such an effect was observed despite the increase in the expression of AhR mRNA (AhR, cytosolic transcription factor responsible for changes in gene transcription). This effect can be explained in several ways. For example, the parent EQ may bind CYP1A1 protein and as a result may inhibit the gene expression and activity of protein [31]. Hepatic antioxidant response elements (ARE) or AhR repressor (AhRR) together with basic-helix-loop-helix-PAS (Per-AhR/ARNT-Sim homology sequence) of transcription factor usually associated with each other to form heterodimers (AhR/ARNT or AhRR/ARNT) may be

TABLE 2: Physical and chemical properties of ethoxyquin.

Properties of EQ	
Molecular formula	$C_{14}H_{19}NO$
Molecular mass	217.34 [g mol^{-1}]
IUPAC name	1,2-dihydro-2,2,4-trimethylquinolin-6-yl ethyl ether
CAS name	6-ethoxy-1,2-dihydro-2,2,4-trimethylquinoline
Chemical class	Quinoline
Melting point	0°C[a]
Boiling point	123–125°C at 2 mm Hg[a]
Solubility	Insoluble in water, soluble in fat and animal and plant oils, and soluble in ethanol, methanol, DMSO, and DMF
Vapor density	7.48 (AIR = 1)[a]
Stability	Stable under ordinary conditions, storage temp. 0–6°C
Spectral properties	Index of refraction: n_D 1.569–1.672 at 25°C[a]
	Max absorption: 354 nm. Intense mass spectral peaks: 202 m/z (100%), 108 m/z (53%), 174 m/z (48%), and 137 m/z (36%)[a]
	^{13}C NMR in CDCl$_3$ EQ (8% solution) δ_C in ppm: 151.2 (EQ-C-6); 137.6 (EQ-C-4); 129.5 (EQ-C-3); 128.5 (EQ-C-9); 122.8 (EQ-C-10); 114.6 (EQ-C-8); 113.6 (EQ-C-7); 111.1 (EQ-C-5); 64.2 (EQ-$\underline{C}H_2$-O); 51.6 (EQ-C-2); 30.4 (EQ-$\underline{C}H_3$-4); 18.5 (EQ-$\underline{C}H_3$-4); 15.0 (EQ-$\underline{C}H_3$-CH$_2$-O-)[b]

[a] According to the National Library of Medicine HSDB Database (last revision date: 2003) [18].
[b] According to Błaszczyk and Skolimowski [19].
DMSO: dimethyl sulfoxide, DMF: N,N-dimethylformamide.

also involved in the CYP1A1 downregulation process. These heterodimers can influence gene expression by binding ARE sequences in the gene promoter regions [32].

However, EQ as other phenolic antioxidants, first of all causes induction of phase II xenobiotic-metabolizing enzymes. Bohne et al. [28, 29] observed elevated dose-related uridine diphosphate glucuronosyl-transferase (UDPGT) mRNA expression after dietary exposure to EQ. As UDPGT reacts with the compounds that have the hydroxyl group (-OH) parent EQ cannot be the potential substrate for glucuronidation, only its metabolites, for example DEQ (6-hydroxy-2,2,4-trimethyl-1,2-dihydroquinoline; Table 3), the metabolite identified by Berdikova Bohne et al. [26] in Atlantic salmon. Changes in the expression of glutathione S-transferase (GST) gene were also observed after feeding animals with EQ containing feed. The alterations in GST activity caused by EQ were documented in Atlantic salmon [28, 29], in rodents [33, 34], and in nonhuman primates [35]. In addition to UDPGT and GST, some other enzymes are involved in phase II metabolism of EQ, for example, NADP(H) : quinone oxidoreductase and epoxide hydrolase [36].

The expression pattern of both phase I and II enzymes involved in EQ metabolism may vary in different animals and should be considered in relation to the ratio of parent EQ and its metabolites (first of all DEQ, QI, and EQDM) in the liver [29, 37]. The research concerning this issue is currently in progress.

3.2. Plants. Ethoxyquin is also registered as an antioxidant to control scald (browning) in apples and pears. The EQ plant metabolites/degradation products were detected, and it was shown that in general they are different from those observed in animals (Table 3). In pears treated with ring-labeled [^{14}C]ethoxyquin the following compounds were detected: N–N and C–N dimers, demethylethoxyquin (DMEQ), dehydrodemethylethoxyquin (DHMEQ), and dihydroethoxyquin (DHEQ) [14, 25]. It was shown that ethoxyquin was rapidly degraded or metabolized but itself it was not translocated into the pulp of fruit where the residues were detected (less than 0.5% of total radioactive residue was EQ). Toxicity of EQ metabolites, MEQ, DHMEQ, and DHEQ was studied in dogs (oral administration, single doses of 50 to 200 mg kg^{-1} bw), and it was found that they did not show any significant toxicity. In the report of Gupta and Boobis [14] the rank order of toxic potency for the plant metabolites and EQ is MEQ < EQ < DHEQ < DHMEQ (the least toxic first). MEQ, DHMEQ, and DHEQ were also evaluated for genotoxicity in *in vitro* and *in vivo* tests. The compounds did not cause gene mutations in *Salmonella typhimurium* and *Escherichia coli* strains, but they induced chromosomal aberrations or/and endoreduplication in Chinese hamster ovary cells. On the other hand, plant metabolites/degradation products did not exhibit genotoxic potential *in vivo*. ADI intake for humans for MEQ, DHMEQ, and DHEQ was estimated at the same level as for EQ (0–0.005 mg kg^{-1} bw).

4. Antioxidant Activity of EQ

EQ possesses high-antioxidative activity. It is very efficient in protecting lipids which are present in food against oxidization [38, 39]. Specifically it is used to retard oxidation of carotene, xanthophylls, and vitamins (like vitamins A or E). In animals treated with ethoxyquin three times higher level of vitamins

TABLE 3: Metabolite/degradation products of EQ detected in different organisms.

	EQ metabolite/degradation product	Organism
	1,2-Dihydro-6-hydroxy-2,2,4-trimethylquinoline sulphate	Rats[1]
	1,2-Dihydro-6-hydroxy-2,2,4-trimethylquinoline glucuronide	Mice[1]
	1,2,3,4-Tetrahydro-3,6-dihydroxy-4-metylene-2,2-dimethylquinoline sulphate	Rats[1]
	1,2-Dihydro-6-ethoxy-8-hydroxy-2,2,4-trimethylquinoline glucuronide	Rats[1] Mice[1]
	8-(S-glutathionyl)-2,2,4-trimethylquinol-6-one	Rats[1] Mice[1]
	1,2,3,4-Tetrahydro-6-ethoxy-3-hydroxy-4-(S-glutathionyl)-2,2,4-trimethylquinoline	Rats[1] Mice[1]
	2,6-Dihydro-2,2,4-trimethyl-6-quinolone (quinone imine (QI))	Rats[1] Atlantic salmon[2]
	6-Hydroxy-2,2,4-trimethyl-1,2-dihydroquinoline (deethylated EQ (DEQ))	Atlantic salmon[2]
	1,8'-Di(1,2-dihydro-6-ethoxy-2,2,4-trimethylquinoline) or 6,6'-diethoxy-2,2,2',2',4,4'-hexamethyl-1',2'-dihydro-2H-1,8'-biquinoline (ethoxyquin dimer (EQDM)) *Instead of the hydrogen atom there is a methyl group, it gives methyl C-N ethoxyquin dimer, 6,6'-Diethoxy-1',2,2,2',2',4,4'-heptamethyl-1',2'-dihydro-2H-1,8'-biquinoline	Atlantic salmon[2] Pears[3,4]

TABLE 3: Continued.

EQ metabolite/degradation product	Organism
N-N Ethoxyquin dimer, 1,1'-di(1,2-dihydro-6-ethoxy-2,2,4-trimethylquinoline) or 6,6'-diethoxy-2,2,2',2',4,4'-hexamethyl-2H,2'H-1,1'-biquinoline	Pears[3,4]
6-Ethoxy-1,2,2,4-tetramethyl-1,2-dihydroquinoline, methylethoxyquin (MEQ)	Pears[3,4]
6-Ethoxy-2,4-dimethylquinoline, dehydrodemethylethoxyquin (DHMEQ)	Pears[3,4]
6-Ethoxy-2,2,4-trimethyl-1,2,3,4-tetrahydroquinoline, dihydroethoxyquin (DHEQ)	Pears[3,4]
6-Ethoxy-2,4-dimethyl-1,2-dihydroquinoline, demethylethoxyquin (DMEQ)	Pears[3,4]

Glu: glucuronide, G: glutathione, Et: ethyl group (C_2H_5).
[1]Burka et al. [21], [2]Berdikova Bohne et al. [26], [3]Gupta and Boobis [14], [4]JMPR [25].

A and E in blood plasma was observed [40]. This finding suggests that an organism is using EQ instead of natural antioxidants. High efficiency of this antioxidant results not only from chemical features of EQ itself but also from the fact that products of its oxidation also possess antioxidative properties [12, 39, 41].

Studies on EQ antioxidant properties were performed by Taimr [39] with the use of alkylperoxyls, and it was shown that the reaction rate of EQ with them is very high. In the presence of high oxygen concentrations EQ reacts with alkylperoxyl molecule to form aminyl radical (6-ethoxy-2,2,4-trimethyl-1,2-dihydroquinolin-1-yl) which subsequently may enter various pathways. In nonoxidizing conditions it can be stabilized both by the loss of methyl group and aromatization of heterocycle to form 2,4-dimethyl-1,2-dihydroquinoline (dehydrodemethylethoxyquin (DHMEQ)) and through dimerization to form EQ dimer (EQDM) [39, 42]. On the other hand, in an oxidizing medium other molecules can be formed, for example, 2,6-dihydro-2,2,4-trimethyl-6-quinolone (QI) or nitroxide radical (6-ethoxy-2,2,4-trimethyl-1,2-dihydroquinolin-N-oxyl) which is also a strong antioxidant [39, 42].

Products of EQ oxidation were detected by different authors in fish oil and meal [22, 43–45]. According to He and Ackman [46] the following oxidization products of EQ dominate in fish meal and fish feed: 2,6-dihydro-2,2,4-trimethyl-6-quinolone (QI) and 1,8'-di(1,2-dihydro-6-ethoxy-2,2,4-trimethylquinoline) (EQDM). At high storage temperature neither QI nor EQDM accumulates; however, another product of EQ oxidation, 2,4-dimethyl-6-ethoxyquinoline, is stable [41]. As it was pointed out earlier, ethoxyquin oxidization products also possess antioxidative properties. The EQDM and QI show 69% and 80% of EQ efficacy, respectively (studies on fish meal) [12]. On the other hand, in the studies of Thorisson et al. [45] quinone imine (QI) and EQ nitroxide were also powerful antioxidants, while

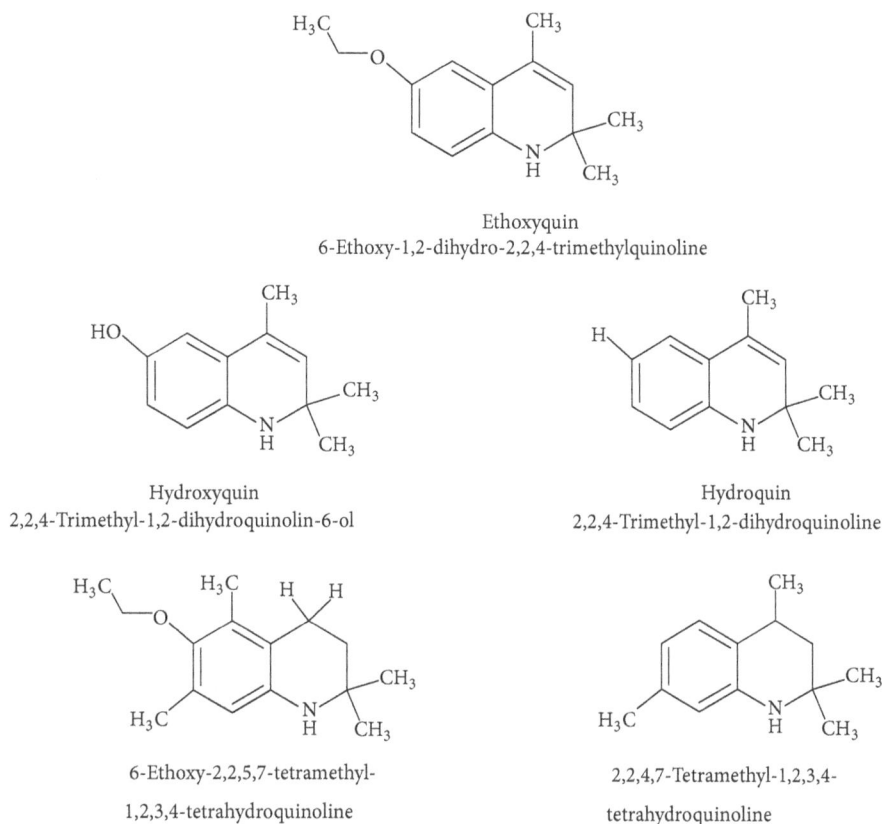

Ethoxyquin
6-Ethoxy-1,2-dihydro-2,2,4-trimethylquinoline

Hydroxyquin
2,2,4-Trimethyl-1,2-dihydroquinolin-6-ol

Hydroquin
2,2,4-Trimethyl-1,2-dihydroquinoline

6-Ethoxy-2,2,5,7-tetramethyl-
1,2,3,4-tetrahydroquinoline

2,2,4,7-Tetramethyl-1,2,3,4-
tetrahydroquinoline

FIGURE 1: Chemical structure of ethoxyquin (EQ) and of some new compounds synthesized on ethoxyquin backbone with promising antioxidant properties (according to de Koning [12], Dorey et al. [15] and Błaszczyk and Skolimowski [17]).

EQDM, the main product of EQ oxidation, showed little or no antioxidant behavior.

Antioxidant activity of EQ was also demonstrated in experiments performed both *in vivo* and *in vitro*. Antimutagenic effect of this antioxidant was observed in mice, rats, and Chinese hamsters treated with cyclophosphamide, an agent widely used in cancer chemotherapy [47–49]. During cyclophosphamide bioactivation reactive oxygen species are formed which can cause damage of genetic material [50, 51]. EQ reduced the number of chromosome aberrations, micronuclei, and dominant lethal mutations induced by the anticancer drug [47–49]. There were also some reports that EQ can modify carcinogenic response to different carcinogens [35, 52, 53]. EQ given to Fischer 344 rats in diet completely prevented the formation of aflatoxin B1-induced preneoplastic liver lesions [52, 53].

In *in vitro* experiments with human lymphocytes, antioxidant activity of EQ was observed in the comet assay (the method used to detect single- and double-strand DNA breaks, cross-links, and alkaline labile sites) and in micronucleus test (the method for the detection of micronuclei induced by clastogens or aneugens). EQ used at the concentrations ranging from $1 \mu M$ to $10 \mu M$ protected human lymphocytes against DNA damage caused by hydrogen peroxide (H_2O_2, $10 \mu M$) [17]. This antioxidant also reduced the number of micronuclei caused by H_2O_2 used at concentration of $75 \mu M$. However, the significant reduction was evident only in the case of lower EQ concentrations ($5 \mu M$, $10 \mu M$) with no effect at higher concentration [17].

5. Adverse Effects of EQ: *In Vivo* and *In Vitro* Studies

Different phenolic antioxidants may be used in animal feed, such as BHA (butylated hydroxyanisole), BHT (butylated hydroxytoluene), and the most efficacious EQ. The levels of the antioxidants in finished feed should not be higher than 150 ppm for EQ and 200 ppm for BHT and BHA (U.S. Food and Drug Administration permissions). The fact that efficient antioxidants work optimally when they are used at low concentrations is their remarkable characteristic. On the other hand, when antioxidants are used at high concentrations they act as prooxidants. The impact of these compounds depends on their concentration as well as on other factors such as metal-reducing potential, chelating behaviour, solubility, and pH. The effect of antioxidants on living organisms also depends on their bioavailability and stability in tissues [54, 55]. Phenolic antioxidants under favorable conditions may be converted to phenoxyl radicals with prooxidant activity [55]. It was shown that dissolved EQ may exist partly in the free radical form which it was also detected in the compound itself [56]. Therefore, ethoxyquin nitroxide which is produced by EQ oxidation similarly as other nitroxide molecules

TABLE 4: Harmful effects of EQ observed after its oral administration in different animals or in humans (contact exposure).

The harmful effect	Animals
Loss of weight	Marmosets[1], rats[2], dogs[2], mice[2], rabbits[3]
Changes in liver	Marmosets[1], rats[1,4], dogs[2], mice[2], broiler chickens[2], tilapia[5]
Changes in kidney	Marmosets[1], rats[2,6,7], dogs[2], broiler chickens[2,8]
Changes in alimentary duct	Marmosets[1], dogs[2], mice[2], broiler chickens[2,8]
Changes in urinary bladder	Rats[2]
Anemia	Marmosets[1]
Changes in mitochondria	Rats[2]
Lethargy	Rabbits[3]
Colored urine, skin, or fur	Dogs[2], rats[2]
Increase in mortality	Broiler chickens[2,8]
Effect on immunity	Tilapia[5]
Condition factor: the final body weight in relation to body length of fish	Large yellow croaker[9]
Allergy (contact exposure)	Humans[10,11,12]

[1] McIntosh et al. [58], [2] Drewhurst [13], [3] Little [10], [4] Ito et al. [59], [5] Yamashita et al. [60], [6] Neal et al. [61], [7] Manson et al. [53], [8] Leong and Brown [62], [9] Wang et al. [63], [10] Rodríguez-Trabado et al. [9], [11] Brandao [5], [12] Savini et al. [6].

(e.g., tempol) may also show prooxidative properties [57]. Formation of free oxygen species as a result of using too high EQ concentrations can cause adverse health effects in animals fed with EQ containing feed or in people consuming meat from farmed animals, for example, different fishes.

The studies on EQ prooxidant activity and toxicity associated with it were performed both *in vivo* and *in vitro*. Dogs are most susceptible to the harmful effects of EQ, and first reports of such effects were received by FDA in 1988. The symptoms observed by dog owners and veterinarians were liver, kidney, thyroid and reproductive dysfunction, teratogenic and carcinogenic effects, allergic reactions, and a host of skin and hair abnormalities [7]. According to the studies on dogs and laboratory animals it was shown that ethoxyquin had little acute toxicity, except when it is administered parenterally. Values of LD_{50} for EQ are $1700\,mg\,kg^{-1}$ bw (rats, oral gavage), $>2000\,mg\,kg^{-1}$ bw (rats, dermal treatment, 24 h), $\sim900\,mg\,kg^{-1}$ bw (mice, intraperitoneal administration), and $\sim180\,mg\,kg^{-1}$ bw (mice, intravenous administration) [13]. Despite species differences in the majority of animals treated with EQ at the concentrations higher than those permitted in animal feed, the same characteristic symptoms and pathologies appeared such as weight loss, liver, and kidney damage, alterations of alimentary duct (Table 4). The concentration of 100 ppm (equivalent to $2.5\,mg\,kg^{-1}$ bw per day) was considered to be a minimal-effect level for clinical signs of toxicity and liver effects in dogs, the most susceptible animals [13, 14].

Detrimental effects of EQ were also seen when the experiments were performed at the cell metabolism level. Hernandez et al. [64] and Reyes et al. [65] analysed the impact of EQ on the metabolic pathways of rat renal and hepatic cells, as well as on mitochondria and submitochondrial particles obtained from bovine heart and kidney. They observed influence of EQ on energy processes in cells. EQ inhibited renal Na^+, K^+-ATPase activity involved in ion transport [64]. The authors suggested that EQ interacted with site I of the mitochondrial respiratory chain, and it resulted in inhibition of oxygen consumption in the mitochondria of kidney and liver cells when glucose was a respiratory substrate. The effect was dose dependent.

More than 30 years ago when EQ began to be more commonly used in animal feed research started to assess its mutagenicity with the use of Ames test which is performed on different *Salmonella typhimurium* strains. The results were equivocal as some results were negative [66–68], but the positive effects were also observed [69, 70]. It was also shown that EQ enhanced the mutagenic activity of DMBA (3,2'-dimethyl-4aminobiphenyl), a compound having carcinogenic properties [70]. Ethoxyquin was reported to both enhance and inhibit genetic changes induced by known carcinogens; on the other hand it can also lead to cancer in exposed animals. Manson et al. [53] observed in Fischer 344 rats that EQ caused severe damage in kidney. Many hyperplastic and putative preneoplastic tubules were found which suggested that EQ may be exerting a carcinogenic effect. Similar effects were observed earlier by Ito et al. [59] in relation not only to the kidney but also to the urinary bladder.

Possible carcinogenicity of EQ is probably connected with its prooxidant activity and induction of reactive oxygen radicals which cause DNA damage. DNA damage is usually repaired by cellular repair system, but if it is severe or there are too many lesions, this leads to programmed cell death (apoptosis). Sometimes, however, the programmed cell death pathway is damaged so when the defense mechanisms fail there is no way to stop a cell from becoming a cancer cell. Some *in vitro* studies showed both cytotoxic effects of EQ leading to cell apoptosis or necrosis and damage of genetic material at DNA or chromosome levels. Cytotoxic effects of pure EQ (purity > 97%) were studied *in vitro* with the use of human lymphocytes. The IC_{50} value (the concentration causing 50% growth inhibition) for EQ determined after 72-hour treatment of the cells in the MTT assay was 0.09 mM [71]. This antioxidant significantly reduced viability of lymphocytes detected with trypan blue exclusion method after 24-hour treatment at the concentrations of 0.25 and 0.5 mM (cell divisions were stimulated by phytohemagglutinin, (PHA)) [19] or of 0.05 mM and higher when 1-hour treatment was performed [72]. EQ-induced apoptosis by observed in *in vitro* cultured human lymphocytes starting from 0.05 mM concentration and the detected number of apoptotic cells depended on the treatment time [71]. Ethoxyquin caused also DNA damage in the comet assay [72] however, most lesions could be repaired by cellular DNA repair systems [73]. On the other hand, the results obtained with the use of chromosome aberration test showed that unrepaired DNA damage induced by EQ could lead to permanent changes in genetic material

[16, 74]. Błaszczyk et al. [16] and Gille et al. [74] showed that this antioxidant induced chromosome aberrations such as breaks, dicentrics, atypical translocated chromosomes, or chromatid exchanges in human lymphocytes and Chinese hamster ovary cells. These aberrations are known to have serious biological consequences [75].

6. Analogues and Derivatives of EQ

Because of adverse health effects caused by EQ it is reasonable to search for new antioxidants as effective in scavenging free radicals as EQ which produce no such problems. In the paper of de Koning [12] nine analogues of EQ prepared to compare their antioxidant efficacy with that of the parent chemical are presented. The compounds have been tested in a refined fish oil and subsequently some of the most promising ones have been also tested in fish meal. It was noted that the results obtained in fish oil were not always the same as in fish meal, for example, hydroxyquin (1,2-dihydro-6-hydroxy-2,2,4-trimethylquinoline; Figure 1) was 3.5 times as effective as EQ in fish oil, while only 3/4 of its efficacy was observed in fish meal. In the case of another compound—hydroquin (1,2-dihydro-2,2,4-trimethylquinoline; Figure 1) antioxidant efficacy in relation to EQ was 101% in fish oil and 52% in fish meal. Despite the lower efficiency of this compound in fish meal the author stated that hydroquin can compete with EQ as an antioxidant of choice [12]. The reason is that preparation of hydroquin based on aniline and acetone is more cost-effective than that of EQ whose production requires p-phenetidine (more expensive than aniline). Hydroquin was earlier patented as an antioxidant in animal feeds in 1997 [76]. The 2-year dermal research with the use of F344/N rats and B6C3F1 mice conducted under the National Toxicology Program [77] showed that the compound was not carcinogenic, but the studies performed by Sitarek and Sapota [78] showed its teratogenic properties.

In 2000 Dorey et al. [15] presented the report concerning the synthesis and biological properties of a new class of antioxidants based on the EQ backbone. The studies were performed to search for new quinolinic derivatives with radical scavenging activity, potential candidates for central nervous system protection. EQ is not suitable for that as it has been shown to exhibit significant hypothermic effect, probably as a result of an inhibition of electron transport in the mitochondrial respiratory chain [65]. Dorey et al. [15] synthesized and studied many 1,2-dihydro and 1,2,3,4-tetrahydroquinolines and then selected for further evaluation a group of antioxidants (5 compounds) with high radical scavenging capacities, relatively low toxicity, and moderate hypothermia. The compounds belonging to the group of 1,2,3,4-tetrahydroquinolines (e.g., 6-ethoxy-2,2,5,7-tetramethyl-1,2,3,4-tetrahydroquinoline, characterized with the lowest toxicity and high radical scavenger capacity) are structurally similar to 2,2,4,7-tetramethyl-1,2,3,4-tetrahydroquinoline synthesized and tested in our laboratory (Figure 1) [17]. The latter compound also had promising features: its antioxidant activity was comparable to that of EQ, but its cytotoxicity and genotoxicity studied with

TABLE 5: Comparison of different activities of ethoxyquin and its derivatives based on the data presented by Błaszczyk and Skolimowski [19, 71, 80–82] and Błaszczyk et al. [79].

Activitity	Activity from the lowest (left) to the highest (right) one
Cytotoxicity	EQ-F < EQ-HCl < EQ-C < EQ-S < EQ-T < EQ-Q < EQ-R < EQ-H < EQ
Genotoxicity	EQ-F < EQ-C < EQ-H < EQ-HCl < EQ-Q < EQ-T < EQ-S < EQ-R < EQ
Antioxidant activity	EQ-F < EQ-HCl < EQ-R < EQ-T < EQ-S < EQ-C < EQ-Q < EQ < EQ-H

EQ-F: ethoxyquin phosphate; EQ-HCl: ethoxyquin hydrochloride; EQ-C: ethoxyquin L-ascorbate; EQ-S: ethoxyquin salicylate; EQ-T: ethoxyquin salt of Trolox C; EQ-Q: ethoxyquin complex with quercetin; EQ-R: ethoxyquin complex with rutin; EQ-H: ethoxyquin n-hexanoate; EQ: ethoxyquin.

the use of human lymphocytes in vitro were significantly lower. We believe that this chemical is worth of detailed studies to confirm its usefulness as a food preservative. Some other EQ derivatives and salts were also studied for cytotoxicity, genotoxicity, and antioxidant activity, namely, the complexes of ethoxyquin with flavonoids (rutin or quercetin), ethoxyquin hydrochloride, ethoxyquin phosphate, ethoxyquin L-ascorbate, ethoxyquin n-hexanoate, ethoxyquin salicylate, and ethoxyquin salt of Trolox C [19, 79–82]. The biological properties of the compounds were analysed with the use of MTT, TUNEL, and trypan blue staining methods (cytotoxicity testing), comet assay (genotoxicity testing), and micronucleus test (mutagenicity testing). From among the compounds tested ethoxyquin phosphate (EQ-F) was the least toxic (Table 5)—its cytotoxic and genotoxic activities in comparison with those of EQ were reduced positively (IC_{50} = 0.8 mM versus 0.09 mM for EQ) [71]. On the other hand, antioxidant activity of EQ-F was observed, but it was the lowest of the tested compounds [82]. The studies showed that all the tested compounds were less toxic to human lymphocytes than EQ, and the antioxidant activity of four of them (ethoxyquin n-hexanoate, ethoxyquin complex with quercetin, ethoxyquin L-ascorbate, and ethoxyquin salicylate) was comparable with that of EQ [79–82]. The results obtained indicate that their use as antioxidants may be considered.

7. Food Safety Aspect

EQ safety has been under consideration for many years. The level of this antioxidant in animal feeds should not be higher than 150 ppm (U.S. Food and Drug Administration permissions). The approved uses of ethoxyquin in animal feeds are addressed in the Code of Federal Regulations (CFR), Title 21, Parts 573.380 and 573.400, and established tolerances are in Part 172.140. On the one hand, the observed adverse health effects (firstly in dogs) could be caused by the fact that the animals ate a lot of feed containing EQ, but on the other hand, it could also be the result of its excessive amounts in the feed. Ethoxyquin is added to animal feed either directly or indirectly as a component of an ingredient. From time to time FDA reminds industry about labeling and safe use requirements for ethoxyquin, but if it is added at the ingredient level this is not always indicated.

Another important safety issue is the presence of EQ oxidation and EQ metabolism products in animal feed or in foods prepared from farmed animal meat. de Koning [12] described main products of EQ oxidation which can be observed in stored feeds or in fish meal: EQ dimer (EQDM, $1,8'$-di(1,2-dihydro-6-ethoxy-2,2,4-trimethylquinoline) and quinone imine (QI, 2,6-dihydro-2,2,4-trimethyl-6-quinolone). Both compounds were shown to be potent antioxidants, but they can also have detrimental effect, especially so because the half-life of the dimer was considerably greater than that of EQ [26, 28]. In the recent studies no adverse toxicological effects of EQDM, in terms of kidney and liver function, were observed in *in vivo* experiments with F344 rats exposed for 90 days to the compound [37]. On the other hand, Augustyniak et al. [83] showed that EQDM, similarly as EQ, was cytotoxic and genotoxic to human lymphocytes. Toxicity of QI has not been studied yet, but the results obtained by the authors indirectly indicated that the compound could be cytotoxic to human cells.

The levels of the parent compound (EQ) in meat of farmed animals are usually lower than MRL (Maximum Residue Level) [20, 84], but EQ oxidation products are usually not controlled. It was shown that EQDM and other EQ residues can be present in different animal tissues [23, 24, 26, 28, 37]. In the studies of Bohne et al. [23] in which Atlantic salmons were fed for 12 weeks with the feed containing this antioxidant, four compounds were identified in their muscles: parent EQ (6-ethoxy-1,2-dihydro-2,2,4-trimethylquinoline), deethylated EQ (6-hydroxy-2,2,4-trimethyl-1,2-dihydroquinoline), quinone imine (2,6-dihydro-2,2,4-trimethyl-6-quinolone, QI), and EQ dimer ($1,8'$-di(1,2-dihydro-6-ethoxy-2,2,4-trimethylquinoline, EQDM). It was also shown that the concentration of EQ in fish muscle was proportional to the duration of exposure and the level of EQ in the feed [23]. The same linear increase was seen for EQDM, the main metabolite of EQ, and the sum of EQ and EQDM. Bohne et al. [23] found that the level of EQ and its metabolites in fish muscle could be predicted from the level of dietary EQ and then controlled, but because it is not the only factor which may affect the levels of EQ and its metabolites in the salmon tissue (others for example are fish size and age), the concentration of EQ and EQDM in fish ready for consumption may be higher than that observed in their studies. In their experiments it was shown that the elimination of EQ from salmon was concurrent with significant increase in the level of EQDM, and they concluded that mandatory 14 days of depuration were not sufficient for elimination of EQ residues—it is mainly because EQDM is characterized by the considerably longer half-life than that of EQ. Moreover, EQDM accounted for 99% of the sum of the two compounds (EQ and EQDM), and its toxicological effects in animals and humans are unknown. EQ and EQ dimer were also detected in similar amounts not only in Atlantic salmon, but also in other commercially important species of farmed fishes (halibut, rainbow trout) by Lundebye et al. [24]. They found that in Atlantic salmon, halibut, and rainbow trout the concentration of EQDM was more than 10-fold

higher than that of EQ. The authors estimated that consumer exposure to EQ from a single portion (300 g) of skinned-fillets of different species of farmed fish could amount up to 15% of the ADI. In the light of data concerning the presence of EQDM in the body of farmed fishes and providing that EQ dimer was included in the ADI, the EQ and EQDM intake from a single portion of Atlantic salmon would be close to ADI [24]. Farmed fish is probably the major source of EQ and its residues for European consumers (its use as a food additive is forbidden). In our opinion, however, both fish and other farmed animals, for example, chickens, should be controlled for the presence of not only EQ, but also EQDM, its main oxidation product.

8. Conclusion

Ethoxyquin has been used as an antioxidant in animal feed for several decades and despite the search for new compounds that could be used as free radical scavengers, it is still the most effective antioxidant. The negative health effects in domestic animals fed with EQ containing feed were observed some years ago, but the presence of its approved doses should not be hazardous. Toxicity and mutagenicity of EQ were observed in *in vivo* and *in vitro* studies showing its potential harmful effects. This makes it very important to label all products and ingredients to which EQ is added and to comply with the recommended doses. Additionally, the results of the studies on products of EQ oxidation, especially EQDM, detected in farmed animal tissues indicate that it should be under control and some regulations should be introduced.

Conflict of Interests

The authors declare that there is no conflict of interests.

References

[1] H. Babich, "Butylated hydroxytoluene (BHT): a review," *Environmental Research*, vol. 29, no. 1, pp. 1–29, 1982.

[2] R. Kahl, "Synthetic antioxidants: biochemical actions and interference with radiation, toxic compounds, chemical mutagens and chemical carcinogens," *Toxicology*, vol. 33, no. 3-4, pp. 185–228, 1984.

[3] G. M. Williams, M. J. Iatropoulos, and J. Whysner, "Safety assessment of butylated hydroxyanisole and butylated hydroxytoluene as antioxidant food additives," *Food and Chemical Toxicology*, vol. 37, no. 9-10, pp. 1027–1038, 1999.

[4] A. A. M. Botterweck, H. Verhagen, R. A. Goldbohm, J. Kleinjans, and P. A. van den Brandt, "Intake of butylated hydroxyanisole and butylated hydroxytoluene and stomach cancer risk: results from analyses in the Netherlands cohort study," *Food and Chemical Toxicology*, vol. 38, no. 7, pp. 599–605, 2000.

[5] F. M. Brandao, "Contact dermatitis to ethoxyquin," *Contact Dermatitis*, vol. 9, no. 3, article 240, 1983.

[6] C. Savini, R. Morelli, E. Piancastelli, and S. Restani, "Contact dermatitis due to ethoxyquin," *Contact Dermatitis*, vol. 21, no. 5, pp. 342–343, 1989.

[7] D. A. Dzanis, "Safety of ethoxyquin in dog foods," *Journal of Nutrition*, vol. 121, no. 11, pp. S163–164, 1991.

[8] K. Alanko, R. Jolanki, T. Estlander, and L. Kanerva, "Occupational 'multivitamin allergy' caused by the antioxidant ethoxyquin," *Contact Dermatitis*, vol. 39, no. 5, pp. 263–264, 1998.

[9] A. Rodríguez-Trabado, J. Miró I Balagué, and R. Guspi, "Hypersensitivity to the antioxidant ethoxyquin," *Actas Dermo-Sifiliograficas*, vol. 98, no. 8, p. 580, 2007.

[10] A. D. Little, "Ethoxyquin, national toxicology program, executive summary of safety and toxicity information," Chemical Committee Draft Report, Ethoxyquin CAS Number 91-53-2, 1990, http://ntp.niehs.nih.gov/ntp/htdocs/chem_background/exsumpdf/ethoxyquin_508.pdf.

[11] U.S Food and Drug Administration, Animal and Veterinary, "FDA requests that ethoxyquin levels be reduced in dog foods," 1997, http://www.fda.gov/AnimalVeterinary/NewsEvents/CVMUpdates/ucm127828.htm.

[12] A. J. de Koning, "The antioxidant ethoxyquin and its analogues: a review," *International Journal of Food Properties*, vol. 5, no. 2, pp. 451–461, 2002.

[13] I. Drewhurst, "Ethoxyquin. JMPR Evaluations," 1998, http://www.inchem.org/documents/jmpr/jmpmono/v098pr09.htm.

[14] P. K. Gupta and A. Boobis, "Ethoxyquin (Addendum)," 2005, http://www.inchem.org/documents/jmpr/jmpmono/v2005pr-10.pdf.

[15] G. Dorey, B. Lockhart, P. Lestage, and P. Casara, "New quinolinic derivatives as centrally active antioxidants," *Bioorganic & Medicinal Chemistry Letters*, vol. 10, no. 9, pp. 935–939, 2000.

[16] A. Błaszczyk, R. Osiecka, and J. Skolimowski, "Induction of chromosome aberrations in cultured human lymphocytes treated with ethoxyquin," *Mutation Research*, vol. 542, no. 1-2, pp. 117–128, 2003.

[17] A. Błaszczyk and J. Skolimowski, "Comparative analysis of cytotoxic, genotoxic and antioxidant effects of 2,2,4,7-tetramethyl-1,2,3,4-tetrahydroquinoline and ethoxyquin on human lymphocytes," *Chemico-Biological Interactions*, vol. 162, no. 1, pp. 70–80, 2006.

[18] National Library of Medicine HSDB Database, "Ethoxyquin CASRN: 91-53-2," 2003, http://toxnet.nlm.nih.gov/cgi-bin/sis/search/r?dbs+hsdb:@term+@DOCNO+400.

[19] A. Błaszczyk and J. Skolimowski, "Synthesis and studies on antioxidants: ethoxyquin (eq) and its derivatives," *Acta Poloniae Pharmaceutica—Drug Research*, vol. 62, no. 2, pp. 111–115, 2005.

[20] A. Hobson Frohock, "Residues of ethoxyquin in poultry tissues and eggs," *Journal of the Science of Food and Agriculture*, vol. 33, no. 12, pp. 1269–1274, 1982.

[21] L. T. Burka, J. M. Sanders, and H. B. Matthews, "Comparative metabolism and disposition of ethoxyquin in rat and mouse. II. Metabolism," *Xenobiotica*, vol. 26, no. 6, pp. 597–611, 1996.

[22] P. He and R. G. Ackman, "Residues of ethoxyquin and ethoxyquin dimer in ocean-farmed salmonids determined by high-pressure liquid chromatography," *Journal of Food Science*, vol. 65, no. 8, pp. 1312–1314, 2000.

[23] V. J. B. Bohne, A. K. Lundebye, and K. Hamre, "Accumulation and depuration of the synthetic antioxidant ethoxyquin in the muscle of Atlantic salmon (*Salmo salar* L.)," *Food and Chemical Toxicology*, vol. 46, no. 5, pp. 1834–1843, 2008.

[24] A. K. Lundebye, H. Hove, A. Måge, V. J. B. Bohne, and K. Hamre, "Levels of synthetic antioxidants (ethoxyquin, butylated hydroxytoluene and butylated hydroxyanisole) in fish feed and commercially farmed fish," *Food Additives and Contaminants A Chemistry, Analysis, Control, Exposure and Risk Assessment*, vol. 27, no. 12, pp. 1652–1657, 2010.

[25] JMPR, "Pesticide residues in food—2005. Report of the Joint Meeting of the FAO Panel of Experts on Pesticide Residues in Food and the Environment and the WHO Core Assessment Group on Pesticide Residues," JMPR Report, Ethoxyquin 035, 2005, http://www.fao.org/ag/AGP/AGPP/Pesticid/JMPR/Download/99_eva/14Ethoxyquin.pdf.

[26] V. J. Berdikova Bohne, H. Hove, and K. Hamre, "Simultaneous quantitative determination of the synthetic antioxidant ethoxyquin and its major metabolite in atlantic salmon (*Salmo salar*, L), ethoxyquin dimer, by reversed-phase high-performance liquid chromatography with fluorescence detection," *Journal of AOAC International*, vol. 90, no. 2, pp. 587–597, 2007.

[27] J. U. Skaare and S. O. Roald, "Ethoxyquin (EMQ) residues in Atlantic salmon measured by fluorimetry and gas chromatography (GLC)," *Nordisk Veterinaermedicin*, vol. 29, no. 4-5, pp. 232–236, 1977.

[28] V. J. B. Bohne, K. Hamre, and A. Arukwe, "Hepatic biotransformation and metabolite profile during a 2-week depuration period in Atlantic Salmon fed graded levels of the synthetic antioxidant, ethoxyquin," *Toxicological Sciences*, vol. 93, no. 1, pp. 11–21, 2006.

[29] V. J. B. Bohne, K. Hamre, and A. Arukwe, "Hepatic metabolism, phase I and II biotransformation enzymes in Atlantic salmon (*Salmo Salar*, L) during a 12 week feeding period with graded levels of the synthetic antioxidant, ethoxyquin," *Food and Chemical Toxicology*, vol. 45, no. 5, pp. 733–746, 2007.

[30] T. D. Bucheli and K. Fent, "Induction of cytochrome P450 as a biomarker for environmental contamination in aquatic ecosystems," *Critical Reviews in Environmental Science and Technology*, vol. 25, no. 3, pp. 201–268, 1995.

[31] A. Arukwe, L. Förlin, and A. Goksøyr, "Xenobiotic and steroid biotransformation enzymes in Atlantic salmon (*Salmo salar*) liver treated with an estrogenic compound, 4-nonylphenol," *Environmental Toxicology and Chemistry*, vol. 16, no. 12, pp. 2576–2583, 1997.

[32] T. M. Buetler, E. P. Gallagher, C. Wang, D. L. Stahl, J. D. Hayes, and D. L. Eaton, "Induction of phase I and phase II drug-detabolizing enzyme mRNA, protein, and activity by BHA, ethoxyquin, and oltipraz," *Toxicology and Applied Pharmacology*, vol. 135, no. 1, pp. 45–57, 1995.

[33] J. D. Hayes, D. J. Pulford, E. M. Ellis et al., "Regulation of rat glutathione S-transferase A5 by cancer chemopreventive agents: mechanisms of inducible resistance to aflatoxin B1," *Chemico-Biological Interactions*, vol. 111-112, pp. 51–67, 1998.

[34] K. L. Henson, G. Stauffer, and E. P. Gallagher, "Induction of glutathione S-transferase activity and protein expression in brown bullhead(*Ameiurus nebulosus*) liver by ethoxyquin," *Toxicological Sciences*, vol. 62, no. 1, pp. 54–60, 2001.

[35] T. K. Bammler, D. H. Slone, and D. L. Eaton, "Effects of dietary oltipraz and ethoxyquin on aflatoxin B1 biotransformation in non-human primates," *Toxicological Sciences*, vol. 54, no. 1, pp. 30–41, 2000.

[36] R. Kahl, "Elevation of hepatic epoxide hydratase activity by ethoxyquin is due to increased synthesis of the enzyme," *Biochemical and Biophysical Research Communications*, vol. 95, no. 1, pp. 163–169, 1980.

[37] R. Ørnsrud, A. Arukwe, V. Bohne, N. Pavlikova, and A. K. Lundebye, "Investigations on the metabolism and potentially adverse effects of ethoxyquin dimer, a major metabolite of the synthetic antioxidant ethoxyquin in salmon muscle," *Journal of Food Protection*, vol. 74, pp. 1574–1580, 2011.

[38] L. Taimr, M. Prusikova, and J. Pospisil, "Antioxidants and stabilizers. 113. Oxidation-products of the antidegradant ethoxyquin," *Angewandte Makromolekulare Chemie*, vol. 190, pp. 53–65, 1991.

[39] L. Taimr, "Study of the mechanism of the antioxidant action of ethoxyquin," *Angewandte Makromolekulare Chemie*, vol. 217, pp. 119–128, 1994.

[40] C. Lauridsen, K. Jakobsen, and T. K. Hansen, "The influence of dietary ethoxyquin on the vitamin E status in broilers," *Archiv fur Tierernahrung*, vol. 47, no. 3, pp. 245–254, 1995.

[41] D. R. Brannegan, *Analysis of ethoxyquin and its oxidation products using supercritical fluid extraction and high performance liquid chromatography with chemiluminescent nitrogen detection [thesis]*, Faculty of the Virginia Polytechnic, Institute and State University in Partial Fulfillment of the Requirements of the Degree of Master of Science in Chemistry, 2000, http://scholar.lib.vt.edu/theses/available/etd-03302000-20440044/.

[42] L. Taimr, M. Smelhausova, and M. Prusikova, "The reaction of 1-cyano-1-methylethyl radical with antidegradant ethoxyquin and its aminyl and nitroxide derivatives," *Angewandte Makromolekulare Chemie*, vol. 206, pp. 199–207, 1993.

[43] A. J. de Koning and G. van der Merwe, "Determination of ethoxyquin and two of its oxidation products in fish meal by gas chromatography," *The Analyst*, vol. 117, no. 10, pp. 1571–1576, 1992.

[44] P. He and R. G. Ackman, "HPLC determination of ethoxyquin and its major oxidation products in fresh and stored fish meals and fish feeds," *Journal of the Science of Food and Agriculture*, vol. 80, pp. 10–16, 2000.

[45] S. Thorisson, F. Gunstone, and R. Hardy, "The antioxidant properties of ethoxyquin and of some of its oxidation products in fish oil and meal," *Journal of the American Oil Chemists' Society*, vol. 69, pp. 806–809, 1992.

[46] P. He and R. G. Ackman, "Purification of ethoxyquin and its two oxidation products," *Journal of Agricultural and Food Chemistry*, vol. 48, no. 8, pp. 3069–3071, 2000.

[47] H. W. Renner, "Antimutagenic effect of an antioxidant in mammals," *Mutation Research*, vol. 135, no. 2, pp. 125–129, 1984.

[48] H. W. Renner and M. Knoll, "Antimutagenic effects on male germ cells of mice," *Mutation Research*, vol. 140, no. 2-3, pp. 127–129, 1984.

[49] D. Guyonnet, C. Belloir, M. Suschetet, M. H. Siess, and A. M. Le Bon, "Antimutagenic activity of organosulfur compounds from Allium is associated with phase II enzyme induction," *Mutation Research*, vol. 495, no. 1-2, pp. 135–145, 2001.

[50] A. Stankiewicz, E. Skrzydlewska, and M. Makieła, "Effects of amifostine on liver oxidative stress caused by cyclophosphamide administration to rats," *Drug Metabolism and Drug Interactions*, vol. 19, no. 2, pp. 67–82, 2002.

[51] S. Ray, B. Pandit, S. D. Ray, S. Das, and S. Chakraborty, "Cyclophosphamide induced lipid peroxidation and changes in cholesterol content: protective role of reduced glutathione," *International Journal of PharmTech Research*, vol. 2, no. 1, pp. 704–718, 2010.

[52] J. R. P. Cabral and G. E. Neal, "The inhibitory effects of ethoxyquin on the carcinogenic action of aflatoxin B1 in rats," *Cancer Letters*, vol. 19, no. 2, pp. 125–132, 1983.

[53] M. M. Manson, J. A. Green, and H. E. Driver, "Ethoxyquin alone induces preneoplastic changes in rat kidney whilst preventing induction of such lesions in liver by aflatoxin B1," *Carcinogenesis*, vol. 8, no. 5, pp. 723–728, 1987.

[54] E. A. Decker, "Phenolics: prooxidants or antioxidants?" *Nutrition Reviews*, vol. 55, no. 11, pp. 396–398, 1997.

[55] Y. Sakihama, M. F. Cohen, S. C. Grace, and H. Yamasaki, "Plant phenolic antioxidant and prooxidant activities: phenolics-induced oxidative damage mediated by metals in plants," *Toxicology*, vol. 177, no. 1, pp. 67–80, 2002.

[56] J. U. Skaare and T. Henriksen, "Free-radical formation in antioxidant ethoxyquin," *Journal of the Science of Food and Agriculture*, vol. 26, pp. 1647–1654, 1975.

[57] C. S. Wilcox and A. Pearlman, "Chemistry and antihypertensive effects of tempol and other nitroxides," *Pharmacological Reviews*, vol. 60, no. 4, pp. 418–469, 2008.

[58] G. H. McIntosh, J. S. Charnock, P. H. Phillips, and G. J. Baxter, "Acute intoxication of marmosets and rats fed high concentrations of the dietary antioxidant "ethoxyquin 66"," *Australian Veterinary Journal*, vol. 63, no. 11, pp. 385–386, 1986.

[59] N. Ito, S. Fukushima, and H. Tsuda, "Carcinogenicity and modification of the carcinogenic response by BHA, BHT and other antioxidants," *Critical Reviews in Toxicology*, vol. 15, no. 2, pp. 109–150, 1985.

[60] Y. Yamashita, T. Katagiri, N. Pirarat, K. Futami, M. Endo, and M. Maita, "The synthetic antioxidant, ethoxyquin, adversely affects immunity in tilapia (*Oreochromis niloticus*)," *Aquaculture Nutrition*, vol. 15, no. 2, pp. 144–151, 2009.

[61] G. E. Neal, D. J. Judah, G. G. Hard, and N. Ito, "Differences in ethoxyquin nephrotoxicity between male and female F344 rats," *Food and Chemical Toxicology*, vol. 41, no. 2, pp. 193–200, 2003.

[62] V. Y. Leong and T. P. Brown, "Toxicosis in broiler chicks due to excess dietary ethoxyquin," *Avian Diseases*, vol. 36, no. 4, pp. 1102–1106, 1992.

[63] J. Wang, Q. Ai, K. Mai et al., "Effects of dietary ethoxyquin on growth performance and body composition of large yellow croaker *Pseudosciaena crocea*," *Aquaculture*, vol. 306, no. 1–4, pp. 80–84, 2010.

[64] M. E. Hernandez, J. L. Reyes, C. Gomez-Lojero, M. S. Sayavedra, and E. Melendez, "Inhibition of the renal uptake of p-aminohippurate and tetraethylammonium by the antioxidant ethoxyquin in the rat," *Food and Chemical Toxicology*, vol. 31, no. 5, pp. 363–367, 1993.

[65] J. L. Reyes, M. Elisabeth Hernández, E. Meléndez, and C. Gómez-Lojero, "Inhibitory effect of the antioxidant ethoxyquin on electron transport in the mitochondrial respiratory chain," *Biochemical Pharmacology*, vol. 49, no. 3, pp. 283–289, 1995.

[66] P. E. Joner, "Butylhydroxyanisol (BHA), butylhydroxytoluene (BHT) and ethoxyquin (EMQ) tested for mutagenicity," *Acta Veterinaria Scandinavica*, vol. 18, no. 2, pp. 187–193, 1977.

[67] T. Ohta, M. Moriya, Y. Kaneda et al., "Mutagenicity screening of feed additives in the microbial system," *Mutation Research*, vol. 77, no. 1, pp. 21–30, 1980.

[68] E. Zeiger, "Mutagenicity of chemicals added to foods," *Mutation Research*, vol. 290, no. 1, pp. 53–61, 1993.

[69] A. Rannug, U. Rannug, and C. Ramel, "Genotoxic effects of additives in synthetic elastomers with special consideration to the mechanism of action of thiurams and dithiocarbamates," *Progress in Clinical and Biological Research*, vol. 141, pp. 407–419, 1984.

[70] B. S. Reddy, D. Hanson, L. Mathews, and C. Sharma, "Effect of micronutrients, antioxidants and related compounds on the mutagenicity of 3,2'-dimethyl-4-aminobiphenyl, a colon and breast carcinogen," *Food and Chemical Toxicology*, vol. 21, no. 2, pp. 129–132, 1983.

[71] A. Błaszczyk and J. Skolimowski, "Apoptosis and cytotoxicity caused by ethoxyquin and two of its salts," *Cellular and Molecular Biology Letters*, vol. 10, no. 1, pp. 15–21, 2005.

[72] A. Błaszczyk, "DNA damage induced by ethoxyquin in human peripheral lymphocytes," *Toxicology Letters*, vol. 163, no. 1, pp. 77–83, 2006.

[73] J. J. Skolimowski, B. Cieślińska, M. Zak, R. Osiecka, and A. Błaszczyk, "Modulation of ethoxyquin genotoxicity by free radical scavengers and DNA damage repair in human lymphocytes," *Toxicology Letters*, vol. 193, no. 2, pp. 194–199, 2010.

[74] J. J. P. Gille, P. Pasman, C. G. M. van Berkel, and H. Joenje, "Effect of antioxidants on hyperoxia-induced chromosomal breakage in Chinese hamster ovary cells: protection by carnosine," *Mutagenesis*, vol. 6, no. 4, pp. 313–318, 1991.

[75] T. H. Rabbitts, "Chromosomal translocations in human cancer," *Nature*, vol. 372, no. 6502, pp. 143–149, 1994.

[76] A. J. de Koning, "An antioxidant for fish meal," Republic of South Africa Patent, 970894, 1997.

[77] National Toxicology Program (NTP), *Toxicology and Carcinogenesis Studies of 1,2-Dihydro-2,2,4-Trimethylquinoline (CAS no. 147-47-7) in F344/N Rats and B6C3DF₁ mice (Dermal Studies) and the Initiation/Promotion (Dermal Study) in Female Sencar Mice*, Technical Report Series no. 456, U.S. Department of Health and Human Services, 1997, http://ntp.niehs.nih.gov/ntp/htdocs/LT_rpts/tr456.pdf.

[78] K. Sitarek and A. Sapota, "Maternal-fetal distribution and prenatal toxicity of 2,2,4-trimethyl-1,2-dihydroquinoline in the rat," *Birth Defects Research Part B—Developmental and Reproductive Toxicology*, vol. 68, no. 4, pp. 375–382, 2003.

[79] A. Błaszczyk, J. Skolimowski, and A. Materac, "Genotoxic and antioxidant activities of ethoxyquin salts evaluated by the comet assay," *Chemico-Biological Interactions*, vol. 162, no. 3, pp. 268–273, 2006.

[80] A. Błaszczyk and J. Skolimowski, "Apoptosis and cytotoxicity caused by ethoxyquin salts in human lymphocytes *in vitro*," *Food Chemistry*, vol. 105, no. 3, pp. 1159–1163, 2007.

[81] A. Błaszczyk and J. Skolimowski, "Evaluation of the genotoxic and antioxidant effects of two novel feed additives (ethoxyquin complexes with flavonoids) by the comet assay and micronucleus test," *Food Additives and Contaminants*, vol. 24, no. 6, pp. 553–560, 2007.

[82] A. Błaszczyk and J. Skolimowski, "Preparation of ethoxyquin salts and their genotoxic and antioxidant effects on human lymphocytes," *Arkivoc*, vol. 2007, no. 6, pp. 217–229, 2007.

[83] A. Augustyniak, A. Niezgoda, J. Skolimowski, R. Kontek, and A. Błaszczyk, "Cytotoxicity and genotoxicity of ethoxyquin dimers," *Bromatologia i Chemia Toksykologiczna*, vol. 45, pp. 228–234, 2012.

[84] Y. Aoki, A. Kotani, N. Miyazawa et al., "Determination of ethoxyquin by high-performance liquid chromatography with fluorescence detection and its application to the survey of residues in food products of animal origin," *Journal of AOAC International*, vol. 93, no. 1, pp. 277–283, 2010.

16

Bioactivity of Nonedible Parts of *Punica granatum* L.: A Potential Source of Functional Ingredients

Nawraj Rummun,[1] **Jhoti Somanah,**[2] **Srishti Ramsaha,**[3]
Theeshan Bahorun,[4] **and Vidushi S. Neergheen-Bhujun**[3]

[1] *Department of Health Sciences, Faculty of Science, University of Mauritius, Réduit, Mauritius*
[2] *Department of Biosciences, and ANDI Centre of Excellence for Biomedical and Biomaterials Research,*
  University of Mauritius, Réduit, Mauritius
[3] *Department of Health Sciences, Faculty of Science and ANDI Centre of Excellence for Biomedical and Biomaterials Research,*
  University of Mauritius, Réduit, Mauritius
[4] *ANDI Centre of Excellence for Biomedical and Biomaterials Research, University of Mauritius, Réduit, Mauritius*

Correspondence should be addressed to Vidushi S. Neergheen-Bhujun; v.neergheen@uom.ac.mu

Academic Editor: Fabienne Remize

Punica granatum L. has a long standing culinary and medicinal traditional use in Mauritius. This prompted a comparative study to determine the bioefficacy of the flower, peel, leaf, stem, and seed extracts of the Mauritian *P. granatum*. The flower and peel extracts resulting from organic solvent extraction exhibited strong antioxidant activities which correlated with the high levels of total phenolics, flavonoids, and proanthocyanidins. The peel extract had the most potent scavenging capacity reflected by high Trolox equivalent antioxidant capacity value (5206.01 ± 578.48 μmol/g air dry weight), very low IC_{50} values for hypochlorous acid (0.004 ± 0.001 mg air dry weight/mL), and hydroxyl radicals scavenging (0.111 ± 0.001 mg air dry weight/mL). Peel extracts also significantly inhibited *S. mutans* ($P < 0.001$), *S. mitis* ($P < 0.001$), and *L. acidophilus* ($P < 0.05$) growth compared to ciprofloxacin. The flower extract exhibited high ferric reducing, nitric oxide scavenging, and iron (II) ions chelation and significantly inhibited microsomal lipid peroxidation. Furthermore, it showed a dose-dependent inhibition of xanthine oxidase with an IC_{50} value of 0.058 ± 0.011 mg air dry weight/mL. This study showed that nonedible parts of cultivated pomegranates, that are generally discarded, are bioactive in multiassay systems thereby suggesting their potential use as natural prophylactics and in food applications.

1. Introduction

Punica granatum L. fruit or fruit juice has for the past decade been advocated as an interesting functional food that can confer health benefits beyond basic nutrition [1, 2]. *P. granatum* L. belongs to the family of Punicaceae and is indigenous to the Himalayas in northern India and to Iran [3] but has grown and been naturalized in a number of Asian and African countries including Mauritius. The edible and nonedible parts (Figure 1) have been reported to treat different pathological conditions in different folklore medicine [4–6]. Documented use of pomegranate in Mauritian folklore medicine includes ingestion of macerated bark extracts to

treat asthma, chronic diarrhea, chronic dysentery, relaxation of the larynx, and intestinal worms [7].

Pomegranate extracts are known for their antidiabetic, antibacterial, anticarcinogenic, antiatherogenic, and antihypertensive potential amongst others [3]. Pomegranate juice is also used as mouthwash in oral hygiene [8]. Consumption of pomegranate juice has been linked with a decrease in inflammatory biomarkers levels and oxidation of both proteins and lipids in a randomized placebo-controlled trial [9]. In the same vein, the beneficial effect of pomegranate juice was reported in an initial phase II clinical trial in patients with prostate cancer [10]. The health benefits of pomegranate have been ascribed to the pluripharmacological effects of

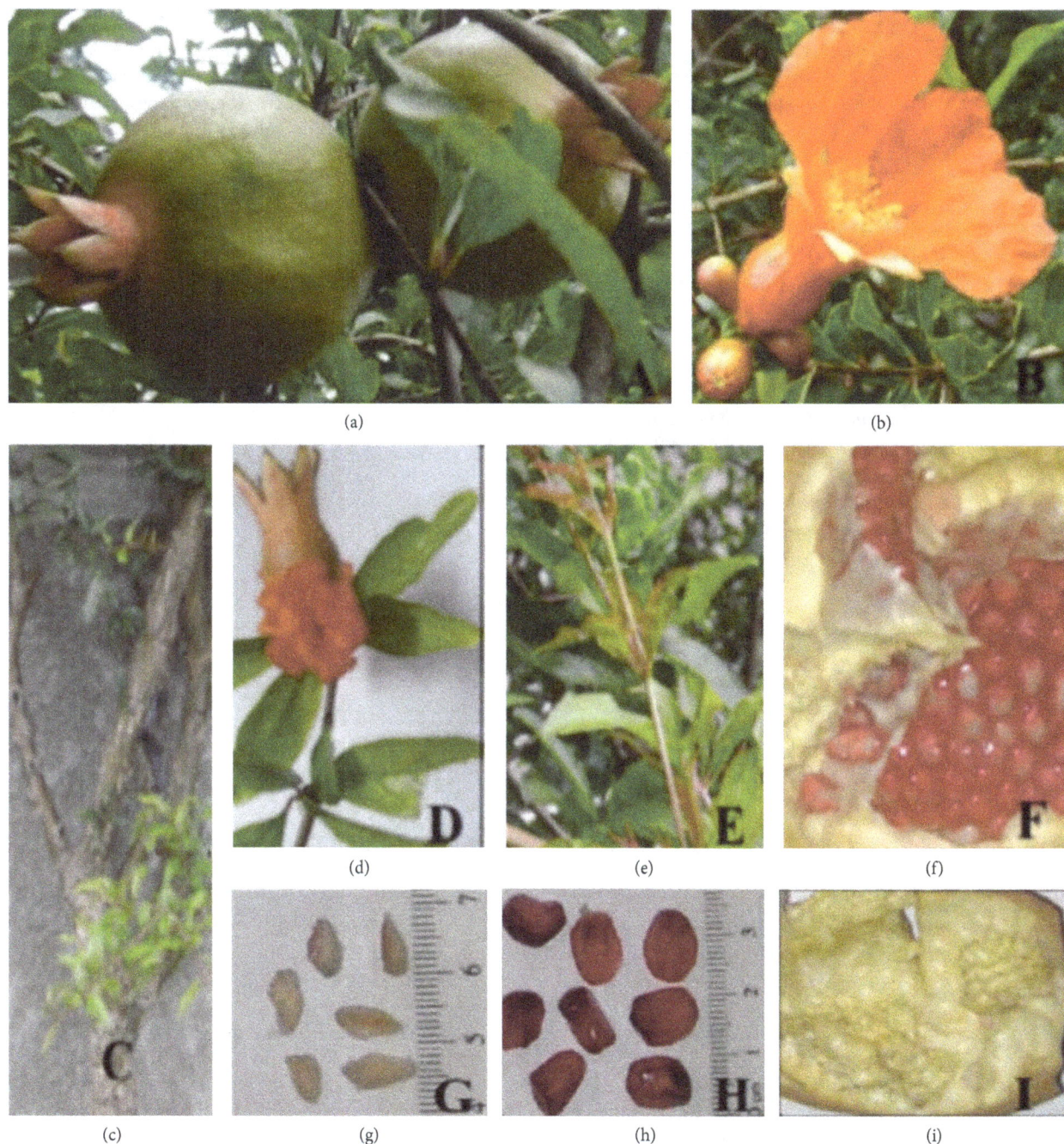

Figure 1: Different anatomical parts of *P. granatum* tree and fruit. (a) Unripe fruit; (b) flower; (c) stem; (d) flower and tubular calyx; (e) young leaves at the branch endings; (f) the fruit's rind with membranous extensions forming compartments which contain the juicy arils; (g) seeds; (h) arils (juicy pulp coating the seed); (i) pomegranate inner membrane.

the secondary metabolites more specifically its polyphenolic compounds present in relatively high concentrations [11–13].

The phytophenolic compositions vary differently in the edible and nonedible parts of the plants and have been widely investigated. Pomegranate fruit (peel, aril, seeds, and juice) has been reported to be rich in phenolic acids, flavanols, flavones, flavonones, anthocyanidins, and anthocyanin [3]. Literature data reported glycated anthocyanins (pelargonidin 3,5-diglucoside, pelargonidin 3-glucoside) apart from

phenolic compounds common to the edible parts like gallic acid in the flowers [14], while the nonedible parts of pomegranate comprising leaves, roots, and stem contained apigenin, punicalin, punicalagin, and luteolin [3, 15, 16]. This rich polyphenolic composition has been intrinsically linked to the pluripharmacological effects of pomegranate extracts. However, it should be noted that sources of variation in the level of phytochemicals and nutrients arising from genetic variability of a naturalized plant in addition to geographical and environmental factors can result in diverse polyphenolic compositions that modulate bioactivity level.

Whilst recent years have witnessed a surge in the scientific evaluation of the ethnopharmacological uses of pomegranate, limited works have been reported on the assessment of the nonedible discarded parts as a source of bioactive ingredients for the functional food industry. Thus, this study aimed at determining the antibacterial, anti-inflammatory, and antioxidant potential of the nonedible parts of the Mauritian cultivar of pomegranate with the view of promoting their utilization in functional health and in potential food applications.

2. Methodology

2.1. Chemicals and Microorganisms.
Streptococcus mutans (ATCC[R] 5175), *Streptococcus mitis* (ATCC[R] 6249), and *Lactobacillus acidophilus* (ATCC[R] 4356) were purchased from ATCC. Brain heart infusion agar (BHI) and de Man, Rogosa and Sharpe agar (MRS) were purchased from Sigma-Aldrich, Germany. BHI was used for growth of *S. mutans* and *S. mitis* while MRS agar was utilized for *L. acidophilus*. All other chemicals used were of analytical grade.

2.2. Plant Materials.
Pomegranate plant and fruit parts were collected from a domesticated plant growing in a backyard in "Triolet" village situated in the Pamplemousses district in the northern part of Mauritius Island, during the month of September 2011 and authenticated by the herbarium of Mauritius Sugar Industry Research Institute, Réduit. *Punica granatum* leaves, stems, flowers, and fruits were collected from the same plant. The latter were air dried and the samples homogenized to a fine powder prior to extraction.

2.3. Preparation of Extracts.
The plant material was extracted thrice with 70% methanol (1 : 3, w/v) and allowed to macerate each time at 4°C for 24 hours. The filtrates were pooled together and concentrated *in vacuo* at 37°C. The concentrated aqueous extract was partitioned in dichloromethane to remove fats and chlorophyll, and the aqueous phase was then collected and lyophilized. The lyophilized powders, thereof derived, were dissolved in deionized water and in 100% methanol to a concentration of 1 g of air dried mass to 5 mL for the subsequent tests.

2.4. Total Phenolic Content (TPC).
The total phenolic content was estimated using the Folin-Ciocalteu assay adapted from Neergheen et al. [17]. The reaction mixture in a final volume of 5 mL contained 0.25 mL of the extracts, 3.50 mL of distilled water, and 0.25 mL of Folin-Ciocalteu reagent. After 3 minutes, 0.75 mL of 20% sodium carbonate solution was added. The tubes were mixed thoroughly and heated for 40 minutes in a water-bath set at 40°C and then allowed to cool. The absorbance of the blue coloration was read at 685 nm against a blank. Total phenolics were calculated with respect to a gallic acid standard curve (stock solution 250 μg/mL) and results expressed in mg of gallic acid equivalent (GAE)/g air dry weight (ADW) of plant material.

2.5. Total Proanthocyanidin Content (TPrC).
A modified HCl/Butan-1-ol assay adapted from Porter et al. [18] was used for the quantification of total proanthocyanidin content of the methanolic plant extracts. The reaction mixture in each tube contained, in a final volume of 3.35 mL, the following in order of addition: 0.25 mL extract, 3 mL of n-BuOH/HCl (95 : 5 v/v), and 0.1 mL of 2% NH_4Fe $(SO_4)_2$·12 H_2O in 2 M HCl, and the tubes were incubated for 40 minutes at 95°C. A red coloration was developed, and the absorbance was read at 550 nm against a blank standard containing 0.25 mL n-BuOH/HCl (95 : 5 v/v) instead of extract. The amount of proanthocyanidins in the extracts was calculated, in triplicates, with respect to a cyanidin chloride standard curve (stock solution 0.1 mg/mL). Results were expressed in mg of cyanidin chloride equivalent (CCE)/g ADW of plant material.

2.6. Total Flavonoid Content (TFC).
Total flavonoids were measured using a colorimetric assay adapted from Zhishen et al. [19]. A total of 150 μL of 5% aqueous $NaNO_2$ was added to 2.50 mL of extract. After 5 min, 150 μL of 10% aqueous $AlCl_3$ was added. A total of 1 mL of 1 M NaOH was added 1 min after the addition of aluminum chloride. The absorbance of the solution was measured at 510 nm. Flavonoid contents were expressed in μg quercetin/g of ADW of plant material.

2.7. Ferric Reducing Antioxidant Power Assay (FRAP).
The reducing power of the extracts was assessed using the method of Benzie and Strain [20]. A total of 100 μL of sample was added to 300 μL of distilled water, followed by 3 mL of FRAP reagent (40 mM HCl and 20 mL of 20 mM ferric chloride in 200 mL of 0.25 M sodium acetate buffer at pH 3.6). The absorbance was read at 593 nm after 4 min of incubation at 37°C. Results were expressed in μmol Fe^{2+} equivalent/g of ADW of plant material.

2.8. Trolox Equivalent Antioxidant Capacity (TEAC) Assay.
The free radical scavenging capacity of the extracts was measured by the TEAC assay according to the method of Campos and Lissi [21]. A total of 0.50 mL of diluted plant extract was added to 3 mL of the $ABTS^{•+}$ solution generated by a reaction between 2,2-azino-bis(3-ethyl-benzthiazoline-6-sulfonic acid) diammonium salt (ABTS, 0.50 mM) and activated MnO_2 (1 mM) in phosphate buffer (0.10 M, pH 7). Decay in absorbance was monitored at 734 nm for 15 min. TEAC values are expressed in μmol Trolox equivalent/g of ADW.

2.9. Iron (II) Chelating Activity.
The method of Neergheen et al. [17] was adapted to assess the iron (II) chelating activity of the extracts. The reaction mixture contained, in order of addition, 200 μL of plant extract (varied concentrations) and 50 μL of $FeCl_2$·$4H_2O$ (0.5 mM). The reaction volume was made up to 1 mL with distilled deionised water and incubated for 5 minutes at room temperature. After incubation, 50 μL of FerroZine (2.5 mM) was added and the purple coloration formed was read at 562 nm. The absorbance of the reaction mixture was read both before and after the addition of

FerroZine to account for possible interferences caused by the plant extract. The controls contained all the reaction reagents and water instead of the extract or the positive control substance. EDTA was used as a positive control. The percentage chelating activity was calculated and results were expressed as mean IC_{50} (mg ADW/mL).

2.10. Scavenging of Hypochlorous Acid (HOCl). The ability of the extracts to scavenge HOCl was assessed essentially as described by Neergheen et al. [17]. Briefly, the reaction mixture contained $100 \mu L$ taurine (10 mM), $100 \mu L$ HOCl (1 mM), $100 \mu L$ plant extract (variable concentrations), and $700 \mu L$ phosphate saline buffer (pH 7.4) in a final volume of 1 mL. The solution was mixed thoroughly and incubated for 10 minutes at ambient temperature. The sample was then assayed for taurine chloramine by adding $10 \mu L$ potassium iodide to the reacting mixture. The I_2 released was determined spectrophotometrically at 350 nm in the presence of excess I^- as I_3^-. The absorbance of the reaction mixture was read both before and after the addition of potassium iodide as to cater for possible interferences caused by the plant extract. The analyses were made in triplicates and the results were expressed as IC_{50} mg ADW/mL for the plant extracts.

2.11. Inhibition of Deoxyribose Degradation. The hydroxyl radical scavenging potential of the extracts was determined using the deoxyribose assay [22]. The reacting mixture contained in a final volume of 1 mL the following reagents, order of addition indicated: $200 \mu L$ of 100 mM KH_2PO_4-KOH, $200 \mu L$ of 0.5 mM $FeCl_3$, $100 \mu L$ of 1 mM EDTA, $100 \mu L$ sample, $200 \mu L$ of 15 mM deoxyribose, $100 \mu L$ of 10 mM H_2O_2, and $100 \mu L$ of 1 mM ascorbic acid. Reaction mixtures were incubated at 37°C for 1 hour.

At the end of the incubation period, 1 mL 1% (w/v) thiobarbituric acid (TBA) was added to each mixture followed by the addition of 1 mL 2.8% (w/v) trichloroacetic acid (TCA). The solutions were heated in a water bath at 80°C for 20 min to develop the pink coloured MDA-$(TBA)_2$ adduct. The MDA-$(TBA)_2$ chromogen was extracted into 3 mL butan-1-ol and its absorbance measured at 532 nm. The analyses were made in triplicates and the results were expressed as IC_{50} g ADW/mL for the plant extracts.

2.12. Inhibition of Microsomal Lipid Peroxidation. Beef liver microsomes were prepared by tissue homogenization as described by Neergheen et al. [17]. The formation of malondialdehyde, measured as thiobarbituric reactive substances (TBARS), was used to monitor microsomal lipid peroxidation. The reaction mixture contained in a final volume of 1 mL the following: $200 \mu L$ of 3.4 mM phosphate buffered saline (pH 7.4), $200 \mu L$ of 0.5 mg/mL microsomal protein, $400 \mu L$ of sample (variable concentrations), $100 \mu L$ of 1 mM $FeCl_3$, and $100 \mu L$ of 1 mM ascorbate. The mixture was incubated for 1 hour at 37°C.

At the end of the incubation period, $100 \mu L$ 2% (w/v) BHT was added followed by 1 mL 1% (w/v) thiobarbituric acid and 2.8% (w/v) trichloroacetic acid. The solutions were heated in a water bath at 80°C for 20 min to develop the pink coloured

MDA-$(TBA)_2$ adduct. As turbidity was encountered, the MDA-$(TBA)_2$ chromogen was extracted into 2 mL butan-1-ol and its absorbance measured at 532 nm. The inhibition of microsomal lipid peroxidation was calculated and results are expressed as mean IC_{50} (mg ADW/mL).

2.13. Superoxide Anion Radical Scavenging Assay. The superoxide anion scavenging activity of the pomegranate extracts was measured according to the modified method of Kumar et al. [23]. One mL of $156 \mu M$ of nitroblue tetrazolium (NBT) aqueous solution and 1 mL of $200 \mu M$ beta-nicotinamide adenine dinucleotide reduced disodium salt hydrate (NADH) aqueous solution were mixed together, followed by the addition of 1 mL of aqueous pomegranate extract (varied concentration). The reaction was started by adding $100 \mu L$ of $60 \mu M$ phenazine methosulphate (PMS) aqueous solution. The reaction mixture was incubated at 25°C for 20 minutes and the absorbance was measured at 560 nm against control sample. Ascorbic acid was used as a positive control. IC_{50} value was calculated from the dose-dependent curve obtained by plotting antioxidant activity (%) against a concentration range for each pomegranate extract. The antioxidant activity was calculated as follows:

$$\text{antioxidant activity \%} = \left[\frac{(A_0 - A_1)}{A_0} \right] \times 100, \quad (1)$$

where A_0 is the absorbance of the control (reaction mixture without test sample), and A_1 is the absorbance of the test sample.

2.14. Nitric Oxide Radical Inhibition Assay. Nitric oxide radical inhibition was evaluated according to the modified method of Sunil et al. [24]. Griess Illosvoy' reagent was modified by using 0.1% (w/v) naphthylethylenediamine dihydrochloride. The reaction mixture contained 0.5 mL of extracts (variable concentrations), 2 mL of 10 mM aqueous sodium nitroprusside, and 0.5 mL phosphate saline buffer. The mixture was incubated at 25°C for 180 minutes. 0.5 mL of the reaction mixture was pipetted out, and 2 mL of Griess Illosvoy's reagent (0.33% sulphanilic acid in 20% glacial acetic acid and 0.1% naphthylethylenediamine dichloride) was added, mixe,d and allowed to stand for 30 minutes. The absorbance of the pink chromophore formed was measured at 546 nm. Ascorbic acid was used as positive control and the percentage antioxidant activity was calculated. IC_{50} value was calculated from the dose-dependent curve obtained by plotting antioxidant activity (%) against a concentration range for each pomegranate extract.

2.15. Xanthine Oxidase (XO) Inhibition Assay. Spectrophotometric determination of XO inhibitory activity measuring uric acid production from xanthine substrate was used. The method was adapted from Havlik et al. [25] with some modifications. The mixture consisted of $250 \mu L$ extract (varied concentrations), $400 \mu L$ 0.12 M phosphate buffer (pH 7.5), and $330 \mu L$ xanthine (8 mM in same buffer). The reaction was initiated by adding $20 \mu L$ of xanthine oxidase (0.5 U/mL

TABLE 1: Total phenolics, total flavonoids, and total proanthocyanidin content of pomegranate parts extracts.

	TPC mg GAE/g ADW ± SD	TFC mg QE/g ADW ± SD	TPrC mg CCE/g ADW ± SD
Flower	336.51 ± 0.70^a	213.54 ± 3.14^a	$1.46 \pm 0.06^{a,b}$
Peel	190.27 ± 0.54^b	180.10 ± 1.31^b	2.48 ± 0.08^a
Leaf	87.81 ± 0.47^c	63.89 ± 0.62^c	0.21 ± 0.01^b
Stem	52.92 ± 0.62^d	41.36 ± 0.52^d	0.32 ± 0.01^b
Seed	0.65 ± 0.00^e	0.33 ± 0.00^e	0.13 ± 0.00^b
LSD value at 5% significance	0.96	2.85	1.29

ADW: air dry weight; CCE: cyanidin chloride equivalent; GAE: gallic acid equivalent; QE: quercetin equivalent; TPC: total phenolic content; TFC: Total Flavonoid content; TPrC: Total proanthocyanidin content. Different superscripts between rows in individual columns represent significant difference between extracts. Data expressed as mean ± standard deviation ($n = 3$).

in same buffer) which was prepared immediately before use. The tubes were incubated at room temperature for 5 minutes and the reaction stopped by the addition of 200 μL 1 M HCl. Absorbance for formation of uric acid was read at 295 nm. Allopurinol was used as the positive control (concentration range: 25–350 μM). The % inhibition was calculated and activity of extract presented as calculated IC$_{50}$ (μM).

2.16. Determination of Minimum Inhibition Concentration of Punica granatum. Sterilized molten agar was dispensed into sterile Petri dishes and allowed to solidify. Microbial suspension (150 μL) containing approximately 1.5×10^8 CFU/mL was spread evenly over the surface of the solidified medium and left to air dry. Meanwhile, 20 μL of sample extracts and ciprofloxacin (2 mg/mL) were loaded separately onto sterile oven-dried paper discs and placed firmly onto medium using forceps. Each plate consisted of four impregnated discs: two extracts, one positive control, and one extract + positive control. The experiment was performed in quadruplicate.

The Petri dishes were inverted and incubated at 37°C for 24 hours. After the incubation period, the diameter of the zone of inhibition, defined as the area which was devoid of or had minimal cell growth, was measured to the nearest millimeter. The antimicrobial activity of the extract was determined by the zone of inhibition of the extracts; a higher inhibition zone indicated a more potent antimicrobial effect of the extract.

2.17. Statistical Analysis. All the antioxidant assays were carried out in triplicate and the results recorded were expressed as mean ± standard deviation. All charts including standard curves, dose response curves, and bar charts were generated using Microsoft Excel software (Version 2010) and GraphPad Prism, version 6.01, from GraphPad Software (San Diego, CA, USA). Correlation between phytoconstituent and antioxidant activity was carried out using the Pearson correlation on SPSS (version 17.0). ANOVA (single factor) was performed in Microsoft Excel software (Version 2010) to test for significant difference in mean values of the different extracts for

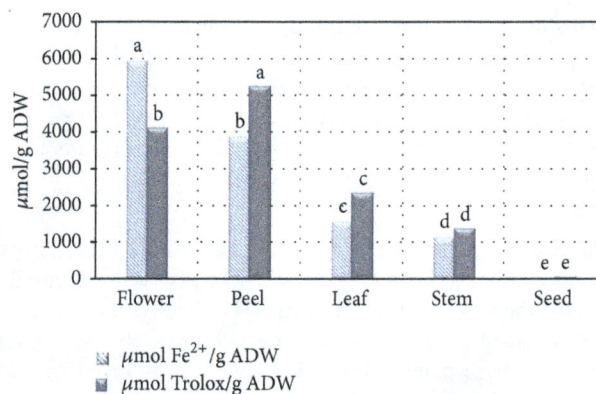

FIGURE 2: TEAC and FRAP values of different pomegranate extracts. ADW: air dry weight. Different superscripts between columns represent significant difference between extracts. Data expressed as mean ± standard deviation μmol Trolox equivalent/g air dry weight for TEAC ($n = 3$); LSD = 920.44, at 5% significance. Data expressed as mean ± standard deviation μmol Fe^{2+}/g ADW for FRAP ($n = 3$); LSD = 101.34, at 5% significance.

each assay. To test for null hypothesis, the least significant difference between extracts, for each independent assay, was calculated.

3. Results

3.1. Polyphenolic Content. Total phenolics of the extracts ranged between 0.65 ± 0.004 mg GAE/g ADW and 336.51 ± 0.70 mg GAE/g ADW with the highest content measured in the flower (Table 1). The amount of total phenolics differed significantly between the extracts ($P < 0.05$). Total flavonoids were between 0.332 ± 0.003 mg QE/g ADW and 213.54 ± 3.14 mg QE/g ADW. Pomegranate flower extract had the most prominent flavonoid level followed by peel, leaf, and stem, with only negligible amount measured in the seed extract. Significant differences were observed in flavonoid content among the extracts ($P < 0.05$). Relatively lower amount of total proanthocyanidins was present in the samples compared to the amount of total phenolics and flavonoid contents (Table 1).

3.2. Antioxidant Activities. TEAC value ranged from 14.04 ± 2.40 μmol Trolox/g ADW to 5206.01 ± 578.48 μmol Trolox/g ADW with the peel exhibiting the highest TEAC value. The ferric reducing potential ranged between 6.29 ± 0.38 μmol Fe^{2+}/g ADW and 5933.00 ± 54.06 μmol Fe^{2+}/g ADW ($P < 0.05$) (Figure 2). The flowers had the highest ferric reducing potential which was statistically different from the activity of the other extracts ($P < 0.05$) (Figure 2).

All extracts showed dose-dependent iron (II) chelating activity. However, pomegranate flower exhibited the highest iron (II) cation chelating activity with the lowest calculated IC$_{50}$ value (Table 2). Statistically significant differences were observed between the IC$_{50}$ values ($P < 0.05$) with pomegranate flower, peel, and stem being more potent than the leaf and the seed. Similarly, most of the extracts except

TABLE 2: Antioxidant activities of different pomegranate parts extract.

Extract	Fe(II) chelating activity	HOCl scavenging activity	Inhibition of deoxyribose degradation	Inhibition of lipid peroxidation	Nitric oxide inhibition	Superoxide scavenging
Flower	0.113 ± 0.006^c	0.012 ± 0.001^b	0.220 ± 0.041^c	0.047 ± 0.006^b	0.396 ± 0.002^e	0.175 ± 0.001^b
Peel	0.157 ± 0.006^c	0.004 ± 0.001^b	0.111 ± 0.001^c	0.333 ± 0.058^b	0.668 ± 0.001^d	0.089 ± 0.001^c
Leaf	0.713 ± 0.006^b	0.017 ± 0.002^b	3.752 ± 0.091^b	0.601 ± 0.100^b	18.155 ± 0.005^b	0.072 ± 0.001^d
Stem	0.397 ± 0.030^c	N.A	0.480 ± 0.031^c	1.700 ± 0.173^b	4.831 ± 0.001^c	0.040 ± 0.001^e
Seed	15.400 ± 0.310^a	5.200 ± 0.400^a	14.300 ± 0.760^a	35.000 ± 2.646^a	48.641 ± 0.001^a	2.523 ± 0.001^a
LSD value at 5% significance	0.300	0.371	0.630	2.162	3.231	1.490

N.A: not available. Data expressed as mean $IC_{50} \pm$ standard deviation (mg/mL) ($n = 3$); different superscripts between rows in individual columns represent significant difference between extracts.

the stem extract showed dose-dependent hypochlorous acid scavenging activity with the peel extract being the most potent HOCl scavenger. The calculated IC_{50} value ranged between 0.004 ± 0.001 mg ADW/mL and 5.200 ± 0.400 mg ADW/mL (Table 2). The calculated IC_{50} value of flower, peel, and leaf extracts differed significantly from that of the seed ($P < 0.05$). However, the flower, peel, and leaf were observed to scavenge hypochlorous acid more efficiently than ascorbic acid, used as a positive control ($IC_{50} = 5.63 \pm 0.21$ mg/mL).

The samples analysed were also strong hydroxyl radical scavengers. The results were regarded as indications of hydroxyl radical scavenging propensity by virtue of their ability to inhibit deoxyribose degradation (Figure 3). Pomegranate peel afforded the highest protection, followed by flower, stem, leaf, and seed extracts (Table 2). Statistically significant differences were observed in IC_{50} values among the extracts ($P < 0.05$).

The degree of microsomal lipid peroxidation inhibition induced by Fe^{3+}/ascorbate was evaluated by measuring the formation of MDA-$(TBA)_2$ adduct spectrophotometrically. All the extracts protected microsome against lipid peroxidation in a dose-dependent manner. Pomegranate flower offered the most prominent protection followed by peel, leaf, and stem extracts. No significant difference was observed among flower, peel, leaf and stem extracts as compared to the seed extract ($P < 0.05$). Gallic acid used as positive control (IC_{50} value of 0.014 ± 0.002 mg/mL) was more potent than the flower extract.

A similar trend was observed for nitric oxide radical inhibition. The flower and peel extracts were the most potent scavenger of $NO^{\bullet}$ with the lowest calculated IC_{50} value (Table 2) and were more effective than ascorbic acid used as positive control (IC_{50} 1253.141 ± 0.002 mg/mL) ($P < 0.05$).

All the extracts exhibited a dose-dependent effect against superoxide radical. However, a different trend of activity for the extracts under study was observed, the stem extract being the most potent followed by leaf, peel, flower, and seed extract. The stem extract was a very powerful scavenger of superoxide (Table 2), 100 folds more powerful than ascorbic acid (IC_{50} 14.191 ± 0.001 mg/mL).

3.3. Anti-Inflammatory Effect of P. granatum Extracts.
The degree of inhibition of xanthine oxidase by the extracts

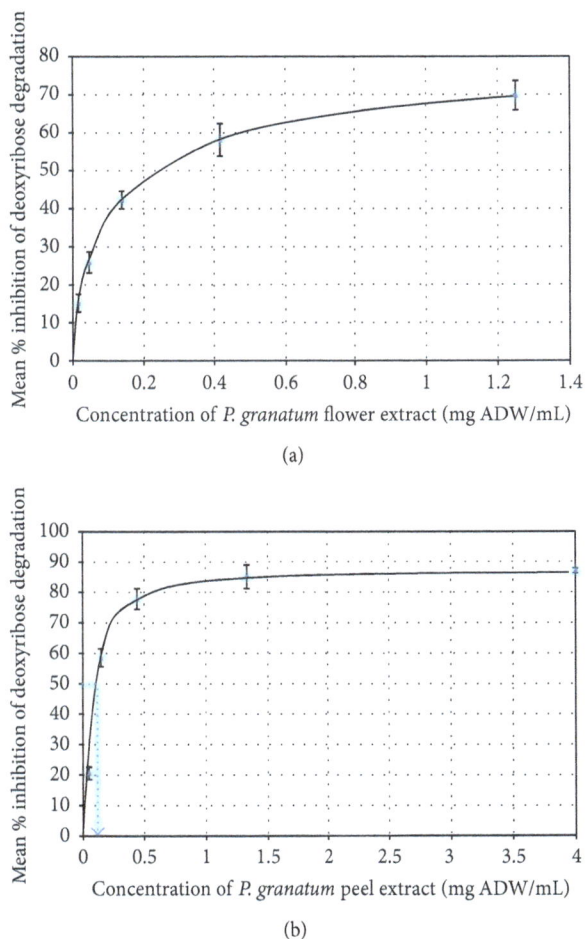

(a)

(b)

FIGURE 3: Dose-dependent hydroxyl radical induced deoxyribose degradation inhibition by pomegranate flower and peel extracts. ADW: air dry weight; data are representative of mean ± standard deviation of three replicates. IC_{50} values were extrapolated from the graphs.

was evaluated by measuring the formation of uric acid spectrophotometrically. Only the flower extract showed xanthine oxidase inhibitory activity, at the concentration range tested. Pomegranate flower extract showed a dose-dependent inhibition of xanthine oxidase with an IC_{50} value

(a)

(b)

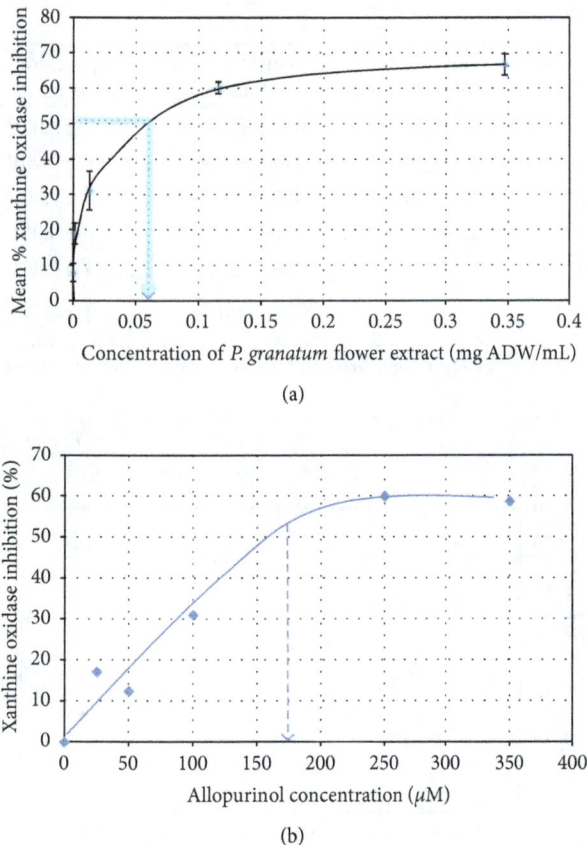

FIGURE 4: Dose-dependent xanthine oxidase inhibitory activity of pomegranate flower extract and allopurinol. ADW: air dry weight. Data are representative of mean ± standard deviation of three replicates. IC$_{50}$ value was extrapolated from the graph.

of 0.058 ± 0.011 mg ADW/mL (Figure 4). Allopurinol, used as a positive control, was however a more potent inhibitor (IC$_{50}$ value: 0.0055 ± 0.0002 mg/mL).

3.4. Antibacterial Activity.

The antimicrobial activity of the extracts from *P. granatum* on some indigenous oral microbiotas known to rapidly colonize smooth surfaces and crevices of the teeth and gums causing dental plaque and tooth caries was evaluated. Using the disc diffusion method, it was noted that bacterial growth was minimal in the presence of all concentrated extracts. Peel extract showed greater antibacterial activity and produced the highest inhibition zones against *S. mutans*, *S. mitis*, and *L. acidophilus* (19.75 mm, 25 mm, and 14.75 mm, resp.) (Table 3). This inhibitory effect was observed to be significantly greater than that of the positive control ciprofloxacin ($P < 0.001$). Leaf extract produced the second highest inhibition zones of 16, 18.25, and 8.75 mm against *S. mutans*, *S. mitis*, and *L. acidophilus* respectively, followed closely by the stem extract. The antibacterial activity effect of the flower extract was much less pronounced compared to its plant counterparts. Although bacterial cell proliferation was minimal, it was still significant compared to the activity of ciprofloxacin ($P < 0.001$).

TABLE 3: Zone of inhibition or minimum growth (mm) by various parts of *Punica granatum* on *S. mutans*, *S. mitis*, and *L. acidophilus*.

	Mean zone of inhibition ± SD (mm)		
	Streptococcus mutans	*Streptococcus mitis*	*Lactobacillus acidophilus*
Extract only			
Peel	19.75 ± 0.50***	25.00 ± 0.00***	14.75 ± 2.22*
Flower	12.25 ± 0.50*	14.75 ± 0.50***	8.38 ± 0.75
Leaf	16.00 ± 0.00***	18.25 ± 0.50***	8.75 ± 0.50
Stem	14.75 ± 0.50***	19.00 ± 1.15***	8.75 ± 0.50
Extract + ciprofloxacin			
Peel	10.75 ± 0.96	22.41 ± 1.41##	11.50 ± 1.00#
Flower	11.25 ± 0.96	13.00 ± 0.00##	8.13 ± 1.03
Leaf	12.50 ± 0.58##	18.82 ± 0.82##	8.25 ± 0.50
Stem	10.0 ± 0.0	12.08 ± 0.58#	9.25 ± 1.26
Ciprofloxacin	10.25 ± 0.50	8.75 ± 1.26	8.75 ± 0.50

Values are expressed as mean ± standard deviation. Significance compared to ciprofloxacin (positive control): $P < 0.05$*,#, $P < 0.01$##, $P < 0.001$***.

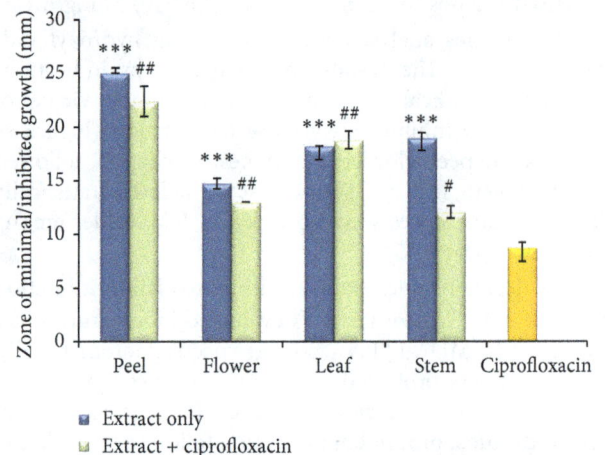

■ Extract only
▨ Extract + ciprofloxacin

FIGURE 5: Zone of inhibition or minimum growth (mm) by various parts of *Punica granatum* on *Streptococcus mitis*. Values are expressed as mean where error bars represent standard deviation. Significance compared to ciprofloxacin (positive control): $P < 0.05$#, $P < 0.01$##, $P < 0.001$***.

In addition to the study of sole plant extracts, bacterial strains were exposed to a combinational treatment (extract + ciprofloxacin). Such treatment produced inhibition zones that ranged between 12.25 and 19.75 mm against *S. mutans*, 14.75 and 25.0 mm against *S. mitis*, and 8.13 and 11.50 mm against *L. acidophilus*. The following trends were observed: leaf > flower > peel > stem, peel > leaf > flower > stem (Figure 5), and peel < stem < leaf < flower, respectively. For the majority of the plant extracts tested, the combinational treatment proved to exhibit a more efficient antibacterial effect that significantly exceeded that of ciprofloxacin, but not that of individual plant extracts.

4. Discussion and Conclusion

In the recent years, *P. granatum* L. has received considerable attention for its pluripharmacological effects and its potent contribution in the maintenance of human health. The efficacy of pomegranate juice has been validated in clinical trials, wherein its ability to decrease inflammatory biomarkers, oxidation of lipids and proteins [9] and to prolong the doubling time of prostate specific antigen in patients with prostate cancer [10] was reported. Phytoconstituents encompassing several phenolic classes [26–28], fatty acids [29], sugars, and organic acids [27, 30, 31] have been characterized in pomegranate fruits and have been ascribed for the diverse pharmacological effects. The edible parts of pomegranate have long been used as food, while the nonedible parts like the roots, rinds, and leaves have a number of applications in ethnomedicine [4, 6]. Thus it can be envisaged that the nonedible parts of pomegranate represent a beneficial source of functional ingredients, a statement warranting indepth investigations. The present study therefore aimed at determining the phytophenolic content and bioefficacy of the pomegranate plant that has been naturalized in the island of Mauritius since 1639 [7]. Different plant parts, namely, the flower, leaf, stem, peel, and seeds were investigated to assess the *in vitro* prophylactic potential.

A significant variation in total phenolics concentration was found among the different parts of pomegranate studied. The flower extract contained the highest phenolic level followed by the peel, leaf, and stem while the seed was relatively poor (Table 1). However, Zhang et al. [32] reported that the ethanolic extracts of pomegranate peel extract from China contained higher levels of total phenolics (508.98 ± 24.19 mg gallic acid equivalent/g DW) compared to the flower receptacles (454.96 ± 18.34 mg gallic acid equivalent/g DW) and leaf extracts (289.76 ± 14.82 mg gallic acid equivalent/g DW). Tehranifar et al. [33] also reported higher total phenolics in the methanolic extracts of peel (423.5 ± 31.8 mg gallic acid equivalent/g DW) followed by seed (384.7 ± 24.2 mg gallic acid equivalent/g DW), while the lowest amounts were measured in the leaf extract (133.3 ± 8.7 mg gallic acid equivalent/g DW).

Using the method of Zhishen et al. [19], the following trend was established for the total flavonoids in the pomegranate extracts: flower > peel > leaf > stem > seed extracts. The HCl/Butan-1-ol assay, on the other hand, indicated low levels of proanthocyanidins in the following order: peel > flower > stem > leaf > seed extracts.

The TPC measured in this study varied considerably with regard to data from the literature [32, 33]. Factors generally contributing to these variations can include treatment mode of samples prior to extraction; in this study plant parts were air dried, extraction methods and solvents [34] and cultivars used [26]. In addition, phenolic and flavonoid contents have been reported to vary due to seasonal changes and the degree of maturation of the plant parts. For instance, the biosynthesis of flavonols has been documented to be light dependent and can also be affected by temperature variation [35, 36]. Plants growing in Mauritius are tolerant of high level of environment stress induced by varying level of sunlight,

ultraviolet radiation, and temperature change throughout the year. This may explain the interesting levels of phenolic compounds in the parts studied.

A multimethod approach was used to determine the antioxidant effect of the extracts since no one method can predict the total antioxidant efficiency of an extract [17, 37]. Thus, several independent methods differing in biological action mechanisms were used to provide a thorough mechanistic insight of the antioxidant actions of the extracts under study. Nevertheless, a very strong correlation was observed between results of each antioxidant test. For instance, HOCl scavenging activity was highly correlated with deoxyribose assay results ($r = 0.968$, $P < 0.050$). A similar relationship was observed between deoxyribose assay and lipid peroxidation assay, while the superoxide anion radical scavenging activity was significantly and highly correlated with the iron chelating activity ($r = 0.997$, $P < 0.001$), the antioxidant activity from HOCl assay ($r = 0.999$, $P < 0.001$), deoxyribose assay ($r = 0.964$, $P < 0.01$), lipid peroxidation assay ($r = 0.997$, $P < 0.001$), and nitric oxide assay ($r = 0.927$, $P < 0.050$). This is further supported by the very high correlation between the calculated IC_{50} values from iron chelation and the inhibition of microsomal lipid peroxidation assay ($r = 0.9991$, $P < 0.0001$).

The TEAC value provided a ranking order of the antioxidant capacity of the extracts mainly peel > flower > leaf > stem > seed extracts. The TEAC value measured in this study was higher than that reported in the literature; for instance, the TEAC value for peel extract was higher than that reported by Shan et al. [38]. A very strong positive correlation between TEAC and proanthocyanidin content ($r = 0.921$, $P < 0.05$) and TFC was observed ($r = 0.936$, $P < 0.05$).

The extracts under study were also potent scavengers of a number of biologically relevant radicals. The HOCl scavenging assay indicated the peel extract as the most potent scavenger of hypochlorous acid. The antioxidant capacity hierarchy based on the HOCl assay of the extracts was in the following order of activity: peel > flower > leaf > seed extracts, the seed extract showing similar efficacy to ascorbic acid (IC_{50} value 5.63 ± 0.21 mg/mL). Similarly, the peel extract was the strongest inhibitor of deoxyribose sugar degradation against hydroxyl radicals generated via the Fenton reaction (Table 2). Only a moderate correlation between polyphenolic content and IC_{50} values was found.

The superoxide anion radical scavenging assay showed a different trend in activities compared to other antioxidant systems. Interestingly, the stem extract was found to be more potent than the leaf, flower, and seed extracts. This finding is in line with data reported by Kaneria et al. [34], whereby pomegranate stem extract exhibited higher antioxidant activity than leaf extract in both DPPH antiradical assay and the superoxide anion radical assay. On the other hand, the flower extract also exhibited interesting antioxidant potential. The latter was the most potent inhibitor of nitric oxide followed by peel, stem, leaf, and seed extracts.

All the pomegranate extracts significantly inhibited Fe^{3+}/ascorbate-induced microsomal lipid peroxidation with a calculated IC_{50} of less than 1.7 mg ADW/mL except for the seed extract. The ability of the pomegranate peel extract to inhibit lipid peroxidation of beef liver microsome was consistent with the findings of Althunibat et al. [39] who reported decreased lipid peroxidation in liver and kidney homogenate of STZ-induced diabetic rat models. The lipid protective ability of the extracts may be partly attributed to its flavonoid content ($r = 0.630$, $P < 0.05$). For instance, the O-dihydroxyl groups in the flavonol ring structure have been reported to be a potent inhibitor of lipid peroxidation in cells [40].

Complex formation with reduced form of transition metals particularly those that can enhance metal-induced free radical generation has been proposed as an alternative antioxidant mechanism of action ascribed to plant phytophenolic. Flavonoids can act by chelating metal ions thereby inhibiting free radical production [41]. In this line, the iron (II) ions chelating ability of the different extracts was investigated. The hierarchy of metal iron chelation of the plant parts was flower > peel > stem > leaf > seed and paralleled the nitric oxide inhibition (Table 2) ($r = 0.947$, $P < 0.05$). The similarity in activity trend in both assays may be attributed to the involvement of the catechol moiety of the flavonoids as part of the mechanism employed in both assays. The structural requirements and the mechanism of nitric oxide production inhibition by flavonoids have been reported [42]. The metal chelating activity correlated with TFC ($r = 0.626$) which may be partly assigned to the chemical structure of flavonoids. The catechol moiety in the ring B, the 3-hydroxyl and 4- oxo groups in the heterocyclic ring C, and the 4- oxo and 5-hydroxyl groups between the C and A rings has been identified as binding sites for metal ions in the flavonoid molecules [41, 43]. In addition, the vital role of Fe^{2+} in inducing and propagating lipid peroxidation has been well documented in the literature [44, 45], and thus iron chelation can be proposed as a mechanism for the potent inhibition of microsomal lipid peroxidation by the flower and peel extracts.

Similarly, the FRAP assay based on the redox reaction involving electron transfer showed the following hierarchy of activity: flower > peel > leaf > stem > seed. The FRAP assay does not detect antioxidant compounds that act by hydrogen atom transfer. The FRAP values of this study were consistent with data from Ardekani et al. [26] who reported FRAP value of peel extract to vary between 3401 and 4788 μmol Fe^{2+}/g DW among different cultivars. Statistically significant positive correlation was obtained between the FRAP and TPC ($r = 0.996$, $P < 0.01$) as well as with TFC ($r = 0.983$, $P < 0.01$). Numerous reports showed similar types of linear relationship between antioxidant activities and phytophenolic contents of fruits [37, 46].

Excess of uric acid in joints has been associated with inflammation [47] leading to pathological conditions. Xanthine oxidase, an important enzyme involved in the conversion of hypoxanthine to xanthine and to uric acid, has been reported as an interesting target against inflammation. In this vein, xanthine oxidase inhibitory activities of the pomegranate plant parts were thus assessed. The pomegranate flower extract exhibited a dose-dependent enzyme inhibition propensity. However, the inhibitory potential of the flower extract as measured by the calculated IC_{50} (0.058 ± 0.011 mg ADW/mL) was weaker than allopurinol (IC_{50}: 0.0055 ± 0.0002 mg/mL) used as control. Other parts extracts showed no inhibitory activity at the tested concentration.

The extracts tested behaved differently in the various experimental systems showing varying hierarchy of activities which were independent of the phenolic content measured. The general trend in bioefficacy demonstrated that pomegranate flower and peel extracts were very potent followed by leaf > stem > seed extracts. The findings indicated that polyphenolic compounds may act synergistically to potentiate the antioxidant activity of extracts. It should also be noted that the pomegranate flower and peel extracts were 100 times more potent than the red and yellow *Psidium cattleianum* Sabine "Chinese guava," Mauritian exotic fruits evaluated using the TEAC and FRAP assays [48]. In addition, pomegranate flower and peel extracts were found to be much more potent than the citrus extracts assayed using similar methodology. For instance, the total phenolic content of pomegranate flower extract was 200 folds higher than *Fortunella margarita* pulp extract ($1694 \pm 19 \mu$g/g FW) [50] and 44 folds higher than *C. reticulata* X *C. sinensis* flavendo extract ($7667 \pm 93 \mu$g/g FW) [38]. Similarly, the total flavonoid content were 200 folds higher than *Citrus maxima* pulp extract ($965 \pm 7 \mu$g/g FW) [50] and 35 folds higher than *C. reticulata* X *C. paradisi* flavendo extract ($5615 \pm 93 \mu$g/g FW) [38]. Likewise, the antioxidant propensities of pomegranate flower and peel extracts were significantly more important than the Mauritian citrus fruits pulp of flavendo extract thereby highlighting the prophylactic potential of the extracts under study.

Furthermore, the growing prevalence of dental caries, gingivitis, periodontitis, and oral microbial infections cases amongst adults prompted the evaluation of the antibacterial effects of the extracts against oral bacterial growth. Pathophysiological mechanisms including deficient nutritional intake, alterations in host response to oral microflora, compromised neutrophil function, and decreased phagocytosis and leukotaxis have been increasingly suggested to account for these disorders. A realistic management plan including regular oral hygiene practice and basic dental treatment can be envisaged for managing dental caries and its associated oral complications. Nowadays, active constituents extracted from plants have been included in the preparation of toothpaste, mouth rinses, dental floss, and chewing gum to ensure a stronger antimicrobial activity [8]. Ongoing studies focusing on the anticariogenic properties of polyphenols isolated from green tea [50], cranberry juice [51], and shiitake mushrooms [52] seemed promising. However despite the numerous studies conducted on such functional foods, only a handful of them can be clinically used to control dental plaque, caries formation, and mouth infections due to their effectiveness, stability, taste, and economic feasibility [53]. George and Sumathy [54] reported effective antibacterial activity of aqueous and ethanolic extracts of pomegranate against *Streptococcus mutans, Staphylococcus* sp., *Escherichia coli, Lactobacillus* sps., and *Candida albicans* isolated from the mouth. In this study, pomegranate was reported to negatively influence the proliferation of Gram Positive bacteria and concurrently to demonstrate potent iron-chelating capabilities. Recently, Kulkarni et al. [55] found that punicalagin, an ellagitannin isolated from the pomegranate peel, could completely suppress iron catalyzing oxidant reactions in vitro. It can therefore be speculated that potent pomegranate extracts under study, in particular, the peel extract, may be liable to remove iron from the broth medium and deprive bacteria of the iron they need for normal growth through chelation in view of the high iron (II) ions chelating efficiency. The presence of both tannins and alkaloids isolated from pomegranate pericarp and seeds has been extensively reviewed for their outstanding ability to block bacterial surface adhesions and inhibit glycosyltransferases thus deterring bacterial attachment to dental surfaces, hence, its colonization [56]. Data from this study provide basic supplementary evidence of the antimicrobial activity of pomegranate and support the imperative need to find new effective bioagents that can avoid a negative impact upon the future oral health of communities affected by dental caries and expenditure on dental services.

While pomegranate' edible parts have common applications in the food and food processing industries due to their excellent nutritional and health values [57], this study showed the prospect of the nonedible parts Mauritian cultivar of pomegranate. An in-depth comparison of the broad classes of phytophenolic and the bioefficacies of the latter indicated the effectiveness of the flower and peel extracts. The available evidence indicates that these extracts might be of therapeutic benefit in bacterial infections and be an ideal candidate for functional food health products. Further investigations need to be directed towards determining the potential toxicity, the phytochemistry of the nonedible parts, and applicability of the extracts in various food matrices. The use of functional foods enriched with pomegranate flower and peel extracts however needs technologies for incorporating these health-promoting ingredients into food without reducing their bioavailability or functionality.

Conflict of Interests

The authors declare that there is no conflict of interests.

References

[1] C. Guo, J. Wei, J. Yang, J. Xu, W. Pang, and Y. Jiang, "Pomegranate juice is potentially better than apple juice in improving antioxidant function in elderly subjects," *Nutrition Research*, vol. 28, no. 2, pp. 72–77, 2008.

[2] E. González-Molina, D. A. Moreno, and C. García-Viguera, "A new drink rich in healthy bioactives combining lemon and pomegranate juices," *Food Chemistry*, vol. 115, no. 4, pp. 1364–1372, 2009.

[3] J. Jurenka, "Therapeutic applications of pomegranate (*Punica granatum* L.): a review," *Alternative Medicine Review*, vol. 13, no. 2, pp. 128–144, 2008.

[4] M. C. Mathabe, R. V. Nikolova, N. Lall, and N. Z. Nyazema, "Antibacterial activities of medicinal plants used for the treatment of diarrhoea in Limpopo Province, South Africa," *Journal of Ethnopharmacology*, vol. 105, no. 1-2, pp. 286–293, 2006.

[5] M.-J. Song and H. Kim, "Ethnomedicinal application of plants in the western plain region of North Jeolla Province in Korea," *Journal of Ethnopharmacology*, vol. 137, no. 1, pp. 167–175, 2011.

[6] P. Tetali, C. Waghchaure, P. G. Daswani, N. H. Antia, and T. J. Birdi, "Ethnobotanical survey of antidiarrhoeal plants of Parinche valley, Pune district, Maharashtra, India," *Journal of Ethnopharmacology*, vol. 123, no. 2, pp. 229–236, 2009.

[7] G. Rouillard and J. Guého, *Les PLantes Et Leur Histoire À L'ILe Maurice*, 1999.

[8] I. E. Alsaimary, "Efficacy of some antibacterial agents against *Streptococcus* mutans associated with tooth decay," *Internet Journal of Microbiology*, vol. 7, no. 2, 2010.

[9] L. Shema-Didi, S. Sela, L. Ore et al., "One year of pomegranate juice intake decreases oxidative stress, inflammation, and incidence of infections in hemodialysis patients: a randomized placebo-controlled trial," *Free Radical Biology & Medicine*, vol. 53, pp. 297–304, 2012.

[10] A. J. Pantuck, J. T. Leppert, N. Zomorodian et al., "Phase II study of pomegranate juice for men with rising prostate-specific antigen following surgery or radiation for prostate cancer," *Clinical Cancer Research*, vol. 12, no. 13, pp. 4018–4026, 2006.

[11] L. S. Adams, Y. Zhang, N. P. Seeram, D. Heber, and S. Chen, "Pomegranate ellagitannin-derived compounds exhibit antiproferative and antiaromatase activity in breast cancer cells in vitro," *Cancer Prevention Research*, vol. 3, no. 1, pp. 108–113, 2010.

[12] E. H. Endo, D. A. Garcia Cortez, T. Ueda-Nakamura, C. V. Nakamura, and B. P. Dias Filho, "Potent antifungal activity of extracts and pure compound isolated from pomegranate peels and synergism with fluconazole against *Candida albicans*," *Research in Microbiology*, vol. 161, no. 7, pp. 534–540, 2010.

[13] M. Haidari, M. Ali, S. Ward Casscells III, and M. Madjid, "Pomegranate (*Punica granatum*) purified polyphenol extract inhibits influenza virus and has a synergistic effect with oseltamivir," *Phytomedicine*, vol. 16, no. 12, pp. 1127–1136, 2009.

[14] L. Zhang, Q. Fu, and Y. Zhang, "Composition of anthocyanins in pomegranate flowers and their antioxidant activity," *Food Chemistry*, vol. 127, no. 4, pp. 1444–1449, 2011.

[15] M. I. Gil, F. A. Tomas-Barberan, B. Hess-Pierce, D. M. Holcroft, and A. A. Kader, "Antioxidant activity of pomegranate juice and its relationship with phenolic composition and processing," *Journal of Agricultural and Food Chemistry*, vol. 48, no. 10, pp. 4581–4589, 2000.

[16] T. Tanaka, G.-I. Nonaka, and I. Nishioka, "Tannins and related compounds. XL. Revision of the structures of punicalin and punicalagin, and isolation and characterization of 2-O-galloylpunicalin from the bark of *Punica granatum* L.," *Chemical and Pharmaceutical Bulletin*, vol. 34, no. 2, pp. 650–655, 1986.

[17] V. S. Neergheen, M. A. Soobrattee, T. Bahorun, and O. I. Aruoma, "Characterization of the phenolic constituents in Mauritian endemic plants as determinants of their antioxidant activities in vitro," *Journal of Plant Physiology*, vol. 163, no. 8, pp. 787–799, 2006.

[18] L. J. Porter, L. N. Hrstich, and B. G. Chan, "The conversion of procyanidins and prodelphinidins to cyanidin and delphinidin," *Phytochemistry*, vol. 25, no. 1, pp. 223–230, 1985.

[19] J. Zhishen, T. Mengcheng, and W. Jianming, "The determination of flavonoid contents in mulberry and their scavenging effects

on superoxide radicals," *Food Chemistry*, vol. 64, no. 4, pp. 555–559, 1999.

[20] I. F. F. Benzie and J. J. Strain, "The ferric reducing ability of plasma (FRAP) as a measure of "antioxidant power": the FRAP assay," *Analytical Biochemistry*, vol. 239, no. 1, pp. 70–76, 1996.

[21] A. M. Campos and E. A. Lissi, "Kinetics of the reaction between 2,2′-azinobis 3-ethylbenzothiazoline-6-sulfonic acid (ABTS) derived radical cations and phenols," *International Journal of Chemical Kinetics*, vol. 29, no. 3, pp. 219–224, 1997.

[22] O. I. Aruoma, "Deoxyribose assay for detecting hydroxyl radicals," *Methods in Enzymology*, vol. 233, pp. 57–66, 1994.

[23] M. Kumar, M. Chandel, S. Kumar, and S. Kaur, "Studies on the antioxidant/ genoprotective activity of extracts of *Koelreuteria paniculata* laxm," *American Journal of Biomedical Sciences*, vol. 1, pp. 177–189, 2011.

[24] C. Sunil, P. Agastian, C. Kumarappan, and S. Ignacimuthu, "In vitro antioxidant, antidiabetic and antilipidemic activities of *Symplocos cochinchinensis* (Lour.) S. Moore bark," *Food and Chemical Toxicology*, vol. 50, no. 5, pp. 1547–1553, 2012.

[25] J. Havlik, R. G. de la Huebra, K. Hejtmankova et al., "Xanthine oxidase inhibitory properties of Czech medicinal plants," *Journal of Ethnopharmacology*, vol. 132, no. 2, pp. 461–465, 2010.

[26] M. R. S. Ardekani, M. Hajimahmoodi, M. R. Oveisi et al., "Comparative antioxidant activity and total flavonoid content of Persian pomegranate (*Punica granatum* l.) cultivars," *Iranian Journal of Pharmaceutical Research*, vol. 10, no. 3, pp. 519–524, 2011.

[27] P. Mena, C. García-Viguera, J. Navarro-Rico et al., "Phytochemical characterisation for industrial use of pomegranate (*Punica granatum* L.) cultivars grown in Spain," *Journal of the Science of Food and Agriculture*, vol. 91, no. 10, pp. 1893–1906, 2011.

[28] A. Tehranifar, M. Zarei, Z. Nemati, B. Esfandiyari, and M. R. Vazifeshenas, "Investigation of physico-chemical properties and antioxidant activity of twenty Iranian pomegranate (*Punica granatum* L.) cultivars," *Scientia Horticulturae*, vol. 126, no. 2, pp. 180–185, 2010.

[29] A. Parashar, N. Sinha, and P. Singh, "Lipid contents and fatty acids composition of seed oil from twenty five pomegranates varieties grown in India," *Advance Journal of Food Science and Technology*, vol. 2, no. 1, pp. 12–15, 2010.

[30] P. Melgarejo, D. M. Salazar, and F. Artés, "Organic acids and sugars composition of harvested pomegranate fruits," *European Food Research and Technology*, vol. 211, no. 3, pp. 185–190, 2000.

[31] N. Hasnaoui, R. Jbir, M. Mars et al., "Organic acids, sugars, and anthocyanins contents in juices of Tunisian pomegranate fruits," *International Journal of Food Properties*, vol. 14, no. 4, pp. 741–757, 2011.

[32] L. Zhang, X. Yang, Y. Zhang, L. Wang, and R. Zhang, "In vitro antioxidant properties of different parts of pomegranate flowers," *Food and Bioproducts Processing*, vol. 89, no. 3, pp. 234–240, 2011.

[33] A. Tehranifar, Y. Selahvarzi, M. Kharrazi, and V. J. Bakhsh, "High potential of agro-industrial by-products of pomegranate (*Punica granatum* L.) as the powerful antifungal and antioxidant substances," *Industrial Crops and Products*, vol. 34, no. 3, pp. 1523–1527, 2011.

[34] M. J. Kaneria, M. B. Bapodara, and S. V. Chanda, "Effect of extraction techniques and solvents on antioxidant activity of pomegranate (*Punica granatum* L.) leaf and stem," *Food Analytical Methods*, vol. 5, no. 3, pp. 396–404, 2012.

[35] E. Braidot, M. Zancani, E. Petrussa et al., "Transport and accumulation of flavonoids in grapevine (*Vitis vinifera* L.)," *Plant Signaling and Behavior*, vol. 3, no. 9, pp. 626–632, 2008.

[36] D. Treutter, "Managing phenol contents in crop plants by phytochemical farming and breeding-visions and constraints," *International Journal of Molecular Sciences*, vol. 11, no. 3, pp. 807–857, 2010.

[37] D. Ramful, T. Bahorun, E. Bourdon, E. Tarnus, and O. I. Aruoma, "Bioactive phenolics and antioxidant propensity of flavedo extracts of Mauritian citrus fruits: potential prophylactic ingredients for functional foods application," *Toxicology*, vol. 278, no. 1, pp. 75–87, 2010.

[38] B. Shan, Y.-Z. Cai, J. D. Brooks, and H. Corke, "The in vitro antibacterial activity of dietary spice and medicinal herb extracts," *International Journal of Food Microbiology*, vol. 117, no. 1, pp. 112–119, 2007.

[39] O. Y. Althunibat, A. H. Al-Mustafa, K. Tarawneh, K. M. Khleifat, B. H. Ridzwan, and H. N. Qaralleh, "Protective role of *Punica granatum* L. peel extract against oxidative damage in experimental diabetic rats," *Process Biochemistry*, vol. 45, no. 4, pp. 581–585, 2010.

[40] I.-W. Peng and S.-M. Kuo, "Flavonoid structure affects the inhibition of lipid peroxidation in Caco-2 intestinal cells at physiological concentrations," *Journal of Nutrition*, vol. 133, no. 7, pp. 2184–2187, 2003.

[41] D. Malešev and V. Kuntić, "Investigation of metal-flavonoid chelates and the determination of flavonoids via metal-flavonoid complexing reactions," *Journal of the Serbian Chemical Society*, vol. 72, no. 10, pp. 921–939, 2007.

[42] H. Matsuda, T. Morikawa, S. Ando, I. Toguchida, and M. Yoshikawa, "Structural requirements of flavonoids for nitric oxide production inhibitory activity and mechanism of action," *Bioorganic and Medicinal Chemistry*, vol. 11, no. 9, pp. 1995–2000, 2003.

[43] D. Procházková, I. Boušová, and N. Wilhelmová, "Antioxidant and prooxidant properties of flavonoids," *Fitoterapia*, vol. 82, pp. 513–523, 2011.

[44] L. Tang, Y. Zhang, Z. Qian, and X. Shen, "The mechanism of Fe^{2+}-initiated lipid peroxidation in liposomes: the dual function of ferrous ions, the roles of the pre-existing lipid peroxides and the lipid peroxyl radical," *Biochemical Journal*, vol. 352, no. 1, pp. 27–36, 2000.

[45] M. Valko, C. J. Rhodes, J. Moncol, M. Izakovic, and M. Mazur, "Free radicals, metals and antioxidants in oxidative stress-induced cancer," *Chemico-Biological Interactions*, vol. 160, no. 1, pp. 1–40, 2006.

[46] R. L. Prior, G. Cao, A. Martin et al., "Antioxidant capacity as influenced by total phenolic and anthocyanin content, maturity, and variety of *Vaccinium* species," *Journal of Agricultural and Food Chemistry*, vol. 46, no. 7, pp. 2686–2693, 1998.

[47] A. Chandrasekara and F. Shahidi, "Antiproliferative potential and DNA scission inhibitory activity of phenolics from whole millet grains," *Journal of Functional Foods*, vol. 3, no. 3, pp. 159–170, 2011.

[48] A. Luximon-Ramma, T. Bahorun, and A. Crozier, "Antioxidant actions and phenolic and vitamin C contents of common Mauritian exotic fruits," *Journal of the Science of Food and Agriculture*, vol. 83, no. 5, pp. 496–502, 2003.

[49] D. Ramful, E. Tarnus, O. I. Aruoma, E. Bourdon, and T. Bahorun, "Polyphenol composition, vitamin C content and antioxidant capacity of Mauritian citrus fruit pulps," *Food Research International*, vol. 44, no. 7, pp. 2088–2099, 2011.

[50] S. Otake, M. Makimura, T. Kuroki, Y. Nishihara, and M. Hirasawa, "Anticaries effects of polyphenolic compounds from Japanese green tea," *Caries Research*, vol. 25, no. 6, pp. 438–443, 1991.

[51] J. Babu, C. Blair, S. Jacob, and O. Itzhak, "Inhibition of *Streptococcus gordonii* metabolic activity in biofilm by cranberry juice high-molecular-weight component," *Journal of Biomedicine and Biotechnology*, vol. 2012, Article ID 590384, 7 pages, 2012.

[52] C. Signoretto, G. Burlacchini, A. Marchi et al., "Testing a low molecular mass fraction of a mushroom (Lentinus edodes) extract formulated as an oral rinse in a cohort of volunteers," *Journal of Biomedicine and Biotechnology*, vol. 2011, Article ID 857987, 2011.

[53] R. A. Bagramian, F. Garcia-Godoy, and A. R. Volpe, "The global increase in dental caries. A pending public health crisis," *American Journal of Dentistry*, vol. 22, no. 1, pp. 3–8, 2009.

[54] S. George and V. J. H. Sumathy, "Anti-bacterial and anti-fungal activity of botanical extracts against microorganisms isolated from mouth flora," *Biosciences International*, vol. 1, pp. 74–77, 2012.

[55] A. P. Kulkarni, H. S. Mahal, S. Kapoor, and S. M. Aradhya, "In vitro studies on the binding, antioxidant, and cytotoxic action of punicalagin," *Journal of Agricultural and Food Chemistry*, vol. 55, no. 4, pp. 1491–1500, 2007.

[56] G. K. Jayaprakasha, P. S. Negi, and B. S. Jena, "Antimicrobial activities of pomegranate," in *From: Ancient Foods to Modern Medicine*, Chapter 11, pp. 168–183, Taylor and Francis, 2006.

[57] L. U. Opara, M. R. Al-Ani, and Y. S. Al-Shuaibi, "Physicochemical properties, vitamin C content, and antimicrobial properties of pomegranate fruit (*Punica granatum* L.)," *Food and Bioprocess Technology*, vol. 2, no. 3, pp. 315–321, 2009.

17

Effect of Extraction Conditions on the Antioxidant Activity of Olive Wood Extracts

Mercedes Pérez-Bonilla,[1] Sofía Salido,[1] Adolfo Sánchez,[1] Teris A. van Beek,[2] and Joaquín Altarejos[1]

[1] Departamento de Química Inorgánica y Orgánica, Facultad de Ciencias Experimentales, Universidad de Jaén, Campus de Excelencia Internacional Agroalimentario, ceiA3, 23071 Jaén, Spain
[2] Laboratory of Organic Chemistry, Natural Products Chemistry Group, Wageningen University, Dreijenplein 8, 6703 HB Wageningen, The Netherlands

Correspondence should be addressed to Joaquín Altarejos; jaltare@ujaen.es

Academic Editor: Philip Cox

An investigation to optimize the extraction yield and the radical scavenging activity from the agricultural by-product olive tree wood (*Olea europaea* L., cultivar Picual) using six different extraction protocols was carried out. Four olive wood samples from different geographical origin, and harvesting time have been used for comparison purposes. Among the fifty olive wood extracts obtained in this study, the most active ones were those prepared with ethyl acetate, either through direct extraction or by successive liquid-liquid partitioning procedures, the main components being the secoiridoids oleuropein and ligustroside. An acid hydrolysis pretreatment of olive wood samples before extractions did not improve the results. In the course of this study, two compounds were isolated from the ethanolic extracts of olive wood collected during the olives' harvesting season and identified as $(7''R)$-7''-ethoxyoleuropein (**1**) and $(7''S)$-7''-ethoxyoleuropein (**2**).

1. Introduction

Since agricultural and industrial residues are attractive sources of natural antioxidants, basically due to their null or low value [1–4], different residues and by-products from fruits [5, 6], vegetables [7, 8], or olive oil manufacturing [9] have been screened for the presence of antioxidants. Due to the large amounts of biomass from pruning generated every year (more than 7 million tonnes per year in Spain), olive tree wood constitutes an important agricultural by-product. During the search of natural antioxidants from *Olea europaea* L. residues and by-products, both solid and liquid residues from olive oil and table olives processing have been studied [2, 10–19].

Our preliminary studies on the radical scavenging activity of olive wood extracts, cultivar Picual, showed that this agricultural by-product could be a source of natural antioxidants [20]. The isolation and radical scavenging activity of the main constituents [21] as well as some minor components present in olive wood extracts have been reported by us [22]. The secoiridoids oleuropein and ligustroside are among the main components. Other compounds present in olive wood are the lignan (+)-cycloolivil, the phenolic alcohol hydroxytyrosol, and several secoiridoids related to oleuropein, such as $(7''S)$-7''-hydroxyoleuropein or oleuropein $3'$-O-β-D-glucoside. Moreover, the human platelet antiaggregant properties of two olive wood components, oleuropein and (+)-cycloolivil, have been evaluated [23]. The cultivar Picual was selected for these studies since it is one of the most important Spanish olive varieties for oil extraction, representing around 860.000 ha in the province of Jaén and other Andalusian areas, and is also cultivated in other regions of Spain and other countries [24].

Solvent extraction is routinely used for the isolation of antioxidants from plant material. Both extraction yield and antioxidant activity of extracts are strongly dependent on the solvent [1]. Hence, a comparative study for selecting optimal extraction conditions to provide the highest antioxidant activity (and proper extraction yield) from olive wood (cultivar Picual) was carried out in this work. For this purpose several extraction processes at room temperature and reflux were designed using solvents of different polarities. Also the influence of an acidic hydrolysis pretreatment of olive wood was investigated, since this methodology has been used sometimes to improve the recovery of phenols [3].

2. Materials and Methods

2.1. Chemicals. The solvents used for extraction (hexane, dichloromethane, ethyl acetate, *n*-butanol, ethanol, methanol, chloroform, and acetone) were glass-distilled prior to use. Methanol used for radical scavenging activity assays was of HPLC grade. Deuterated methanol was used to prepare solutions of purified compounds for NMR analysis. The 2,2-diphenyl-1-picrylhydrazyl radical (DPPH•, 95%) was purchased from Sigma-Aldrich Chemie (Steinheim, Germany). A commercial rosemary oleoresin was obtained from Evesa (Cádiz, Spain).

2.2. General Experimental Procedures. Optical rotations ($[\alpha]_D$) were recorded in MeOH on a Perkin-Elmer 241 automatic polarimeter (Perkin-Elmer Instruments, Norwalk, CT, USA), in a 10 cm 2 mL cell. Ultraviolet (UV) spectra were recorded in MeOH on a Perkin-Elmer UV/Vis spectrophotometer Lambda 19 (Perkin-Elmer Instruments, Norwalk, CT, USA). Infrared (IR) spectra were recorded on a FT-IR Perkin-Elmer 1760X spectrometer (Perkin-Elmer Instruments, Norwalk, CT, USA). NMR spectra (^{1}H NMR, ^{13}C NMR, DQF-COSY, HSQC, HMBC) were recorded on a Bruker Avance AMX 500 spectrometer (Bruker Daltonik GmbH, Rheinstetten, Germany), using CD_3OD as solvent and tetramethylsilane (TMS) as internal reference. Mass spectra (MS) were recorded on an Finnigan MAT LCQ ion trap mass spectrometer (Waters Integrity System, Milford, MA, USA). The ESI interface was used in both positive and negative modes, with the capillary temperature at 200°C and a spray voltage of 4.5 kV.

High-performance liquid chromatography (HPLC) analyses were performed on an analytical RP-HPLC Spherisorb ODS-2 column (250 mm × 3 mm i.d., 5 μm) (Waters Chromatography Division, Milford, MA, USA) on a Waters 600E instrument (Waters Chromatography Division, Milford, MA, USA) equipped with a diode array detector, scan range: 190–800 nm (Waters CapLC 2996 Photodiode Array Detector, Waters Chromatography Division, Milford, MA, USA), and operating at 30°C. Samples of the extracts were prepared in MeOH at a concentration of 10 mg mL^{-1}, and the injection volume was 10 μL. The best separation was performed with H_2O : CH_3COOH, 99.8 : 0.2, v/v (solvent A) and CH_3OH : CH_3COOH, 99.8 : 0.2, v/v (solvent B), at a flow rate of 0.7 mL min^{-1}, using a linear gradient from 20% to 70% B for 55 min. The HPLC analyses were recorded at 230 nm, since most of the compounds present in olive extracts have an intense absorption at that wavelength.

Preparative HPLC separations were performed on an Alltima C18 column (250 mm × 22 mm i.d., 5 μm) (Alltech Associates Inc., Deerfield, IL, USA) with a Shimadzu preparative HPLC instrument (Shimadzu, Kyoto, Japan), equipped with a diode array detector, scan range: 190–600 nm (SPD-M10Ap Photodiode Array Detector, Shimadzu, Kyoto, Japan) and a sample collector FRC-10 A (Shimadzu, Kyoto, Japan), and operating at 30°C and a flow rate of 12 mL min^{-1} with H_2O : CH_3OH : CH_3COOH (59.9 : 39.9 : 0.2, v/v/v).

HPLC–DAD–MS analyses were performed on an Spherisorb ODS-2 column (125 mm × 3 mm i.d., 5 μm) (Waters Chromatography Division, Milford, MA, USA) with an Agilent 1100 HPLC instrument (Agilent Technologies, Santa Clara, CA, USA) equipped with a diode array detector, scan range: 190–600 nm (G1315B Photodiode Array Detector, Agilent Technologies, Santa Clara, CA, USA) and an ion trap mass spectrometer Esquire 6000 (Bruker Daltonics, Bremen, Germany). The sample preparation and gradient were the same as those in the HPLC analysis. The flow rate was 0.4 mL min^{-1}. The ESI source parameters were as follows: capillary voltage: 4 kV; cap exit: –100 V; skimmer: –40 V; trap drive: 70; nebulizer: 50 psi; dry gas: 10 mL min^{-1}; dry temperature: 350°C; scan range: m/z 50–1000.

2.3. Plant Material and Collection Data. Four samples of olive wood (*Olea europaea* L., cultivar Picual) were collected at two olive groves located in Jaén province (southern Spain) during the pruning period (March and April) and at the beginning of the olives' harvesting season (November) from 2003 until 2006. The samples collected were labelled as **A**, **B**, **C**, and **D**, and the location and collection date were as follows: **A** (Fuensanta village; April, 2003), **B** (Fuensanta village; March, 2005), **C** (Mogón village; March, 2005), and **D** (Fuensanta village; November, 2006). In each case, the plant material consisted of a single piece of *ca.* 10 cm diameter and 50 cm length from the pruning of the same olive grove near Fuensanta (**A**, **B**) and a different olive grove near Mogón (**C**). Another similar single piece was cut in the same olive grove near Fuensanta during the olives' harvesting season for comparison (**D**). Each sample was stored in a dry and dark place at room temperature with passive ventilation for 3 months. Just before starting the extraction process, the samples (including bark and heartwood) were scraped in a local sawmill (wood shavings: length 3–5 cm, thickness 0.1–0.3 mm).

2.4. Extraction Protocols. Olive wood samples **A**, **B**, **C**, and **D** were extracted by the following procedures (*i–iv*).

(i) Procedures E1, E2, and E3. These procedures involved the sequential extraction of olive wood samples with solvents of increasing polarity at room temperature for 24 h or under reflux for 2 h. The procedure E1 employed the sequence of solvents CH_2Cl_2 and EtOH at room temperature (Figure 1). The procedure E2 used the sequence of solvents *n*-hexane, CH_2Cl_2, EtOAc, and EtOH at room temperature (Figure 2). The procedure E3 employed CH_2Cl_2 at room temperature and then EtOAc under reflux (Figure 3). The olive wood sample **A** (35 g each) was extracted by the procedures E1 and E2 using 250 mL of each solvent (extracts **A1**–**A3**; see Table 2). Olive wood samples **A**, **B**, **C**, and **D** (35 g each) were extracted by the procedure E3 using 500 mL of each solvent (extracts **A4**, **B1**, **C1**, and **D1**; see Tables 2 and 4). The extracts prepared

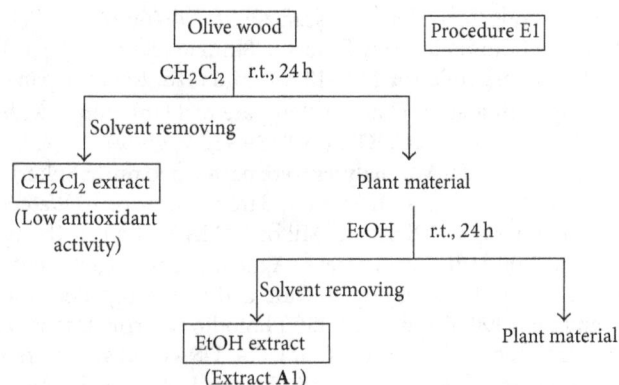

FIGURE 1: Solvent extractions of olive wood sample **A** following procedure E1.

with *n*-hexane and CH_2Cl_2 were discarded since their radical scavenging activities were low.

(ii) Procedure E4. This procedure involved the direct extraction of olive wood samples with different solvents at room temperature for 24 h and under reflux for 2 h (Figure 4). The solvents used were EtOAc, EtOH, EtOH : H_2O (3 : 2, v/v), H_2O, and H_2O : HCOOH (4 : 1, v/v). The olive wood sample **A** (35 g) was extracted by the procedure E4 at room temperature (**A5**–**A9**) and under reflux using 500 mL of each solvent (**A10**–**A14**) (Table 2). The olive wood samples **B**, **C**, and **D** (35 g each) were extracted by the procedure E4 using 500 mL of EtOH under reflux (extracts **B2**, **C2**, and **D2**) (Table 4).

(iii) Procedures E5 and E6. These procedures involved the direct extraction of olive wood samples with a polar solvent under reflux for 2 h, followed by a liquid-liquid partitioning of the resulting extract with solvents of increasing polarity (Figures 5 and 6). A mixture of EtOH : H_2O (3 : 2, v/v) was used for the direct extraction in procedure E5 (Figure 5), while MeOH was used in procedure E6 (Figure 6). Olive wood sample **A** (30 g) was extracted by the procedure E5 using 350 mL of EtOH : H_2O (3 : 2). The resulting EtOH : H_2O extract was evaporated to dryness under vacuum, suspended in water (200 mL), and successively liquid-liquid partitioned with EtOAc (150 mL) and *n*-BuOH (150 mL) to yield extracts **A15** and **A16**, respectively (Figure 5 and Table 2). The remaining aqueous phase was also evaporated to yield extract **A17**. Another olive wood sample **A** (30 g) was extracted by the procedure E6 using 350 mL of MeOH. In a similar manner, the resulting MeOH extract was evaporated and partitioned with Et_2O (150 mL), $CHCl_3$ (150 mL), and *n*-BuOH (150 mL) to yield extracts **A18**, **A19**, and **A20**, respectively (Figure 6, Table 2).

(iv) Procedure E7. This procedure involved an acidic hydrolysis pretreatment of the olive wood samples using different acids for 1 h, 3 h, and 5 h at 130°C, followed by a solvent extraction of both the resulting liquid acidic extract and solid pretreated wood (Figure 7). Every olive wood sample **A** (5 g each) was hydrolysed with 40 mL of 0.5 M H_2SO_4 in H_2O, 40 mL of 0.5 M H_2SO_4 in EtOH : H_2O (1 : 1, v/v), 40 mL

of 1 M HCl in H_2O, or 40 mL of 1 M HCl in EtOH : H_2O (1 : 1, v/v) under the conditions described above. Thus, for example, an olive wood sample **A** (5 g) was pretreated with 40 mL of 0.5 M H_2SO_4 in H_2O for 1 h at 130°C. Then, the mixture was filtered, and the liquid acidic extract diluted with water (200 mL) and its pH adjusted to 3 with Na_2CO_3. EtOAc (100 mL) were added to the acidic aqueous phase and refluxed for 0.5 h. Then, the EtOAc layer was separated and dried over anhydrous Na_2SO_4, and the EtOAc extract evaporated to dryness under vacuum to yield extract **A21** (Figure 7 and Table 3). This procedure generated, after the initial filtering, a solid pretreated wood, which was also extracted first with CH_2Cl_2 at room temperature for 24 h and then with EtOAc under reflux for 2 h (Figure 7). The CH_2Cl_2 extract was discarded, due to its low radical scavenging activity, and the EtOAc extract **A22** was kept (Figure 7 and Table 3). Another olive wood sample **A** (5 g) was pretreated with 40 mL of 0.5 M H_2SO_4 in H_2O for 3 h at 130°C, yielding finally extracts **A23** and **A24** (Figure 7 and Table 3). Since the hydrolysis pretreatments of olive wood samples were carried out with five different acid conditions and three different times (1 h, 3 h, and 5 h) and each pretreatment yielded two extracts to be investigated, the procedure E7 afforded twenty-four extracts (Figure 7 and Table 3).

The solvent of the extracts obtained in the different procedures was evaporated under vacuum at temperatures not higher than 40°C. The resulting dry extracts were stored under argon in sealed vials at −20°C until analysis. Extraction yields were calculated as grams of the dry extract per kilogram of olive wood sample.

2.5. DPPH Radical Scavenging Assay. Radical scavenging activity of extracts was determined spectrophotometrically with the stable DPPH radical [25, 26]. Methanolic solutions (2.4 mL) of DPPH$^{\bullet}$ ($\sim 7 \times 10^{-5}$ mol L^{-1}) with an absorbance at 515 nm of 0.80 ± 0.03 AU were mixed with methanolic solutions (1.2 mL) of extracts at 50 μg mL^{-1} by dissolving the dry extracts in methanol. The experiment was carried out in triplicate. The samples were shaken and kept in the dark for 15 min at room temperature, and then the decrease of absorbance was measured at 515 nm. Radical scavenging activity of extracts is expressed as radical scavenging percentage (RSP) and was calculated using the following equation [26]:

$$\text{RSP (\%)} = \left[\frac{A_B - A_A}{A_B} \right] \times 100, \tag{1}$$

where A_B is the absorbance of the blank ($t = 0$ min) and A_A is the absorbance of the tested extract solution ($t = 15$ min).

2.6. Isolation and Structure Elucidation of Purified Compounds. An aliquot (73 mg) of extract **D2** (Figure 4 and Table 4) was chromatographed by preparative RP-HPLC (see Section 2.2) to afford compounds **1** and **2**. Pure compounds **1** (21 mg) and **2** (15 mg) were obtained after removing the solvents with a rotary evaporator and the remaining H_2O with a freeze dryer. The structures of purified compounds

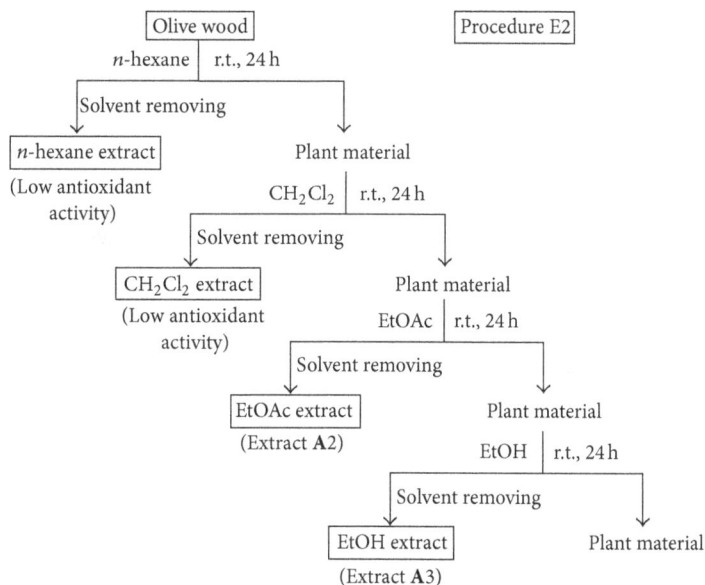

FIGURE 2: Solvent extractions of olive wood sample **A** following procedure E2.

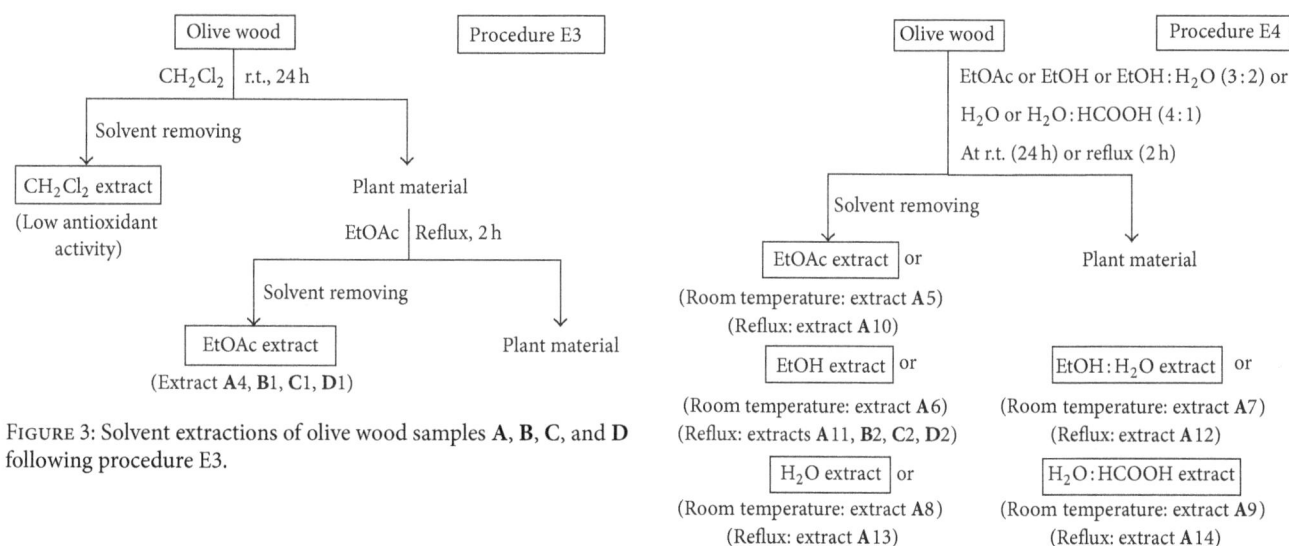

FIGURE 3: Solvent extractions of olive wood samples **A**, **B**, **C**, and **D** following procedure E3.

FIGURE 4: Solvent extractions of olive wood samples **A**, **B**, **C**, and **D** following procedure E4.

were elucidated by various spectroscopic methods and specific optical rotations measurements (see Section 2.2).

(7″R)-7″-Ethoxyoleuropein (**1**). Colourless syrup; $[\alpha]_D^{25}$ −112° (*c* 0.10, methanol); UV (methanol) λ_{max} (log ε) 231 (4.04), 282 nm (3.36); IR (film) ν_{max} 3384 (OH), 1705 (C=O), 1629 (α,β-unsaturated C=O), 1384, 1076, 1045 (C–O–C) cm^{-1}; ESIMS (positive), *m/z* 607.2 ([M+Na]$^+$), and 1190.7 ([2 M+Na]$^+$), ESIMS (negative), *m/z* 583.2 ([M–H]$^-$); for ^{1}H and ^{13}C NMR data see Table 1.

(7″S)-7″-Ethoxyoleuropein (**2**). Colourless syrup; $[\alpha]_D^{25}$ −95° (*c* 0.05, methanol); UV (methanol) λ_{max} (log ε) 231 (4.17), 282 nm (3.50); IR (film) ν_{max} 3385 (OH), 1703 (C=O), 1628 (α,β-unsaturated C=O), 1384, 1076, 1045 (C–O–C) cm^{-1}; ESIMS (positive), *m/z* 607.2 ([M+Na]$^+$), and 1190.8 ([2M+Na]$^+$), ESIMS (negative), *m/z* 583.3 ([M–H]$^-$); for ^{1}H and ^{13}C NMR data see Table 1.

3. Results and Discussion

Following up our preliminary results on the radical scavenging activity of dichloromethane and ethanol extracts of olive (*O. europaea*) wood [20], several extraction procedures using different sequences of solvents with different polarities, at different temperatures and times were investigated in this work. Moreover, the influence of using both (a) acidified solvents (e.g. mixture of water and formic acid) to extract olive wood and (b) acidic hydrolysis pretreatments of the plant material on the yield and antioxidant activity of the resulting extracts was also studied (Figures 1–7). Four samples of olive tree wood, cultivar Picual, have been used to prepare all

TABLE 1: NMR data (400 MHz, CD$_3$OD) of (7″R)-7″-ethoxyoleuropein (**1**) and (7″S)-7″-ethoxyoleuropein (**2**).

Position	1		2	
	δ_H, mult. (J in Hz)	δ_C, mult.	δ_H, mult. (J in Hz)	δ_C, mult.
1	5.92, bs	95.0, CH	5.92, bs	95.1, CH
3	7.51, s	155.2, CH	7.52, s	155.2, CH
4	—	109.3, qC	—	109.4, qC
5	3.99, dd (4.5, 9.0)	31.8, CH	3.94–4.00, m	31.7, CH
6a	2.72, dd (4.5, 14.2)	41.1, CH$_2$	2.72, dd (4.4, 14.4)	41.1, CH$_2$
6b	2.47, dd (9.0, 14.2)		2.47, dd (9.1, 14.4)	
7	—	173.1, qC	—	173.0, qC
8	6.08, bq (6.9)	124.9, CH	6.09, bq (6.6)	124.9, CH
9	—	130.5, qC	—	130.5, qC
10	1.69 bd (6.9)	13.6, CH$_3$	1.69, bd (6.6)	13.6, CH
11	—	168.7, qC	—	168.7, qC
OCH$_3$	3.71, s	51.9, CH$_3$	3.71, s	51.9, CH$_3$
1′	4.80, d (7.8)	100.8, CH	4.80, d (7.9)	100.8, CH
2′	3.34–3.42, m	74.8, CH	3.34–3.44, m	74.8, CH
3′	3.34–3.42, m	77.9, CH	3.34–3.44, m	77.9, CH
4′	3.34–3.42, m	71.5, CH	3.34–3.44, m	71.5, CH
5′	3.34–3.42, m	78.4, CH	3.34–3.44, m	78.4, CH
6′a	3.89, bd (11.9)	62.7, CH$_2$	3.89, bd (11.8)	62.7, CH$_2$
6′b	3.67, dd (4.9, 11.9)		3.66, dd (4.9, 11.8)	
1″	—	131.3, qC	—	131.4, qC
2″	6.77, bs	114.9, CH	6.77, bs	114.8, CH
3″	—	146.5, qC	—	146.6, qC
4″	—	146.4, qC	—	146.4, qC
5″	6.75, d (7.6)	116.3, CH	6.75, d (8.3)	116.3, CH
6″	6.65, bd (7.6)	119.7, CH	6.65, bd (8.3)	119.7, CH
7″	4.39, dd (4.1, 7.2)	80.7, CH	4.39, dd (4.0, 7.8)	80.6, CH
8″a	4.02–4.10, m	69.3, CH$_2$	4.14, dd (7.8, 11.2)	69.2, CH$_2$
8″b	4.02–4.10, m		3.94–4.00, m	
OC<u>H</u>$_2$CH$_3$	3.34–3.42, m	65.2, CH$_2$	3.34–3.44, m	65.2, CH$_2$
OCH$_2$C<u>H</u>$_3$	1.15, t (7.0)	15.6, CH$_3$	1.15, t (7.0)	15.6, CH$_3$

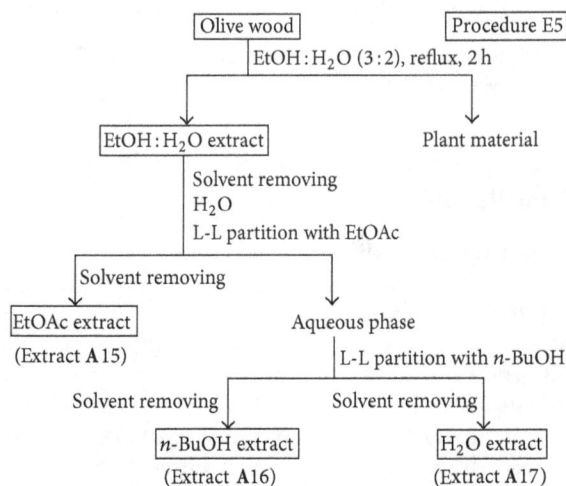

FIGURE 5: Solvent extractions of olive wood sample **A** following procedure E5.

the extracts; three of them were collected in the same olive grove during the pruning period (**A, B**) or the harvesting season (**D**) and the other one in a different olive grove (**C**). All extracts obtained have been evaluated for their radical scavenging activities, except the hexane and dichloromethane extracts. The two latter ones showed low antioxidant activity in our previous works but allowed the removal of nonpolar components from polar extracts [20, 21]. Tables 2–4 show the extraction yields and the DPPH radical scavenging activity of the fifty extracts obtained in this work. The purpose of this work was to find improved extraction conditions to yield olive wood extracts with high antioxidant activities and appropriate extraction yields. The design of every extraction procedure was based on our previous experience and that of others working on optimization of extraction processes, taking into account general considerations on cost, easiness, and suitable scaling-up. Thus, procedure E1 involved a simple sequential extraction with dichloromethane and ethanol at

TABLE 2: Extraction yields and radical scavenging percentages of several extracts prepared by the extraction procedures E1–E6 from the olive wood sample **A**.

Extract[a]	Procedure[b]	Solvent	Temperature	Yield[c]	RSP ± SD[d]
A1	E1	EtOH	r.t.	54.1	48.8 ± 0.2
A2	E2	EtOAc	r.t.	11.0	63.2 ± 0.8
A3	E2	EtOH	r.t.	51.4	42.1 ± 0.4
A4	E3	EtOAc	Reflux	14.2	64.9 ± 0.1
A5	E4	EtOAc	r.t.	8.6	48.5 ± 0.8
A6	E4	EtOH	r.t.	40.0	42.1 ± 1.1
A7	E4	$EtOH : H_2O$ 3 : 2	r.t.	111.4	27.1 ± 1.9
A8	E4	H_2O	r.t.	80.0	17.1 ± 1.4
A9	E4	$H_2O : HCOOH$ 4 : 1	r.t.	122.9	25.9 ± 0.5
A10	E4	EtOAc	Reflux	11.4	50.9 ± 1.5
A11	E4	EtOH	Reflux	94.3	52.4 ± 1.7
A12	E4	$EtOH : H_2O$ 3 : 2	Reflux	145.7	39.6 ± 1.1
A13	E4	H_2O	Reflux	108.6	44.0 ± 0.2
A14	E4	$H_2O : HCOOH$ 4 : 1	Reflux	211.4	33.7 ± 1.8
A15	E5	EtOAc	Reflux	22.6	54.3 ± 0.3
A16	E5	n-BuOH	Reflux	54.9	45.7 ± 1.3
A17	E5	H_2O	Reflux	42.0	11.1 ± 0.8
A18	E6	Et_2O	Reflux	4.0	48.5 ± 0.1
A19	E6	$CHCl_3$	Reflux	4.0	26.1 ± 0.4
A20	E6	n-BuOH	Reflux	33.7	39.6 ± 0.1
Rosemary oleoresin (reference extract)[d]					95.0 ± 0.3

[a] Extracts **A1**–**A20** were prepared from the olive wood sample **A**, collected in April, 2003 (during the pruning period) at the village of Fuensanta, Jaén province, Spain.

[b] Procedures E1–E6 are detailed in Figures 1–6.

[c] Yield is expressed as grams of extract per kilogram of olive wood sample.

[d] Radical scavenging percentage (RSP) is expressed as DPPH$^\bullet$ scavenging (%). Values are means of three replicates ± SD (standard deviation).

[d] Commercially available rosemary extract was used as reference, at the same concentration (50 μg mL^{-1}).

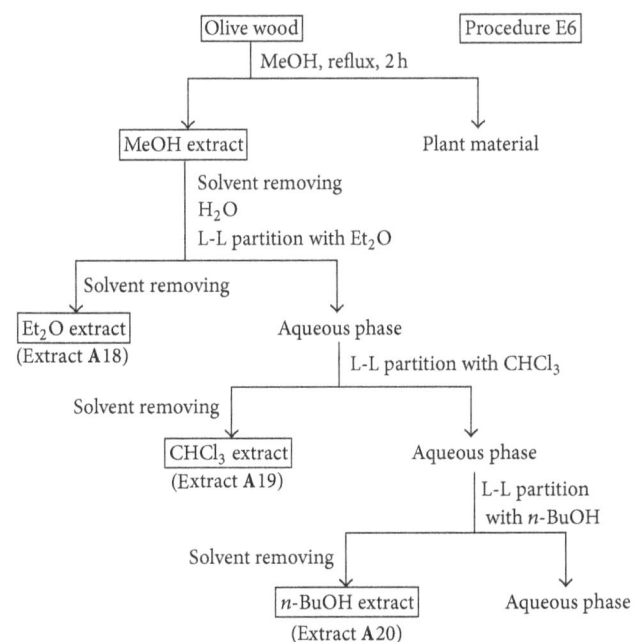

FIGURE 6: Solvent extractions of olive wood sample **A** following procedure E6.

room temperature (Figure 1). This protocol was used in our preliminary work [20] and allowed us to conclude that ethanol yielded the largest amounts of extracts, with the highest radical scavenging activities. Procedure E2 used the sequence of solvents n-hexane, dichloromethane, ethyl acetate, and ethanol at room temperature (Figure 2) in order to perform a better separation of metabolites by groups and choose the better solvent (ethyl acetate or ethanol) to extract antioxidants. As ethyl acetate seemed to be a more selective solvent to extract antioxidants from olive wood, procedure E3 used only two solvents to simplify the protocol (Figure 3): dichloromethane at room temperature to remove nonpolar components, and ethyl acetate under reflux to increase extract yields in active compounds. Procedure E4 involved the direct extraction of olive wood with ethyl acetate at room temperature or under reflux, without previous removal on nonpolar components, and the direct extraction with several more polar solvents, including acidified water (water-formic acid), for comparison purposes (Figure 4). Procedures E5 and E6 involved the initial extraction under reflux of olive wood with aqueous ethanol (Figure 5) and methanol (Figure 6), respectively, followed by several liquid-liquid partitioning with solvents of increasing polarity. These "inverse" procedures were designed to compare results with those of the

TABLE 3: Extraction yields and radical scavenging percentages of several extracts prepared by the extraction procedure E7 from the olive wood sample **A**.

Extract[a]	Pretreatment[b]	Time[c]	Reextracted material[d]	Yield[e]	RSP ± SD[f]
A21	H_2SO_4 in H_2O	1 h	Acidic extract	29.2	49.7 ± 0.6
A22	H_2SO_4 in H_2O	1 h	Pre-treated wood	8.8	37.4 ± 0.1
A23	H_2SO_4 in H_2O	3 h	Acidic extract	23.6	49.6 ± 0.8
A24	H_2SO_4 in H_2O	3 h	Pre-treated wood	4.7	59.6 ± 0.4
A25	H_2SO_4 in H_2O	5 h	Acidic extract	34.8	49.2 ± 0.9
A26	H_2SO_4 in H_2O	5 h	Pre-treated wood	5.5	59.6 ± 1.6
A27	H_2SO_4 in H_2O : EtOH	1 h	Acidic extract	25.7	54.0 ± 1.3
A28	H_2SO_4 in H_2O : EtOH	1 h	Pre-treated wood	5.2	29.9 ± 1.9
A29	H_2SO_4 in H_2O : EtOH	3 h	Acidic extract	71.2	48.1 ± 2.9
A30	H_2SO_4 in H_2O : EtOH	3 h	Pre-treated wood	5.9	38.4 ± 1.7
A31	H_2SO_4 in H_2O : EtOH	5 h	Acidic extract	74.5	45.7 ± 1.2
A32	H_2SO_4 in H_2O : EtOH	5 h	Pre-treated wood	2.3	63.1 ± 0.6
A33	HCl in H_2O	1 h	Acidic extract	27.5	56.6 ± 0.5
A34	HCl in H_2O	1 h	Pre-treated wood	4.9	38.1 ± 0.9
A35	HCl in H_2O	3 h	Acidic extract	26.4	58.5 ± 2.1
A36	HCl in H_2O	3 h	Pre-treated wood	6.1	68.7 ± 1.1
A37	HCl in H_2O	5 h	Acidic extract	23.5	48.8 ± 0.4
A38	HCl in H_2O	5 h	Pre-treated wood	9.0	64.5 ± 0.4
A39	HCl in H_2O : EtOH	1 h	Acidic extract	40.0	43.0 ± 1.2
A40	HCl in H_2O : EtOH	1 h	Pre-treated wood	7.5	53.8 ± 0.3
A41	HCl in H_2O : EtOH	3 h	Acidic extract	58.5	41.6 ± 0.8
A42	HCl in H_2O : EtOH	3 h	Pre-treated wood	7.1	56.4 ± 1.8
A43	HCl in H_2O : EtOH	5 h	Acidic extract	94.1	45.9 ± 2.4
A44	HCl in H_2O : EtOH	5 h	Pre-treated wood	3.6	60.9 ± 1.9
Rosemary oleoresin (reference extract)[g]					95.0 ± 0.3

[a] Extracts **A21**–**A44** were prepared from the olive wood sample **A**, collected in April, 2003 (during the pruning period) at the village of Fuensanta, Jaén province, Spain.

[b] The olive wood sample **A** was subjected to a hydrolysis pre-treatment with 0.5 M H_2SO_4 in H_2O (extracts **A21**–**A26**), 0.5 M H_2SO_4 in H_2O : EtOH (50 : 50, v/v) (extracts **A27**–**A32**), 1 M HCl in H_2O (extracts **A33**–**A38**), and 1 M HCl in H_2O : EtOH (50 : 50, v/v) (extracts **A39**–**A44**) (see Figure 7).

[c] Time of the hydrolysis pre-treatment on the olive wood sample **A** at 130°C.

[d] The hydrolysis pre-treatment of olive wood sample **A** afforded a liquid acidic extract and a solid pre-treated wood on which further extractions were performed (see Figure 7).

[e] Yield is expressed as grams of extract per kilogram of olive wood sample.

[f] Radical scavenging percentage (RSP) is expressed as DPPH• scavenging (%). Values are means of three replicates ± SD (standard deviation).

[g] Commercially available rosemary extract was used as reference, at the same concentration (50 μg mL^{-1}).

direct extraction in sequence included in procedures E1, E2, and E3. Finally, procedure E7 (Figure 7) was designed to explore the influence on yield and antioxidant activity of the extracts obtained after an acidic treatment of olive wood (see the following).

3.1. Temperature Effect. The influence of temperature on the extraction was investigated since it affects both the equilibrium and mass transfer rate. Higher temperatures could produce the breakage of bonds between analytes and plant matrix and could thus increase the yield of the extraction [27] or could favour the reaction of compounds like phenols with other plant components, impeding their extraction [1]. Higher temperatures increase the solubility of the compounds, although they may also affect their stability, and chemical transformations may happen; the changes in extract composition usually involve changes in radical scavenging

activity. In this study, olive wood shavings from sample **A** were subjected to extraction with solvent systems at two temperatures, room temperature, and reflux temperature of each solvent (see Section 2), and as expected, higher yields were obtained under reflux (from 11.4 to 211.4 g extract kg^{-1} wood) than at room temperature (from 8.6 to 122.9 g extract kg^{-1} wood) (Table 2). For instance, the increase of the yield of extracts obtained under reflux with respect to those obtained at room temperature is considerable; from a 30% (extract **A4** versus extract **A2**) up to a 136% (extract **A11** versus extract **A6**). In contrast, the increases observed for the radical scavenging percentage (RSP) of the same extracts are comparatively lower: from a 3% (extract **A4** versus extract **A2**) up to a 24% (extract **A11** versus extract **A6**). In general, it can be said that extract yields reach a considerable increase with the temperature while RSP values increase moderately. Indeed, the highest value for radical scavenging activity at

TABLE 4: Extraction yields and radical scavenging percentages of several extracts prepared by the extraction procedures E3 and E4 from olive wood samples **A**, **B**, **C**, and **D**.

Olive wood sample[a]	Extract[a]	Procedure[b]	Solvent	Yield[c]	RSP ± SD[d]
A	A4	E3	EtOAc	14.2	64.9 ± 0.1
B	B1	E3	EtOAc	46.8	63.7 ± 1.3
C	C1	E3	EtOAc	91.8	59.1 ± 2.6
D	D1	E3	EtOAc	14.2	40.5 ± 0.7
A	A11	E4	EtOH	94.3	52.4 ± 1.7
B	B2	E4	EtOH	117.3	38.3 ± 2.6
C	C2	E4	EtOH	172.5	42.4 ± 0.6
D	D2	E4	EtOH	81.7	42.9 ± 2.2
Rosemary oleoresin (reference extract)[e]					95.0 ± 0.3

[a]Extracts were prepared from (a) the olive wood sample **A**, collected in April, 2003 (during the pruning period) at the village of Fuensanta, Jaén province, Spain; (b) the olive wood sample **B**, collected in March, 2005 (during the pruning period) at the same location of sample **A**; (c) the olive wood sample **C**, collected in March, 2005 (during the pruning period) at the village of Mogón, Jaén province, Spain; (d) the olive wood sample **D**, collected in November, 2006 (during the harvesting season) at the same location of sample **A**.

[b]Procedures E3 and E4 are detailed in Figures 3 and 4.

[c]Yield is expressed as grams of extract per kilogram of olive wood sample.

[d]Radical scavenging percentage (RSP) is expressed as DPPH[•] scavenging (%). Values are means of three replicates ± SD (standard deviation).

[e]Commercially available rosemary extract was used as reference, at the same concentration (50 μg mL^{-1}).

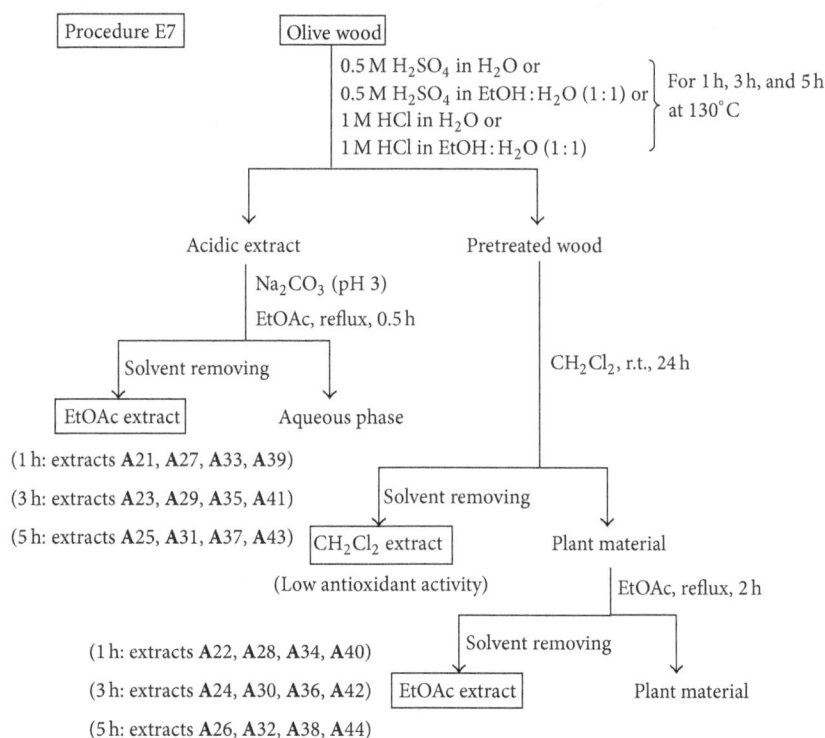

FIGURE 7: Hydrolysis pretreatment and solvent extractions of olive wood sample **A** following procedure E7.

room temperature (63.2%, extract **A2**) was similar to the highest one for extractions under reflux (64.9%, extract **A4**) (Table 2). This means that the extraction protocols using refluxing solvents are not necessarily better than those using solvents at room temperature.

3.2. Solvent Composition. Since the radical scavenging activity depends on the extract composition, comparative studies for selecting the optimal solvents providing maximum antioxidant activity are required for each plant material. Methanol, mixtures of ethanol (or methanol) and water, ethyl acetate, and diethyl ether have been the most common extraction solvents reported in the literature for the extraction of phenols from wood samples [28]. Considering our previous work on olive wood extracts [20–22], where nonoptimized extraction protocols were used, a more comprehensive study of the extraction yields and radical scavenging activities of olive wood extracts obtained from sample **A** with different neat solvents or mixtures of them was carried out in this work (Figures 1–6). The results are shown in Table 2. Hexane, dichloromethane, ethyl acetate, and ethanol were chosen as solvents to extract olive tree wood shavings in three different manners (procedures E1, E2, and E3). The ethyl acetate extract obtained under reflux of an olive wood sample (A4), previously extracted with dichloromethane at room temperature, afforded the best radical scavenging activity (64.9%). Ethyl acetate, ethanol, ethanol-water, water, and water-formic acid mixtures have also been used for each extraction of a fresh and non previously extracted olive tree wood sample, following procedure E4 (Figure 4). Ethanolic and aqueous extracts yields were around 6-fold and 9-fold higher than those of ethyl acetate extracts, respectively (Table 2). It is documented that an addition of water to solvents causes swelling of the plant material, thereby allowing the solvent to penetrate the solid matrix more easily, which leads to higher yields [27]. However, the radical scavenging percentages of ethanolic and aqueous extracts were up to 3-fold lower than those of ethyl acetate extracts. Moreover, a pH effect on the extraction yield has also been reported [1]. To study the pH influence, two additional extracts obtained with a water-formic acid mixture have also been evaluated in this work (extracts A9 and A14). The yields of these extracts were higher than those of aqueous extracts without acid (extracts A8 and A13), but there were no significant changes in their radical scavenging percentages. Two other extraction procedures (E5 and E6) have been checked according to those described in the literature [29, 30]. In both procedures, an aqueous ethanolic extract (procedure E5, Figure 5) and a methanolic extract (procedure E6, Figure 6) were partitioned using successively ethyl acetate and *n*-butanol (procedure E5) and diethyl ether, chloroform, and *n*-butanol (procedure E6) as solvents. The yields of alcoholic (extracts A16 and A20) and aqueous extracts (extract A17) were the highest ones, and the ethyl acetate extract (extract A15) was the most active one again. Thus, ethyl acetate extracts from olive wood, obtained either through direct extraction (A4) or by successive liquid-liquid partitioning (A15) procedures, are the most active ones. This solvent has been used on other occasions to separate low molecular weight polyphenols from other wood sources [31, 32].

3.3. Acid Hydrolysis Pretreatment Effect on Wood Shavings.

It is known that during the conversion of hemicellulose into sugar and sugar oligomers by mild hydrolysis of lignocellulosic materials, cell wall-linked phenolic compounds are also solubilized [12]. A number of technologies are available for the hydrolysis of these materials. The autohydrolysis is the simplest one, where the lignocellulosic material is contacted with water or steam [3, 33]. In hardwoods, acid hydrolysis processes have been extensively studied as well as the effect of the operational conditions on the yield and the antioxidant activity of the phenolic fraction recovered [3]. Solvent extraction with ethyl acetate has successfully been applied for purification or refining purposes, since saccharides remain in the aqueous phase, whereas the nonsaccharide compounds (part of them, of phenolic nature) are transferred to the organic phase. To evaluate the effect of this pretreatment on olive wood samples, several experiments were designed based

on the literature using sulfuric acid [34, 35] or hydrochloric acid [36]. In this work, after acidic hydrolysis pretreatment of olive wood shavings, the filtration and extraction of both liquid and solid phases were carried out using procedure E7 (Figure 7). Table 3 shows the extraction yields and the DPPH radical scavenging activity of the ethyl acetate extracts obtained according to this procedure. Yields of the ethyl acetate extracts obtained from the liquid phase of procedure E7 ("acidic extract" in Figure 7) by partitioning against ethyl acetate (extracts A21, A23, A25, A27, A29, A31, A33, A35, A37, A39, A41, A43) were up to 10-fold higher than those of ethyl acetate extracts obtained following a consecutive extraction of the solid phase of procedure E7 ("pretreated wood" in Figure 7) with dichloromethane and ethyl acetate (extracts A22, A24, A26, A28, A30, A32, A34, A36, A38, A40, A42, and A44) (Table 3). However, among the radical scavenging percentage data of the latter extracts (from 29.9 to 68.7%), the best result (extract A36, 68.7%) was only 10% higher than the highest one (extract A35, 58.5%) of the former ones (from 41.6 to 58.5%). Besides, it can be said that there are no important differences between the radical scavenging activities obtained with the acidic pretreatment of olive wood using sulfuric acid (63.1% is the highest radical scavenging value; extract A32) or hydrochloric acid (68.7% is the highest one; extract A36), neither in aqueous-alcoholic solutions nor in aqueous solutions. In terms of pretreatment time, the radical scavenging activity was in general higher after 3 or 5 h than after 1 h. In conclusion, the best radical scavenging activities were those corresponding to extracts A36 and A38, obtained from the ethyl acetate extraction of the pretreated olive wood shavings with 1 M HCl after 3 h and 5 h, whose radical scavenging percentages were 68.7% and 64.5%, respectively. These results were similar to that obtained for extract A4 (64.9% of radical scavenging activity), which was prepared by extraction with ethyl acetate under reflux of the same wood sample A previously extracted with dichloromethane according to procedure E3 (Figure 3 and Table 2). However, if yield values are compared, it can be said that the acid hydrolysis pretreatment of olive wood does not improve the simplest and cheapest procedure E3, since extract A4 was obtained with a 14.2% yield while extracts A36 and A38 were obtained with 6.1% and 9.0% yields, respectively.

3.4. Location and Season Collection Data.

The geographic origin, as well as climatic condition, harvesting date, storage, environmental, and technological factors, affects the composition of plant material samples and consequently their antioxidant activities [1, 24, 37]. Four different woods collected at two different locations and seasons have been studied (samples A, B, C, and D; see Section 2 for details). Taking into consideration the best results previously found for sample A, procedure E3 was chosen as the extraction protocol for extracting the other three samples. Later, these samples were also extracted by procedure E4 (with EtOH under reflux), which also showed good behaviour with sample A. Yields of the ethyl acetate extracts (procedure E3) ranged from 14.2 g extract kg^{-1} wood for A to 91.8 g extract kg^{-1} wood for C, and the yields of the ethanol ones (procedure E4) from 81.7 g extract kg^{-1} wood for D to 172.5 g

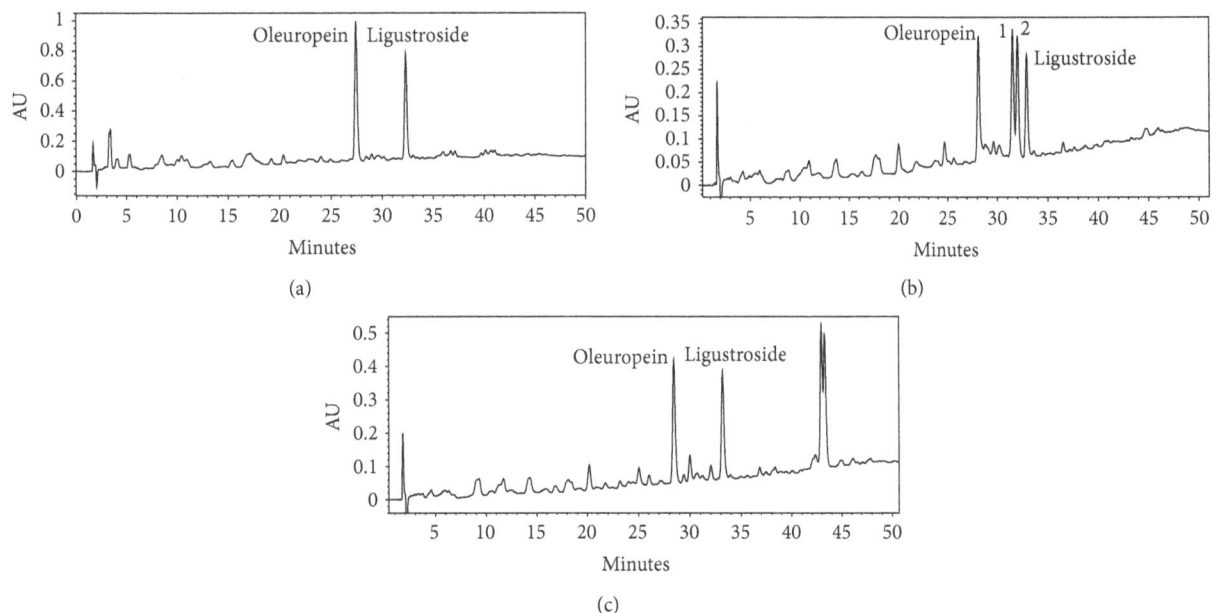

FIGURE 8: HPLC profiles of olive wood extracts at 230 nm: (a) ethyl acetate extract **A**4, (b) ethanol extract **D**2, and (c) direct *n*-butanol extract from olive wood sample **D**.

extract kg^{-1} wood for **C** (Table 4). Hence, yields of the ethyl acetate extracts were lower than those of the ethanol ones which is in agreement with previous results obtained for sample **A**. However, some differences in antioxidant activity and composition have been found among the four olive wood samples studied; regarding the radical scavenging activity, the ethyl acetate extract of the sample collected during the olives' harvesting season (sample **D**) showed a slightly lower radical scavenging percentage than that of the ethanol extract, which is in contrast with the results obtained for sample **A**. The HPLC–DAD–MS analyses of seven of these eight extracts (both ethyl acetate and ethanol extracts) showed similar chromatographic profiles (Figure 8(a)), where oleuropein and ligustroside were the main components as described before by us for other olive wood samples [21, 22]. However, the chromatogram of the ethanolic extract from the sample collected during the olives' harvesting season (extract **D**2) showed four major peaks: oleuropein, ligustroside, and two compounds not identified previously by us (Figure 8(b)).

3.5. Structure Elucidation of the Unidentified Compounds. An aliquot of the ethanolic extract from the sample collected in autumn (extract **D**2) was submitted to further preparative RP-HPLC separations, and two pure secoiridoids, compounds **1** and **2**, were therefore isolated (Figure 9). Compounds **1** and **2** were characterized by UV, IR, MS, ^{1}H NMR, ^{13}C NMR, 2D NMR, and specific optical rotation measurements. These spectroscopic data indicate that compounds **1** and **2** were two stereoisomers of 7″-ethoxyoleuropein. The spectral data of **1** are in agreement with earlier published data for lucidumoside C, which was isolated for the first time from an ethanolic extract of *Ligustrum lucidum* fruit [38]. However, the exact configuration at C-7″ was not given in that paper. In order to establish the stereochemistry at C-7″ of **1** and **2**,

a comparative study of the NMR spectra of both compounds with those of (7″*R*)- and (7″*S*)-7″-methoxyoleuropein was carried out. These methoxyoleuropein derivatives were isolated for the first time from the methanolic extract of *Jasminun officinale* L. var. *grandiflorum* leaves and stems [39]. The 7″*R*-epimer has signals for H-8″a and H-8″b differing by 0.06 ppm while in the 7″*S*-epimer the signals for H-8″a and H-8″b differ by 0.20 ppm. Since compounds **1** and **2** showed Δδ values of 0.03 and 0.17, respectively, the stereochemistry at carbon C-7″ for **1** is assigned as 7″*R* and for **2** as 7″*S*. Both diastereoisomers of 7″-ethoxyoleuropein seem to be artefacts of 7″-hydroxyoleuropein produced by the extraction with ethanol. In order to prove this hypothesis, the same olive wood sample **D** was extracted following the same extraction procedure (procedure E4; 2 h at reflux) with acetone, in one case, and with *n*-butanol in the other case. The corresponding acetone and *n*-butanol extracts were analysed by HPLC–DAD–MS and only the second extracts presented two new peaks with an [M–H]$^-$ ion at *m/z* 611.2 for the two diastereoisomers. These were assigned as the corresponding artefacts produced by *n*-butanol (Figure 8(c)), although no further efforts were made to isolate them. It is well documented that olive drupes contain a hydroxylated oleuropein derivative, with a hydroxyl group at the elenoic moiety, known as 10-hydroxyoleuropein [40]. However, the presence of 7″-hydroxyoleuropein, with the hydroxyl group located at the phenylethanolic moiety, has only been detected in some occasion in olive drupes [41]. Recently, we reported the presence of (7″*S*)-7″-hydroxyoleuropein in olive wood [22]. This molecule, never found previously in *O. europaea*, is a secondary metabolite in other genera of the Oleaceae family, such as *Fraxinus* and *Ligustrum* [42]. It is known that a hydroxyl group located at a benzylic position, such as in the case of 7″-hydroxyoleuropein, is endowed with

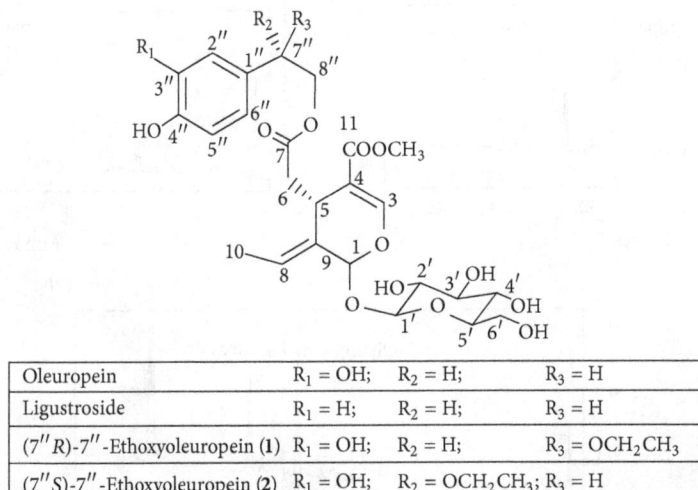

Oleuropein	$R_1 = OH$;	$R_2 = H$;	$R_3 = H$
Ligustroside	$R_1 = H$;	$R_2 = H$;	$R_3 = H$
$(7''R)$-$7''$-Ethoxyoleuropein (**1**)	$R_1 = OH$;	$R_2 = H$;	$R_3 = OCH_2CH_3$
$(7''S)$-$7''$-Ethoxyoleuropein (**2**)	$R_1 = OH$;	$R_2 = OCH_2CH_3$;	$R_3 = H$

FIGURE 9: Structures of the isolated compounds from the olive wood sample **D**.

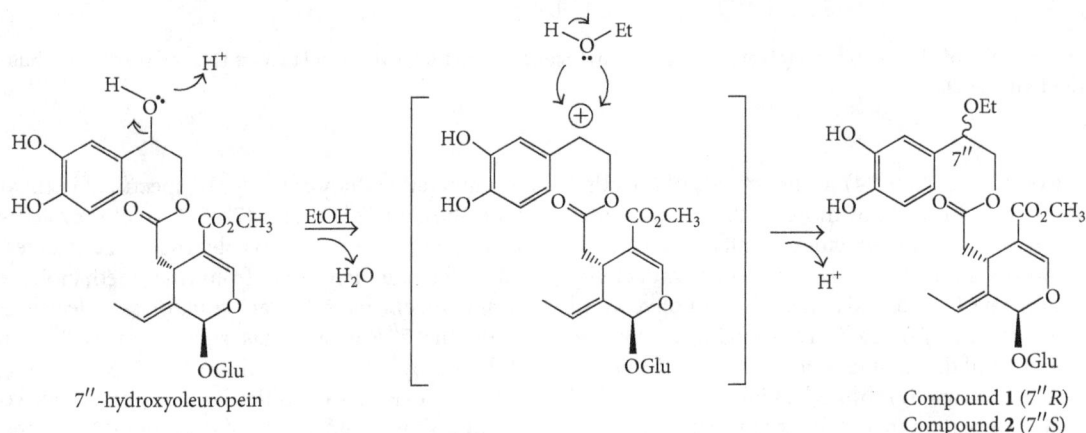

SCHEME 1: Proposed reaction pathway for the conversion of the natural product $7''$-hydroxyoleuropein into the artefacts **1** and **2** during the extraction of olive wood (sample **D**) with ethanol.

a special reactivity. Indeed, the acid-catalysed synthesis of ethers from benzylic alcohols and aliphatic alcohols has been described [43]. Thus, we can postulate that when the olive wood extracts were prepared under reflux with ethanol as solvent, a catalytical substitution of the hydroxyl group of $7''$-hydroxyoleuropein took place, yielding the related $7''$-ethoxyoleuropein derivatives (Scheme 1).

4. Conclusions

Fifty extracts of olive (*Olea europaea* L., cultivar Picual) wood have been prepared following seven different solvent extraction protocols in order to find the best conditions to optimize yield and radical scavenging activity. It was observed that the yields of the ethanolic, aqueous, and acid-aqueous extracts were higher than those of the ethyl acetate extracts, while the opposite was observed for the antioxidant activity. Indeed, the most active extracts were obtained with ethyl acetate either through direct extraction or by successive liquid-liquid partitioning procedures. When the extracts were obtained under

reflux, the yields were higher than at room temperature, although the radical scavenging activities were similar. There are no significant differences between the results obtained from the pretreatment of olive wood with sulfuric acid or hydrochloric acid, neither in aqueous-alcoholic solutions nor in aqueous solutions. Pretreatment times of 3 and 5 h gave higher radical scavenging activities than those of 1 h. The best result for the hydrolysis pretreatments (with yields of 9.0 g extract kg^{-1} wood and radical scavenging percentages of 68.7%) was similar to that obtained for ethyl acetate extractions without pretreatment (from 8.6 to 14.2 g extract kg^{-1} wood for yield and from 48.5 to 64.9% for radical scavenging activity). Significant differences were observed for the extraction yields and radical scavenging activity from those olive wood samples collected at two different geographical origins, years, and seasons. The HPLC–DAD–MS analysis of the ethyl acetate and ethanol extracts showed similar profiles, where oleuropein and ligustroside were the main components. However, the chromatogram of the ethanolic extract from the sample collected during the olives' harvesting season (extract

D2) showed four major peaks: oleuropein, ligustroside, and two compounds identified as $(7''R)$-$7''$-ethoxyoleuropein (**1**) and $(7''S)$-$7''$-ethoxyoleuropein (**2**). Compounds **1** and **2** were shown to be artefacts formed from the natural product $7''$-hydroxyoleuropein during the extraction process with ethanol.

Acknowledgments

The authors wish to thank the Spanish *Ministerio de Educación y Ciencia* for financial support (R + D Project CTQ2005-07005/PPQ; partial financial support from the FEDER funds of the European Union) and the Andalusian Government, *Junta de Andalucía*, for a predoctoral fellowship to M. Pérez-Bonilla. They are grateful to Dr. P. de Waard for his technical assistance in recording NMR spectra and also to the *Centro de Instrumentación Científico-Técnica* of the University of Jaén for the financial support.

References

[1] A. Moure, J. M. Cruz, D. Franco et al., "Natural antioxidants from residual sources," *Food Chemistry*, vol. 72, no. 2, pp. 145–171, 2001.

[2] F. Visioli and C. Galli, "Olives and their production waste products as sources of bioactive compounds," *Current Topics in Nutraceutical Research*, vol. 1, no. 1, pp. 85–88, 2003.

[3] G. Garrote, J. M. Cruz, A. Moure, H. Domínguez, and J. C. Parajó, "Antioxidant activity of byproducts from the hydrolytic processing of selected lignocellulosic materials," *Trends in Food Science and Technology*, vol. 15, no. 3-4, pp. 191–200, 2004.

[4] N. Balasundram, K. Sundram, and S. Samman, "Phenolic compounds in plants and agri-industrial by-products: antioxidant activity, occurrence, and potential uses," *Food Chemistry*, vol. 99, no. 1, pp. 191–203, 2006.

[5] R. Carle, P. Keller, A. Schieber et al., "Method for obtaining useful materials from the by-products of fruit and vegetable processing," WO Patent 2001078859, 2001.

[6] J. L. Torres, B. Varela, M. T. García et al., "Valorization of grape (*Vitis vinifera*) byproducts. Antioxidant and biological properties of polyphenolic fractions differing in procyanidin composition and flavonol content," *Journal of Agricultural and Food Chemistry*, vol. 50, no. 26, pp. 7548–7555, 2002.

[7] W. Peschel, F. Sánchez-Rabaneda, W. Diekmann et al., "An industrial approach in the search of natural antioxidants from vegetable and fruit wastes," *Food Chemistry*, vol. 97, no. 1, pp. 137–150, 2006.

[8] S. Savatovic, G. Cetkovic, J. Canadanovic-Brunet, and S. Djilas, "Tomato waste: a potential source of hydrophilic antioxidants," *International Journal of Food Sciences and Nutrition*, vol. 63, no. 2, pp. 129–137, 2012.

[9] V. Papadimitriou, G. A. Maridakis, T. G. Sotiroudis, and A. Xenakis, "Antioxidant activity of polar extracts from olive oil and olive mill wastewaters: an EPR and photometric study," *European Journal of Lipid Science and Technology*, vol. 107, no. 7-8, pp. 513–520, 2005.

[10] F. Visioli, A. Romani, N. Mulinacci et al., "Antioxidant and other biological activities of olive mill waste waters," *Journal of Agricultural and Food Chemistry*, vol. 47, no. 8, pp. 3397–3401, 1999.

[11] O. Benavente-García, J. Castillo, J. Lorente, A. Ortuño, and J. A. del Río, "Antioxidant activity of phenolics extracted from *Olea europaea* L. leaves," *Food Chemistry*, vol. 68, no. 4, pp. 457–462, 2000.

[12] B. Felizón, J. Fernández-Bolaños, A. Heredia, and R. Guillén, "Steam explosion pretreatment of olive cake," *Journal of the American Oil Chemists Society*, vol. 77, no. 1, pp. 15–22, 2000.

[13] H. Domínguez, J. Torres, and M. J. Núñez, "Antioxidant phenolics as food additives from agricultural wastes," *Polyphénols Actualités*, vol. 21, pp. 25–30, 2001.

[14] B. Amro, T. Aburjai, and S. Al-Khalil, "Antioxidative and radical scavenging effects of olive cake extract," *Fitoterapia*, vol. 73, no. 6, pp. 456–461, 2002.

[15] A. Ranalli, L. Lucera, and S. Contento, "Antioxidizing potency of phenol compounds in olive mill wastewater," *Journal of Agricultural and Food Chemistry*, vol. 51, no. 26, pp. 7636–7641, 2003.

[16] H. K. Obied, M. S. Allen, D. R. Bedgood, P. D. Prenzler, K. Robards, and R. Stockmann, "Bioactivity and analysis of biophenols recovered from olive mill waste," *Journal of Agricultural and Food Chemistry*, vol. 53, no. 4, pp. 823–837, 2005.

[17] G. Rodríguez, R. Rodríguez, J. Fernández-Bolaños, R. Guillén, and A. Jiménez, "Antioxidant activity of effluents during the purification of hydroxytyrosol and 3,4-dihydroxyphenylglycol from olive oil waste," *European Food Research and Technology*, vol. 224, no. 6, pp. 733–741, 2007.

[18] G. Rodríguez, A. Lama, M. Trujillo, J. L. Espartero, and J. Fernández-Bolaños, "Isolation of powerful antioxidant from *Olea europaea* fruit-mill waste: 3,4-dihydroxyphenylglycol," *Food Science and Technology*, vol. 42, no. 2, pp. 483–490, 2009.

[19] I. González-Hidalgo, S. Bañón, and J. M. Ros, "Evaluation of table olive by-product as a source of natural antioxidants," *International Journal of Food Science and Technology*, vol. 47, no. 4, pp. 674–681, 2012.

[20] J. Altarejos, S. Salido, M. Pérez-Bonilla et al., "Preliminary assay on the radical scavenging activity of olive wood extracts," *Fitoterapia*, vol. 76, no. 3-4, pp. 348–351, 2005.

[21] M. Pérez-Bonilla, S. Salido, T. A. van Beek et al., "Isolation and identification of radical scavengers in olive tree (*Olea europaea*) wood," *Journal of Chromatography A*, vol. 1112, no. 1-2, pp. 311–318, 2006.

[22] M. Pérez-Bonilla, S. Salido, T. A. van Beek et al., "Isolation of antioxidative secoiridoids from olive wood (*Olea europaea* L.) guided by on-line HPLC-DAD-radical scavenging detection," *Food Chemistry*, vol. 124, no. 1, pp. 36–41, 2011.

[23] H. Zbidi, S. Salido, J. Altarejos et al., "Olive tree wood phenolic compounds with human platelet antiaggregant properties," *Blood Cells, Molecules, and Diseases*, vol. 42, no. 3, pp. 279–285, 2009.

[24] G. Beltrán, C. del Río, S. Sánchez, and L. Martínez, "Seasonal changes in olive fruit characteristics and oil accumulation during ripening process," *Journal of the Science of Food and Agriculture*, vol. 84, no. 13, pp. 1783–1790, 2004.

[25] W. Brand-Williams, M. E. Cuvelier, and C. Berset, "Use of a free radical method to evaluate antioxidant activity," *Food Science and Technology*, vol. 28, no. 1, pp. 25–30, 1995.

[26] A. von Gadow, E. Joubert, and C. F. Hansmann, "Comparison of the antioxidant activity of aspalathin with that of other plant phenols of rooibos tea (*Aspalathus linearis*), α-tocopherol, BHT, and BHA," *Journal of Agricultural and Food Chemistry*, vol. 45, no. 3, pp. 632–638, 1997.

[27] S. Mukhopadhyay, D. L. Luthria, and R. J. Robbins, "Optimization of extraction process for phenolic acids from black cohosh (*Cimicifuga racemosa*) by pressurized liquid extraction," *Journal*

of the Science of Food and Agriculture, vol. 86, no. 1, pp. 156–162, 2006.

[28] B. F. de Simón, E. Cadahía, E. Conde, and M. C. García-Vallejo, "Low molecular weight phenolic compounds in Spanish oakwoods," *Journal of Agricultural and Food Chemistry*, vol. 44, no. 6, pp. 1507–1511, 1996.

[29] H. Tsukamoto, S. Hisada, and S. Nishibe, "Lignans from bark of the *Olea* plants. I," *Chemical and Pharmaceutical Bulletin*, vol. 32, no. 7, pp. 2730–2735, 1984.

[30] I. Parejo, F. Viladomat, J. Bastida et al., "Comparison between the radical scavenging activity and antioxidant activity of six distilled and nondistilled mediterranean herbs and aromatic plants," *Journal of Agricultural and Food Chemistry*, vol. 50, no. 23, pp. 6882–6890, 2002.

[31] J. M. Cruz, J. M. Domínguez, H. Domínguez, and J. C. Parajó, "Antioxidant and antimicrobial effects of extracts from hydrolysates of lignocellulosic materials," *Journal of Agricultural and Food Chemistry*, vol. 49, no. 5, pp. 2459–2464, 2001.

[32] A. Moure, H. Domínguez, and J. C. Parajó, "Antioxidant activity of liquors from aqueous treatment of *Pinus radiata* wood," *Wood Science and Technology*, vol. 39, no. 2, pp. 129–139, 2005.

[33] J. M. Cruz, H. Domínguez, and J. C. Parajó, "Anti-oxidant activity of isolates from acid hydrolysates of *Eucalyptus globulus* wood," *Food Chemistry*, vol. 90, no. 4, pp. 503–511, 2005.

[34] J. Fernández-Bolaños, B. Felizón, M. Brenes, R. Guillén, and A. Heredia, "Hydroxytyrosol and tyrosol as the main compounds found in the phenolic fraction of steam-exploded olive stones," *Journal of the American Oil Chemists Society*, vol. 75, no. 11, pp. 1643–1649, 1998.

[35] I. Romero, E. Ruiz, E. Castro, and M. Moya, "Acid hydrolysis of olive tree biomass," *Chemical Engineering Research and Design*, vol. 88, no. 5-6, pp. 633–640, 2010.

[36] M. Bouaziz and S. Sayadi, "Isolation and evaluation of antioxidants from leaves of a Tunisian cultivar olive tree," *European Journal of Lipid Science and Technology*, vol. 107, no. 7-8, pp. 497–504, 2005.

[37] M. M. Torres, P. Pierantozzi, M. E. Cáceres, P. Labombarda, G. Fontanazza, and D. M. Maestri, "Genetic and chemical assessment of Arbequina olive cultivar grown in Córdoba province, Argentina," *Journal of the Science of Food and Agriculture*, vol. 89, no. 3, pp. 523–530, 2009.

[38] Z. D. He, P. P. H. But, T. W. D. Chan et al., "Antioxidative glucosides from the fruits of *Ligustrum lucidum*," *Chemical and Pharmaceutical Bulletin*, vol. 49, no. 6, pp. 780–784, 2001.

[39] T. Tanahashi, T. Sakai, Y. Takenaka, N. Nagakura, and C. C. Chen, "Structure elucidation of two secoiridoid glucosides from *Jasminum officinale* L. var. *grandiflorum* (L.) Kobuski," *Chemical and Pharmaceutical Bulletin*, vol. 47, no. 11, pp. 1582–1586, 1999.

[40] D. Caruso, R. Colombo, R. Patelli, F. Giavarini, and G. Galli, "Rapid evaluation of phenolic component profile and analysis of oleuropein aglycon in olive oil by atmospheric pressure chemical ionization-mass spectrometry (APCI-MS)," *Journal of Agricultural and Food Chemistry*, vol. 48, no. 4, pp. 1182–1185, 2000.

[41] L. Di Donna, F. Mazzotti, A. Napoli, R. Salerno, A. Sajjad, and G. Sindona, "Secondary metabolism of olive secoiridoids. New microcomponents detected in drupes by electrospray ionization and high-resolution tandem mass spectrometry," *Rapid Communications in Mass Spectrometry*, vol. 21, no. 3, pp. 273–278, 2007.

[42] Y. Takenaka, T. Tanahashi, M. Shintaku, T. Sakai, N. Nagakura, and Parida, "Secoiridoid glucosides from *Fraxinus americana*," *Phytochemistry*, vol. 55, no. 3, pp. 275–284, 2000.

[43] S. W. Wright, D. L. Hageman, A. S. Wright, and L. D. McClure, "Convenient preparations of t-butyl esters and ethers from t-butanol," *Tetrahedron Letters*, vol. 38, no. 42, pp. 7345–7348, 1997.

18

Microbiological Safety Assessment of Fermented Cassava Flour "*Lafun*" Available in Ogun and Oyo States of Nigeria

A. O. Adebayo-Oyetoro,[1] O. B. Oyewole,[2] A. O. Obadina,[2] and M. A. Omemu[3]

[1] Department of Food Technology, Yaba College of Technology, PMB 2011, Yaba, Lagos 101212, Nigeria
[2] Department of Food Science and Technology, Federal University of Agriculture, PMB 2240, Ogun State, Abeokuta 110001, Nigeria
[3] Department of Food Service and Tourism, Federal University of Agriculture, PMB 2240, Ogun State, Abeokuta 110001, Nigeria

Correspondence should be addressed to A. O. Adebayo-Oyetoro; wonunext@yahoo.com

Academic Editor: Rosana G. Moreira

The microorganisms involved in the fermentation and spoilage of fermented cassava flour were investigated. The water samples used at the different processing sites were also investigated to determine their safety status. There was predominance of *Staphylococcus aureus*, *Aspergillus* spp., and *Escherichia coli* in all samples. Coliforms were observed to be present in all of the processing water. In the fermented cassava flour, the total bacterial count ranged between 4.9×10^6 cfu/mL from Eleso, Bakatari, and Oja Odan processing sites and 8.10×10^6 cfu/mL in Eruku processing site. The majority of the microorganisms involved in the spoilage of "*lafun*" were found to be *Aspergillus niger* which ranged between 4.6×10^5 cfu/mL in Eleso and 8.1×10^5 cfu/mL in Kila. The control sample prepared in the laboratory had a low microbial load compared to samples collected from various sites and markets.

1. Introduction

Cassava (*Manihot esculenta* Crantz) is a major root crop in the tropics and its starchy roots are significant sources of calories for more than 500 million people worldwide [1]. It is the most important root crop in Nigeria in terms of food security, employment creation, and income generation for crop-producing households [2]. It supplies about 70% of the daily calories of over 50 million people in Nigeria [3].

Nigeria is the largest producer of cassava in the world [1] with about 45 million metric tonnes and its cassava transformation is the most advanced in Africa [4]. Cassava is grown throughout the tropic and could be regarded as the most important root crop, in terms of area cultivated and total production [5]. It is a major food crop in Nigeria [6]. It is essentially a carbohydrate food with low protein and fat. The edible part of fresh cassava root contains 32–35% carbohydrate, 2-3% protein, 75–80% moisture, 0.1% fat, 1.0% fibre, and 0.70–2.50% ash [3]. The production of cassava for human consumption has been estimated to be 65% of cassava products, while 25% is for industrial use, mostly starch 6% or animal feed 19% and 10% lost as waste [7].

The major uses of cassava in Nigeria include flour, gari, fufu, livestock feeds, confectionaries, monosodium glutamate processing, sweeteners, glues, textiles, and pharmaceuticals. In spite of the desirability of cassava for consumption as food and animal feed, it contains some toxic compounds such as cyanogenic glycosides, linamarin, and lotaustralin which are highly toxic. Thus, the consumption of an inadequately processed cassava product for prolonged periods may result in chronic toxicity. However, the toxicity of the cyanogens is a result of inadequate processing [8]. Cassava roots are highly perishable and a lot of postharvest losses occur to this commodity during storage due to high physiological activities and activities of microorganisms that enter bruises received during harvesting as well as the inherent high moisture content of fresh roots, which promote both microbial deterioration and unfavourable biochemical changes in the commodity [9].

Lafun is a fine flour obtained from the traditional fermentation of cassava (Figure 1). It is usually prepared as a stiff porridge using boiling water, prior to being consumed with soup. The processing involves peeling, cutting, submerged fermentation, dewatering, sun drying, and milling. One of

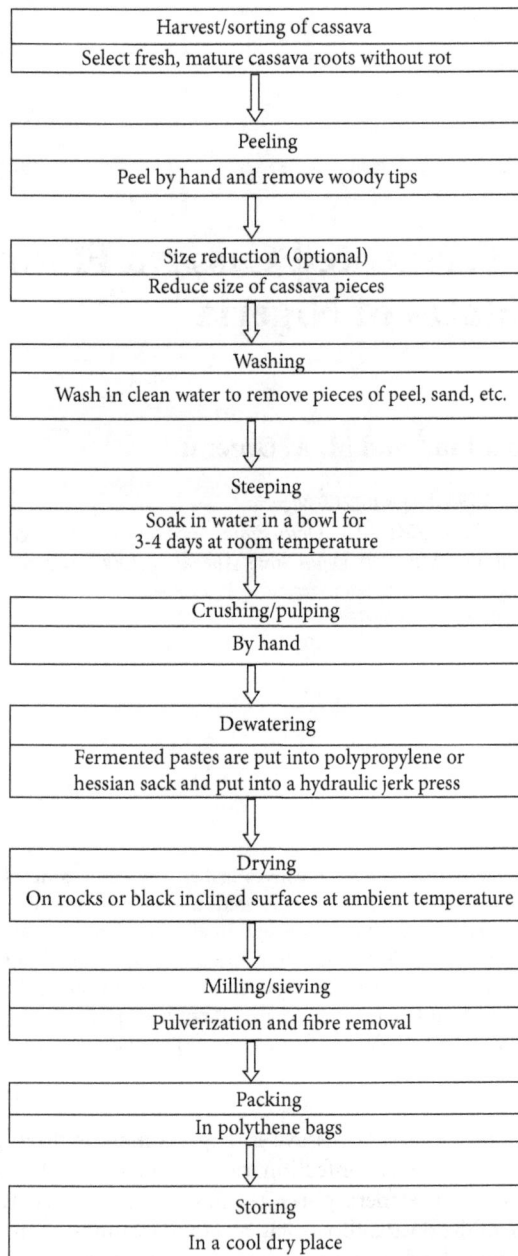

FIGURE 1: Process diagram for *lafun*.

The diagram contains the following sequential steps:

Harvest/sorting of cassava	Select fresh, mature cassava roots without rot
Peeling	Peel by hand and remove woody tips
Size reduction (optional)	Reduce size of cassava pieces
Washing	Wash in clean water to remove pieces of peel, sand, etc.
Steeping	Soak in water in a bowl for 3-4 days at room temperature
Crushing/pulping	By hand
Dewatering	Fermented pastes are put into polypropylene or hessian sack and put into a hydraulic jerk press
Drying	On rocks or black inclined surfaces at ambient temperature
Milling/sieving	Pulverization and fibre removal
Packing	In polythene bags
Storing	In a cool dry place

two states were selected for being the major producers and consumers of this product. The samples after collection were labeled and taken to the laboratory for analysis which is usually not later than 24 hours after collection. Processing water samples from boreholes, rain, and streams were also collected in sterile containers for analysis.

2.1. Microbiological Analysis. The analysis involved total plate count, fungi count, and coliform count.

2.2. Isolation and Identification. Samples of the *lafun* collected were serially diluted tenfold in which ten grams of each sample was diluted in 90 mL peptone water followed by homogenization by horizontal and vertical agitations for a few minutes to obtain 10^{-1} dilution. Further tenfold serial dilutions were made up to 10^{-5} for colony count. 1 mL of volume of each dilution was spread plated in triplicate on de Man Rogosa Sharpe Agar (MRS; Oxoid CM 361) and incubated anaerobically at 35°C for 48 h for the enumeration of lactic acid bacteria, and Plate Count Agar (PCA; Oxoid CM 325) incubated at 32°C for 48 h was used for the enumeration of aerobic bacteria [12]. 0.1 mL of each of the samples was also plated on Potato Dextrose Agar (Oxoid) supplemented with 60 μg mL^{-1} chloramphenicol for fungal isolation. This was incubated at 28°C for 5 days. All plates were prepared in triplicate. The colonies were counted and recorded followed by isolation, purification, and storage on Nutrient Agar (Lab M) slants and kept at 4°C for further characterization and identification [13, 14].

2.3. Characterization of Isolates. Bacterial isolates were characterized and identified using series of cultural and biochemical tests such as Gram staining, glucose, indole reaction coagulase, and catalase as described by [15].

2.4. pH Determination. The pH was determined using Kent pH meter (Kent Ind. Measurement Ltd., UK) model 7020 equipment with a glass electrode as described by [16]. Ten grams of the *lafun* sample was weighed and dissolved in 100 mL sterile distilled water. The solution was later decanted and the pH was measured. The pH of the processing water was also determined using the same equipment.

2.5. Statistical Analysis. The mean of the total viable bacterial count obtained from the fermented cassava flour was subjected to analysis of variance (ANOVA) and the Duncan multiple range test to separate the means.

3. Results and Discussion

Table 1 shows the mean values for the total colony count for bacterial isolates from fermented cassava flour (*lafun*) obtained from the processing sites and markets. Among the organisms isolated are *Escherichia coli*, *Klebsiella oxytoca*, *Bacillus cereus*, *Staphylococcus aureus*, and *Clostridium sporogenes*.

the constraints in the commercialization of locally fermented cassava products is that the quality of the products varies from one processor to the other and even from one processing batch to the other by the processor [10, 11].

Meanwhile, a report on high rate of spoilage as well as toxicity resulting from the consumption of *lafun* has necessitated this research. This study is aimed at evaluating the microbiological safety status of the *lafun* produced in south west Nigeria.

2. Materials and Methods

Lafun samples were collected aseptically from the processors and sellers in 10 locations in Ogun and Oyo States. The

TABLE 1: Mean values for the total aerobic colony count for bacterial isolates.

Sampling points/code	Total colony count for aerobic bacteria (cfu/g) $\times 10^6$ (processing sites)	Total colony count for aerobic bacteria (cfu/g) $\times 10^6$ (markets)
ERK	4.0 ± 0.07	4.0 ± 0.07
OJO	3.0 ± 0.02	4.0 ± 0.07
OWO	4.3 ± 0.23	4.8 ± 0.73
BTS	3.1 ± 0.09	3.2 ± 0.08
EOS	4.5 ± 0.43	5.3 ± 1.23
SOP	4.2 ± 0.13	4.8 ± 0.73
MNI	4.8 ± 0.73	4.8 ± 0.73
TVE	4.5 ± 0.43	4.7 ± 0.63
PTJ	4.2 ± 0.13	4.8 ± 0.73
OSJ	4.1 ± 0.03	4.5 ± 0.43
CTR	1.1 ± 0.11	

Results are presented as mean values of triplicate samples.

TABLE 2: Mean values for the total colony count for fungal isolates.

Sampling points/code	Total colony count for fungi (cfu/g) $\times 10^3$ (processing sites)	Total colony count for fungi (cfu/g) $\times 10^3$ (markets)
ERK	2.8 ± 0.27	3.4 ± 0.34
OJO	3.0 ± 0.29	3.3 ± 0.33
OWO	2.3 ± 0.22	2.2 ± 0.22
BTS	3.4 ± 0.34	4.0 ± 0.39
EOS	2.2 ± 0.22	3.3 ± 0.32
SOP	3.7 ± 0.37	4.0 ± 0.39
MNI	4.0 ± 0.39	3.9 ± 0.39
TVE	3.9 ± 0.40	4.0 ± 0.39
PTJ	3.9 ± 0.38	4.0 ± 0.39
OSJ	3.5 ± 0.30	3.4 ± 0.34
CTR	NIL	

Results are presented as mean values of triplicate samples.

TABLE 3: Biochemical characteristics of bacterial isolates.

Bacteria found	Gram staining	Glucose	Indole	Coagulase	Catalase
Escherichia coli	−	+	+	NA	+
Klebsiella oxytoca	−	+	+	NA	+
Bacillus cereus	+	+	+	NA	+
Staphylococcus aureus	+	+	−	+	+
Clostridium sporogenes	+	+	−	NA	−

Table 2 shows the results of the fungal count for the samples obtained from both the processing sites and the markets. It was observed that some of the values were higher than 1.9×10^3 cfu/g to 3.9×10^3 cfu/g as reported by [17] and 0.5–2.5×10^3 cfu/g as reported by [18].

TABLE 4: Microbiological status of water used at the processing sites.

Location	Aerobic cfu/mL	Coliform cfu/mL
ERK	7.3 ± 0.17	7.0 ± 0.2
OJO	8.5 ± 1.37	8.0 ± 1.13
OWO	7.7 ± 0.57	7.2 ± 0.33
BTS	6.4 ± 0.07	6.1 ± 0.08
EOS	8.3 ± 1.17	8.0 ± 1.13
SOP	8.1 ± 0.97	7.9 ± 1.03
MNI	7.5 ± 0.37	7.1 ± 0.23
TVE	6.0 ± 0.01	5.8 ± 0.01
PTJ	5.5 ± 0.01	5.3 ± 0.01
OSJ	6.0 ± 0.11	6.3 ± 0.57
CTR	1.2 ± 0.1	1.1 ± 0.1

Bacterial isolates from the samples collected were *E. coli, B. cereus, S aureus, K. oxytoca,* and *C. sporogenes* as shown in Table 3. Some of the organisms were Gram positive, catalase positive, and glucose positive. *Bacillus cereus* have been reported to be pathogenic especially if found to be present at 10^5–10^7 cfu/g [19]. Meanwhile, sample MNI was found to have the highest amount of aerobic bacterial count of 4.8×10^6 cfu/g, while sample EOS had the highest count of 5.3×10^6 cfu/g among the samples from the markets. These values were higher than the findings of microbial count of the fermented flour obtained by [20] who reported 1.5×10^6 cfu/g. It was also higher than the findings of [17] who reported 2.7×10^3 to 1.2×10^7 cfu/g.

More so, Table 4 shows the aerobic count of stream water used for processing which was found to be the highest in OJO with value of 8.5 ± 1.37 cfu/mL and the lowest in PTJ with 5.5 ± 0.01 cfu/mL. The coliform counts (Table 4) from sites EOS and OJO were found to be the highest with 8.0 ± 1.13 cfu/mL and the lowest in PTJ with 5.3 ± 0.01 cfu/mL. This could be as a result of the same stream water being used by processors for washing and bathing as well as for processing. Some people even defecate in the water thereby reducing the quality of the product. In addition, lactic acid bacteria, *Bacillus* spp. yeast, and filamentous fungi have been shown to be present in traditional fermented cassava (*lafun*) [21].

Although water samples used for fermenting previous batches of "*lafun*" have been suggested to be used as back slopping by the processors in the production area [10], this may also account for the high microbial load obtained in these samples. This is due to the possibility of the safety being compromised. The isolation of *Staphylococcus aureus* and *Escherichia coli* from the fermented cassava flour is attributed to postprocessing handling and exposure both at the processing sites and in the markets [22]. This is because the drying is usually done in open air where animals are reared. In some locations, the toilet is constructed near the processing area which could facilitate cross-contamination.

Table 5 shows the results of the pH values for the various samples obtained from the processing sites and markets. The value ranged from 4.05 ± 0.16 in sample EOS to 5.51 ± 0.42 in sample MNI for those samples collected from the processing sites, while the value ranged from 4.08 ± 0.16 in sample EOS

TABLE 5: Results of the pH values for *lafun* obtained from processing sites and markets.

Sample codes	Processing site	Market
ERK	4.51 ± 0.34	4.54 ± 0.33
OJO	4.42 ± 0.32	4.42 ± 0.32
OWO	4.24 ± 0.24	4.20 ± 0.22
BTS	4.15 ± 0.20	4.16 ± 0.21
EOS	4.05 ± 0.16	4.08 ± 0.16
SOP	5.50 ± 0.45	5.55 ± 0.42
MNI	5.51 ± 0.42	5.50 ± 0.40
TVE	5.23 ± 0.40	5.52 ± 0.41
PTJ	4.63 ± 0.36	5.53 ± 0.42
OSJ	4.55 ± 0.34	5.51 ± 0.42

to 5.55 ± 0.42 in sample SOP for those collected from the markets. It was observed that the result was comparable to the findings of [23]. Meanwhile, the increase in pH observed in some locations may be due to short period of fermentation by some of these processors thereby limiting the activities of lactic acid bacteria. This is because the product is always in high demand which allows some of these processors to compromise.

4. Conclusion

Our results had shown that although *lafun* is a fermented product, there is possibility of contamination resulting due to poor handling either at the processing site or from the market. In the course of our investigation, we observed that quality is often compromised as some processors make use of back slopping to reduce the length of time for fermentation so as to meet high customers' demand. The displayed pattern in the market also encourages cross-contamination. We suggest provision of pipe borne water or borehole for constant supply of water for processing. We also suggest the provision of modern driers to each community so as to prevent sun drying of *lafun*.

Finally, it is recommended that close monitoring of the steps involved in processing fermented cassava flour should be carried out by health officials so as to ensure strict compliance by the processors and the sellers in the market.

Conflict of Interests

A. O. Adebayo-Oyetoro wishes to state on behalf of the other authors that the commercial identity (OXOID) mentioned in this paper has no direct financial relation with the research work. The media were rather purchased and used for the study. There is therefore no conflict of interests.

Acknowledgments

The authors would like to acknowledge the Tertiary Education Trust Fund in Nigeria as well as Yaba College of Technology, Lagos, Nigeria.

References

[1] FAO, "Corporate Document Repository," The impact of HIV/AIDS on the agricultural sector, 2008, http://www.fao.org/docrep/005/Y4636E/y4636e05.htm.

[2] B. O. Ugwu and U. J. Ukpabi, "Potential of soy-cassava flour processing to sustain increasing cassava production in Nigeria," *Outlook on Agriculture*, vol. 31, no. 2, pp. 129–133, 2002.

[3] O. B. Oluwole, O. O. Olatunji, and S. A. Odunfa, "A process technology for conversion of dried cassava chips into gari," *Journal of Food Science and Technology*, vol. 22, pp. 65–77, 2004.

[4] C. Egesi, E. Mbanaso, F. Ogbe, E. Okogbenin, and M. Fregene, "Development of cassava varieties with high value root quality through induced mutations and marker-aided breeding," NRCRI, Umudike Annual Report, 2006.

[5] A. O. Ano, "Studies on the effect of Liming on the Yield of two cassava cultivars," NRCRI Annual Report, 2003.

[6] F. O. Ogbe, J. K. U. Emehute, and J. Legg, "Screening of cassava varieties for whitefly populations," NRCRI Annual Report, 2007.

[7] D. M. Fish and D. S. Trim, "A review of research in the drying of cassava," *Tropical Science*, vol. 33, no. 2, pp. 191–203, 1993.

[8] J. H. Bradbury, A. Cumbana, E. Mirione, and J. Cliff, "Reduction of cyanide content of cassava flour in Mozambique by the wetting method," *Food Chemistry*, vol. 101, no. 3, pp. 894–897, 2006.

[9] J. E. Wenham, "Post-harvest deterioration of cassava. A Biotechnology perspective," F.A.O. Plant Production and Protection Paper 130, Food and Agriculture Organization of the United Nations, Rome, Italy, 1995.

[10] S. W. Padonou, J. D. Hounhouigan, and M. C. Nago, "Physical, chemical and microbiological characteristics of *lafun* produced in Benin," *African Journal of Biotechnology*, vol. 8, no. 14, pp. 3320–3325, 2009.

[11] O. B. Oyewole and L. O. Sanni, "Constraints in traditional cassava food processing: the case of fufu production," in *Cassava Food Processing*, T. Agbor Egbe, A. Bauman, T. Griffon, and S. Treche, Eds., pp. 523–529, ORSTOM, Paris, France, 1995.

[12] J. Owusu-Kwarteng, K. Tano-Debra, R. L. K. Glover, and F. Akabanda, "Process characteristics and Microbiology of *fura* produced in Ghana," *Nature and Science*, vol. 8, no. 8, pp. 41–51, 2010.

[13] Y. M. Somorin, M. O. Bankole, A. M. Omemu, and O. O. Atanda, "Impact of milling on the microbiological quality of yam flour iii southwestern Nigeria," *Research Journal of Microbiology*, vol. 6, no. 5, pp. 480–487, 2011.

[14] A. O. Obadina, O. B. Oyewole, and A. O. Odusami, "Microbiological safety and quality assessment of some fermented cassava products (*lafun, fufu , gari*)," *Scientific Research and Essays*, vol. 4, no. 5, pp. 432–435, 2009.

[15] J. Ochei and A. Kolhatkar, *Medical Laboratory Science: Theory and Practice*, McGraw-Hill, New York, NY, USA, 2008.

[16] O. B. Oyewole and S. A. Odunfa, "Characterization and distribution of lactic acid bacteria in cassava fermentation during fufu production," *Journal of Applied Bacteriology*, vol. 68, no. 2, pp. 145–152, 1990.

[17] J. A. Tsav-Wua, C. U. Inyang, and M. A. Akpapunam, "Microbiological quality of fermented cassava flour 'kpor umilin'," *International Journal of Food Sciences and Nutrition*, vol. 55, no. 4, pp. 317–324, 2004.

[18] C. O. Eleazu, J. U. Amajor, A. I. Ikpeama, and E. Awa, "Studies on the nutrient composition, antioxidant activities,

functional properties and microbial load of the flours of 10 Elite cassava (*Manihot esculenta*) varieties," *Asian Journal of Clinical Nutrition*, vol. 3, no. 1, pp. 33–39, 2011.

[19] N. Michelet, P. E. Granea, and J. Mahillon, "*Bacillus cereus* enterotoxins, bi- and tri-component cytolysins, and other hemolysins," in *The Comparative Sourcebook of Bacterial Protein Toxins.*, J. E. Alouf and M. R. Popoff, Eds., pp. 779–790, Elsievier, Amsterdam, The Netherlands, 2006.

[20] A. O. Ijabadeniyi, "Microbiological safety of gari, *lafun* and ogiri in Akure metropolis, Nigeria," *African Journal of Biotechnology*, vol. 6, no. 22, pp. 2633–2635, 2007.

[21] O. B. Oyewole and S. Ayo Odunfa, "Effects of fermentation on the carbohydrate, mineral, and protein contents of cassava during "fufu" production," *Journal of Food Composition and Analysis*, vol. 2, no. 2, pp. 170–176, 1989.

[22] A. O. Obadina, O. B. Oyewole, L. O. Sanni, K. I. Tomlins, and A. Westby, "Identification of hazards and critical control points (CCP) for cassava fufu processing in South-West Nigeria," *Food Control*, vol. 19, no. 1, pp. 22–26, 2008.

[23] F. Hongbete, C. Mestres, N. Akissoe, and M. C. Nago, "Effect of processing conditions on cyanide content and colour of cassava flours from West Africa," *African Journal of Food Science*, vol. 3, no. 1, pp. 1–6, 2009.

19

Mineral Properties and Dietary Value of Raw and Processed Stinging Nettle (*Urtica dioica* L.)

Laban K. Rutto,[1] Yixiang Xu,[2] Elizabeth Ramirez,[3] and Michael Brandt[1]

[1] *Alternative Crops Program, Agriculture Research Station, Virginia State University, Petersburg, VA 23806, USA*
[2] *Food Processing and Engineering Program, Agriculture Research Station, Virginia State University, Petersburg, VA 23806, USA*
[3] *College of Agriculture and Life Sciences, Virginia Polytechnic Institute and State University, Blacksburg, VA 24061, USA*

Correspondence should be addressed to Laban K. Rutto; lrutto@vsu.edu

Academic Editor: Fernanda Fonseca

Stinging nettle (*Urtica dioica* L.) has a long history of usage and is currently receiving attention as a source of fiber and alternative medicine. In many cultures, nettle is also eaten as a leafy vegetable. In this study, we focused on nettle yield (edible portion) and processing effects on nutritive and dietary properties. Actively growing shoots were harvested from field plots and leaves separated from stems. Leaf portions (200 g) were washed and processed by blanching (1 min at 96–98°C) or cooking (7 min at 98-99°C) with or without salt (5 g·L^{-1}). Samples were cooled immediately after cooking and kept in frozen storage before analysis. Proximate composition, mineral, amino acid, and vitamin contents were determined, and nutritive value was estimated based on 100 g serving portions in a 2000 calorie diet. Results show that processed nettle can supply 90%–100% of vitamin A (including vitamin A as β-carotene) and is a good source of dietary calcium, iron, and protein. We recommend fresh or processed nettle as a high-protein, low-calorie source of essential nutrients, minerals, and vitamins particularly in vegetarian, diabetic, or other specialized diets.

1. Introduction

Stinging nettle (*Urtica dioica* L.) has a long history as one among plants foraged from the wild and eaten as a vegetable [1, 2]. Although not fully domesticated, the species remains popular even in the current era for food and medicine as reported, for example, in Nepal [2] and Poland [3].

Despite *U. dioica* being recognized as an edible and highly nutritious vegetable, research attention has focused more on its value as a source of alternative medicine and fiber. Clinical trials have confirmed the effectiveness of nettle root and saw palmetto (*Serenoa repens* (Bart.) Small) fruit extracts in the treatment of benign prostatic hyperplasia [4]. Dried nettle leaf preparations are also known to alleviate symptoms associated with allergic rhinitis [5], and a technology for granulating lipophilic leaf extracts for medicine has been developed [6]. A recent report from ongoing work in Italy confirms the potential of *U. dioica* as a sustainable source of textile fiber [7].

There are a number of reports that address the role of *U. dioica* in human nutrition. Fatty acid and carotenoid content in leaf, stem, root, and seed samples have been measured [8], and the properties of phenolic compounds in leaves, stalks, and fibers have been reported [9]. Furthermore, the quality and safety [10] and microbiological properties [11] of sucuk, a Turkish dry-fermented sausage, incorporating dried *U. dioica* leaf have been studied, and the capacity of nettle extracts to improve oxidative stability in brined anchovies has been reported [12]. In the Basque region of Spain, young shoots are reportedly eaten raw or included in omelets [13]. In terms of postharvest processing for long-term storage, microwave drying at 850 W was found to be the best method for preservation of leaf color, energy consumption, and processing time [14]. Mineral content [15] and trace metal concentrations [16] in nettle leaf tea made by infusion or decoction have also been determined.

However, nettle is consumed primarily as a fresh vegetable whereby it is added to soups, cooked as a pot herb, or used as a vegetable complement in dishes. In this sense, more work needs to be done on nutritive value of fresh nettle, and the fate of minerals and bioactive compounds in processed products. This information is essential because the capacity of fresh nettle to irritate bare skin may discourage potential consumers and postharvest processing methods that make it safe to handle, while maintaining nutritive value will benefit the development of *U. dioica* as a specialty vegetable.

In this study, we report dietary values, mineral properties, and other quality attributes of raw, blanched, and cooked stinging nettle.

2. Materials and Methods

2.1. Plant Materials. Plant samples were obtained from field plots planted as a part of an ongoing agronomic study on *U. dioica* at Randolph Farm (37.1°N; 77.3°W), Virginia State University (VSU). Samples from fall and spring growth were collected in October 2011 and May 2012, respectively, by harvesting actively growing shoots (20 ± 2 cm) before the onset of flowering. Individual shoots were clipped with a pair of shears and consolidated in vented plastic bags before transfer to a demonstration kitchen located at the VSU Farm Pavilion for further processing.

2.2. Sample Processing. In the kitchen, the shoots were washed, and twelve 200 ± 5 g units were weighed before separating leaves and tender shoot tips from the woody stem. The edible portion (leaves and tender shoot tips) was weighed, and mean yield was determined by presenting the weight of edible portion as a percentage of total unit mass. Treatments, each replicated three times, were applied as follows: raw samples were packaged and frozen without further processing, blanched samples were immersed in boiling water (98-99°C) for 1 min, and cooked samples were boiled (98–100°C) with or without salt (5 g·L^{-1} H$_2$O) for 7 min. Both blanched and cooked samples were cooled to 0°C with shaved ice immediately after treatment. All samples were kept in frozen storage (−4°C) before analysis. Samples for proximate composition analysis were submitted frozen, while those for fatty and amino acid analysis were freeze-dried and ground to a fine powder before analysis.

2.3. Proximate Analysis. All analysis was done according to the Association of Analytical Chemists (AOAC) methods (AOAC, 2000). Moisture content was determined by drying samples to constant weight using a convection oven. Nitrogen (N) content was measured using a CN analyzer (LECO 528, LECO Corp., St. Joseph, MI), and protein content was derived by multiplying N values with 6.25. Total fat was determined by gas chromatography (Agilent 5890, Agilent Technologies, Santa Clara, CA, USA) after extraction of saponifiable and unsaponifiable fractions, and ash content was measured by ignition at 550°C to constant weight. Carbohydrate content and calorie values were calculated by difference. Total dietary fiber was determined following methods described by the American Association of Cereal Chemists (AACCI method 32-07.01).

2.4. Vitamin and Mineral Analysis. Total vitamin A and vitamin A as β-carotene were determined by colorimetry after alkaline digestion followed by extraction with hexane. Vitamin C was extracted in acid and sample content determined by titration. For mineral analysis, samples were subjected to wet digestion before calcium, iron, and sodium content was determined using an ICP spectrometer (AOAC, 2000).

2.5. Amino Acid Analysis. For amino acid analysis, a ground subsample of nettle tissue was hydrolyzed with 6 M HCl at 100°C for 24 hr as previously described [17]. Acid hydrolyzed amino acids were derivatized with phenyl isothiocyanate (Acros Organics, Geel, Belgium) and separated using a 2695 Alliance HPLC equipped with a 15-cm Pico-Tag column, 2487 UV/Vis detector, and Empower software (all from Waters Corp., Milford, MA) using previously described conditions [18]. Amino acid concentrations are expressed in g/100 g of nettle leaf.

2.6. Fatty Acid Analysis. Fatty acid methyl esters (FAMEs) were prepared by treating raw and processed samples with ethyl chloride and absolute methanol as described [19]. Fatty acid methyl esters were analyzed by gas chromatography using an Agilent 6890 N GC system (Agilent Technologies), equipped with a HP-INNOWax column (30 m × 0.32 mm I.D. × 0.5 μm film thickness) and flame ionization detector. Peaks were identified against retention times for a known FAME and quantified by the aid of heptadecanoic acid (17:0) included as an internal standard. The concentration of each fatty acid is presented as a percentage of total saponifiable oil in sample.

2.7. Statistical Analysis. One-way analysis of variance (ANOVA) using the Analyst function in SAS (version 9.2 for Windows, SAS Institute, Cary, NC) was performed to compare the effects of blanching and cooking on stinging nettle quality and nutritive value. Treatments were treated as independent variables, and data for fall 2011 and spring 2012 were analyzed separately. Tukey's HSD ($P < 0.05$) was used to separate treatment means within season.

3. Results and Discussion

3.1. Yield of Edible Portion in U. dioica. Actively growing stinging nettle shoots are ideally harvested before flowering for consumption as a potherb or spinach alternative. Leaves on stems were found to be tender enough for use as a vegetable up to 25 cm from the growing point, but stems become woody about 4 cm away from the growing point necessitating destemming after harvest to separate the tender tip (approx. 4 cm and leaves) from the woody stem. Our results show that the woody stem portion accounts for 23%–30% of total biomass with edible portion comprising of 70% or more of harvested material (Table 1). Yield (edible portion) was higher in fall than in spring samples because of seasonal differences in *U. dioica* growth characteristics. Consistent with published observations [20], *U. dioica* displays two distinct phenological stages when grown in south-central

TABLE 1: Edible portion (leaf) yield as a percentage of total biomass in stinging nettle (*Urtica dioica* L.) harvested from field plots in the fall of 2011 and spring of 2012. Actively growing shoots (20 ± 2 cm) were harvested and processed by de-stemming.

Season	Shoot wt. (g)	Stem wt. (g)	Leaf wt. (g)	Loss (%)
Fall 2011	203 ± 1.73[a]	46 ± 3.5	157 ± 4.69	23 ± 1.8
Spring 2012	199 ± 5.5	55 ± 7.9	144 ± 10.3	28 ± 4.2

[a]Mean (*n* = 3) ± standard deviation.

Virginia: reproductive growth up to late spring, limited development during summer, and mostly vegetative growth in the fall.

3.2. Effect of Blanching and Boiling on Proximate Composition, Vitamin, and Mineral Content in U. dioica. After draining, there was not much difference in moisture content between raw and processed samples in the fall of 2011, while there was slightly more moisture in processed samples in the spring of 2012, likely due to differences in draining time. There was a slight reduction in crude protein, ash, and fat after blanching or cooking in both fall and spring samples. In both cases, the most significant reductions were observed with longer exposure to heat and also to salt. The same applies to dietary fiber, carbohydrate content, and calorie value. Samples harvested in the spring contained significantly higher values for all parameters measured and showed higher decline after processing (Table 2). Preparation and cooking generally result in deterioration of vegetable quality. For example, cooking significantly reduces ash, carbohydrate content, and calorific value in Cocoyam (*Colocasia esculenta*) leaves [21], while chopping amaranth (*Amaranthus* sp.) leaves before cooking can result in increased loss of vitamins and minerals [22]. Our results show that vitamin A, calcium, and iron contents in *U. dioica* leaf are similarly affected by cooking. Sodium content was low and was not affected by cooking, but the salt added to cooking water in one of the treatments significantly ($P < 0.05$) increased sodium content in drained samples (Table 2). Salt addition for seasoning or preservation has been reported to affect vegetable quality through dilution of minerals and other chemical changes [23]. Cooking led to changes in the fatty acid profile of *U. dioica* with more saturated fat being converted into mono-unsaturated and polyunsaturated forms (Table 3) or lost into solution. Saponifiable oil content in raw and processed *U. dioica* samples (3.2%–4.7% in the spring; 3.2%–4.1% in the spring) was comparable to that in wild asparagus (*Asparagus acutifolius*) and black bryony (*Tamus communis*), edible wild greens common to Mediterranean diets [24].

3.3. Effect of Cooking on Fatty and Amino Acid Composition in U. dioica Tissue Samples. Data on individual amino and fatty acid content in stinging nettle shows that the species can supply significant quantities of oleic (18:1), linoleic (18:2), and α-linoleic (18:3) acids and is a good source of unsaturated fatty acids. Considerable amounts of palmitic acid (16:0), a saturated fatty acid, were found in the leaf (Table 3; Figure 1). There were no significant differences in fatty acid content

between samples collected from fall and spring growth. Similarly blanching and cooking with or without salt did not affect fatty acid content within season except for a general trend showing an increase in unsaturated fatty acid content and a corresponding decrease in the concentration of saturated fatty acids (Table 3). Similarly, high levels of linoleic and α-linoleic acids in young and mature leaves and the presence of relatively high concentrations of the same oils in *U. dioica* seed, stem, and roots portions have been reported [8], with the seed containing up to 15% saponifiable oil.

In terms of omega-3 fatty acid content, *U. dioica* compares favorably with frozen spinach (*Spinacia oleracea* L.) pre-treated by steaming, blanching, or autoclaving [25]. Relative to other commonly consumed wild plants, it contains a higher concentration of omega-3 fatty acids than borage (*Borago officinalis*), and about the same level as water-blinks (*Montia fontana*) [26], watercress (*Rorippa nasturtium-aquaticum*), sheep sorrel (*Rumex acetosella*), and sorrel (*Rumex induratus*) [27]. However, carbohydrate content (including total sugars) was significantly lower in raw and processed *U. dioica* (4.2%–16.5%) than in the four species above reported to constitute 66.6%–78.9% total carbohydrates [27]. These results show that processing by blanching and cooking has a minimal impact on *U. dioica* fatty acid composition, implying that it can be a good source of essential fatty acids when eaten as a leafy vegetable.

With regard to individual amino acids, tissue content was similarly not affected by season. Our results show that *U. dioica* can supply considerable amounts of essential amino acids including threonine, valine, isoleucine, leucine, phenylalanine, and lysine, along with lower concentrations of histidine and methionine (Table 4; Figure 1). Amino acid content was largely unchanged in the spring as compared with fall growth though asparagine, glutamine, leucine, and histidine levels were generally lower in samples from spring growth. There were slight to significant increases in amino acid content after blanching or cooking in fall samples, but no similar observation was made for samples collected in the spring (Table 4). There may be differences between and within species in response to postharvest handling and processing conditions. In one study, a significant increase in amino acid content was recorded after cooking relative to raw spinach [28], while the opposite was true for cooked and frozen versus raw Brussels sprouts [29].

Data from this experiment show that both raw and cooked *U. dioica* can be important sources of dietary protein. The species can supply higher concentrations of essential amino acids than Brussels sprouts [29] and has a better amino acid profile than most other leafy vegetables. Although similar to *S. oleracea* in terms of total amino acid content, *U. dioica* contains higher levels of all essential amino acids except leucine and lysine. Some of the published recipes incorporating *U. dioica* leaf flour in bread, pasta, and noodle dough suggest that it can be used as a protein-rich supplement in starchy diets associated with poor and undernourished populations. This is because on a dry weight basis, *U. dioica* leaf is better than almond (dry) and is comparable to common bean (*Phaseolus vulgaris*) and chicken (*Gallus gallus*) as a source of essential amino acids [30]. The agronomic

TABLE 2: Proximate composition, vitamins, minerals, and fatty acid profile of raw and processed stinging nettle (*Urtica dioica* L.) shoots harvested from field plots in the fall of 2011 and spring of 2012.

	Fall 2011				Spring 2012			
	Raw	Blanched	Cooked	Cooked + salt	Raw	Blanched	Cooked	Cooked + salt
Proximate analysis								
Moisture (%)	89.0 ± 1.4[a]	87.2 ± 0.9[a]	87.7 ± 0.7[a]	88.6 ± 0.5[a]	75.1 ± 1.5[c]	84.6 ± 2.5[b]	85.6 ± 0.8[b]	91.7 ± 0.9[a]
Protein (%)	3.7 ± 0.5[a]	3.6 ± 0.4[ab]	3.6 ± 0.3[a]	2.7 ± 0.2[b]	6.3 ± 0.3[a]	4.1 ± 0.2[b]	3.8 ± 0.3[b]	2.2 ± 0.2[c]
Fat (%)	0.6 ± 0.1[a]	0.4 ± 0.1[b]	0.4 ± 0.0[b]	0.2 ± 0.0[b]	1.4 ± 0.3[a]	1.1 ± 0.1[a]	1.1 ± 0.2[a]	0.6 ± 0.1[b]
Ash (%)	2.1 ± 0.3[a]	1.8 ± 0.3[ab]	1.5 ± 0.3[b]	1.5 ± 0.1[b]	3.4 ± 0.2[a]	1.4 ± 0.1[b]	1.2 ± 0.1[c]	1.0 ± 0.1[c]
Fiber, total dietary (%)	6.4 ± 0.4[a]	4.2 ± 0.1[b]	3.5 ± 0.3[c]	3.6 ± 0.3[bc]	9.7 ± 1.0[a]	5.4 ± 0.9[b]	4.9 ± 1.0[b]	4.2 ± 0.2[c]
Carbohydrates, total (%)	7.1 ± 1.7[a]	6.6 ± 1.4[ab]	6.3 ± 0.8[b]	6.2 ± 1.2[b]	16.5 ± 1.6[a]	8.9 ± 0.7[b]	8.1 ± 1.1[b]	4.2 ± 0.6[c]
Other carbohydrates (%)	2.7 ± 0.2[ab]	2.9 ± 0.3[a]	2.5 ± 0.1[b]	2.7 ± 0.1[a]	6.2 ± 1.0[a]	3.5 ± 0.7[b]	3.3 ± 0.5[b]	2.0 ± 0.1[c]
Calories, total (kcal/100 g)	45.7 ± 3.1[a]	42.6 ± 2.1[a]	44.7 ± 2.5[a]	36.5 ± 2.3[b]	99.7 ± 2.5[a]	62.0 ± 1.0[b]	57.3 ± 1.5[c]	32.0 ± 1.0[d]
Calories from fat (kcal/100 g)	5.0 ± 1.0[a]	4.3 ± 0.6[ab]	2.7 ± 0.5[bc]	2.3 ± 0.6[c]	12.3 ± 1.6[a]	10.0 ± 1.0[ab]	8.7 ± 3.1[b]	4.0 ± 1.0[c]
Vitamins and minerals								
Vitamin A, total (IU/100 g)	4935 ± 104[a]	4851 ± 56[a]	4548 ± 53[b]	4362 ± 78[b]	11403 ± 1333[a]	6470 ± 222[bc]	6021 ± 90[c]	7872 ± 354[b]
Vitamin A, as β-carotene (IU/100 g)	5035 ± 213[a]	4689 ± 37[b]	4549 ± 130[b]	4062 ± 39[c]	7860 ± 460[a]	4811 ± 88[b]	5028 ± 65[b]	4154 ± 148[c]
Vitamin C (mg/100 g)	1.1 ± 0.1[a]	0.6 ± 0.1[b]	0.6 ± 0.1[b]	0.5 ± 0.1[b]	0.5 ± 0.0[a]	0.5 ± 0.0[a]	0.5 ± 0.0[a]	0.5 ± 0.0[a]
Calcium (mg/100 g)	278 ± 9[c]	441 ± 12[a]	376 ± 9[ab]	318 ± 52[bc]	788 ± 41[a]	464 ± 10[b]	430 ± 10[b]	316 ± 7[c]
Iron (mg/100 g)	1.2 ± 0.1[c]	1.8 ± 0.2[b]	2.6 ± 0.1[a]	2.5 ± 0.3[a]	3.4 ± 0.3[a]	2.1 ± 0.2[b]	2.1 ± 0.3[b]	1.6 ± 0.1[c]
Sodium (mg/100 g)	5.7 ± 0.1[b]	6.3 ± 0.4[b]	6.5 ± 0.3[b]	87.7 ± 6.0[a]	5.5 ± 0.6[b]	7.0 ± 0.2[b]	6.7 ± 0.2[b]	81.1 ± 2.9[a]
Fatty acid profile								
Saturated fat (%)	35.5 ± 2.6[a]	25.7 ± 2.5[b]	23.6 ± 4.1[c]	21.7 ± 1.9[d]	32.7 ± 2.8[a]	16.5 ± 1.5[bc]	17.3 ± 1.2[b]	15.7 ± 1.4[c]
Monounsaturated (%)	2.7 ± 0.2[c]	3.3 ± 0.2[a]	4.8 ± 0.3[a]	3.2 ± 0.1[b]	7.5 ± 0.6[a]	5.3 ± 0.3[b]	5.8 ± 1.1[b]	4.6 ± 0.2[c]
Polyunsaturated (%)	61.8 ± 3.5[c]	71.0 ± 2.0[b]	71.6 ± 1.2[c]	75.1 ± 1.9[a]	59.8 ± 2.7[d]	78.2 ± 4.4[b]	76.9 ± 2.5[c]	79.7 ± 2.6[a]
Cholesterol (mg/100 g)	1.0 ± 0.0[a]	1.0 ± 0.0[a]	1.0 ± 0.0[a]	1.0 ± 0.0[a]	1.0 ± 0.0[a]	1.0 ± 0.0[a]	1.0 ± 0.0[a]	1.0 ± 0.0[a]

[a]Mean (n = 3) ± standard deviation. Values within a year followed by different letters are significantly different at $P < 0.05$ (Tukey's HSD).

TABLE 3: Fatty acid content[a] in raw and processed stinging nettle (*Urtica dioica* L.) shoots harvested from field plots in the fall of 2011 and spring of 2012.

| | Total fat (%) | Fatty acid[b] (% of total fat) | | | | | | | | | | |
		16:0	16:1	18:0	18:1	18:2	18:3	20:0	20:1	22:0	22:1	24:0
Fall 2011												
Raw	3.15 ± 0.12^c	17.06 ± 0.05^a	2.54 ± 0.04^b	1.86 ± 0.01^a	2.18 ± 0.01^c	23.30 ± 0.20^a	49.55 ± 0.10^d	0.83 ± 0.01^a	0.03 ± 0.01^a	1.37 ± 0.02^a	0.06 ± 0.01^b	1.23 ± 0.03^a
Blanched	4.72 ± 0.05^{ab}	14.91 ± 0.12^b	2.54 ± 0.02^b	1.41 ± 0.02^c	2.23 ± 0.02^b	21.58 ± 0.20^b	54.42 ± 0.37^c	0.67 ± 0.01^c	0.06 ± 0.01^a	1.11 ± 0.01^a	0.09 ± 0.02^{ab}	0.98 ± 0.01^b
Cooked	4.65 ± 0.10^b	14.83 ± 0.09^b	2.45 ± 0.02^c	1.60 ± 0.01^b	1.91 ± 0.03^d	20.96 ± 0.10^c	55.48 ± 0.20^b	0.69 ± 0.01^b	0.03 ± 0.01^a	1.13 ± 0.01^b	0.05 ± 0.01^b	0.88 ± 0.01^d
Cooked + salt	4.78 ± 0.14^a	14.22 ± 0.11^c	2.62 ± 0.01^a	1.35 ± 0.01^d	2.54 ± 0.01^a	19.67 ± 0.2^d	56.70 ± 0.34^a	0.67 ± 0.01^{bc}	0.05 ± 0.01^a	1.13 ± 0.01^b	0.14 ± 0.01^a	0.91 ± 0.01^c
Spring 2012												
Raw	3.17 ± 0.01^d	16.30 ± 0.04^a	1.88 ± 0.01^b	1.76 ± 0.01^a	2.99 ± 0.02^a	23.89 ± 0.07^a	48.06 ± 0.09^d	1.12 ± 0.01^a	0.25 ± 0.01^a	1.63 ± 0.02^a	0.53 ± 0.01^a	1.61 ± 0.01^a
Blanched	4.27 ± 0.04^b	14.58 ± 0.06^b	1.70 ± 0.01^d	1.64 ± 0.01^c	2.95 ± 0.01^a	21.56 ± 0.27^b	53.25 ± 0.30^c	1.05 ± 0.02^b	0.20 ± 0.01^c	1.48 ± 0.01^b	0.38 ± 0.01^b	1.26 ± 0.01^d
Cooked	4.50 ± 0.15^a	14.07 ± 0.21^d	1.93 ± 0.03^a	1.64 ± 0.03^c	2.30 ± 0.07^c	20.78 ± 0.35^c	55.59 ± 0.41^a	0.99 ± 0.04^c	0.21 ± 0.01^b	1.47 ± 0.03^b	0.38 ± 0.01^b	1.35 ± 0.12^c
Cooked + salt	3.58 ± 0.06^c	14.29 ± 0.05^c	1.84 ± 0.05^c	1.67 ± 0.03^b	2.39 ± 0.01^b	20.19 ± 0.47^d	54.44 ± 0.37^b	1.04 ± 0.02^d	0.18 ± 0.01^d	1.50 ± 0.01^b	0.36 ± 0.01^c	1.50 ± 0.03^b

[a]Methylated samples were analyzed for total fatty acid content using gas chromatography.

[b]Palmitic acid (16:0); palmitoleic acid (16:1); stearic acid (18:0); oleic acid (18:1); linoleic acid (18:2); α-linoleic acid (18:3); gadoleic acid (20:1); behenic acid (22:0); erucic acid (22:1); lignoceric acid (24:0). Column values followed by different letters within season are significantly different at $P < 0.05$ (Tukey's HSD).

[c]Mean ($n = 3$) ± standard deviation.

TABLE 4: Amino acid content in raw and processed stinging nettle (*Urtica dioica* L.) shoots harvested from field plots in the fall of 2011 and spring of 2012.

Amino acid (g/100 g)	Fall 2011				Spring 2012			
	Raw	Blanched	Cooked	Cooked + salt	Raw	Blanched	Cooked	Cooked + salt
Isoleucine	0.90 ± 0.17^b	1.13 ± 0.20^{ab}	1.30 ± 0.10^a	1.39 ± 0.06^a	1.04 ± 0.08^a	1.04 ± 0.08^a	1.06 ± 0.09^a	0.97 ± 0.05^a
Leucine	1.65 ± 0.27^b	2.09 ± 0.033^{ab}	2.37 ± 0.18^a	2.56 ± 0.18^a	1.79 ± 0.38^a	1.91 ± 0.06^a	1.91 ± 0.08^a	1.75 ± 0.03^a
Lysine	1.11 ± 0.21^a	1.37 ± 0.11^a	1.37 ± 0.30^a	1.48 ± 0.17^a	1.16 ± 0.38^a	1.33 ± 0.20^a	1.19 ± 0.30^a	1.10 ± 0.19^a
Methionine	0.24 ± 0.05^a	0.31 ± 0.04^a	0.33 ± 0.05^a	0.35 ± 0.06^a	0.23 ± 0.15^a	0.19 ± 0.13^a	0.17 ± 0.07^a	0.20 ± 0.13^a
Tyrosine	0.75 ± 0.13^b	0.95 ± 0.13^{ab}	1.11 ± 0.10^{ab}	1.18 ± 0.14^a	0.97 ± 0.20^a	0.90 ± 0.10^a	0.93 ± 0.12^a	0.91 ± 0.13^a
Phenylalanine	1.03 ± 0.19^b	1.27 ± 0.17^{ab}	1.43 ± 0.15^a	1.51 ± 0.03^a	1.15 ± 0.23^a	1.14 ± 0.05^a	1.13 ± 0.04^a	1.06 ± 0.04^a
Threonine	1.00 ± 0.17^a	1.08 ± 0.05^a	1.12 ± 0.15^a	1.24 ± 0.08^a	1.03 ± 0.24^a	0.75 ± 0.07^a	0.84 ± 0.11^a	0.75 ± 0.14^a
Valine	1.11 ± 0.19^b	1.40 ± 0.23^{ab}	1.60 ± 0.11^a	1.72 ± 0.16^a	1.30 ± 0.24^a	1.28 ± 0.12^a	1.32 ± 0.15^a	1.22 ± 0.10^a
Histidine	0.42 ± 0.09^b	0.53 ± 0.11^{ab}	0.64 ± 0.06^{ab}	0.68 ± 0.11^a	0.32 ± 0.15^a	0.30 ± 0.12^a	0.37 ± 0.08^a	0.22 ± 0.12^a
Total essential amino acids	8.23 ± 1.36^b	10.13 ± 1.39^{ab}	11.26 ± 1.00^a	12.11 ± 1.60^a	8.95 ± 2.14^a	8.83 ± 0.39^a	8.93 ± 0.29^a	8.20 ± 0.59^a
Arginine	1.22 ± 0.21^b	1.57 ± 0.27^{ab}	1.79 ± 0.16^a	1.97 ± 0.14^a	1.55 ± 0.42^a	1.43 ± 0.26^a	1.56 ± 0.21^a	1.52 ± 0.24^a
Aspartic acid + asparagine	0.85 ± 0.32^a	1.01 ± 0.25^a	0.88 ± 0.40^a	1.01 ± 0.04^a	0.60 ± 0.37^a	0.47 ± 0.09^a	0.49 ± 0.10^a	0.39 ± 0.14^a
Glutamic acid + glutamine	1.69 ± 0.39^a	2.13 ± 0.19^a	1.97 ± 0.62^a	2.22 ± 0.26^a	1.49 ± 0.72^a	1.25 ± 0.26^a	1.42 ± 0.13^a	1.14 ± 0.27^a
Serine	0.85 ± 0.14^b	1.06 ± 0.15^{ab}	1.14 ± 0.13^{ab}	1.26 ± 0.10^a	1.00 ± 0.29^a	0.82 ± 0.12^a	0.96 ± 0.15^a	0.82 ± 0.20^a
Proline	0.90 ± 0.15^b	1.11 ± 0.17^{ab}	1.31 ± 0.13^{ab}	1.41 ± 0.20^a	1.24 ± 0.17^a	1.06 ± 0.22^a	1.19 ± 0.28^a	1.07 ± 0.16^a
Glycine	0.92 ± 0.15^b	1.13 ± 0.16^{ab}	1.26 ± 0.10^{ab}	1.39 ± 0.17^a	1.14 ± 0.23^a	0.98 ± 0.22^a	1.12 ± 0.22^a	0.97 ± 0.12^a
Alanine	1.20 ± 0.19^b	1.40 ± 0.13^{ab}	1.54 ± 0.11^{ab}	1.66 ± 0.16^a	1.54 ± 0.29^a	1.24 ± 0.22^a	1.38 ± 0.26^a	1.21 ± 0.14^a
Total amino acids	17.46 ± 2.88^b	21.58 ± 3.40^{ab}	22.87 ± 2.21^{ab}	24.76 ± 0.96^a	19.40 ± 5.00^a	17.77 ± 1.83^a	18.73 ± 1.44^a	16.97 ± 1.63^a
Dry matter (g/100 g edible portion)	11.0	12.8	12.3	11.4	14.9	15.4	14.4	8.3

[a] Mean ($n = 3$) ± standard deviation. Row values followed by different letters within season are significantly different at $P < 0.05$ (Tukey's HSD).

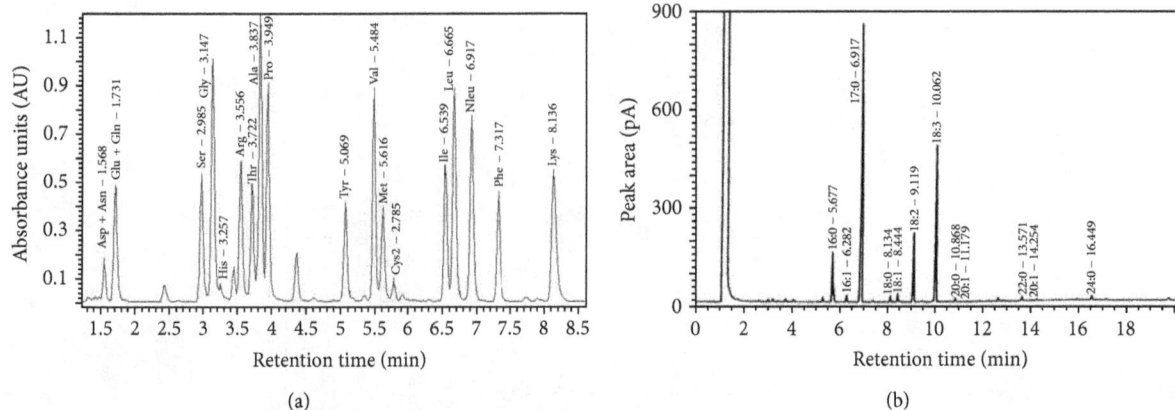

FIGURE 1: Representative chromatograms showing peaks and retention times for different amino (a) and fatty (b) acids in raw and processed stinging nettle (*Urtica dioica* L.) leaf samples.

FIGURE 2: Suggested food labeling information for raw and processed stinging nettle (*Urtica dioica* L.).

properties of *U. dioica* including perennial growth, quick response to fertilization, and high biomass yield make it an excellent candidate for low-cost mass production for such a purpose.

3.4. Labeling Information for Processed U. dioica. Results from this study show that *U. dioica* retains a significant portion of minerals, vitamins, and essential nutrients after

pre-treatment by blanching or cooking prior to frozen storage. Processing may be the most effective approach to availing the nutritional benefits of *U. dioica* to consumers discouraged by the stinging quality of live or fresh nettle. The nutritional information in Figure 2, representing means of data from both spring and fall growth, can be used to label frozen raw and processed *U. dioica* leaf. However, lower vitamin A and higher carbohydrate content and other data reported for blanched *U. dioica* samples collected from the wild [31] show that more work is required to evaluate the properties of *U. dioica* products as affected by interactions between landrace, environment, harvesting time, and processing conditions.

4. Conclusions

Although the usage of *U. dioica* as a leafy vegetable is widespread, there is little information on processing potential, and the impact of different processing methods on nutritive and functional value. The results presented in this report show that *U. dioica* retains significant amounts of minerals, vitamins, and other functional values after blanching or cooking. We recommend processing and selling of *U. dioica* leaf as a highly functional and nutritive food.

Conflict of Interests

The authors declare that they have no conflict of interests.

Acknowledgments

The authors are grateful to Mr. Robert Kraemer and Mr. Landon West, VSU Farm Manager and Assistant Farm Manager, respectively, for field support and to Dr. Ngowari Jaja for assistance with sample preparation. This is a contribution of Virginia State University Research Station Article No. 303.

References

[1] A. di Tizio, J. Ł. Łuczaj, C. L. Quave, S. Redzic, and A. Pieroni, "Traditional food and herbal uses of wild plants in the ancient South-Slavic diaspora of Mundimitar/Montemitro (Southern Italy)," *Journal of Ethnobiology and Ethnomedicine*, vol. 8, article 21, 2012.

[2] Y. Uprety, R. C. Poudel, K. K. Shreshta et al., "Diversity of use and local knowledge of wild edible plant resources in Nepal," *Journal of Ethnobiology and Ethnomedicine*, vol. 8, article 16, 2012.

[3] Ł. Łuczaj, A. Pieroni, J. Tardío et al., "Wild food plant use in 21st century Europe: the disappearance of old traditions and the search for new cuisines involving wild edibles," *Acta Societatis Botanicorum Poloniae*, vol. 81, no. 4, pp. 359–370, 2012.

[4] N. Lopatkin, A. Sivkov, C. Walther et al., "Long-term efficacy and safety of a combination of sabal and urtica extract for lower urinary tract symptoms—a placebo-controlled, double-blind, multicenter trial," *World Journal of Urology*, vol. 23, no. 2, pp. 139–146, 2005.

[5] P. Mittmann, "Randomized, double-blind study of freeze-dried *Urtica dioica* in the treatment of allergic rhinitis," *Planta Medica*, vol. 56, no. 1, pp. 44–47, 1990.

[6] S. A. Minina and M. G. Ozhigova, "Development of granulation technology for lipophilic extracts," *Pharmaceutical Chemistry Journal*, vol. 44, no. 4, pp. 213–215, 2010.

[7] L. Bacci, S. Baronti, S. Predieri, and N. di Virgilio, "Fiber yield and quality of fiber nettle (*Urtica dioica* L.) cultivated in Italy," *Industrial Crops and Products*, vol. 29, no. 2-3, pp. 480–484, 2009.

[8] J. L. Guil-Guerrero, M. M. Rebolloso-Fuentes, and M. E. Torija Isasa, "Fatty acids and carotenoids from Stinging Nettle (*Urtica dioica* L.)," *Journal of Food Composition and Analysis*, vol. 16, no. 2, pp. 111–119, 2003.

[9] P. Pinelli, F. Ieri, P. Vignolini, L. Bacci, S. Baronti, and A. Romani, "Extraction and HPLC analysis of phenolic compounds in leaves, stalks, and textile fibers of *Urtica dioica* L.," *Journal of Agricultural and Food Chemistry*, vol. 56, no. 19, pp. 9127–9132, 2008.

[10] S. Karabacak and H. Bozkurt, "Effects of *Urtica dioica* and *Hibiscus sabdariffa* on the quality and safety of sucuk (Turkish dry-fermented sausage)," *Meat Science*, vol. 78, no. 3, pp. 288–296, 2008.

[11] G. Kaban, M. I. Aksu, and M. Kaya, "Behavior of Staphylococcus aureus in sucuk with nettle (*Urtica dioica* L.)," *Journal of Food Safety*, vol. 27, no. 4, pp. 400–410, 2007.

[12] S. Turhan, I. Sagir, and H. Temiz, "Oxidative stability of brined anchovies (*Engraulis encrasicholus*) with plant extracts," *International Journal of Food Science and Technology*, vol. 44, no. 2, pp. 386–393, 2009.

[13] G. Menendez-Baceta, L. Aceituno-Mata, J. Tardío, V. Reyes-García, and M. Pardo-de-Santayana, "Wild edible plants traditionally gathered in Gorbeialdea (Biscay, Basque Country)," *Genetic Resources and Crop Evolution*, vol. 59, pp. 1329–1347, 2012.

[14] I. Alibas, "Energy consumption and color characteristics of nettle leaves during microwave, vacuum and convective drying," *Biosystems Engineering*, vol. 96, no. 4, pp. 495–502, 2007.

[15] M. M. Özcan, A. Ünver, T. Uçar, and D. Arslan, "Mineral content of some herbs and herbal teas by infusion and decoction," *Food Chemistry*, vol. 106, no. 3, pp. 1120–1127, 2008.

[16] D. Kara, "Evaluation of trace metal concentrations in some herbs and herbal teas by principal component analysis," *Food Chemistry*, vol. 114, no. 1, pp. 347–354, 2009.

[17] D. M. Albin, J. E. Wubben, and V. M. Gabert, "Effect of hydrolysis time on the determination of amino acids in samples of soybean products with ion-exchange chromatography or precolumn derivatization with phenyl isothiocyanate," *Journal of Agricultural and Food Chemistry*, vol. 48, no. 5, pp. 1684–1691, 2000.

[18] S. A. Cohen, M. Meys, and T. L. Tarvin, *The Pico-Tag Method: A Manual of Advanced Techniques for Amino Acid Analysis*, Waters Division of Millipore, Milford, Mass, USA, 1989.

[19] G. Lepage and C. C. Roy, "Improved recovery of fatty acid through direct transesterification without prior extraction or purification," *Journal of Lipid Research*, vol. 25, no. 12, pp. 1391–1396, 1984.

[20] M. Šrůtek, "Growth responses of *Urtica dioica* to nutrient supply," *Canadian Journal of Botany*, vol. 73, no. 6, pp. 843–851, 1995.

[21] M. N. Lewu, P. O. Adebola, and A. J. Afolayan, "Effect of cooking on the proximate composition of the leaves of some accessions of *Colocasia esculenta* (L.) Schott in KwaZulu-Natal province of South Africa," *African Journal of Biotechnology*, vol. 8, no. 8, pp. 1619–1622, 2009.

[22] O. M. Funke, "Evaluation of nutrient contents of amaranth leaves prepared using different cooking methods," *Food and Nutrition Sciences*, vol. 2, pp. 249–252, 2011.

[23] J. C. Rickman, C. M. Bruhn, and D. M. Barrett, "Nutritional comparison of fresh, frozen, and canned fruits and vegetables II. Vitamin A and carotenoids, vitamin E, minerals and fiber," *Journal of the Science of Food and Agriculture*, vol. 87, no. 7, pp. 1185–1196, 2007.

[24] D. Martins, L. Barros, A. M. Carvalho, and I. C. F. R. Ferreira, "Nutritional and *in vitro* antioxidant properties of edible wild greens in Iberian Peninsula traditional diet," *Food Chemistry*, vol. 125, no. 2, pp. 488–494, 2011.

[25] E. Cho, J. Lee, K. Park, and S. Lee, "Effects of heat pretreatment on lipid and pigments of freeze-dried spinach," *Journal of Food Science*, vol. 66, no. 8, pp. 1074–1079, 2001.

[26] J. Tardío, M. Molina, L. Aceituno-Mata et al., "*Montia fontana* L. (Portulacaceae), an interesting wild vegetable traditionally consumed in the Iberian Peninsula," *Genetic Resources and Crop Evolution*, vol. 58, pp. 1105–1118, 2011.

[27] C. Pereira, L. Barros, A. M. Carvalho, and I. C. F. R. Ferreira, "Nutritional composition and bioactive properties of commonly consumed wild greens: potential sources for new trends in modern diets," *Food Research International*, vol. 44, no. 9, pp. 2634–2640, 2011.

[28] Z. Lisiewska, W. Kmiecik, P. Gebczynski, and L. Sobczynska, "Amino acid profile of raw and as-eaten products of spinach (*Spinacia oleracea* L.)," *Food Chemistry*, vol. 126, no. 2, pp. 460–465, 2011.

[29] Z. Lisiewska, J. Słupski, R. Skoczeń-Słupska, and W. Kmiecik, "Content of amino acids and the quality of protein in Brussels sprouts, both raw and prepared for consumption," *International Journal of Refrigeration*, vol. 32, no. 2, pp. 272–278, 2009.

[30] FAO, "Amino-acid content of foods and biological data on proteins," *FAO Food and Nutrition* Series no. 21, Rome, Italy, 1970.

[31] U.S. Department of Agriculture, Agricultural Research Service, "USDA Nutrient Database for Standard Reference," Release 24, 2011, http://ndb.nal.usda.gov/ndb/foods/show/7593.

20

UHPLC/MS-MS Analysis of Six Neonicotinoids in Honey by Modified QuEChERS: Method Development, Validation, and Uncertainty Measurement

Michele Proietto Galeano, Monica Scordino, Leonardo Sabatino, Valentina Pantò, Giovanni Morabito, Elena Chiappara, Pasqualino Traulo, and Giacomo Gagliano

Dipartimento dell'Ispettorato Centrale della Tutela della Qualità e della Repressione Frodi dei Prodotti Agroalimentari (ICQRF), Laboratory of Catania, Ministero delle Politiche Agricole Alimentari e Forestali (MIPAAF), Via A. Volta 19, 95122 Catania, Italy

Correspondence should be addressed to Monica Scordino; m.scordino@mpaaf.gov.it

Academic Editor: Yao-wen Huang

Rapid and reliable multiresidue analytical methods were developed and validated for the determination of 6 neonicotinoids pesticides (acetamiprid, clothianidin, imidacloprid, nitenpyram, thiacloprid, and thiamethoxam) in honey. A modified QuEChERS method has allowed a very rapid and efficient single-step extraction, while the detection was performed by UHPLC/MS-MS. The recovery studies were carried out by spiking the samples at two concentration levels (10 and 40 μg/kg). The methods were subjected to a thorough validation procedure. The mean recovery was in the range of 75 to 114% with repeatability below 20%. The limits of detection were below 2.5 μg/kg, while the limits of quantification did not exceed 4.0 μg/kg. The total uncertainty was evaluated taking the main independent uncertainty sources under consideration. The expanded uncertainty did not exceed 49% for the 10 μg/kg concentration level and was in the range of 16–19% for the 40 μg/kg fortification level.

1. Introduction

Neonicotinoids are a relatively new class of insecticides that share a common mode of action that affect the central nervous system of insects, resulting in paralysis and death [1]. They possess either a nitromethylene, nitroimine, or cyanoimine group [2]. They include acetamiprid, clothianidin, imidacloprid, nitenpyram, thiacloprid, and thiamethoxam. Studies suggested that neonicotinoids residues can accumulate in pollen and nectar of treated plants and represent a potential risk to pollinators [3]. Therefore, neonicotinic pesticides may play a role in recent pollinator declines. The Honey Italian Observatory stated that in 2008 more than half of Italian hives, and that 600,000 of a total of 1,100,000 have been put out of production for the depopulation of entire apiaries. The honey production in 2008 fell by 50% reduced to 7,000 tons. One result might be expected given that the previous year, the European Food Safety Authority (EFSA)

stated that the bee die-off had hit the 50% bee population, compared to the annual average of 15%.

Neonicotinoids can also be persistent in the environment and, when used as seed treatments, translocate to residues in pollen and nectar of treated plants. The potential for these residues to affect bees and other pollinators remains uncertain. Despite these uncertainties, neonicotinoids are beginning to dominate the market place because of their high systemicity, the broad spectrum of action, and the reduced dose. In light of these findings, the Italian Ministry of Agriculture has asked the Ministry of Health to suspend action. The Ministry of Health, after consultation with the Pesticides Committee, issued the ministerial decree of September 17, 2008 that stated the precautionary suspension of the authorized use for the seeds tanning of plant protection products containing the active substances clothianidin, thiamethoxam, imidacloprid, and fipronil [4]. On June 25, 2012, a decree of the Ministry of Health extended to January 31, 2013 stating the neonicotinoids suspension for seeds treatment [5]. Similar measures have already been taken by other European states.

Recently, many researchers detected these insecticides in honey bees, honey, soil, pollen, and treated seeds for agriculture [6–12]. Measurement of pesticide residues in different matrices involves two basic steps, namely, sample preparation (extraction and clean up) and instrumental analysis. Ideally, a sample preparation should be rapid, simple, cheap, and environment friendly and provide clean extracts. After extraction, clean up is the most important process for multiresidue analysis. QuEChERS (Quick Easy Cheap Effective Rugged Safe) technique, which was developed between 2000 and 2002 and first reported in 2003 [13], is a fast and complete extraction and clean up procedure and also employs the use of dispersive-solid phase extraction (d-SPE) for sample clean up.

In this paper, we report a rapid modified QuEChERS method for multiresidue analysis for 6 neonicotinoids (acetamiprid, clothianidin, imidacloprid, nitenpyram, thiacloprid, and thiamethoxam) in honey with good selectivity, sensitivity, and cost effectiveness. In order to demonstrate the suitability of the method for routine regulatory purposes, the method was validated and the statistical parameters are discussed.

2. Materials and Methods

2.1. Reagents and Standards. The certified analytical standards of all the 6 pesticides (acetamiprid, clothianidin, imidacloprid, nitenpyram, thiacloprid, and thiamethoxam) and internal standard Tris(1-chloro-2-propyl)phosphate (TCPP) were purchased from Ultra Scientific (Bologna, Italy) ($100.0 \pm 0.5 \mu g/mL$ each) in acetonitrile. All the solvents and chemicals used in the study were of analytical reagent (AR) grade, ethanol was supplied by Romil (Milan, Italy), and formic acid, ammonium formiate, and acetonitrile were by Carlo Erba (Milan, Italy). Distilled water was purified at $18.2 M\Omega$ with a MilliQ ULTRA (Millipore, Vimodrone (MI), Italy) purification system. A mixture of dispersive SPE Citrate Extraction Tube Supelco (4 g magnesium sulphate, 1 g sodium chloride, 0.5 g sodium citrate dibasic sesquihydrate, and 1 g sodium citrate tribasic dihydrate) was used, supplied by Sigma-Aldrich (Milan, Italy).

2.2. Instrumentation. Ultra high-performance liquid chromatography UHPLC-MS/MS (Thermo Scientific, TSQ Quantum Access Max) equipped with Thermo hypersilgold column (50 mm × 2.1 mm, 1.9 μm) was used for quantification of neonicotinoids. The flow rate was 400 μL/min, the column temperature 30°C, and the injection loop volume 5 μL. A binary gradient of 0.05% HCOOH and HCOONH$_4$ 2 mM in water (A) and 0.05% HCOOH and HCOONH$_4$ 2 mM in CH$_3$OH (B) was employed. The mobile-phase gradient was programmed as follows: 0 min, 10% B; 7 min, 95% B; 8 min, 95% B; 9 min, 10% B; and 10 min, 10% B. Mass spectral analyses were performed using an LC-TSQ Quantum Access Max operating in the positive ion mode

TABLE 1: UHPLC-MS/MS fragmentation of studied neonicotinoids.

MW	Pesticide	Precursor ion (m/z)	Product ions (m/z)	Collision energy (eV)
222	Acetamiprid	223.0	90.3	31
			99.2	34
			126.2	19
249	Clothianidin	250.0	113.2	25
			132.1	16
			169.2	12
252	Thiacloprid	253.0	90.3	36
			99.2	37
			126.2	21
255	Imidacloprid	256.0	175.2	17
			209.1	14
270	Nitenpyram	271.1	126.2	27
			225.2	9
			237.2	21
291	Thiamethoxam	292.0	181.2	20
			210.2	7
			211.2	10

using a h-ESI interface. The electrospray ionization (ESI) needle spray voltage was 4000. The heated capillary was 270°C. Flush volume was 700 μL and Collision Gas Pressure was 1.3 mTorr. The neonicotinoids and the internal standard TCPP were detected in MS/MS conditions, programming the chromatographic run in SRM mode (selected reaction monitoring) as reported in Table 1. Preliminary tunings were carried out with continuous introduction of a dilute solution of certified standards. Flow rate of syringe pump infusion of 5 μL/min and the voltages on the lenses were optimized in TSQ Tune Master (Excalibur software).

2.3. Reference Solution. The standard mix solution at 5 μg/mL of standard pesticides was diluted by transferring 500 μL ($100.0 \pm 0.5 \mu g/mL$) into a volumetric flask (10 mL, Class A certified).

The standard mix solution at 1 μg/mL of standard pesticides was diluted by transferring 100 μL ($100.0 \pm 0.5 \mu g/mL$) into a volumetric flask (10 mL, Class A certified).

The standard mix solution at 0.1 μg/mL of standard pesticides was diluted by transferring 200 μL of solution at 5 μg/mL into a volumetric flask (10 mL, Class A certified). All mix solutions are making up at volume with acetonitrile. Stock solutions stored at −18°C were stable for at least 3 months.

2.4. Method Validation

2.4.1. Specificity. The specificity of the analytical method for neonicotinoids detection was confirmed by obtaining positive results from honey containing the analyte, coupled with negative results from samples which do not contain

UHPLC/MS-MS Analysis of Six Neonicotinoids in Honey by Modified QuEChERS: Method Development, Validation, and Uncertainty Measurement

161

it (negative controls). The matrix effect was assessed by preparing pesticide standards in blank matrix extracted from untreated honey. The matrix extracts were analyzed before spiking to confirm the absence of the test pesticides in them.

2.4.2. Linearity.
The quantification of pesticide was based on a six-point matrix-matched calibration graph by plotting the detector response (SRM area ratio with respect to internal standard TCPP) against concentration of the calibration standards within the range 1–50 μg/L making three replicates for each concentration. A linear regression of six calibration points for each component was used to determine the relationship with the analyte concentrations calculated for each component on the basis of their occurrence in the reference material. The regression equations with slope, y-intercept, and coefficient of correlation (r^2) were evaluated for acetamiprid, clothianidin, imidacloprid, nitenpyram, thiacloprid, and thiamethoxam. Statistical test (Mandel and residual analysis with normal distribution of the calibration points) were performed to prove the linearity of regression lines.

2.4.3. Limit of Detection (LOD) and Limit of Quantification (LOQ).
The LOD and LOQ were determined by signal-to-noise approach [14]. The noise and signal are measured experimentally on the chromatogram printout. LOQ was estimated by the response of method noise level by approximately ten and LOD is, therefore, 3.3-fold lower.

2.4.4. Method Accuracy (Recovery) and Precision (Repeatability).
Method recovery studies were performed at two spiking concentration levels (10 μg/kg and 40 μg/kg). The sample matrix was prepared by homogenizing a series of different honeys in order to develop a highly specific method. The samples were prepared by weighing 5.0 ± 0.5 g of honey spiked in 50 mL tube (Meus srl, Piove di Sacco (PD), Italy). These sample tubes were vortexed (Velp, Usmate (MB), Italy) for 30 seconds after adding 10 mL of water and 10 mL of acetonitrile, in order to homogenize and fluidize the sample, and 50 μL of Tris(1-chloro-2-propyl)phosphate (TCPP) at 50 mg/L. In each tube was added a mixture of salts (4 g magnesium sulphate, 1 g sodium chloride, 0.5 g sodium citrate dibasic sesquihydrate, and 1 g sodium citrate tribasic dihydrate). The extract was stirred for 1 minute in vortex, in order to maximize the distribution of the analytes in the organic phase. The samples were centrifuged at 3000 rpm for 5 minutes and the supernatant was filtered at 0.45 μm PTFE filters (VWR, Milan, Italy). The extract was analyzed by UHPLC-MS/MS, making 6 replicates for each concentration. The average percentage of recovery and the relative standard deviation (RSD, repeatability) were evaluated.

2.4.5. Determination of Uncertainties.
Combined uncertainty in estimation was determined for all the neonicotinoids at the two fortification levels studied (10 and 40 μg/kg) as the statistical procedure of the EURACHEM/CITAC Guide CG 4 [15]. Individual sources of uncertainty were taken into account as described below.

Uncertainty of Analytical Standard Solutions. As the uncertainty of standard concentration declared in the supplier's certificate was given without any confidence level, rectangular distribution was assumed for calculating standard uncertainty

$$U1 = \frac{u(x)/C(x)}{\sqrt{3}}, \tag{1}$$

where $u(x)$ represents the uncertainty value given in the certificate and $C(x)$ the concentration of the standard solution.

Uncertainty of Weighing. The relative uncertainty due to honey weighing was calculated using normal distribution given by

$$U2 = \frac{(0.00005)}{Wi}, \tag{2}$$

where Wi is the weight of the sample, and 0.00005 is the value of uncertainty of the balance at 95% confidence level as reported in the certificate.

Uncertainty of Calibration Linearity. Uncertainty associated with the calibration curve, was calculated according to

$$U3 = \left(\frac{s}{b_1}\right)\left(\left\{\frac{1}{p}\right\} + \left\{\frac{1}{n}\right\} + \left\{\frac{(c_0 - c')^2}{s_{xx}}\right\}\right)^{1/2}, \tag{3}$$

where s is the standard deviation of the residuals of the calibration curve, b_1 is the slope of the calibration curve, p is the number of measurements of the unknown, n is the number of points used to form the calibration curve, c_0 is the calculated concentration of the analyte from the calibration curve, c' is the arithmetic mean of the concentrations of the standards used to make the calibration curve, and s_{xx} is calculated as given in

$$s_{xx} = \sum (cj - c')^2, \tag{4}$$

where $j = 1, 2, \ldots, n$. cj is the concentration of each calibration standard used to build up the calibration curve.

Uncertainty Associated with Precision. In the present study, the random errors of extraction, clean up, and UHPLC analyses steps were approximated by standard deviations which were calculated from repeated determinations of analytes expressed as repeatability. The precision was calculated according to

$$U4 = \frac{s}{(\sqrt{n} \times x)}, \tag{5}$$

where s is the standard deviation of the results obtained from the recovery study, n is the number of assays and x is the mean value of the concentration recovered.

Uncertainty of Volume. The volume of the solution is subject to 3 sources of uncertainty: calibration, repeatability, and temperature effects. (a) Calibration: the uncertainty in the certified internal volume of the flask and of the pipettes. For

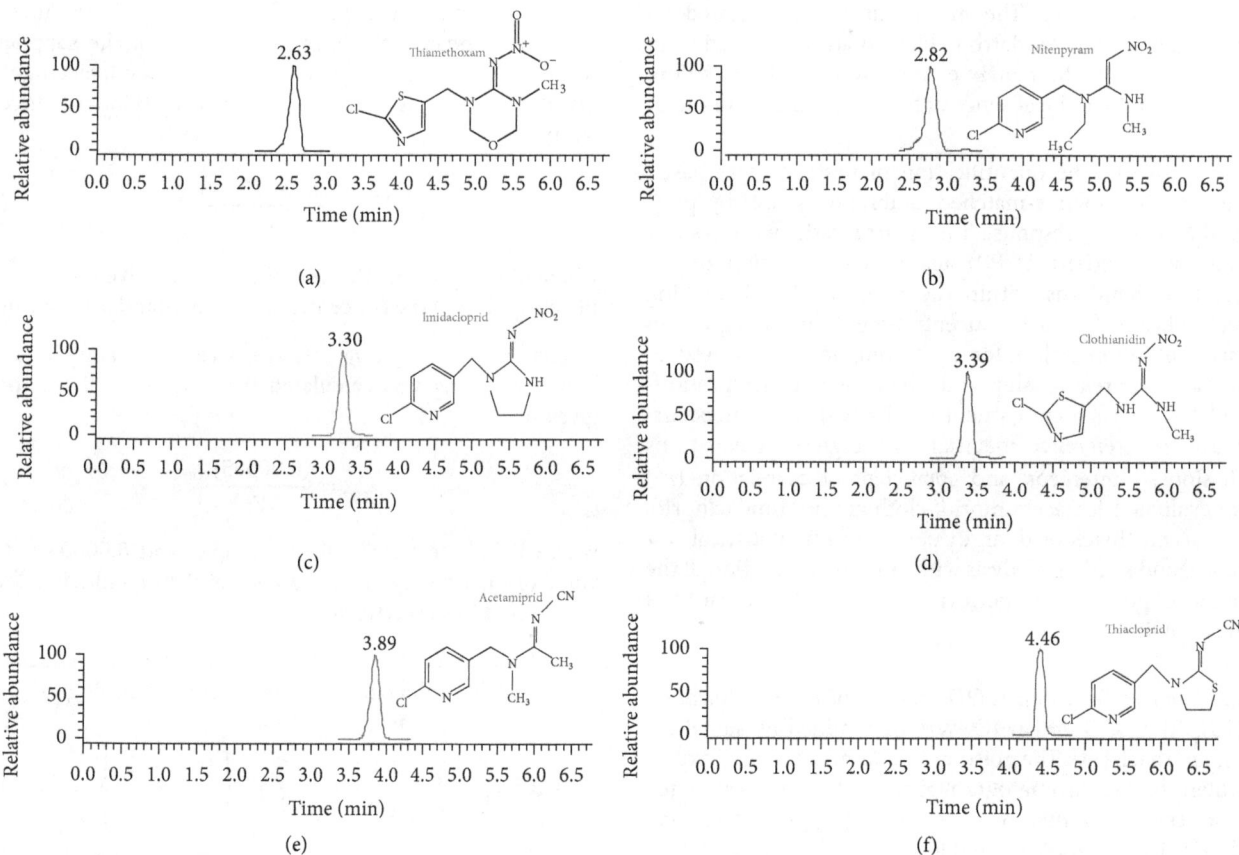

FIGURE 1: Representative UHPLC/MS-MS chromatogram of studied neonicotinoids for matrix-matched standards 40 μg/kg.

example, the manufacturer gives a volume of 10.00 ± 0.02 mL ($V \pm a$) for the flask, when measured at a temperature of $20°$C. Because the value of the uncertainty is given without a confidence level or distribution information, an assumption is necessary. In this work, the standard uncertainty is calculated by assuming a triangular distribution according to

$$U5 = \frac{(a/\sqrt{3})}{V}. \qquad (6)$$

In the same way, the volumes of the pipettes used to prepare the solutions at different levels are calculated by assuming a triangular distribution. The contributions due to the dilution operations performed for each concentration level are calculated separately and combined to give the standard uncertainty of the volume. (b) Repeatability: the uncertainty due to variations in filling is considered in the repeatability experiments. (c) Temperature: the temperatures of the flask and solution differ from the temperature at which the volume of the flask was calibrated. According to the manufacturer, the flask was calibrated at a temperature of $20°$C, whereas the laboratory temperature varies by $\pm 2°$C. The uncertainty from this effect can be calculated from the estimate of the temperature range and the coefficient of the volume expansion. In the case of acetonitrile as a solvent, this effect is negligible.

The combined uncertainty (U) was calculated as $U = x[(U1^2 + U2^2 + U3^2 + U4^2)^{1/2}]$, where Cx is the mean neonicotinoids concentration, and reported as expanded uncertainty ($2U$) which is twice the value of the combined uncertainty at 95% confidence level.

3. Results and Discussion

3.1. Method Development. In order to identify the major species produced in collisional experimental fragmentation of MS/MS analysis, a mass characterization study was firstly performed for direct infusion of each investigated neonicotinoids. Mass scans in positive ions mode were performed with h-ESI source ionization; all investigated molecules showed a good fragmentation. The collision energy was modulated from 5 to 50 of instrumental maximum to obtain the better fragmentation pattern. The ESI spectrum is characterized by the parent ion $[M + H]^+$ for all molecules. The neutral losses of NO_2 and/or HCl were observed for clothianidin, imidacloprid, nitenpyram, and thiamethoxam. The fragment at m/z 126, corresponding to $[C_6H_5\text{-}OCl]^+$ was a characteristic for acetamiprid, nitenpyram, and thiacloprid (Table 1). The discussed SRM data were in agreement with what reported by Sabatino et al. [10] and Ferrer et al. [16].

UHPLC/MS-MS Analysis of Six Neonicotinoids in Honey by Modified QuEChERS: Method Development, Validation, and Uncertainty Measurement

163

TABLE 2: Method validation results.

Compound	R_t (min)	Linearity range (μg/L)	r^2	LOQ (μg/kg)	Recovery % (10 μg/kg)[†]	Recovery % (40 μg/kg)[†]
Thiamethoxam	2.63	1–50	0.999	0.50	101 ± 11	100 ± 12
Nitenpyram	2.82	1–50	0.998	4.00	75 ± 20	97 ± 9
Imidacloprid	3.30	1–50	0.998	2.80	114 ± 3	109 ± 7
Clothianidin	3.39	1–50	0.999	3.20	111 ± 8	105 ± 9
Acetamiprid	3.89	1–50	0.995	0.12	107 ± 5	105 ± 6
Thiacloprid	4.46	1–50	0.997	0.10	89 ± 6	92 ± 2

[†] Mean value of six determinations; relative standard deviations (precision) in parenthesis.

The chromatographic method has been developed on the results of preliminary studies carried out on matrix-fortified standards. Different solvents were used for the chromatographic separation and several chromatographic separations were evaluated. The best results were obtained using an elution gradient starting with a binary gradient of 0.05% HCOOH and $HCOONH_4$ 2 mM in water and 0.05% HCOOH and $HCOONH_4$ 2 mM in CH_3OH combined with the Thermo hypersil GOLD 50 × 2.1 mm (1.9 μm i.d.) column. Under the described chromatographic conditions, the studied molecules were resolved in less than 5 minutes (Figure 1) and well recognizable on the basis of m/z signals, and good sensitivities were obtained; each analyte showed a typical mass spectrum profile previously identified by direct infusion.

The concept of a single extraction and dilution of the extracts was chosen in this study to achieve good results in the shortest time. In 2011, Tanner and Czerwenka [11] applied two steps of purification with d-SPE applying the QuEChERS methodology to the honey. Our protocol eliminated the second purification step, limiting the extraction to the use of d-SPE citrate extraction tube and reducing times and costs of analyses. Nevertheless, results were satisfactory in terms of statistical parameters, the selectivity for the analytes of interest, and reduction of the matrix effect (see paragraph below). This protocol permitted to analyze a high number of samples per day and is, therefore, suitable for a routine application in control laboratories. The proposed analytical protocol is currently applied in ICQRF Catania laboratory in the frame of Italian Ministry quality control investigation.

3.2. Method Validation. Analytical parameters of the proposed method were evaluated according to the criteria given in Section 2. Results are reported in Table 2.

3.2.1. Specificity. The specificity of the method toward the studied analytes was good. No interferences due to matrixes were found. Hence, no further time-consuming concentration/cleanup pretreatments were required.

3.2.2. Linearity of Calibration Curve. The linearity of each pesticide was established by plotting UHPLC response area ratio versus concentration. The analytes showed linear behavior in the studied concentration range of 1–50 μg/L. The correlation coefficient (r^2) was found to be ≥0.995 for all pesticides.

3.2.3. LOD and LOQ. LOD and LOQ were estimated as the lowest concentrations of pesticide injected that yielded a signal/noise ratio of 3 and 10, respectively. LOQs evaluation showed the lowest value 0.10 μg/kg for thiacloprid to the higher value of 4.00 μg/kg for nitenpyram. The LOQs attained in the proposed method fit with maximum residue limits (MRLs) of 10 μg/kg for nonallowed pesticides [17].

3.2.4. Recovery and Precision. The single-step extraction method adopted for honey samples provided satisfactory recovery which ranged from 75% (nitepyram) to 114% (imidacloprid) for the fortification level of 10 μg/kg and from 92 (thiacloprid) to 109% (imidacloprid) for the fortification level of 40 μg/kg. The precision of the method was good, not exceeding a coefficient of variation of 12%, with the exception of nitenpyram at the lowest fortification level. These data are in agreement with the criteria of document no. SANCO/12495/2011, that recommend general recovery limits of 70–120% within laboratory repeatability ≤20% [18]. Therefore, the method could be considered sufficiently accurate and precise for the purpose.

3.2.5. Uncertainty of Measurement. The study of uncertainty was performed at 2 concentration levels (10 and 40 μg/kg), identifying and studying the most important parameters that determined the uncertainty of the analytical method. The parameters selected were point calibration, standard solution, weigh, volume, and precision; their contributions to method uncertainty were calculated as indicated in the experimental section. The different contributions of uncertainty for each concentration level, together with the relative combined standard uncertainty, are shown in Tables 3 and 4 for each neonicotinoid. Results showed that the contribution to uncertainty due to the dilution operations and the standard purities was constant for each concentration level and for each analyte. The same value of uncertainty concerning the amount of weighed sample was used for each level and for all pesticides because the quantity of analyzed sample did not change among the experiments; moreover, this contribution could be considered negligible. The uncertainty associated with repeatability has a moderate contribution to the expanded uncertainties, showing the higher value for nitenpyram, thiamethoxam, and clothianidin. The 10 μg/kg level showed the uncertainty of calibration point as the main constituent of total uncertainty, followed by the volume contribution. On the contrary, the volume uncertainty was

TABLE 3: Results of individual and combined uncertainties for each pesticide calculated at 10 μg/kg concentration level.

Compound	Standard solution	Weighting	Calibration curve	Precision	Dilution operations	Combined uncertainty	Expanded uncertainty	Relative expanded uncertainty
	$U1$	$U2$	$U3$	$U4$	$U5$	U	$2U$	$U\%$
Thiamethoxam	0.003	0.00001	0.064	0.044	0.070	1.1	2.2	21
Nitenpyram	0.003	0.00001	0.164	0.082	0.070	1.5	3.0	39
Imidacloprid	0.003	0.00001	0.161	0.011	0.070	2.0	4.0	35
Clothianidin	0.003	0.00001	0.089	0.033	0.070	1.3	2.6	23
Acetamiprid	0.003	0.00001	0.232	0.019	0.070	2.6	5.2	49
Thiacloprid	0.003	0.00001	0.220	0.024	0.070	2.1	4.2	46

TABLE 4: Results of individual and combined uncertainties for each pesticide calculated at 40 μg/kg concentration level.

Compound	Standard solution	Weighting	Calibration curve	Precision	Dilution operations	Combined uncertainty	Expanded uncertainty	Relative expanded uncertainty
	$U1$	$U2$	$U3$	$U4$	$U5$	U	$2U$	$U\%$
Thiamethoxam	0.003	0.00001	0.017	0.040	0.070	3.3	6.6	16
Nitenpyram	0.003	0.00001	0.036	0.032	0.070	3.3	6.6	16
Imidacloprid	0.003	0.00001	0.040	0.029	0.070	3.7	7.4	17
Clothianidin	0.003	0.00001	0.023	0.037	0.070	3.5	7.0	16
Acetamiprid	0.003	0.00001	0.061	0.023	0.070	4.0	8.0	19
Thiacloprid	0.003	0.00001	0.054	0.009	0.070	3.3	6.6	17

the major source to total uncertainty at the 40 μg/kg level, while the uncertainty of repeatability and calibration point had approximately similar values.

When the uncertainty of the result is reported, the combined standard uncertainty is multiplied with a so-called coverage factor, yielding an expanded uncertainty. A factor $k = 2$ was used because of the resemblance of the expanded uncertainty to a 95% confidence interval. The document no. SANCO/12495/2011 recommended a default expanded uncertainty of 50% to be used by regulatory authorities in cases of enforcement decisions (MRL exceedances) [18]. Our results showed a relative uncertainty ($U\%$) ranging from 21 (thiamethoxam) to 49% (acetamiprid) at levels of 10 μg/kg. Lower values were obtained for the 40 μg/kg level. At this level, all pesticides had $U\%$ ranging from 16 to 19%.

References

[1] S. Kagabu, "Studies on the synthesis and insecticidal activity of neonicotinoid compounds," *Journal of Pesticide Science*, vol. 21, pp. 237–239, 1996.

[2] K. Matsuda, S. D. Buckingham, D. Kleier, J. J. Rauh, M. Grauso, and D. B. Sattelle, "Neonicotinoids: insecticides acting on insect nicotinic acetylcholine receptors," *Trends in Pharmacological Sciences*, vol. 22, no. 11, pp. 573–580, 2001.

[3] T. Iwasa, N. Motoyama, J. T. Ambrose, and R. M. Roe, "Mechanism for the differential toxicity of neonicotinoid insecticides in the honey bee, *Apis mellifera*," *Crop Protection*, vol. 23, no. 5, pp. 371–378, 2004.

[4] Italian Government, 2008, Ministerial Decree of 17. 09. 2008 relating the precautionary suspension of the authorization of plant protection products containing active substances clothianidin, thiamethoxam, imidacloprid and fipronil for use in the seeds tanning. Italian Republic Official Gazzette of 20. 09. 2008 No. 221.

[5] Italian Government, 2012, Ministerial Decree of 25 giugno 2012 relating the prorogation of precautionary suspension of the authorization of plant protection products containing active substances clothianidin, thiamethoxam, imidacloprid and fipronil for use in the seeds tanning. Italian Republic Official Gazzette of 30.06.2012 No. 151.

[6] P. Fidente, S. Seccia, F. Vanni, and P. Morrica, "Analysis of nicotinoid insecticides residues in honey by solid matrix partition clean-up and liquid chromatography-electrospray mass spectrometry," *Journal of Chromatography A*, vol. 1094, no. 1-2, pp. 175–178, 2005.

[7] A. Kamel, "Refined methodology for the determination of neonicotinoid pesticides and their metabolites in honey bees and bee products by liquid chromatography-tandem mass spectrometry (LC-MS/MS)," *Journal of Agricultural and Food Chemistry*, vol. 58, no. 10, pp. 5926–5931, 2010.

[8] N. O. Z. Abaga, P. Alibert, S. Dousset, P. W. Savadogo, M. Savadogo, and M. Sedogo, "Insecticide residues in cotton soils of Burkina Faso and effects of insecticides on fluctuating asymmetry in honey bees (*Apis mellifera* Linnaeus)," *Chemosphere*, vol. 83, no. 4, pp. 585–592, 2011.

[9] S. Rossi, A. G. Sabatini, R. Cenciarini, S. Ghini, and S. Girotti, "Use of high-performance liquid chromatography-UV and gas chromatography-mass spectrometry for determination of the imidacloprid content of honeybees, pollen, paper filters, grass, and flowers," *Chromatographia*, vol. 61, no. 3-4, pp. 189–195, 2005.

[10] L. Sabatino, M. Scordino, V. Pantò, E. Chiappara, P. Traulo, and G. Gagliano, "Survey of neonicotinoids and fipronil in corn seeds for agriculture," *Food Additives and Contaminants B*, no. 1, pp. 11–16, 2012.

UHPLC/MS-MS Analysis of Six Neonicotinoids in Honey by Modified QuEChERS: Method Development, Validation, and Uncertainty Measurement

165

[11] G. Tanner and C. Czerwenka, "LC-MS/MS analysis of neonicotinoid insecticides in honey: methodology and residue findings in Austrian honeys," *Journal of Agricultural and Food Chemistry*, vol. 59, no. 23, pp. 12271–12277, 2011.

[12] S. Totti, M. Fernández, S. Ghini et al., "Application of matrix solid phase dispersion to the determination of imidacloprid, carbaryl, aldicarb, and their main metabolites in honeybees by liquid chromatography-mass spectrometry detection," *Talanta*, vol. 69, no. 3, pp. 724–729, 2006.

[13] M. Anastassiades, S. J. Lehotay, D. Štajnbaher, and F. J. Schenck, "Fast and easy multiresidue method employing acetonitrile extraction/partitioning and "dispersive solid-phase extraction" for the determination of pesticide residues in produce," *Journal of AOAC International*, vol. 86, no. 2, pp. 412–431, 2003.

[14] J. Vial and A. Jardy, "Experimental comparison of the different approaches to estimate LOD and LOQ of an HPLC method," *Analytical Chemistry*, vol. 71, no. 14, pp. 2672–2677, 1999.

[15] EURACHEM/CITAC Guide CG 4, EURACHEM/CITAC Guide (2000), Quantifying uncertainty in analytical measurement, 2nd edition, http://www.measurementuncertainty.org.

[16] I. Ferrer, E. M. Thurman, and A. M. Fernàndez-Alba, "Quantitation and accurate mass analysis of pesticides in vegetables by LC/TOF-MS," *Analytical Chemistry*, vol. 77, no. 9, pp. 2818–2825, 2005.

[17] European Commission Regulation (EC), "No. 396/2005 of 23 February 2005 on maximum residue levels of pesticides in or on food and feed of plant and animal origin and amending Council Directive 91/414/EEC," *Official Journal of the European Union*, vol. L70, pp. 1–16.

[18] Document No. SANCO/12495/2011, Method Validation and Quality Control Procedures for Pesticide Residues Analysis in Food and Feed, http://ec.europa.eu/food/plant/protection/pesticides/docs/qualcontrol_en.pdf.

21

Effect of Maltodextrins on the Rheological Properties of Potato Starch Pastes and Gels

Lesław Juszczak,[1] **Dorota Gałkowska,**[1] **Teresa Witczak,**[2] **and Teresa Fortuna**[1]

[1] Department of Analysis and Evaluation of Food Quality, University of Agriculture in Krakow, Balicka 122, 30-149 Krakow, Poland
[2] Department of Engineering and Machinery for Food Industry, University of Agriculture in Krakow, Balicka 122, 30-149 Krakow, Poland

Correspondence should be addressed to Lesław Juszczak; rrjuszcz@cyf-kr.edu.pl

Academic Editor: Kiyoshi Ebihara

The study examines the effects of maltodextrins saccharified to various degrees on some rheological properties of potato starch dispersions. Pasting characteristics, flow curves, and mechanical spectra were determined for native potato starch and for its blends with potato maltodextrins having dextrose equivalents (DE) of 10.5, 18.4, and 26.5. The results showed that medium-saccharified maltodextrin (DE = 18.4) gave the strongest effect, manifesting itself as a considerable reduction in the viscosity at pasting, a decrease in apparent viscosity during flow, and a decrease in the storage and loss moduli. Addition of high-(DE = 26.5) or low-(DE = 10.5) saccharified maltodextrins had a markedly smaller effect on the rheological properties of starch. The differences in the effects produced by the maltodextrins are closely connected to the degree of polymerisation of the maltooligosaccharides in the systems.

1. Introduction

Starch, one of the most common polysaccharides, has a number of specific properties that make it highly useful in the food industry and other sectors of the economy. It belongs to the cheapest thickening agents, texturizers, filling agents, and stabilizers. Heating starch granules in water environment causes them to paste. When they release amylose in the process, starch loses its specific granular structure [1–3]. The characteristic pasting temperature depends on botanical origin of starch as well as on the presence of other substances in the system. Sugars present in a starch suspension reduce the water activity of the system and stabilize the amorphous regions of the granules. As a result, starch pasting temperature increases and rheological properties of the system change [4–7]. The greater this effect, the higher the concentration of the solution and the greater the molecular weight of the substance added [3, 8, 9].

In the temperate climatic zone countries, such as Poland, potatoes are an important source of starch. Potato starch differs from cereal starches mainly in the size and structure of granules [10, 11], amylose content, phosphorus content,

and manner of phosphorus bonding, as well as in fat and protein contents [1, 12–14]. Since natural properties of starch are not always advantageous in terms of technology and application, it is often subjected to various modifications, among them hydrolysis. Such a modification may produce maltodextrins, that is, carbohydrate polymers built of D-glucose units having a dextrose equivalent (DE) of under 20 [15, 16]. In Poland, starch hydrolysates with a DE of 20 to 30 are called high-saccharified maltodextrins. Due to their properties, maltodextrins are widely applied in the food industry [15]. One of the interesting issues is the influence of maltodextrins on starch polymers. For example, Smits et al. [17] observed that the presence of maltooligosaccharides with polymerization degrees (DP) of 2 to 5 hinders the formation of amylose helices, thus reducing the retrogradation degree of wheat starch, while those with a DP exceeding 6 may form by themselves small helices that co-crystallise with starch polymers, thus accelerating retrogradation. An increase in the level of retrogradation of starch at temperature of 2°C in the presence of high-molecular-weight maltooligosaccharides has been reported by Wang and Jane [18]. As found by Durán et al. [19] adding oligosaccharides with DP of 3 to 5 delays the gelatinisation of starch and reduces the enthalpy of its retrogradation. Such a phenomenon may be used for inhibiting the staling of bread [17, 20–22].

Knowledge of the rheological properties of starch pastes and gels is of vital importance to the food industry and other sectors utilizing starches as a raw material [3, 22]. Since in complex food systems starch coexists with a wide range of other compounds, it is useful to understand the influence of individual components of foods on the properties of starch. The present study was designed to determine the effect of maltodextrins of different dextrose equivalents (called low-, medium-, and high-saccharified maltodextrins) on chosen rheological properties of potato starch.

2. Materials and Methods

Potato starch was obtained from PZZ Piła, Poland, and potato maltodextrins were provided by CLPZ Luboń, Poland. The maltodextrins were saccharified to different degrees: low (DE = 10.5, DP = 10.6), medium (DE = 18.4, DP = 6.0), and high (DE = 26.5, DP = 4.2). The dextrose equivalent (DE) was determined by Lane-Eynon's method according to the relevant Polish Standard (PN-EN ISO 5377:2001). The mean degree of polymerization (DP) was calculated on the basis of dextrose equivalent values: DP = 111/DE.

Rheological studies were conducted at constant concentration of starch (5 g d.w./100 g). The starch-maltodextrin systems were produced by dissolving an appropriate amount of maltodextrins (1, 2, or 3 g d.w./100 g) in distilled water and then adding starch.

The pasting characteristics of both native starch and starch-maltodextrin blends were determined in a Brabender viscograph, type 801201 (Germany) with a measuring cup of 250 cmg at a rotation speed of 75 rpm. The systems studied were heated and then cooled at a rate of 1.5°C/min using the following procedure: raising temperature from 25 to 96°C, maintaining constant temperature of 96°C during 20 minutes, reducing temperature from 96 to 50°C, and maintaining constant temperature of 50°C during 10 minutes. The viscograms obtained were used to read pasting temperature, peak viscosity, temperature at peak viscosity, viscosity at 96°C, viscosity after 20 minutes at 96°C, viscosity at 50°C, and viscosity after 10 minutes at 50°C.

Samples for rheometric investigations were prepared by heating the suspension of starch or starch with each maltodextrin at temperature of 95°C for 30 minutes while stirring it continuously at a rate of 250 rpm. Next the hot paste was placed in the measuring element of the rheometer, relaxed, and thermostated during 15 minutes at the temperature of measurement. Flow curves at 50°C were obtained by using a rotational rheometer Rheolab MC1 (Physica, Germany) with a coaxial cylinders system (cup diameter 27.12 mm, bob diameter 25.00 mm) for the shear rate range of 1–300 s^{-1}. The experimental curves were described employing Herschel-Bulkley equation:

$$\tau = \tau_0 + K \cdot \dot{\gamma}^n, \tag{1}$$

where τ is the shear stress (Pa), $\dot{\gamma}$ is the shear rate (s^{-1}), τ_0 is the yield stress (Pa), K is the consistency coefficient (Pa $\cdot$ s^n), and n is the flow behavior index.

Mechanical spectra at 25°C were determined by using a Rheostress RS rheometer (Haake, Germany) with a cone-plate system (cone diameter 35 mm, angle 2°, gap width 0.105 mm). The measurements were made in the linear viscoelasticity range at a constant strain of 0.03 in the frequency range of 0.1–10 Hz.

Statistical assessment was done by performing a one-way analysis of variance and calculating the least significant difference (LSD) at $\alpha = 0.05$.

3. Results and Discussion

Figure 1 shows the pasting curves of native potato starch and starch-maltodextrin systems, and Table 1 provides the pasting characteristics. Maltodextrins added to starch altered its viscosity at pasting. The changes depended on the kind of maltodextrin and its amount in the system. They did not have any influence on the pasting temperature of starch, except for low-saccharified maltodextrin added in the amount of 3 g/100 g, in which case this temperature slightly (by 1.5°C) but significantly increased (Table 1). Low-saccharified maltodextrin blended with starch brought about a marked fall in peak viscosity which was increasing with maltodextrin content in the system. In addition, such systems reached peak viscosity at a slightly higher temperature than native starch (Table 1). Potato starch is characterized by significantly higher values of peak viscosity as compared to cereal starches, that is, due to its high swelling capacity at relatively low temperature [14]. The presence of maltodextrins in the system reduces swelling capacity of the starch due to restriction of the amount of water available for starch granules, in the way depending on a DE of maltodextrin. Maltodextrins with low DE and thus with high DP values can also swell, however, to a lower degree than the native starch granules. High-saccharified maltodextrins swell to a low degree but more easily solubilize and thus thicken the continuous phase of the system. In the present study, the viscosity at 96°C of the systems containing low-saccharified maltodextrin was significantly decreased compared to starch paste. Maintaining the pastes at that temperature caused a sharp fall in viscosity both for native starch and the blends. The system with low-saccharified maltodextrin added at a level of 1 g/100 g displayed similar viscosity to that of the paste of native starch, while at higher maltodextrin levels the viscosity of the systems was significantly reduced. Similarly, at cooling, the viscosity of the system containing the smallest amount of low-saccharified maltodextrin did not differ from that of the native starch paste, while adding a greater amount of maltodextrin caused the viscosity of the pastes to decrease. No differences in viscosity were observed between the systems with 2 and 3 g/100 g maltodextrin. The fall in paste viscosity due to the addition of maltodextrin was the most pronounced for medium-saccharified one and was larger when maltodextrin content was higher (Figure 1(b), Table 1). What is more, the systems containing this kind of maltodextrin reached peak

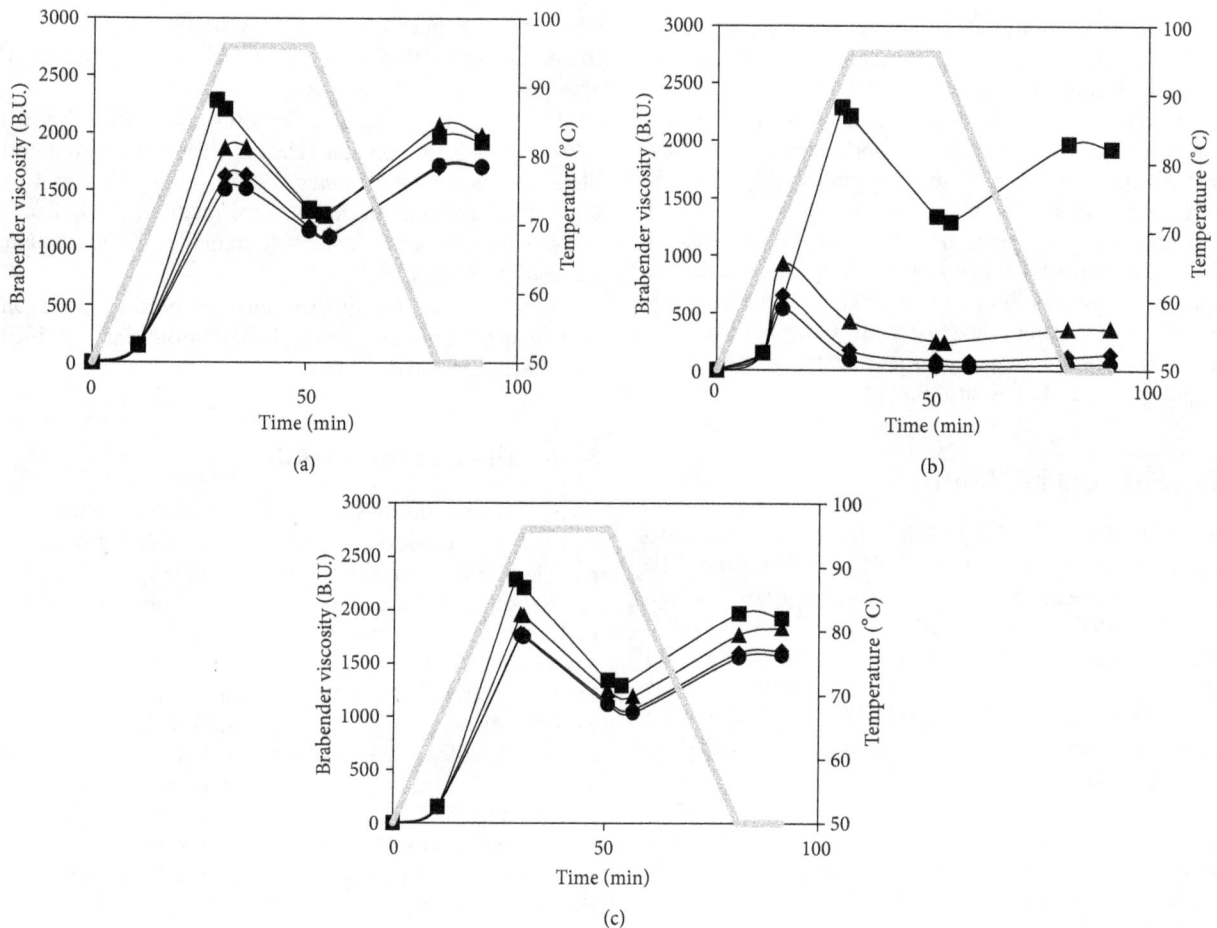

FIGURE 1: Pasting curves of native starch and blends with (a) low-saccharified maltodextrin, (b) medium-saccharified maltodextrin, (c) high-saccharified maltodextrin. Maltodextrin concentration: 0—□, 1—△, 2—◊, 3—○ g/100g.

viscosity at much lower temperature (72.3–73.0) than the other systems. High-saccharified maltodextrin (Figure 1(c), Table 1) also significantly reduced the peak viscosity of the paste, but to a much smaller degree than medium-saccharified one. The systems containing 2 and 3 g/100 g of the maltodextrin in question showed similar viscosity. The peak viscosity and the viscosity at 96°C were higher for the systems with high-saccharified maltodextrin than for the corresponding systems with low-saccharified maltodextrin, while after cooling, that pattern became reversed. The final viscosity of the starch paste results from a structure of two-phase gel-like system formed after cooling stage, in which the continuous phase is composed of associated linear amylose chains, while the dispersed phase is made of fragments of starch granules consisted mainly of amylopectin. The process of the association of the linear amylose chains is an initial stage of the retrogradation phenomenon. According to the literature data [19, 20], low-saccharified maltodextrins, that is, with higher DP, can be involved in forming the structures of the continuous phase, while medium-saccharified maltodextrins with medium-length chains are too small for cocreation of the gel-like structures; however, they have

enough long chains in order to restrict amylose association and weaken the structure of the system.

Figure 2 shows the flow curves of native starch and its blends with maltodextrins. The experimental curves were described using the parameters of Herschel-Bulkley model (Table 2). Addition of low-saccharified maltodextrin resulted in reduced shear stresses, especially at higher shear rates ($>50 s^{-1}$) (Figure 2(a)). The flow curves of starch-low-saccharified maltodextrin systems were similar for all amounts of the maltodextrin added. The pastes with low-saccharified maltodextrin exhibited lower values of the yield stress than the paste of native starch and, except for a system with 1 g of maltodextrin per 100 g, smaller values of the flow behavior index (Table 2). The consistency coefficient decreased for the latter system and increased for the others. According to the pasting curves, the effect of maltodextrins on the rheological properties of starch pastes was the largest for the pastes containing medium-saccharified maltodextrin. In this case the flow curves showed a considerable reduction in shear stresses, as compared to the paste of native starch (Figure 2(b)). The greater the decrease, the higher was the amount of maltodextrin in the system. The same was true

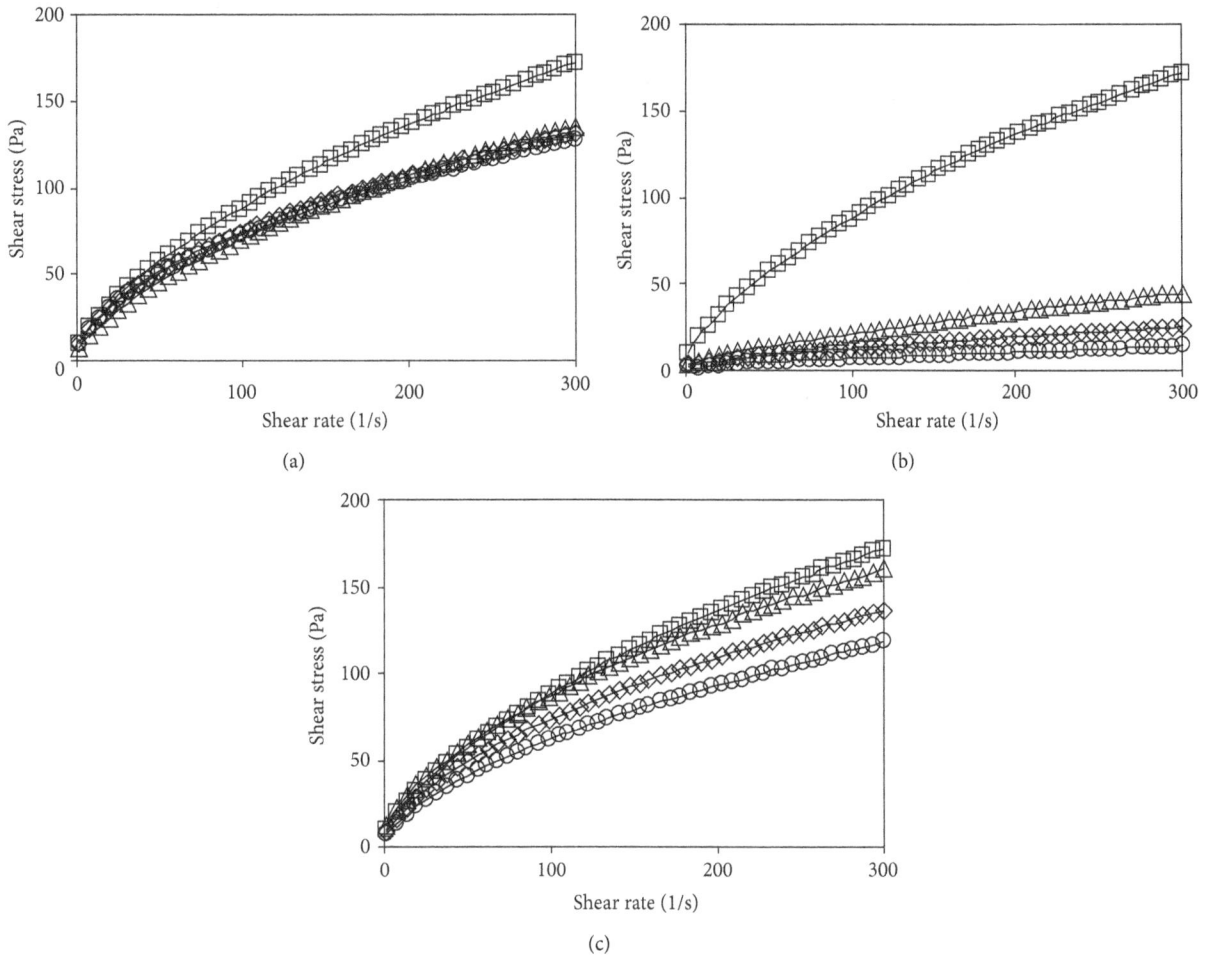

FIGURE 2: Flow curves of native starch and blends with (a) low-saccharified maltodextrin, (b) medium-saccharified maltodextrin, (c) high-saccharified maltodextrin. Maltodextrin concentration: 0—□, 1—△, 2—◊, 3—○ g/100g.

TABLE 1: Pasting characteristics of native starch and blends with maltodextrins.

Sample	Pasting temperature (°C)	Peak viscosity (B.U.)	Temperature at peak viscosity (°C)	Viscosity at 96°C (B.U.)	Viscosity after 20 min at 96°C (B.U.)	Viscosity at 50°C (B.U.)	Viscosity after 10 min at 50°C (B.U.)
NS	65.8 ± 0.3	2285 ± 15	93.3 ± 0.3	2210 ± 10	1340 ± 0	1970 ± 90	1925 ± 0
NS/LSM 1	65.8 ± 0.3	1875 ± 5	96.0 ± 0.0	1870 ± 10	1320 ± 10	2070 ± 0	1980 ± 0
NS/LSM 2	65.8 ± 0.3	1635 ± 5	96.0 ± 0.0	1630 ± 0	1180 ± 0	1705 ± 15	1705 ± 15
NS/LSM 3	67.3 ± 0.3	1515 ± 5	96.0 ± 0.0	1510 ± 10	1150 ± 20	1720 ± 40	1710 ± 30
NS/MSM 1	65.5 ± 0.0	925 ± 15	73.0 ± 0.0	435 ± 15	253 ± 13	358 ± 23	360 ± 20
NS/MSM 2	65.8 ± 0.3	650 ± 10	72.8 ± 0.0	175 ± 5	90 ± 0	120 ± 0	140 ± 0
NS/MSM 3	66.0 ± 0.0	535 ± 15	72.3 ± 0.0	95 ± 5	43 ± 8	50 ± 10	55 ± 5
NS/HSM 1	65.8 ± 0.3	1955 ± 25	95.0 ± 1.0	1945 ± 15	1245 ± 15	1765 ± 65	1830 ± 30
NS/HSM 2	66.0 ± 0.0	1775 ± 15	95.0 ± 0.0	1765 ± 15	1140 ± 15	1595 ± 15	1615 ± 15
NS/HSM 3	66.0 ± 0.0	1760 ± 0	94.0 ± 0.5	1750 ± 0	1115 ± 0	1555 ± 5	1575 ± 5
LSD$_{0.05}$	0.5	33	1.0	26	26	125	56

Mean values from three repetitions ± standard deviation.
NS: native starch, NS/LSM: native starch/low-saccharified maltodextrin (1, 2, and 3 g/100 g), NS/MSM: native starch/medium-saccharified maltodextrin (1, 2, and 3 g/100 g), NS/HSM: native starch/high-saccharified maltodextrin (1, 2, and 3 g/100 g).
LSD: least significant differences.

TABLE 2: Herschley-Bulkel model parameters of native starch paste and blends with maltodextrins.

Sample	Yield stress (Pa)	Consistency coefficient (Pa s^n)	Flow behaviour index (−)	R^2
NS	6.07 ± 0.55	3.67 ± 0.01	0.67 ± 0.00	0.9993
NS/LSM 1	4.45 ± 0.04	2.88 ± 0.11	0.68 ± 0.01	0.9992
NS/LSM 2	4.49 ± 0.38	4.79 ± 0.02	0.58 ± 0.00	0.9993
NS/LSM 3	5.40 ± 0.27	4.63 ± 0.08	0.58 ± 0.01	0.9995
NS/MSM 1	3.01 ± 0.09	0.56 ± 0.02	0.76 ± 0.01	0.9999
NS/MSM 2	2.50 ± 0.20	0.35 ± 0.02	0.74 ± 0.02	0.9994
NS/MSM 3	1.38 ± 0.10	0.17 ± 0.03	0.76 ± 0.03	0.9982
NS/HSM 1	7.32 ± 0.01	4.61 ± 0.06	0.62 ± 0.00	0.9996
NS/HSM 2	4.89 ± 0.14	3.71 ± 0.02	0.63 ± 0.00	0.9994
NS/HSM 3	4.24 ± 0.23	2.95 ± 0.07	0.65 ± 0.01	0.9976
LSD$_{0.05}$	0.65	0.14	0.03	

Mean values from three repetitions ± standard deviation.
NS: native starch, NS/LSM: native starch/low-saccharified maltodextrin (1, 2, and 3 g/100 g), NS/MSM: native starch/medium-saccharified maltodextrin (1, 2, and 3 g/100 g), NS/HSM: native starch/high-saccharified maltodextrin (1, 2, and 3 g/100 g).
LSD: least significant differences.

for the yield stress and the consistency coefficient of these systems (Table 2). In contrary, values of the flow behavior indices of these systems were markedly greater than those of the native starch paste; however, they were not significantly dependent on the amount of maltodextrin. The presence of the high-saccharified maltodextrin at a level of 1 g/100 g caused a rise in the values of the yield stress and the consistency coefficient and a significant decrease in the values of the flow behavior index (Table 2). When the level of the high-saccharified maltodextrin was increased, the yield stress of the paste decreased, as did the consistency coefficient. There were no significant differences in the flow behavior indices between the systems containing different amounts of the maltodextrin (Table 2). During shearing of the starch paste, the destruction and the following reconstruction of its structure take place. The presence of maltodextrins in the starch pastes affected in a different way their flow behavior. Similarly to the pasting characteristic, the medium-saccharified maltodextrins with medium DP value had the greatest effect on the flow behavior of the starch pastes. It results presumably from the length of the maltodextrin chains, which are too short in order to cocreate the structure of starch paste but enough long in order to disturb formation of the continuous phase consisting of the linear amylose.

The mechanical spectra shown in Figure 3 demonstrate that all the starch-maltodextrin systems behaved as weak gels. In the whole range of the frequencies studied, the values of the storage modulus (G') were higher than those of the loss modulus (G''). However, the storage modulus did not display a plateau that is characteristic for strong gels and depended on frequency over whole study range, with the values of tg δ = G''/G' amounting to about 0.48. The various maltodextrins added to the starch differently affected its viscoelastic properties. Low-saccharified maltodextrin in the amount of 1 g/100 g caused an apparent decrease

in both moduli, as compared to the gel of native starch (Figure 3), while the other systems exhibited similar values of the storage modulus and slightly lower values of the loss modulus comparing to the gel of the native starch. Medium-saccharified maltodextrin in the amount of 1 g/100 g did not affect the storage modulus but decreased the loss modulus (Figure 3(b)). Increased amount of that maltodextrin in the system resulted in a marked decrease in the values of both moduli. For the loss modulus, the larger the decrease, the greater the amount of the maltodextrin. The values of the storage modulus for the systems with 2 and 3 g/100 g of medium-saccharified maltodextrin were similar. The gels of the systems containing high-saccharified maltodextrin showed lower values of both moduli, as compared to the gel of native starch (Figure 3(c)). The decrease was the biggest when the maltodextrin was added at the amount of 1 g/100 g. Increasing maltodextrin content in the system resulted in a much smaller decrease in both moduli, with the values of the loss modulus being similar for the blends containing high-saccharified maltodextrin at the level of 2 and 3 g/100 g, and those of the storage modulus being slightly higher for the system with 3 g/100 g of the maltodextrin. Due to the fact that starch gel forming is closely related to the association of amylose chains and retrogradation of the starch polymers [20], presence of any compounds which prevent that phenomenon results in weakening the gel structure and, consequently, decreasing G' and G'' moduli. In the present study, similarly to the pasting characteristic and flow behaviour, the greatest effect on the weakening gel structure and reduction of starch retrogradation had medium-saccharified maltodextrin with DP = 6. Due to a possibility of medium-saccharified starch polymers to reduce starch retrogradation, addition of them to native starches can be an alternative way to the use of stabilized starches and can be a factor that reduces bread staling.

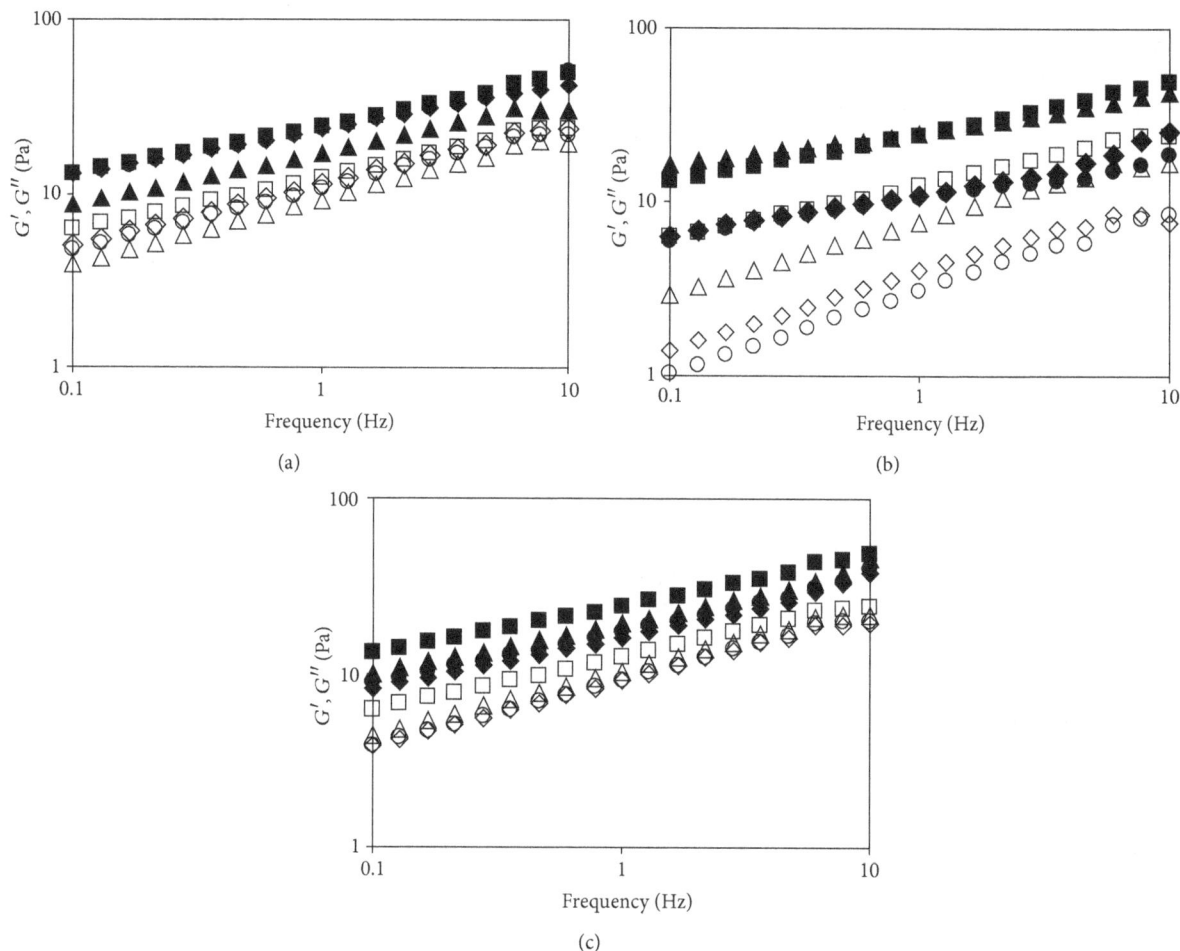

FIGURE 3: Mechanical spectra (G': black markers, G'': white markers) of native starch and blends with (a) low-saccharified maltodextrin, (b) medium-saccharified maltodextrin, and (c) high-saccharified maltodextrin. Maltodextrin concentration: 0—□, 1—△, 2—◇, 3—○ g/100g.

4. Conclusions

Maltodextrins with varied dextrose equivalents showed different effects on the rheological properties of potato starch pastes. Medium-saccharified maltodextrin (DE = 18.4, DP = 6.0) had the greatest effect on starch pasting characteristics, flow behavior, and viscoelastic properties. The contribution of the maltodextrins to the formation of starch pastes and gels was closely associated with their degree of polymerization. High-(DP = 4.2) and medium-(DP = 6.0) saccharified maltodextrins hindered the formation of the structure of starch pastes and gels. Low-saccharified maltodextrin (DE = 10.5, DP = 10.6) added to the starch affected its rheological properties to a much smaller extent than medium-saccharified maltodextrin. This could be attributed to the fact that maltooligosaccharides of DP exceeding 6, which are able by themselves to form amylose helices, participated in the formation of the structure of starch pastes and gels.

References

[1] R. Parker and S. G. Ring, "Aspects of the physical chemistry of starch. Mini review," *Journal of Cereal Science*, vol. 34, no. 1, pp. 1–17, 2001.

[2] S. Lagarrigue, G. Alvarez, G. Cuvelier, and D. Flick, "Swelling kinetics of waxy maize and maize starches at high temperatures and heating rates," *Carbohydrate Polymers*, vol. 73, no. 1, pp. 148–155, 2008.

[3] N. Singh, N. Isono, S. Srichuwong, T. Noda, and K. Nishinari, "Structural, thermal and viscoelastic properties of potato starches," *Food Hydrocolloids*, vol. 22, no. 6, pp. 979–988, 2008.

[4] B. Abu-Jdayil, M. O. J. Azzam, and K. I. M. Al-Malah, "Effect of glucose and storage time on the viscosity of wheat starch dispersions," *Carbohydrate Polymers*, vol. 46, no. 3, pp. 207–215, 2001.

[5] P. A. Perry and A. M. Donald, "The effect of sugars on the gelatinisation of starch," *Carbohydrate Polymers*, vol. 49, no. 2, pp. 155–165, 2002.

[6] V. M. Acquarone and M. A. Rao, "Influence of sucrose on the rheology and granule size of cross-linked waxy maize starch dispersions heated at two temperatures," *Carbohydrate Polymers*, vol. 51, no. 4, pp. 451–458, 2003.

[7] P. J. Torley and F. van der Molen, "Gelatinization of starch in mixed sugar systems," *LWT—Food Science and Technology*, vol. 38, no. 7, pp. 762–771, 2005.

[8] D. Yoo and B. Yoo, "Rheology of rice starch-sucrose composites," *Starch—Stärke*, vol. 57, no. 6, pp. 254–261, 2005.

[9] D. Gałkowska, "Effect of saccharides on gelatinization and retrogradation of modified potato starch," *Electronic Journal of Polish Agricultural Universities*, vol. 11, no. 1, p. 19, 2008.

[10] L. Juszczak, T. Fortuna, and F. Krok, "Non-contact atomic force microscopy of starch granules surface—part I: potato and tapioca starches," *Starch—Stärke*, vol. 55, no. 1, pp. 1–7, 2003.

[11] M. Sujka and J. Jamroz, "Starch granule porosity and its changes by means of amylolysis," *International Agrophysics*, vol. 21, no. 1, pp. 107–113, 2007.

[12] R. Hoover, "Composition, molecular structure, and physic-ochemical properties of tuber and root starches: a review," *Carbohydrate Polymers*, vol. 45, no. 3, pp. 253–267, 2001.

[13] N. Singh, J. Singh, L. Kaur, N. S. Sodhi, and B. S. Gill, "Morphological, thermal and rheological properties of starches from different botanical sources," *Food Chemistry*, vol. 81, no. 2, pp. 219–231, 2003.

[14] K. Pycia, L. Juszczak, D. Gałkowska, and M. Witczak, "Physic-ochemical properties of starches obtained from Polish potato cultivars," *Starch—Stärke*, vol. 64, no. 2, pp. 105–144, 2012.

[15] I. S. Chronakis, "On the molecular characteristics, compositional properties, and structural-functional mechanisms of maltodextrins: a review," *Critical Reviews in Food Science and Nutrition*, vol. 38, no. 7, pp. 599–637, 1998.

[16] U. Uthumporn, I. S. M. Zaidul, and A. A. Karim, "Hydrolysis of granular starch at sub-gelatinization temperature using a mixture of amylolytic enzymes," *Food and Bioproducts Processing*, vol. 88, no. 1, pp. 47–54, 2010.

[17] A. L. M. Smits, P. H. Kruiskamp, J. J. G. van Soest, and J. F. G. Vliegenthart, "The influence of various small plasticisers and malto-oligosaccharides on the retrogradation of (partly) gelatinised starch," *Carbohydrate Polymers*, vol. 51, no. 4, pp. 417–424, 2003.

[18] Y. J. Wang and J. Jane, "Correlation between glass transition temperature and starch retrogradation in the presence of sugars an maltodextrins," *Cereal Chemistry*, vol. 71, pp. 527–531, 1994.

[19] E. Durán, A. León, B. Barber, and C. Benedito de Barber, "Effect of low molecular weight dextrins on gelatinization and retrogradation of starch," *European Food Research and Technology*, vol. 212, no. 2, pp. 203–207, 2001.

[20] J. A. Rojas, C. M. Rosell, and C. Benedito de Barber, "Role of maltodextrins in the staling of starch gels," *European Food Research and Technology*, vol. 212, no. 3, pp. 364–368, 2001.

[21] M. Witczak, J. Korus, R. Ziobro, and L. Juszczak, "The effects of maltodextrins on gluten-free dough and quality of bread," *Journal of Food Engineering*, vol. 96, no. 2, pp. 258–265, 2010.

[22] B. Wang, D. Li, L. J. Wang, and N. Özkan, "Anti-thixotropic properties of waxy maize starch dispersions with different pasting conditions," *Carbohydrate Polymers*, vol. 79, no. 4, pp. 1130–1139, 2010.

22

Grape Seed Procyanidin Extract Improves Insulin Production but Enhances Bax Protein Expression in Cafeteria-Treated Male Rats

Lídia Cedó, Anna Castell-Auví, Victor Pallarès, Mayte Blay, Anna Ardévol, and Montserrat Pinent

Nutrigenomics Research Group, Departament de Bioquímica i Biotecnologia, Universitat Rovira i Virgili, Marcel·lí Domingo s/n, 43007 Tarragona, Spain

Correspondence should be addressed to Montserrat Pinent; montserrat.pinent@urv.cat

Academic Editor: Kiyoshi Ebihara

In a previous study, the administration of a grape seed procyanidin extract (GSPE) in female Wistar rats improved insulin resistance, reduced insulin production, and modulated apoptosis biomarkers in the pancreas. Considering that pharmacokinetic and pharmacodynamic parameters in females are different from these parameters in males, the aim of the present study was to evaluate the effects of GSPE on male Wistar cafeteria-induced obese rats. The results have confirmed that the cafeteria model is a robust model mimicking a prediabetic state, as these rats display insulin resistance, increased insulin synthesis and secretion, and increased apoptosis in the pancreas. In addition, GSPE treatment (25 mg/kg of GSPE for 21 days) in male rats improves insulin resistance and counteracts the cafeteria-induced effects on insulin synthesis. However, the administration of the extract enhances the cafeteria-induced increase in Bax protein levels, suggesting increased apoptosis. This result contradicts previous results from cafeteria-fed female rats, in which GSPE seemed to counteract the increased apoptosis induced by the cafeteria diet.

1. Introduction

Procyanidins are the second most abundant natural phenolic after lignin, and they are widely distributed in fruits, berries, beans, nuts, cocoa, and wine [1]. They are potent antioxidants that possess biological properties that may protect against cardiovascular diseases [1]. They participate in glucose homeostasis [2] and modulate insulin synthesis, secretion, and degradation [3]. Moreover, changes in β-cell insulin production may also be due to variations in the number of insulin-producing cells. β-cell mass adapts to increased metabolic demands caused, for example, by obesity, pregnancy, or insulin resistance. However, when β-cells are unable to compensate for increased insulin demand, there is a decrease in β-cell mass characteristic of the onset of type 2 diabetes mellitus (T2DM) [4]. Procyanidins modulate apoptotic and proliferation processes, mainly reported in cancerous cell lines [5]. Moreover, they protect cells from

diverse drug- or chemical-induced toxic assaults by decreasing apoptosis and inducing cell growth [5]. However, there is little information available regarding their effects on β-cells. Other studies by our research group have reported that procyanidins modulate proliferation and apoptosis of the pancreatic β-cell line INS-1E under altered conditions [6]. Procyanidins also alter the protein and/or gene expression of factors involved in apoptosis in Zucker fatty rats [7].

Obesity has become a worldwide problem, leading to an explosion of obesity-related health issues [8]. Obese individuals develop resistance to the cellular actions of insulin, a key etiological factor for T2DM, which is also becoming an epidemic [9]. T2DM is characterised by peripheral insulin resistance as well as pancreatic β-cell dysfunction and decreased β-cell mass [10]. It involves a combination of genetic and environmental or lifestyle factors. These lifestyle changes, involving high-energy diets and reduced physical activity, are linked to the pandemics of obesity and T2DM

[11]. Given the high prevalence of the disease, obtaining knowledge about natural compounds with potential beneficial effects on glucose homeostasis is of great interest.

Several animal models have been used to study obesity, including both genetic and diet-induced obesity models. However, the cafeteria diet is a more robust model to reproduce the diet in Western society [12]. In a previous study, we analysed the effects of procyanidins in an insulin resistance model induced by cafeteria diet administration in female Wistar rats. We also considered the effects of the cafeteria diet on insulin production and apoptosis in the pancreas. The study showed that the cafeteria diet increased insulin production as well as activated apoptosis biomarkers [13]. Furthermore, procyanidin administration caused a reduction in the Homeostasis Model Assessment for Insulin Resistance (HOMA-IR) index, suggesting improved insulin resistance [14]. Moreover, in the pancreas, procyanidins caused a decrease in insulin production [15] and modulated pro- and antiapoptosis markers [6].

The pharmacokinetics and pharmacodynamics in females are different from the same parameters in males because of the female's unique anatomy and physiology [16, 17]. Thus, the aim of the present study was to evaluate the effects of GSPE in male Wistar cafeteria-induced obese rats and to compare the results on insulin synthesis, apoptosis, and proliferation in the pancreas with those observed in the previous study of female rats.

2. Materials and Methods

2.1. GSPE.
The procyanidin extract was derived from grape seed and contained the following: catechin ($58\,\mu$mol/g), epicatechin ($52\,\mu$mol/g), epigallocatechin ($5.50\,\mu$mol/g), epicatechin gallate ($89\,\mu$mol/g), epigallocatechin gallate ($1.40\,\mu$mol/g), dimeric procyanidins ($250\,\mu$mol/g), trimeric procyanidins ($15.68\,\mu$mol/g), tetrameric procyanidins ($8.8\,\mu$mol/g), pentameric procyanidins ($0.73\,\mu$mol/g), and hexameric procyanidins ($0.38\,\mu$mol/g) [18].

2.2. Animal Experimental Procedures.
Wistar male rats weighting between 250–330 g were purchased from Charles River Laboratories (Barcelona, Spain) and housed in animal quarters at 22°C with a 12 h light/dark cycle. After 1 week in quarantine, the animals were divided in two groups, a diet-control group (7 animals) fed a standard diet (Panlab A03) and a cafeteria group (21 animals) fed a cafeteria diet (bacon, biscuits with pâté, biscuits with cheese, muffins, carrots, and milk with sugar) in addition to standard chow and water. Every day at 9 AM, food was withdrawn, and it was replaced at 6 PM. Obesity was induced in the animals on the cafeteria diet for 52 days. Afterwards, the diet-control group and 7 animals from the cafeteria-fed group were sacrificed, as a reference for the state of the animals before the beginning of treatment. The rest of the cafeteria-fed rats were divided in two subgroups (7 animals/group). These two groups were the (i) cafeteria group: rats treated with a vehicle (gum arabic 5% w/v) and the (ii) GSPE-treated group: rats treated with 25 mg of GSPE/kg of body weight (bw) per day. The treatment was administrated every evening for 21 days before the replacement of the food. Three days before the end of the treatment and after 8 h of fasting, blood was collected from the tails of the rats to measure glucose and insulin levels. At the end of the treatment regimen and after 3 h of fasting, the animals were anesthetised using sodium pentobarbital (50 mg/kg of bw, Sigma-Aldrich, St. Louis, MO) and sacrificed by abdominal aorta exsanguination. The pancreas was isolated from all of the animals, frozen immediately in liquid nitrogen, and stored at −80°C until analysis. All of the procedures were approved by the Experimental Animals Ethics Committee of the Universitat Rovira i Virgili.

2.3. Plasmatic and Pancreatic Measurements.
Insulin plasma levels were assayed using an ELISA method following the manufacturer's instructions (Mercodia, Uppsala, Sweden). Glucose plasma levels were determined using an enzymatic colorimetric kit (QCA, Amposta, Spain).

The HOMA-IR and HOMA-β were calculated using the fasting values of glucose and insulin and the following formulas:

$$\text{HOMA-IR} = \frac{\text{insulin } (\mu\text{U/mL}) \times \text{glucose (mM)}}{22.5},$$

$$\text{HOMA-}\beta = \frac{20 \times \text{insulin } (\mu\text{U/mL})}{\text{glucose (mM)} - 3.5}. \qquad (1)$$

Triglycerides (TAG) and nonesterified fatty acids (NEFAs) from the pancreas were extracted by homogenising the tissue with PBS containing 0.1% triton X-100 (Sigma-Aldrich, St. Louis, MO), and their concentrations were determined using enzymatic colorimetric kits (QCA, Amposta, Spain for TAG and Wako chemicals GmbH, Neuss, Germany for NEFAs).

The pancreas was homogenised with six volumes of PBS containing 50 mM EDTA at pH 7.4 and centrifuged at 3000 g for 5 min at 4°C. Reactive oxygen species (ROS) in the supernatants were quantified using 20 μM DCFH-DA ($2',7'$-dichlorofluorescin diacetate) (Sigma-Aldrich, St. Louis, MO), and the fluorescence was measured after 50 minutes at 37°C at $\lambda_{\text{ex}} = 485$ nm and $\lambda_{\text{em}} = 530$ nm. The values were normalised to the protein content, which was analysed by the Bradford method [19].

2.4. Quantitative Real-Time PCR.
Total RNA from the pancreas was extracted using the RNeasy Mini Kit (Qiagen, Barcelona, Spain). cDNA was generated with the High-Capacity cDNA Reverse Transcription Kit (Applied Biosystems, Madrid, Spain) and was subjected to quantitative Real-Time PCR amplification using the TaqMan Master Mix (Applied Biosystems, Madrid, Spain). Specific TaqMan probes (Applied Biosystems, Madrid, Spain) were used for each gene: Rn99999125_m1 for Bcl2, Rn01480160_g1 for Bax, Rn01492401_m1 for Ccnd2 (Cyclin D2), Rn01774648_g1 for Ins, and Rn00565544_m1 for Cpe. Actb was used as the reference gene (Rn00667869_m1). Reactions were run on a quantitative RT-PCR 7300 system (Applied Biosystems, Madrid, Spain) according to the manufacturer's instructions.

TABLE 1: Effects of the cafeteria diet and GSPE treatment on plasmatic glucose and insulin levels, HOMA-IR and HOMA-β index. $^{**}P \leq 0.01$, $^{***}P \leq 0.001$, and $^{\#}P \leq 0.1$ for cafeteria diet versus standard diet. $^{\ddagger}P \leq 0.05$ and $^{\dagger}P \leq 0.1$ for cafeteria + GSPE versus cafeteria + vehicle.

	Standard diet	Cafeteria diet	Cafeteria + vehicle	Cafeteria + GSPE
Glucose (mM)	4.13 ± 0.2	4.49 ± 0.3	4.76 ± 0.2	4.88 ± 0.2
Insulin (ng/mL)	1.23 ± 0.1	$2.47 \pm 0.2^{***}$	2.78 ± 0.2	$1.80 \pm 0.3^{\ddagger}$
Insulin/Glucose	6.02 ± 0.6	$12.40 \pm 1.6^{**}$	13.25 ± 0.7	$8.39 \pm 1.2^{\ddagger}$
HOMA-IR	4.97 ± 0.5	$10.75 \pm 1.0^{***}$	14.66 ± 2.8	$8.37 \pm 1.2^{\dagger}$
HOMA-β	609.45 ± 93.4	$1508.98 \pm 388.9^{\#}$	1084.1 ± 13.7	$750.58 \pm 121.2^{\ddagger}$

The relative mRNA expression levels were calculated using the $\Delta\Delta$Ct method.

2.5. Western Blot.

The protein levels of Bax and Bcl-2 were quantified by Western Blot as previously described [13]. Primary antibodies were purchased from Cell Signalling Technology (Beverly, MA). 25 μg of protein was loaded onto the gel, and the antibody dilution was 1 : 1500 for Bax and Bcl-2. After incubation with peroxidase-conjugated monoclonal anti-rabbit secondary antibody (Sigma-Aldrich, Madrid, Spain) at a 1 : 10000 dilution, immunoreactive proteins were visualised with the ECL Plus Western Blotting Detection System (GE Healthcare, Buckinghamshire, UK). Chemiluminescence and densitometric analysis of the immunoblots was performed using ImageJ 1.44p software, and all proteins were quantified relative to the loading control.

2.6. Calculations and Statistical Analysis.

The results are expressed as the mean $\pm$ SEM. Effects were assessed using Student's t-test. All calculations were performed with SPSS software v19.

3. Results

3.1. Cafeteria Diet Increases Insulin Production in the Pancreas and GSPE Treatment Counteracts This Diet.

We first examined the effects of the cafeteria diet on pancreatic insulin production after 52 days of diet administration. Insulin and glucose plasma levels were quantified at day 49, and the cafeteria diet-fed rats showed significantly higher plasma insulin levels and no changes in glucose levels (Table 1). The HOMA-IR index indicated that the cafeteria-fed animals had peripheral insulin resistance, and their HOMA-β index tended to increase (Table 1). Therefore, there was a tendency to increase pancreatic functionality response to glucose in order to counteract peripheral insulin resistance. The increased plasma insulin levels agree with an increase in the insulin gene expression, as well as with increased gene expression of carboxypeptidase E (Cpe) (Table 2).

After induction of obesity via the cafeteria diet, rats were treated with 25 mg/kg of bw GSPE for 21 days, concomitantly with cafeteria diet administration. The animals treated with the procyanidin extract had lower insulinemia and decreased HOMA-IR and HOMA-β indexes (Table 1), counteracting the effects observed in the cafeteria-fed rats. Moreover,

TABLE 2: Effects of cafeteria diet on gene expression in the pancreas. $^{*}P \leq 0.05$ and $^{\#}P \leq 0.1$ versus standard diet.

	Standard diet	Cafeteria diet
Ins2	1.22 ± 0.3	$5.37 \pm 1.4^{*}$
Cpe	1.16 ± 0.3	$3.71 \pm 0.8^{*}$
Bcl-2	1.15 ± 0.3	$0.28 \pm 0.1^{*}$
Bax	1.08 ± 0.2	1.64 ± 0.4
Bcl-2/Bax	1.27 ± 0.5	$0.24 \pm 0.1^{\#}$
Ccnd2	1.31 ± 0.4	1.12 ± 0.3

TABLE 3: Effects of GSPE treatment of cafeteria-fed rats on gene expression in the pancreas. $^{**}P \leq 0.01$ and $^{\#}P \leq 0.1$ versus vehicle-treated group.

	Cafeteria + vehicle	Cafeteria + GSPE
Ins2	1.16 ± 0.3	$0.55 \pm 0.1^{\#}$
Cpe	1.04 ± 0.2	$0.2 \pm 0.1^{**}$
Bcl-2	1.12 ± 0.2	0.96 ± 0.3
Bax	1.13 ± 0.3	1.18 ± 0.2
Bcl-2/Bax	1.67 ± 0.4	0.95 ± 0.4
Ccnd2	1.15 ± 0.3	1.80 ± 0.3

insulin gene expression tended to decrease in these rats, and decreased expression of Cpe was also observed (Table 3).

3.2. Effects of Cafeteria Diet and GSPE on Apoptosis Biomarkers.

To examine the effects of the cafeteria diet and GSPE on apoptosis and proliferation in the pancreas, several markers were analysed at the gene and protein level.

The cafeteria-fed rats showed a decrease in the antiapoptotic marker Bcl-2 at both the gene (Table 2) and protein levels (Figure 1(a)). For the pro-apoptotic marker Bax, the mRNA levels of this gene were not significantly altered (Table 2), but we did observe an increase in the protein levels of Bax in the cafeteria-diet-fed group (Figure 1(a)). Therefore, the ratio of Bcl-2/Bax was reduced both at the gene and protein levels, suggesting an increase in apoptosis in the pancreas (Table 2 and Figure 1(a)).

The administration of GSPE had no effect on Bcl-2 and Bax at gene expression compared to the cafeteria diet (Table 3) and on Bcl-2 at protein expression (Figure 1(b)). In contrast, Bax protein levels were increased by the GSPE treatment, enhancing the effects observed in the cafeteria-fed rats (Figure 1(b)). The ratio of Bcl-2/Bax was significantly reduced

(a)

(b)

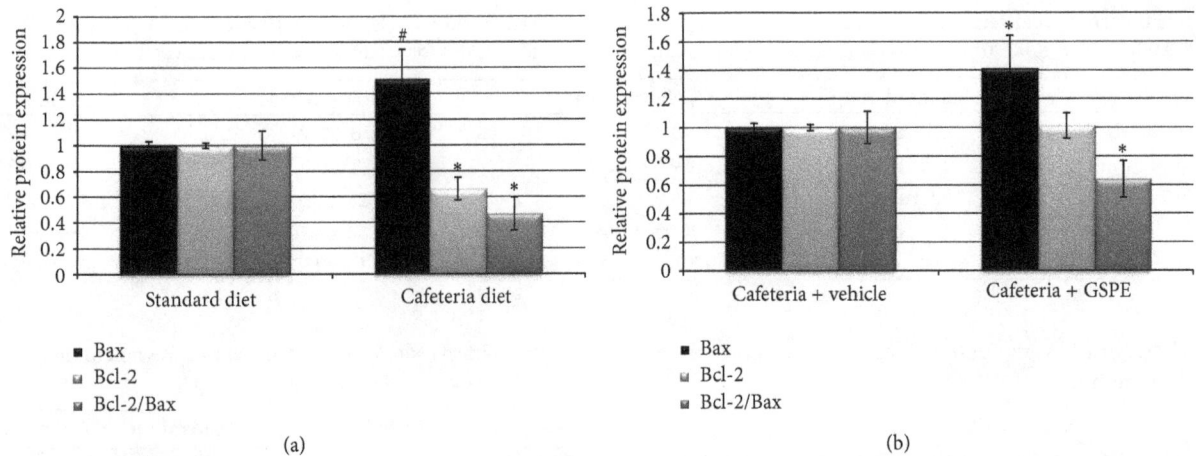

FIGURE 1: Protein expression of the apoptosis markers Bcl-2 and Bax and the calculated ratio of Bcl-2/Bax in: (a) standard-diet-fed rats and cafeteria-diet-fed rats and (b) in GSPE-treated rats and vehicle-treated rats assessed by Western Blot. Data are shown as the mean ± SEM. $^{*}P \leq 0.05$ and $^{\#}P \leq 0.1$.

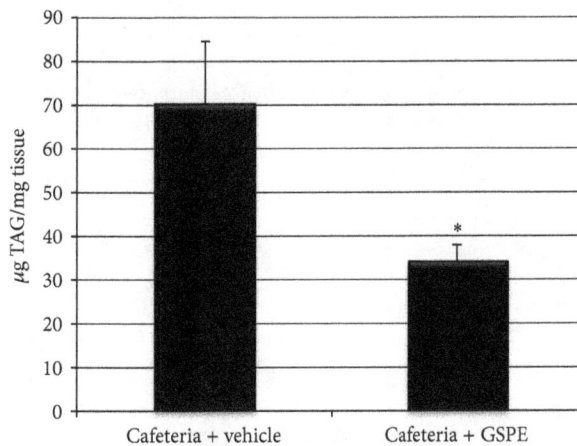

FIGURE 2: TAG content in the pancreas of cafeteria-fed rats treated with GSPE or vehicle, expressed as μg/mg of pancreatic tissue. Data are shown as the mean ± SEM. $^{*}P \leq 0.05$.

FIGURE 3: ROS content in the pancreas, expressed as fluorescence arbitrary units/(mg/mL) of protein and normalised to the respective control group. Data are shown as the mean ± SEM.

at protein level (Figure 1(b)) compared to the cafeteria-fed animals.

Finally, we also analysed Cyclin D2, a proliferation marker, but no changes were observed in the cafeteria-fed animals or in the GSPE-treated rats (Table 2).

3.3. GSPE Treatment Avoids the Increase of TAG in the Pancreas Induced by Cafeteria Diet. Pancreas malfunction is in part due to the accrual of TAG in its cells. To measure it, we examined the TAG content in this tissue and found that TAG triplicated its levels in the pancreas of cafeteria-fed rats compared to the standard-diet-fed rats (33.15 ± 2.7 versus 11.42 ± 0.3 μg TAG/mg tissue, $P \leq 0.01$). After 21 days of treatment, the TAG contents in the pancreas increased, due to cafeteria diet, but GSPE avoided this (Figure 2).

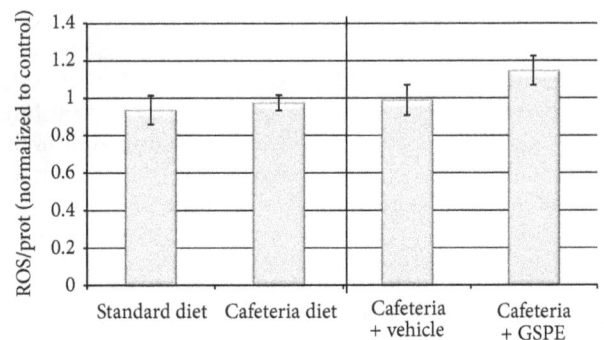

In contrast, the content of NEFAs was not modified neither in the cafeteria group compared to the standard-diet-fed rats (3.37 ± 0.9 versus 4.40 ± 0.4 μg NEFA/mg tissue) nor in the GSPE-treated rats compared to the vehicle-treated rats (2.43 ± 0.3 versus 10.29 ± 4.6 μg NEFA/mg tissue).

ROS content in the pancreas was also analysed, and no significant differences were observed in either the cafeteria-fed rats or in the GSPE-treated animals (Figure 3).

4. Discussion

This study was designed to examine the effects of the cafeteria diet on insulin production in male Wistar rats by evaluating pancreas functionality, apoptosis, and proliferation. We have also evaluated the effects of procyanidins on these processes, since procyanidins were shown to have positive effects on glucose metabolism under conditions of slightly disrupted homeostasis [2].

We had previously shown that 17 weeks of a cafeteria diet led to insulin resistance, high plasma insulin levels, and increased insulin synthesis and secretion in female Wistar rats [13]. It has been reported that female rats are more sensitive to cafeteria-induced obesity than males [20, 21]. However, with respect to insulin resistance, we now show that 52 days of cafeteria diet (nearly 7 and a half weeks) administrated to male Wistar rats confirms the effects observed in the previous study of female rats. In the males, we have observed high insulin plasma levels and elevated HOMA-IR index, indicating peripheral insulin resistance. Additionally, we have seen an elevated HOMA-β index, which indicates an increase in pancreatic functionality in terms of glucose response to counteract peripheral insulin resistance [22]. We have also found an increase in the expression of the insulin gene in the cafeteria-fed rats and an increase in the gene expression of Cpe. Cpe is the enzyme thought to be involved in the cleavage of proinsulin, which results in insulin and C-peptide molecules [23]. Therefore, the pancreas of cafeteria diet-fed rats is still functional, and it tries to counteract peripheral insulin resistance despite the increased lipid accumulation in the pancreas.

Increased deposits of fat are associated with obesity and lead to an increase in free fatty acids (FFAs). The induction of apoptosis in *in vivo* high-fat diet models and *in vitro* models of FFA-induced apoptosis is important evidence for β-cell lipotoxicity [24]. In mice, administration of a high-fat diet for 12 weeks led to increased β-cell mass, despite showing an increase in β-cell apoptosis [25]. In our study, we have found a decrease in the Bcl-2/Bax ratio both at the gene and protein levels in the pancreas of the cafeteria-fed rats compared to the standard chow-fed rats which also suggest an increase in apoptosis in the cafeteria-fed animals. Palmitate was reported to induce β-cell endoplasmic reticulum stress and death mediated by Cpe degradation [26, 27]. Thus we have checked the expression levels of this gene. We have found that Cpe is not decreased but increased by the cafeteria diet, suggesting that the apoptosis observed in the cafeteria-fed rats is not mediated by lipotoxicity. In fact, the levels of NEFA in the pancreas are not altered in the cafeteria-fed rats when compared to the standard-diet-fed rats.

The results from the apoptosis markers are in accordance with the data obtained in the previous study which evaluated the cafeteria diet in Wistar female rats [13]. We have also analyzed the expression of the proliferation marker Cyclin D2 and found no changes due to the cafeteria treatment suggesting that at the time of the analysis, β-cell mass had likely already increased. This also agrees with the pervious results in females [6], as well as with those found in rats fed high fat diets, in which no changes in Ki67 (a proliferation marker) expression were observed [25].

Taken together, the data suggests that the effects of the cafeteria diet on insulin synthesis, secretion, and apoptosis are not influenced by gender or treatment duration.

Once the effects of the cafeteria have been established, we have analyzed the effects of a GSPE treatment on the cafeteria-fed animals. After 52 days of cafeteria diet administration, male Wistar rats have been treated with 25 mg/kg of GSPE for 21 days concomitant with the cafeteria diet. The GSPE treatment has decreased insulin production, plasma insulin levels, and pancreatic insulin Cpe gene expression compared to the cafeteria-vehicle-fed rats. Moreover, the

decreased insulin production could be at least in part explained through GSPE's lipid-lowering effect, since the triglyceride content has also been reduced in the pancreas of GSPE-treated rats. Previously, we showed that 25 mg of GSPE/kg of bw administered to female Wistar rats for 30 days resulted in decreased insulin production and reduced triglyceride content in the pancreas, likely via decreased fatty acid synthesis and increased β-oxidation [15]. Despite this lower insulinemia, GSPE improved glycemia in female Wistar cafeteria-fed rats acting peripherally on adipose tissue [14]. Present results reinforce the effects of GSPE decreasing insulin production and ameliorating lipid accumulation in cafeteria-fed rats and support that these effects are not gender dependent.

GSPE has counteracted the effects of the cafeteria diet reducing the accumulation of triglycerides in the pancreas but has not counteracted the cafeteria-diet effects on the apoptosis markers. Instead, GSPE-treated rats have shown an increase in Bax protein levels and a decreased ratio of Bcl-2/Bax, suggesting an enhancement in the apoptosis. Therefore, the lipid-lowering effects of GSPE do not involve a reduction in apoptosis in the pancreas of cafeteria-fed male rats. Actually, *in vitro* GSPE does not modulate fatty acid-induced apoptosis [6]. In fact, GSPE increases glucose uptake in β-cells under high-glucose conditions and impairs mitochondrial and cellular membrane potentials [3]. Moreover, GSPE is reported to enhance the pro-apoptotic effects of high glucose *in vitro* [6]. Therefore, the enhanced apoptosis observed in the GSPE-treated rats could be due to increased glucose uptake in the β-cells that potentiate glucotoxicity. ROS is one of the players in glucose-induced apoptosis in β-cells; ROS is increased as a consequence of chronically increased glucose metabolism. β-cells have relatively low expression of antioxidant enzymes and are more sensitive to ROS attack when they are exposed to oxidative stress [28]. However, the levels of ROS in the pancreas have not been modified by the cafeteria diet or by GSPE treatment, suggesting that glucose toxicity could be mediated by another mechanism.

The apoptosis marker results conflict with those in the previous study of female Wistar rats, which showed that 25 and 50 mg/kg of GSPE seemed to counteract the deleterious effects of the cafeteria diet by inhibiting the down-regulation of Bcl-2 protein expression after 10 and 30 days of treatment [6]. In addition, 50 mg/kg bw of GSPE also counteracted the decrease in the Bcl-2/Bax ratio at the protein level after 10 days of administration. However, no GSPE effects were observed with respect to the ratio of Bcl-2/Bax gene expression at any dose or treatment duration [6]. Therefore, the modulation of apoptosis biomarkers by GSPE in cafeteria-fed rats is clearly dependent on the dose and treatment period; these effects may also be dependent on gender.

In conclusion, the present study has confirmed that the cafeteria model is a suitable reproduction of the pre-diabetic state. This model induces an insulin resistance state, shows increased insulin synthesis and secretion, and exhibits increased apoptosis in the pancreas. Moreover, GSPE treatment in male rats treated with 25 mg/kg of GSPE for 21 days improves the insulin resistance state and counteracts the cafeteria-induced effects on insulin synthesis. However, procyanidins enhance the elevated levels of Bax, a proapoptotic protein observed in the cafeteria-fed rats, potentially suggesting an increase in apoptosis. This result indicates that the effects of GSPE on apoptosis markers are dose, time, and/or gender dependent.

Conflict of Interests

The authors declare no conflict of interests.

Acknowledgments

The authors thank the members of the Nutrigenomics group with whom they have collaborated to care for and sample the animals. They also thank Pol Andres for assistance in analysing gene and protein expressions. This study was supported by Grants AGL2011-23879 and AGL2008-00387/ALI from the Ministerio de Educación y Ciencia (MEC) of the Spanish Government. Fellowships have been given to L. Cedó from the Generalitat de Catalunya; A. Castell-Auví from the Ministerio de Educación of Spanish Government; and V. Pallarès from de Universitat Rovira i Virgili.

References

[1] S. E. Rasmussen, H. Frederiksen, K. S. Krogholm, and L. Poulsen, "Dietary proanthocyanidins: occurrence, dietary intake, bioavailability, and protection against cardiovascular disease," *Molecular Nutrition and Food Research*, vol. 49, no. 2, pp. 159–174, 2005.

[2] M. Pinent, L. Cedó, G. Montagut, M. Blay, and A. Ardévol, "Procyanidins improve some disrupted glucose homoeostatic situations: an analysis of doses and treatments according to different animal models," *Critical Reviews in Food Science and Nutrition*, vol. 52, no. 7, pp. 569–584, 2012.

[3] A. Castell-Auví, L. Cedó, V. Pallarès et al., "Procyanidins modify insulinemia by affecting insulin production and degradation," *Journal of Nutritional Biochemistry*, vol. 23, no. 12, pp. 1565–1572, 2012.

[4] C. J. Rhodes, "Type 2 diabetes—a matter of β-cell life and death?" *Science*, vol. 307, no. 5708, pp. 380–384, 2005.

[5] D. Bagchi, M. Bagchi, S. J. Stohs, S. D. Ray, C. K. Sen, and H. G. Preuss, "Cellular protection with proanthocyanidins derived from grape seeds," *Annals of the New York Academy of Sciences*, vol. 957, pp. 260–270, 2002.

[6] L. Cedó, A. Castell-Auví, V. Pallarès et al., "Grape seed procyanidin extract modulates proliferation and apoptosis of pancreatic β-cells," *Food Chemistry*, vol. 138, no. 1, pp. 524–530, 2012.

[7] L. Cedó, A. Castell-Auví, V. Pallarès et al., "Pancreatic islet proteome profile in Zucker fatty rats chronically treated with a grape seed procyanidin extract," *Food Chemistry*, vol. 135, no. 3, pp. 1948–1956, 2012.

[8] Z. Li, S. Bowerman, and D. Heber, "Health ramifications of the obesity epidemic," *Surgical Clinics of North America*, vol. 85, no. 4, pp. 681–701, 2005.

[9] M. Qatanani and M. A. Lazar, "Mechanisms of obesity-associated insulin resistance: many choices on the menu," *Genes and Development*, vol. 21, no. 12, pp. 1443–1455, 2007.

[10] S. E. Kahn, "The relative contributions of insulin resistance and beta-cell dysfunction to the pathophysiology of type 2 diabetes," *Diabetologia*, vol. 46, no. 1, pp. 3–19, 2003.

[11] C. J. Nolan, P. Damm, and M. Prentki, "Type 2 diabetes across generations: from pathophysiology to prevention and management," *The Lancet*, vol. 378, no. 9786, pp. 169–181, 2011.

[12] B. P. Sampey, A. M. Vanhoose, H. M. Winfield et al., "Cafeteria diet is a robust model of human metabolic syndrome with liver and adipose inflammation: comparison to high-fat diet," *Obesity*, vol. 19, no. 6, pp. 1109–1117, 2011.

[13] A. Castell-Auví, L. Cedó, V. Pallarès, M. Blay, A. Ardévol, and M. Pinent, "The effects of a cafeteria diet on insulin production and clearance in rats," *The British Journal of Nutrition*, vol. 108, no. 7, pp. 1155–1162, 2012.

[14] G. Montagut, C. Bladé, M. Blay et al., "Effects of a grapeseed procyanidin extract (GSPE) on insulin resistance," *Journal of Nutritional Biochemistry*, vol. 21, no. 10, pp. 961–967, 2010.

[15] A. Castell-Auví, L. Cedó, V. Pallarès, M. Blay, M. Pinent, and A. Ardévol, "Grape seed procyanidins improve β-cell functionality under lipotoxic conditions due to their lipid-lowering effect," *Journal of Nutritional Biochemistry*, 2012.

[16] E. Tanaka, "Gender-related differences in pharmacokinetics and their clinical significance," *Journal of Clinical Pharmacy and Therapeutics*, vol. 24, no. 5, pp. 339–346, 1999.

[17] M. Gochfeld, "Framework for gender differences in human and animal toxicology," *Environmental Research*, vol. 104, no. 1, pp. 4–21, 2007.

[18] A. Serra, A. Macià, M. P. Romero et al., "Bioavailability of procyanidin dimers and trimers and matrix food effects in in vitro and in vivo models," *The British Journal of Nutrition*, vol. 103, no. 7, pp. 944–952, 2010.

[19] M. M. Bradford, "A rapid and sensitive method for the quantitation of microgram quantities of protein utilizing the principle of protein dye binding," *Analytical Biochemistry*, vol. 72, no. 1-2, pp. 248–254, 1976.

[20] A. M. Rodríguez and A. Palou, "Uncoupling proteins: gender dependence and their relation to body weight control," *International Journal of Obesity and Related Metabolic Disorders*, vol. 28, pp. 500–502, 2004.

[21] A. Nadal-Casellas, A. M. Proenza, M. Gianotti, and I. Lladó, "Brown adipose tissue redox status in response to dietary-induced obesity-associated oxidative stress in male and female rats," *Stress*, vol. 14, no. 2, pp. 174–184, 2011.

[22] D. R. Matthews, J. P. Hosker, A. S. Rudenski, B. A. Naylor, D. F. Treacher, and R. C. Turner, "Homeostasis model assessment: Insulin resistance and β-cell function from fasting plasma glucose and insulin concentrations in man," *Diabetologia*, vol. 28, no. 7, pp. 412–419, 1985.

[23] D. F. Steiner, S. Y. Park, J. Støy, L. H. Philipson, and G. I. Bell, "A brief perspective on insulin production," *Diabetes, Obesity and Metabolism*, vol. 11, supplement 4, pp. 189–196, 2009.

[24] A. Giacca, C. Xiao, A. I. Oprescu, A. C. Carpentier, and G. F. Lewis, "Lipid-induced pancreatic β-cell dysfunction: focus on in vivo studies," *The American Journal of Physiology*, vol. 300, no. 2, pp. E255–E262, 2011.

[25] A. M. Owyang, K. Maedler, L. Gross et al., "XOMA 052, an anti-IL-1β monoclonal antibody, improves glucose control and β-cell function in the diet-induced obesity mouse model," *Endocrinology*, vol. 151, no. 6, pp. 2515–2527, 2010.

[26] K. D. Jeffrey, E. U. Alejandro, D. S. Luciani et al., "Carboxypeptidase E mediates palmitate-induced β-cell ER stress

and apoptosis," *Proceedings of the National Academy of Sciences of the United States of America*, vol. 105, no. 24, pp. 8452–8457, 2008.

[27] J. D. Johnson, "Proteomic identification of carboxypeptidase E connects lipid-induced β-cell apoptosis and dysfunction in type 2 diabetes," *Cell Cycle*, vol. 8, no. 1, pp. 38–42, 2009.

[28] Y. Kajimoto and H. Kaneto, "Role of oxidative stress in pancreatic β-cell dysfunction," *Annals of the New York Academy of Sciences*, vol. 1011, pp. 168–176, 2004.

23

Peanut Allergy, Allergen Composition, and Methods of Reducing Allergenicity: A Review

Yang Zhou,[1] **Jin-shui Wang,**[2] **Xiao-jia Yang,**[1] **Dan-hua Lin,**[1] **Yun-fang Gao,**[1] **Yin-jie Su,**[1] **Sen Yang,**[2] **Yan-jie Zhang,**[2] **and Jing-jing Zheng**[1]

[1] *College of Food Science and Technology, Henan University of Technology, Zhengzhou 450001, China*
[2] *College of Bioengineering, Henan University of Technology, Zhengzhou 450001, China*

Correspondence should be addressed to Jin-shui Wang; jinshuiw@163.com

Academic Editor: Kiyoshi Ebihara

Peanut allergy affects 1-2% of the world's population. It is dangerous, and usually lifelong, and it greatly decreases the life quality of peanut-allergic individuals and their families. In a word, peanut allergy has become a major health concern worldwide. Thirteen peanut allergens are identified, and they are briefly introduced in this paper. Although there is no feasible solution to peanut allergy at present, many methods have shown great promise. This paper reviews methods of reducing peanut allergenicity, including physical methods (heat and pressure, PUV), chemical methods (tannic acid and magnetic beads), and biological methods (conventional breeding, irradiation breeding, genetic engineering, enzymatic treatment, and fermentation).

1. Introduction

Food allergy is a worldwide health problem. It affects approximately 5% of young children and 3% to 4% of adults in westernized countries [1], and it becomes more and more common in developing countries. Although virtually any food can cause allergy, over 90% of food allergy is triggered by eight food sources: milk, egg, peanut, tree nuts, shellfish, fish, wheat, and soy [2]. Among them, peanut is one of the most allergenic. Peanut allergy affects many individuals and its prevalence is increasing rapidly (the prevalence of peanut allergy in some countries is summarized in Table 1). In western countries, the prevalence of peanut allergy in children in the USA increased from 0.4% in 1997 to 1.4% in 2008 [3]; the prevalence of sensitization to peanuts of 3-year olds in the UK rose from 1.3% to 3.2% between 1989 and 1995 [4]; over 1% of Canadian children are allergic to peanuts [5]; the prevalence of peanut allergy in Denmark and France is 0.2–0.4% and 0.3–0.75%, respectively [6, 7]. In Asia, although few epidemic studies of peanut allergy have been carried out, a study suggests that 0.47% of 14–16-year-old local Singapore schoolchildren and 0.43% of 14–16-year-old Philippine schoolchildren are allergic to peanuts [8].

Considering that 76.8% of Singapore residents are Chinese [9], peanut allergy is likely to become serious in China in the future. Moreover, peanut allergy can sometimes be life-threatening and usually cannot be outgrown, and it is almost impossible to avoid accidental ingestion [10, 11]. Therefore, peanut allergy greatly reduces the life quality of the patient [12] and brings trouble to food industry in allergen labeling. Solving this problem has a great significance not only to the peanut-allergic individuals but also to the food industry.

2. Peanut Allergen

To date, 13 peanut allergens (Ara h 1 through h 13) have been recognized by the Allergen Nomenclature Sub-Committee of the International Union of Immunological Societies. These allergens come from 7 protein families. Except for Ara h 1 (150 kD) and Ara h 3 (360–380 kD), the molecular weight of the other allergens ranges from 5 to 17 kD [13]. The genes corresponding to the 13 allergens have already been elucidated, the sequence of many linear epitopes of peanut allergens has been identified (Table 2), and the 3D models of Ara h 1-Ara h 6 have been built. Although the allergenicity of these allergens has not been thoroughly studied and there is still some debate

TABLE 1: The prevalence of peanut allergy in some countries [3–8].

Countries	Prevalence
US children	1.40%
Britain children	3.2%
Canadian children	1.03%
Denmark	0.2–0.4%
France	0.3–0.75%
Local Singapore schoolchildren (14–16 years old)	0.47%
Philippine schoolchildren (14–16 years old)	0.43%

TABLE 2: Sequence of linear epitopes of peanut allergens [17, 19, 26, 30].

Allergen	Epitope number	Epitope sequence
Ara h 1 core region	7[a]	PGQFEDFF
	8[a]	YLQGFSRN
	9[a]	FNAEFNEIRR
	10[a]	QEERGQRR
	11[a]	DITNPINLRE
	12[a]	NNFGKLFEVK
	13[a]	GNLELV
	14[a]	RRYTARLKEG
	15[a]	ELHLLGFGIN
	16[a]	HRIFLAGDKD
	17[a]	IDQIEKQAKD
	18[a]	KDLAFPGSGE
	19[a]	KESHFVSARP
	21[b]	NEGVIVKVSKEHVEELTKHAKSVSK
Ara h 2	1	HASARQQWEL
	2	QWELQGDRRC
	3	DRRCQSQLER
	4	LRPCEQHLMQ
	5	KIQR.DEDSYE
	6	YERDPYSPSQ
	7	SQDPYSPSPY
	8	DRLQ..GRQQEQ
	9	KRELRNLPQQ
	10	QRCDLDVESG
Ara h 3	1	IETWNPNNQEFECAG
	2	GNIFSGFTPEFLAQA
	3	VTVRGGLRILSPDRK
	4	DEDEYEYDE--EDRRRG

[a]Determined by [17]. [b]Determined by [19].

about the definition of major allergens, the major peanut allergens that are most widely accepted are Ara h 1, Ara h 2, and Ara h 3.

2.1. Ara h 1. Ara h 1 is a glycoprotein and belongs to the vicilin (7S) family. It comprises 12–16% of the total peanut protein [14] and affects 35–95% of peanut-allergic patients in different populations [15]. Native Ara h 1 exists as a trimer formed by three identical monomers, and the crystal structure of its core region has been elucidated (Figure 1) [16]. The topology and basic structure of its core region are very similar to other known structures of 7S globulins. Those similarities indicate that there is a high possibility of cross-reactivity between 7S globulins [16]. To date, 21 linear epitopes have been identified on the mature Ara h 1 [17–20], and 14 epitopes were found in the core region [16]. It is found that most epitopes on the core region become either slightly (<50% burial) or significantly (≥50% burial) buried upon trimer formation [16]. The burial of those epitopes likely explains the relatively weak activity of native (trimer) Ara h 1 in cross-linking IgE and the strong binding of IgE to denatured monomers [18, 21].

2.2. Ara h 2. Ara h 2 (16-17 kDa) is also a glycoprotein and accounts for 5.9–9.3% of the total peanut protein [22]. It is a 2S albumin, also known as conglutin, and functions as a trypsin inhibitor [23]. More than 95% of peanut-allergic individuals in the USA have specific IgE to Ara h 2, and Ara h 2 was found to be a more potent allergen than Ara h 1 [21, 24, 25]. The structure of Ara h 2 is five α-helices arranged in a right-handed superhelix and connected by several extended loops (Figure 2). This three-dimensional conformation is stabilized by four conserved disulphide bridges. Ten epitopes have been mapped on Ara h 2, and these epitopes show a fairly well exposition on the molecular surface [26].

2.3. Ara h 3. Ara h 3 is a seed storage protein and belongs to the legumin (11S) family [27]. It is recognized by 50% of peanut-allergic individuals and also functions as a trypsin inhibitor [28, 29]. Ara h 3 and soybean glycinin result in a sequence identity of 47.2% [30]. Mature Ara h 3 is a hexamer (360–380 kD) formed by a head-to-head association of two trimers (Figure 3) [30]. Each monomer was found to have 4 linear epitopes [31]. In the natural form of Ara h 3, epitope 4 is fully exposed, while the side chains of most of the critical residues of the other three epitopes are completely or nearly completely buried. This suggests that linear epitopes 1 and 2 may not be recognized by IgE antibodies in the intact form, while epitope 4 and part of epitope 3 may be allergic in the natural form of Ara h 3 [30].

2.4. Ara h 4. Ara h 4 is actually an isoform of Ara h 3. Now, it is no longer thought to be a distinct allergen and renamed to Ara h 3.02 [13, 32].

2.5. Ara h 5. Ara h 5 (15 kD) belongs to the profilin family and regulates the polymerization of actin [13, 32]. It is presented at low levels in peanut extracts and is recognized by 13% of 40 patients' sera [27, 33]. The structure of Ara h 5 is shown in Figure 4.

2.6. Ara h 6. Ara h 6 is a 15 kD protein and belongs to the conglutin family [13]. It is 59% homologous to Ara h 2 and has similar allergenicity [34, 35]. Ara h 6 is a heat and digestion stable protein and showed resistance to proteolytic treatment [36, 37]. The structure of Ara h 6 is shown in Figure 5.

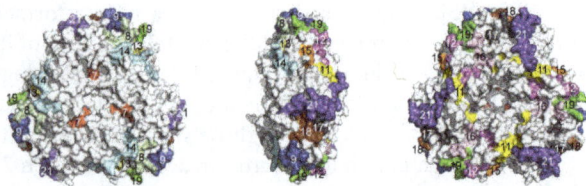

FIGURE 1: IgE epitopes are mapped on the surface of the 3D model of Ara h 1 core region [16].

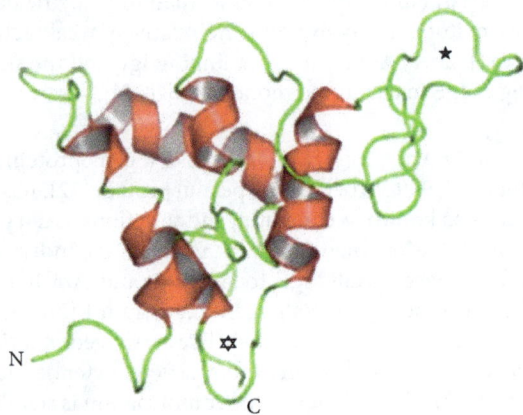

FIGURE 2: Ribbon diagram of Ara h 2 [26].

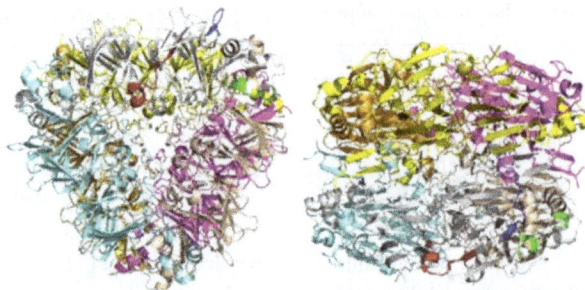

FIGURE 3: Ara h 3 is represented with each of the monomers shown in a different color. In the gray monomer, linear epitope 1, 2, and 3 are shown in red, green, and blue, respectively [30].

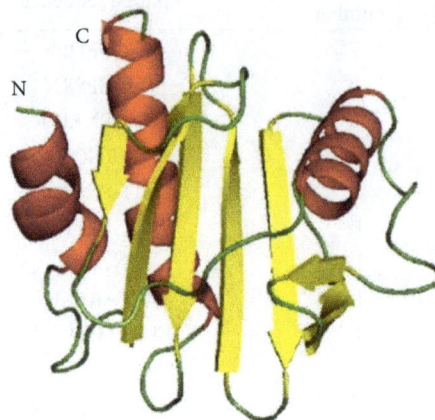

FIGURE 4: Ribbon diagram of the three-dimensional model of Ara h 5. Strands of b-sheet and stretches of a-helix are in yellow and red, respectively. Coil structures or loops are in green, N and C indicate the N- and C-terminus of the polypeptide, respectively [38].

2.7. Ara h 7. Ara h 7 is also a 15 kD protein and belongs to the conglutin family [13]. The sequence identity between Ara h 2 and Ara h 6 is 35%, and it is recognized by 13% of 40 patients' sera [23].

2.8. Other Peanut Allergens. Ara h 8 (17 kD) is a Pathogenesis-related protein. Ara h 9 (9.8 kD, 2 isoforms) is a nonspecific lipid-transfer protein. Ara h 10 (16 kD, 2 isoforms) and Ara h 11 (14 kD) belong to oleosin. Ara h 12 and Ara h 13 are defensin, with molecular weight ranging from 5 to 12 kD [13].

3. Harm of Peanut Allergy

Peanut-allergic reactions involve the skin, the respiratory tract, and the gastrointestinal tract [39]. The common symptoms include acute urticaria, acute vomiting, laryngeal oedema, hypotension, and dysrhythmia [40, 41]. Peanut allergy is very dangerous. Ingestion of even a trace amount of peanut may elicit life-threatening reactions within minutes [42]. Peanut, together with tree nuts, causes most of the fatal or near-fatal food-related anaphylaxis, and peanut allergy leads to 100–200 deaths each year in the USA [43, 44]. Moreover, peanut allergy is usually lifelong, with only 10% of peanut-allergic children outgrowing it [10]. Last but not least, due to the ubiquitous use of peanut in food industry, it is almost impossible for a peanut-allergic patient to completely avoid peanut even if he/she strictly obeys the doctor's guidance. Studies suggest that up to 75% of individuals with known peanut allergy experience reactions caused by accidental exposure [11]. Thus, peanut allergy gives enormous pressure to the patients and their families and greatly impairs their life

quality [12]. In addition, the US law demands that allergen content be labeled on any product sold in the USA [45], and tracing and determining peanut allergens in food products increase the cost and bring inconvenience to the food trade.

4. Methods of Reducing Allergenicity

Although there is now no feasible solution to peanut allergy, many methods have shown great prospect, including oral immunotherapy and some methods to reduce peanut allergenicity. The methods of reducing peanut allergic potential are reviewed as follows.

4.1. Physical Methods

4.1.1. Heat and Pressure Treatment. There are three ways to decrease peanut allergenicity by heat treatment. The first is roasting. Roasting has been recognized as a process that can increase peanut allergenicity [46]. However, Vissers et al. found that after heating Ara h 2/6 (purified from raw peanuts) in a dry form for 20 min at 145°C, the IgE-binding capacity and the degranulation capacity of Ara h 2/6 were 600–700-fold lower than those in the native form [47].

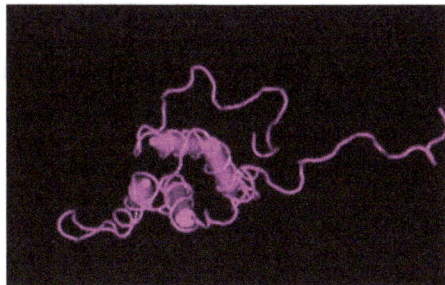

FIGURE 5: Ara h 6 (PDB Entry 1W2Q, first molecule in the entry).

The second is boiling. Boiling native Ara h 2/6 (15 min, 110°C) and boiling native Ara h 1 (15 min, 100°C) resulted in decreased IgE reactivity and mediator-releasing capacity; but for Ara h 2/6 and Ara h 1 extracted from roasted peanut, boiling had no effect [48, 49].

The third one is autoclaving. Cabanillas et al. discovered that IgE-binding capacity of peanut allergens is significantly decreased by autoclaving at 2.56 atm, for 30 min [50]. However, this method obviously comes with high energy consumption and expensive devices.

4.1.2. PUV.

Pulsed ultraviolet light (PUV) is another effective method in reducing peanut allergenicity. Yang et al. treated protein extracts from raw and roasted peanuts and peanut butter slurry in a Xenon Steripulse XL 3000 PUV system. The treatment time was 2, 4, and 6 min for protein extracts and 1, 2, and 3 min for peanut butter slurry. The distance from the central axis of the lamp was varied at 10.8, 14.6, and 18.2 cm. The research found that PUV treatment resulted in reduction in the level of Ara h 1, Ara h 2, and Ara h 3 and decreased IgE binding ability by 12.9% to 6.7% [51]. However, like all the other irradiation technologies, this method comes with concern of food safety.

4.2. Chemical Methods

4.2.1. Tannic Acids.

Chung and Reed reduced the allergenicity of peanut butter by adding tannic acid. The principal is that tannic acid interacts with allergens to form indigestible complex, and epitopes on the allergens are covered during complex formation, making the epitopes inaccessible to antibodies and resulting in reduced allergenicity. Chung and Reed added tannic acid to a peanut butter extract (5 mg/mL; pH = 7.2) and discovered that when pH = 2 and pH = 8, the complexes do not release Ara h 1, or Ara h 2, and the IgE binding ability is decreased; and when concentration of tannic acid is 1-2 mg/mL, the IgE binding ability of the complex is reduced substantially [52]. Since tannic acid interacts with both allergen and non-allergen peanut proteins, such treatment has two obvious deficiencies: first, peanut nutrition is reduced to a great extent, and second, intake of much indigestible food may cause stomach discomfort and thus greatly limit consumption of peanut products.

4.2.2. Magnetic Beads.

Magnetic beads can also be used to remove peanut allergens. The principle is that phenolic compounds and ferric ions (Fe^{3+}) can bind to peanut allergens; thus, one can reduce peanut allergenicity by using magnetic beads attached with or without phenolics to capture peanut allergens or allergen-Fe^{3+} complexes and then separate the beads by a magnetic device. Chung and Champagne found the following: treating peanut extracts by CHL beads (magnetic beads covalently attached with chlorogenic acid, a phenolic) resulted in marked decrease of Ara h 1 and small reduction of Ara h 2; when using magnetic beads without phenolic compounds to treat peanut extracts that have been incubated with Fe^{3+} and dialyzed, both Ara h 1 and Ara h 2 were markedly reduced; those two methods reduced IgE binding ability of the treated extracts by 28–47%. Chung and Champagne believed that the magnetic beads system was a simple way to partially remove peanut allergens from peanut extracts, and it could be a potential approach to produce hypoallergenic peanut products and beverages [53].

4.3. Biological Methods

4.3.1. Conventional Breeding.

The rationale of conventional breeding is crossing hypoallergenic varieties to produce a variety that is more hypoallergenic. Perkins et al. crossbred peanuts that were missing either an Ara h 2 or Ara h 3 isoform and produced a variety lacking both isoforms. The observed numbers of the new variety conformed to the 15 : 1 Mendelian dihybrid ratio [54]. However, considering the large amount of peanut allergens, the progress of this method seems very slow.

4.3.2. Irradiation Breeding.

As for mutation breeding, a type of technology is well worth mentioning. It is the heavy-ion beam irradiation (HIBI). This technology leads to mutation and inactivation of a single gene or multiple genes in a plant, thus inducing stable knockout mutants [55, 56]. Cabanos et al. treated a Japanese peanut variety—Nakateyutaka—with either N or C heavy-ion beams at a dose of 100 Gy and obtained seventeen knockout mutants from 11,335 screened M2 seeds. Among the seventeen mutants, eight lacked either one of the two isoforms of Ara h 2, and the other nine are missing one of the isoforms of Ara h 3 [57]. Cabanos et al. believe that HIBI is a powerful means of producing knockout hypoallergenic peanuts and has many advantages [57], including low radiation exposure levels, less cellular damage, no need for tedious tissue culture or regenerative procedures, no severe growth inhibition, and, in general, less plant death and a high rate of mutation producing diverse kinds of mutants [55, 56, 58]. However, like all the other irradiation technologies, HIBI comes with the concern of food safety.

4.3.3. Genetic Engineering.

Great advance has been made in removing peanut allergens by genetic technology. Chu et al. silenced Ara h 2 and Ara h 6 by RNA interference and produced three independent transgenic lines. All the three lines were featured by significant reduction in human IgE binding to Ara h 2 and Ara h 6 as well as the level of Ara h 2, whereas the level of Ara h 6 was only reduced in two lines. In addition, there were no significant differences between the seed weight and germination data of transgenic and nontransgenic plants [59]. Another research comes from Ananga et al. who tried to produce hypoallergenic peanuts by silencing Ara h 1, Ara h 2, and Ara h 3 with RNA interference. Ananga et al. have found the following: the percentages of transgenic peanut that showed reduction in Ara h 1, Ara h 2, and Ara h 3 were 9%, 10%, and 16%, respectively 3% transgenic seeds were free of all three allergens; the IgE-binding capacity was significantly reduced in at least nine transgenic seeds with reduction in Ara h 1 or Ara h 2, Ara h 3 [60]. Although genetic technology has shown great promise to produce allergen-free peanut, this technology also has two big drawbacks. One is people's increasing repulsion to transgenic food. The other is the fact that peanut allergens account for 20–30% of total peanut proteins, and if all the allergens are removed, peanuts may not taste like peanuts.

4.3.4. Enzymatic Treatment. Enzymatic treatment is full of potential to produce allergen-free peanut and there are two types of enzymatic treatment.

One is using enzymes to cross-link allergen proteins, resulting in the burial of epitopes. Chung and Champagne treated protein extract from roasted and raw peanuts with peroxidase (POD) and transglutaminase (TGA) at 37°C. Both enzymes catalyze cross-links between proteins. Chung and Champagne found the following: POD treatment of roasted peanut resulted in partial loss of Ara h 1 and Ara h 2 along with reduced IgE binding ability and formation of new polymers; on the other hand, TGA treatment of roasted peanut had no effect on the content of Ara h 1 and Ara h 2 as well as IgE binding ability; both POD and TGA had no effect on the IgE binding ability of protein extract from raw peanut. Chung and Champagne believed that POD may be useful in desensitizing peanut while TGA should be useless [61].

The other is using enzymes to break down allergens, destroying their epitopes. Cabanillas et al. studied the effect of hydrolysis with alcalase and flavourzyme on the allergenicity of the soluble protein fraction of roasted peanut. Parameters for alcalase hydrolysis were $S = 2\%$, $E/S = 0.4\,AU/g$ of protein, $T = 50°C$, and pH 8.0; parameters for flavourzyme hydrolysis were $S = 2\%$, $E/S = 100\,LAPU/g$ of protein, $T = 50°C$, and pH 7.0. Cabanillas et al. discovered the following: 30 min alcalase treatment resulted in an important decrease of Ara h 1, Ara h 2, and Ara h 3 levels and reduced IgE binding reactivity by 98%; 90 min alcalase treatment could fully eliminate IgE binding reactivity; while 30 min flavourzyme treatment caused an increase in IgE reactivity, hydrolyzing with flavourzyme for 300 min led to a 65% inhibition of IgE reactivity [62]. Although Cabanillas et al. demonstrated that enzymatic treatment with alcalase or flavourzyme could reduce IgE reactivity in peanuts, Guo et al. found that the allergenicity was retained after treating roasted peanut protein extract in a similar way [63]. The two researches adopted different methods to assess the allergenicity of enzymatic products, and the liability of such assessment methods is still in debate. It is unclear whether alcalase and flavourzyme have an effect or not.

The most promising enzymatic method for desensitizing peanut is the research of Ahmedna et al. Mohamed's team has been working on the subject for over 7 years. They have developed a method which is very likely to completely eliminate peanut allergenicity in a quick, simple, and inexpensive way, without greatly changing the flavor and texture of natural peanuts. Since Mohamed's team are using their research to apply a patent, only a little detail of the method can be obtained. It is only told that peanut allergenicity is reduced by direct application of enzymatic solution to either raw, blanched, or roasted peanuts, or peanut products or derivatives (including but not limited to peanut butter, peanut kernels, peanut skins, peanut protein isolate, peanut flour, or peanut milk); the enzymatic solution used in this method contains at least one endopeptidase whose hypoallergenically effective amount is at least 0.001% (w/w) [64].

4.3.5. Fermentation. Few studies have been reported on reducing peanut allergenicity by fermentation. It is only reported that Dr. Ahmedna et al. found that fermenting whole or ground peanuts with an edible fungus reduced the detectable level of major allergenic proteins Ara h 1 and Ara h 2 by as much as 70 percent, and this study is still in the early stages [65]. Although fermentation method has rarely been reported, this method has already successfully reduced the allergic potential of soybean meal and bovine whey proteins [66, 67] and is very likely to reduce the allergenicity of peanuts. The major principle of fermentation is almost the same as that of enzymatic treatment, and fermentation has all the merits of the enzyme method. Furthermore, this method is usually much cheaper. Therefore, fermentation is still a very promising method to produce hypoallergenic peanuts.

5. Prospect

Breeding is an effective way, but some problems cannot be ignored. The advance of conventional breeding seems very slow, and mutation breeding is always involved with the problem of food safety; all of these impede the development of the breeding method.

Heat and pressure treatment is another effective approach to reduce peanut allergenicity. However, it still has some deficiencies: roasting can only be an assistant method; the effect of boiling is very limited; autoclaving requires high energy consumption and expensive devices. All of these constitute obstacles to the development of this method.

Transgenic technology is very promising to produce allergen-free variety in the near future. However, with people's repulsion to transgenic food and flavor problems, it has a long way to go to have transgenic allergen-free peanuts in the market.

Tannic acid is a useful agent, but considering its obvious drawbacks, it can only be an assistant approach. PUV has the same problem of mutation breeding which hinders its development. Magnetic beads capture is a promising way to decrease peanut allergenicity.

At present, enzymatic treatment is the most promising way to produce nonallergic peanuts. Compared with breeding and gene technology, enzymatic treatment is mild, is natural, usually does not produce harmful substance and can be readily accepted by the public. Compared with autoclaving, the approach of Mohamed's team is very cheap. Moreover, enzymatic treatment does not impair peanut nutrition value. Therefore, the authors believe that it is the most promising approach nowadays.

Fermentation is full of potential to reduce peanut allergenicity. Moreover, it has all the merits of the enzyme method and is usually much cheaper. Although this method has rarely been reported, it may be the best way to reduce peanut allergenicity.

References

[1] S. H. Sicherer and H. A. Sampson, "Food allergy," *Journal of Allergy and Clinical Immunology*, vol. 125, supplement 2, pp. S116–S125, 2010.

[2] S. L. Hefle, J. A. Nordlee, and S. L. Taylor, "Allergenic foods," *Critical Reviews in Food Science and Nutrition*, vol. 36, supplement 1, pp. S69–S89, 1996.

[3] S. H. Sicherer, A. Muñoz-Furlong, J. H. Godbold, and H. A. Sampson, "US prevalence of self-reported peanut, tree nut, and sesame allergy: 11-year follow-up," *Journal of Allergy and Clinical Immunology*, vol. 125, no. 6, pp. 1322–1326, 2010.

[4] J. Grundy, S. Matthews, B. Bateman, T. Dean, and S. H. Arshad, "Rising prevalence of allergy to peanut in children: data from 2 sequential cohorts," *Journal of Allergy and Clinical Immunology*, vol. 110, no. 5, pp. 784–789, 2002.

[5] M. Ben-Shoshan, R. S. Kagan, R. Alizadehfar et al., "Is the prevalence of peanut allergy increasing? A 5-year follow-up study in children in Montreal," *Journal of Allergy and Clinical Immunology*, vol. 123, no. 4, pp. 783–788, 2009.

[6] M. Morisset, D.-A. Moneret-Vautrin, and G. Kanny, "Prevalence of peanut sensitizion in a population of 4,737 subjects-an allergo-vigilance network enquiry carried out in 2002," *European Annals of Allergy and Clinical Immunology*, vol. 37, no. 2, pp. 54–57, 2005.

[7] M. Osterballe, T. K. Hansen, C. G. Mortz, A. Høst, and C. Bindslev-Jensen, "The prevalence of food hypersensitivity in an unselected population of children and adults," *Pediatric Allergy and Immunology*, vol. 16, no. 7, pp. 567–573, 2005.

[8] L. P. Shek, E. A. Cabrera-Morales, S. E. Soh et al., "A population-based questionnaire survey on the prevalence of peanut, tree nut, and shellfish allergy in 2 Asian populations," *Journal of Allergy and Clinical Immunology*, vol. 126, no. 2, pp. 324–331, 2010.

[9] Central Intelligence Agency, Library, Publications, The World Factbook, Singapore, people and society, ethnic groups, 2000, https://www.cia.gov/library/publications/the-world-factbook/geos/sn.html.

[10] Berger and Smith, "Science commentary: why do some children grow out of peanut allergy?" *British Medical Journal*, vol. 316, p. 1275, 1998.

[11] R. S. Kagan, L. Joseph, C. Dufresne et al., "Prevalence of peanut allergy in primary-school children in Montreal, Canada," *Journal of Allergy and Clinical Immunology*, vol. 112, no. 6, pp. 1223–1228, 2003.

[12] R. M. King, R. C. Knibb, and J. O. Hourihane, "Impact of peanut allergy on quality of life, stress and anxiety in the family," *Allergy*, vol. 64, no. 3, pp. 461–468, 2009.

[13] "Allergen Nomenclature (IUIS Allergen Nomenclature Sub-Committee)," http://www.allergen.org/search.php?Allergen source=Arachis+hypogaea.

[14] E. C. de Jong, M. Van Zijverden, S. Spanhaak, S. J. Koppelman, H. Pellegrom, and A. H. Penninks, "Identification and partial characterization of multiple major allergens in peanut proteins," *Clinical and Experimental Allergy*, vol. 28, no. 6, pp. 743–751, 1998.

[15] A. Mari, E. Scala, P. Palazzo, S. Ridolfi, D. Zennaro, and G. Carabella, "Bioinformatics applied to allergy: allergen databases, from collecting sequence information to data integration. The Allergome platform as a model," *Cellular Immunology*, vol. 244, no. 2, pp. 97–100, 2006.

[16] C. Cabanos, H. Urabe, M. R. Tandang-Silvas, S. Utsumi, B. Mikami, and N. Maruyama, "Crystal structure of the major peanut allergen Ara h 1," *Molecular Immunology*, vol. 49, no. 1-2, pp. 115–123, 2011.

[17] A. W. Burks, D. Shin, G. Cockrell, J. S. Stanley, R. M. Helm, and G. A. Bannon, "Mapping and mutational analysis of the IgE-binding epitopes on Ara h 1, a legume vicilin protein and a major allergen in peanut hypersensitivity," *European Journal of Biochemistry*, vol. 245, no. 2, pp. 334–339, 1997.

[18] D. S. Shin, C. M. Compadre, S. J. Maleki et al., "Biochemical and structural analysis of the IgE binding sites on Ara h1, an abundant and highly allergenic peanut protein," *Journal of Biological Chemistry*, vol. 273, no. 22, pp. 13753–13759, 1998.

[19] W. G. Shreffler, K. Beyer, T. T. Chu, A. W. Burks, and H. A. Sampson, "Microarray immunoassay: association of clinical history, in vitro IgE function, and heterogeneity of allergenic peanut epitopes," *Journal of Allergy and Clinical Immunology*, vol. 113, no. 4, pp. 776–782, 2004.

[20] H. J. Wichers, T. De Beijer, H. F. J. Savelkoul, and A. Van Amerongen, "The major peanut allergen Ara h 1 and its cleaved-off N-terminal peptide; possible implications for peanut allergen detection," *Journal of Agricultural and Food Chemistry*, vol. 52, no. 15, pp. 4903–4907, 2004.

[21] G. W. Palmer, D. A. Dibbern Jr., A. W. Burks et al., "Comparative potency of Ara h 1 and Ara h 2 in immunochemical and functional assays of allergenicity," *Clinical Immunology*, vol. 115, no. 3, pp. 302–312, 2005.

[22] S. J. Koppelman, R. A. A. Vlooswijk, L. M. J. Knippels et al., "Quantification of major peanut allergens Ara h 1 and Ara h 2 in the peanut varieties Runner, Spanish, Virginia, and Valencia, bred in different parts of the world," *Allergy*, vol. 56, no. 2, pp. 132–137, 2001.

[23] S. J. Maleki, O. Viquez, T. Jacks et al., "The major peanut allergen, Ara h 2, functions as a trypsin inhibitor, and roasting enhances this function," *Journal of Allergy and Clinical Immunology*, vol. 112, no. 1, pp. 190–195, 2003.

[24] A. M. Scurlock and A. W. Burks, "Peanut allergenicity," *Annals of Allergy, Asthma and Immunology*, vol. 93, supplement 5, pp. S12–S18, 2004.

[25] S. J. Koppelman, M. Wensing, M. Ertmann, A. C. Knulst, and E. F. Knol, "Relevance of Ara h1, Ara h2 and Ara h3 in peanut-allergic patients, as determined by immunoglobulin E Western blotting, basophil-histamine release and intracutaneous testing: Ara h2 is the most important peanut allergen," *Clinical and Experimental Allergy*, vol. 34, no. 4, pp. 583–590, 2004.

[26] A. Barre, J. Borges, R. Culerrier, and P. Rougé, "Homology modelling of the major peanut allergen Ara h 2 and surface mapping of IgE-binding epitopes," *Immunology Letters*, vol. 100, no. 2, pp. 153–158, 2005.

[27] S. J. Koppelman, E. F. Knol, R. A. A. Vlooswijk et al., "Peanut allergen Ara h 3: isolation from peanuts and biochemical characterization," *Allergy*, vol. 58, no. 11, pp. 1144–1151, 2003.

[28] H. W. Wen, W. Borejsza-Wysocki, T. R. DeCory, and R. A. Durst, "Peanut allergy, peanut allergens, and methods for the detection of peanut contamination in food products," *Comprehensive Reviews in Food Science and Food Safety*, vol. 6, no. 2, pp. 47–58, 2007.

[29] H. W. Dodo, O. M. Viquez, S. J. Maleki, and K. N. Konan, "cDNA clone of a putative peanut (Arachis hypogaea L.) trypsin inhibitor has homology with peanut allergens Ara h 3 and Ara h 4," *Journal of Agricultural and Food Chemistry*, vol. 52, no. 5, pp. 1404–1409, 2004.

[30] T. C. Jin, A. F. Guo, Y. W. Chen, A. Howard, and Y. Zhang, "Crystal structure of Ara h 3, a major allergen in peanut," *Molecular Immunology*, vol. 46, no. 8-9, pp. 1796–1804, 2009.

[31] P. Rabjohn, E. M. Helm, J. S. Stanley et al., "Molecular cloning and epitope analysis of the peanut allergen Ara h 3," *Journal of Clinical Investigation*, vol. 103, no. 4, pp. 535–542, 1999.

[32] H. Breiteneder and C. Radauer, "A classification of plant food allergens," *Journal of Allergy and Clinical Immunology*, vol. 113, no. 5, pp. 821–830, 2004.

[33] T. Kleber-Janke, R. Crameri, U. Appenzeller, M. Schlaak, and W. Becker, "Selective cloning of peanut allergens, including profilin and 2S albumins, by phage display technology," *International Archives of Allergy and Immunology*, vol. 119, no. 4, pp. 265–274, 1999.

[34] S. J. Koppelman, G. A. H. De Jong, M. Laaper-Ertmann et al., "Purification and immunoglobulin E-binding properties of peanut allergen Ara h 6: evidence for cross-reactivity with Ara h 2," *Clinical and Experimental Allergy*, vol. 35, no. 4, pp. 490–497, 2005.

[35] X. Chen, Q. Wang, R. El-Mezayen et al., "Ara h 2 and Ara h 6 have similar allergic effector activity and are substantially redundant," *International Archives of Allergy and Immunology*, vol. 160, no. 3, pp. 251–258, 2013.

[36] M. Suhr, D. Wicklein, U. Lepp, and W. Becker, "Isolation and characterization of natural Ara h 6: evidence for a further peanut allergen with putative clinical relevance based on resistance to pepsin digestion and heat," *Molecular Nutrition and Food Research*, vol. 48, no. 5, pp. 390–399, 2004.

[37] K. Lehmann, K. Schweimer, G. Reese et al., "Structure and stability of 2S albumin-type peanut allergens: implications for the severity of peanut allergic reactions," *Biochemical Journal*, vol. 395, no. 3, pp. 463–472, 2006.

[38] C. Cabanos, M. R. Tandang-Silvas, V. Odijk et al., "Expression, purification, cross-reactivity and homology modeling of peanut profilin," *Protein Expression and Purification*, vol. 73, no. 1, pp. 36–45, 2010.

[39] S. H. Sicherer, A. W. Burks, and H. A. Sampson, "Clinical features of acute allergic reactions to peanut and tree nuts in children," *Pediatrics*, vol. 102, no. 1, p. e6, 1998.

[40] S. A. Bock, A. Muoz-Furlong, and H. A. Sampson, "Fatalities due to anaphylactic reactions to foods," *Journal of Allergy and Clinical Immunology*, vol. 107, no. 1, pp. 191–193, 2001.

[41] H. A. Sampson, L. Mendelson, and J. P. Rosen, "Fatal and near-fatal anaphylactic reactions to food in children and adolescents," *New England Journal of Medicine*, vol. 327, no. 6, pp. 380–384, 1992.

[42] H. A. Sampson, "Peanut allergy," *New England Journal of Medicine*, vol. 346, no. 17, pp. 1294–1299, 2002.

[43] P. L. Jackson, "Peanut allergy: an increasing health risk for children," *Pediatric Nursing*, vol. 28, no. 5, pp. 496–504, 2002.

[44] C. W. Lee and A. L. Sheffer, "Peanut allergy," *Allergy and Asthma Proceedings*, vol. 24, no. 4, pp. 259–264, 2003.

[45] USDA, "Get the Facts: New Food Allergen Labeling Laws," 2006, http://www.fns.usda.gov/fdd/facts/nutrition/FoodAllergenFactSheet.pdf.

[46] R. A. Kopper, N. J. Odum, M. Sen, R. M. Helm, J. S. Stanley, and A. W. Burks, "Peanut protein allergens: the effect of roasting on solubility and allergenicity," *International Archives of Allergy and Immunology*, vol. 136, no. 1, pp. 16–22, 2005.

[47] Y. M. Vissers, M. Iwan, K. Adel-Patient et al., "Effect of roasting on the allergenicity of major peanut allergens Ara h 1 and Ara h 2/6: the necessity of degranulation assays," *Clinical and Experimental Allergy*, vol. 41, no. 11, pp. 1631–1642, 2011.

[48] Y. M. Vissers, F. Blanc, P. S. Skov et al., "Effect of heating and glycation on the allergenicity of 2S albumins (Ara h 2/6) from peanut," *PLoS ONE*, vol. 6, no. 8, Article ID e23998, 2011.

[49] F. Blanc, Y. M. Vissers, K. Adel-Patient et al., "Boiling peanut Ara h 1 results in the formation of aggregates with reduced allergenicity," *Molecular Nutrition and Food Research*, vol. 55, no. 12, pp. 1887–1894, 2011.

[50] B. Cabanillas, S. J. Maleki, J. Rodríguez et al., "Heat and pressure treatments effects on peanut allergenicity," *Food Chemistry*, vol. 132, no. 1, pp. 360–366, 2012.

[51] W. W. Yang, N. R. Mwakatage, R. Goodrich-Schneider, K. Krishnamurthy, and T. M. Rababah, "Mitigation of major peanut allergens by pulsed ultraviolet light," *Food and Bioprocess Technology*, vol. 5, no. 7, pp. 2728–2738, 2012.

[52] S. Y. Chung and S. Reed, "Removing peanut allergens by tannic acid," *Food Chemistry*, vol. 134, no. 3, pp. 1468–1473, 2012.

[53] S. Y. Chung and E. Champagne, "Using magnetic beads to reduce peanut allergens from peanut extracts," *Journal of Allergy Clinical Immunology*, vol. 125, no. 2, supplement 1, p. AB223, 2010.

[54] T. Perkins, D. A. Schmitt, T. G. Isleib et al., "Breeding a hypoallergenic peanut," *The Journal of Allergy and Clinical Immunology*, vol. 117, supplement 2, p. S328, 2006.

[55] S. Kikuchi, Y. Saito, H. Ryuto et al., "Effects of heavy-ion beams on chromosomes of common wheat, Triticum aestivum," *Mutation Research*, vol. 669, no. 1-2, pp. 63–66, 2009.

[56] Y. Kazama, H. Saito, M. Fujiwara et al., "An effective method for detection and analysis of DNA damage induced by heavy-ion beams," *Bioscience, Biotechnology and Biochemistry*, vol. 71, no. 11, pp. 2864–2869, 2007.

[57] C. S. Cabanos, H. Katayama, H. Urabe et al., "Heavy-ion beam irradiation is an effective technique for reducing major allergens in peanut seeds," *Molecular Breeding*, vol. 30, no. 2, pp. 1037–1044, 2011.

[58] T. Abe, T. Matsuyama, S. Sekido, I. Yamaguchi, S. Yoshida, and T. Kameya, "Chlorophyll-deficient mutants of rice demonstrated the deletion of a DNA fragment by heavy-ion irradiation," *Journal of Radiation Research*, vol. 43, pp. S157–S161, 2002.

[59] Y. Chu, P. Faustinelli, M. L. Ramos et al., "Reduction of IgE binding and nonpromotion of Aspergillus flavus fungal growth by simultaneously silencing Ara h 2 and Ara h 6 in peanut," *Journal of Agricultural and Food Chemistry*, vol. 56, no. 23, pp. 11225–11233, 2008.

[60] A. Ananga, H. Dodo, and K. Konan, "Elimination of the three major allergens in transgenic peanut (Arachis hypogea L)," in *Vitro Cellular & Developmental Biology-Animal*, vol. 44, pp. S36–S37, 2008.

[61] S. Chung and E. T. Champagne, "Effect of enzyme treatment on the allergenic properties of peanuts," *Journal of Allergy and Clinical Immunology*, vol. 111, no. 2, 2003.

[62] B. Cabanillas, M. M. Pedrosa, J. Rodríguez et al., "Influence of enzymatic hydrolysis on the allergenicity of roasted peanut protein extract," *International Archives of Allergy and Immunology*, vol. 157, no. 1, pp. 41–50, 2011.

[63] R. Guo, X. Shi, B. White et al., "Allergenicity of peanut proteins is retained following enzymatic hydrolysis," *Journal of Allergy and Clinical Immunology*, vol. 129, no. 2, p. AB367, 2011.

[64] M. Ahmedna, J. M. Yu, and I. Goktepe, "Process for preparing hypoallergenic and non-allergenic peanut butter and associated products," 2010, United States Patent Application Publication, Pub. No: US2010/0080870 A1.

[65] "New process removes allergy proteins from peanuts," *Magazine of the Agricultural Research Program at North Carolina Agricultural and Technical State University*, vol. 2, p. 3, 2005.

[66] Y. S. Song, J. Frias, C. Martinez-Villaluenga, C. Vidal-Valdeverde, and E. G. de Mejia, "Immunoreactivity reduction of soybean meal by fermentation, effect on amino acid composition and antigenicity of commercial soy products," *Food Chemistry*, vol. 108, no. 2, pp. 571–581, 2008.

[67] G. H. Bu, Y. K. Luo, Y. Zhang, and F. Chen, "Effects of fermentation by lactic acid bacteria on the antigenicity of bovine whey proteins," *Journal of the Science of Food and Agriculture*, vol. 90, no. 12, pp. 2015–2020, 2010.

24

Microwave Heating as an Alternative Quarantine Method for Disinfestation of Stored Food Grains

Ipsita Das,[1] **Girish Kumar,**[1] **and Narendra G. Shah**[2]

[1] Department of Electrical Engineering, Indian Institute of Technology, Bombay, Powai, Mumbai 400076, India
[2] Centre for Technology Alternatives for Rural Areas, Indian Institute of Technology, Bombay, Powai, Mumbai 400076, India

Correspondence should be addressed to Ipsita Das; ipsitdas@gmail.com

Academic Editor: Mitsuru Yoshida

Insects and pests constitute a major threat to food supplies all over the world. Some estimates put the loss of food grains because of infestation to about 40% of the world production. Contemporary disinfestation methods are chemical fumigation, ionizing radiation, controlled atmosphere, conventional hot air treatment, and dielectric heating, that is, radio frequency and microwave energy, and so forth. Though chemical fumigation is being used extensively in stored food grains, regulatory issues, insect resistance, and environmental concerns demand technically effective and environmentally sound quarantine methods. Recent studies have indicated that microwave treatment is a potential means of replacing other techniques because of selective heating, pollution free environment, equivalent or better quality retention, energy minimization, and so forth. The current paper reviews the recent advances in Microwave (MW) disinfestation of stored food products and its principle and experimental results from previous studies in order to establish the usefulness of this technology.

1. Importance of Disinfestation

Agricultural commodities produced on the fields have to undergo a series of operations such as harvesting, threshing, winnowing, bagging, transportation, storage, and processing before they reach the consumer, and there are appreciable losses in crop output at all these stages. Various estimates have been made to assess the postharvest food grain losses. The losses are caused either by environmental factors such as temperature, moisture, and type of storage structure or by biological agents, namely, insects, rodents, birds, and fungi. The major losses during production, storage and marketing of food grain are being attributed to infestation by insect pests, microbiological contamination, and physiological changes. Insect infestations can occur just prior to harvest or during storage or in-transit in a variety of carriers. The occurrence and numbers of stored grain insect pests are directly related to geographical and climatic conditions [1]. Almost all species have remarkably high rates of multiplication and, within one season, may destroy 10–15% of the grain and contaminate the rest with undesirable odors and flavors. Insect pests also play a pivotal role in transportation of storage fungi [2]. Therefore, preventing economic losses caused by stored-product insects is important from the field to the consumer's table [3]. The losses during storage are classified as quantity losses and quality losses. Quantity losses occur when the grain is consumed by insects, rodents, mites, birds and microorganisms. Quality losses are reflected as reduced economic value of the crop. The stored-grain insects affect not only the quantity of grain but also the quality of grain. Infestation causes decreased nutritional value, reduced seed germination, and lower economic value and also causes changes in chemical compositions such as increase in moisture, free fatty acid levels, nonprotein nitrogen content, and decrease in pH and protein contents in food grain [4]. They reduce the product quality directly by damage through feeding and indirectly by producing webbing and frass [5]. Also the grain quality has been found to decrease with time with increasing levels of infestation [6]. It is estimated that more than 20,000 species of field and storage pests destroy approximately one-third of the world's food production, valued annually at more than 100 billion dollar [7]. The quantitative and qualitative damage to stored grains and grain product from the insect pests may amount to 20–30% in the tropical zone and 5–10% in the temperate zone [8]. Food grain production in India has reached 250 million tons in the year 2010-2011, in which nearly 20–25% food grains are damaged by stored grain insect pests [7]. With population growth and the amount of cultivable land shrinking, grain losses will continue to be a problem in the developing countries. Control of stored-product pests is one of the major tasks because the damage inflicted to foodstuff is irreversible. Also with progressive increase in the quantity of food grains and necessity for longer storage periods, these losses will escalate unless disinfestation measures are improved. International organizations such as FDA [9] and FGIS [10] have set tolerances and grade standards regulating the number of insects and insect fragments above specified tolerances to make the product illegal for human consumption. In developed countries even the mere presence of a few insects in a bulk, at densities of considerably less than one insect per kg grain, can cause a serious loss in its market value. In some developed countries grain can be downgraded or rejected completely if even a single live insect is found [11]. The efficient control and removal of stored grain pests from food commodities have long been the goal of entomologists throughout the world. Various methods of insect control have been practiced to save the grain. In recent years, technology has made marked progress in the study of disinfestation of stored-grain insects and of finding improved ways to control them especially to detect latent forms of infestation. The implementation of an insect-disinfestation method requires detailed analysis of all the elements of an infestation problem: the insects, their age, species, and distribution and their survival and developmental rates under different environmental conditions [12]. Conventional chemicals, grain protectants, and fumigants are extensively used around the world to control insect pests in stored commodities because of low cost, fast processing, and easy application. Greater regulation and restriction of methyl bromide use will likely increase the cost of the fumigant, as well as reduce its availability [13]. With the concerns about health hazards of chemical pesticides and their resulting environmental pollution, there is interest in developing alternative, nonchemical process protocol to control insect pests while retaining acceptable product quality. These include conventional hot air or water heating, controlled atmosphere, and dielectric heating (radio frequency (RF) and microwave (MW)). Currently, the use of chemical fumigation remains widespread and the efficient use of RF and MW methods for disinfestation is still in research stage. The current paper reviews the various disinfestation methods of stored food grains with special emphasis on recent advances in microwave disinfestation of stored food grains. The principle of microwave disinfestation, experimental results of quality characteristics of microwave-treated grains, and the challenges of microwave disinfestation have also been described.

2. Various Methods of Insect Control

The control of stored-product insects is very important if the quality of foodstuff is to be maintained. The goal of the control measure is to render the habitat unsuitable for the growth and reproduction of stored-product insects. The five major potential quarantine treatment methods used for disinfestation of several insect pests for both the domestic and international markets are chemical fumigation, ionizing radiation, controlled atmosphere, conventional hot air/water heating, and dielectric heating using radio frequency (RF) and microwave (MW) energy which have been described in this paper.

2.1. Chemical Method. Since the 1950s, chemical insecticides have been used extensively in grain storage facilities to control stored-product insect pests due to low cost, fast speed in processing, and ease of use. Most postharvest pest management programs, therefore, rely heavily on fumigants. Currently, over 2.5 million tons of chemicals, worth over 30 billion US dollars, are applied to crops in the world [14]. The chemicals used to control insects in the bulk stored-grains are composed of two classes, namely, contact insecticides and fumigants. Contact insecticides, such as malathion, chlorpyrifos-methyl, or deltamethrin, are sprayed directly on grain or structures which provide protection from infestation for several months [15]. Fumigants are gaseous insecticides applied to control insects in grains that are inaccessible by contact insecticide. Some of the commonly used fumigants are methyl bromide (MeBr) and phosphine which rapidly kill all life stages of stored-product insects in food product [16]. Methyl bromide is now under the threat of withdrawal because it apparently depletes the Earth's ozone layer [17]. It has already been established that the use of MeBr has led to serious environmental damage and hazards to people's health. For these reasons, the Montreal Protocol constituted by the United Nations Environment Programme agreed on a phasing out of methyl bromide in the developed countries by 2005 and in the developing countries by 2015 [18]. Phosphine has been used as a replacement of methyl bromide for a long time [19]. Conventional use of phosphine has failed frequently to control insects [20]. Another limiting factor is that insects develop resistance to some particular chemical fumigants. Some of the contact insecticides have become ineffective because of wide-spread resistance in insect populations. A worldwide survey of stored-product insects revealed that 87% of 505 strains of the red flour beetle, *Tribolium castaneum*, collected from 78 countries were resistant to malathion [15]. Resistance to malathion is widespread in Canada, USA, and Australia [21] while resistance to phosphine is great in Australia and India, which may cause control failures [22]. The other major problem associated with the chemical methods is that, even if they are applied with care and in limited quantity, there is a possibility that these chemicals may remain in the food grains and have adverse effects on humans. Fumigation often only kills live larvae or adult insects but does not sterilize the eggs which are still alive in the grain kernels [23]. The consumption of organic products is also increasing each year. There is, therefore, interest in developing an alternative, nonchemical process method to control insect pests in food grains so as to minimize the environmental hazards associated with chemical insecticides while retaining acceptable product quality.

2.2. Ionizing Radiation. It is a process where infested food products are being exposed to ionizing radiation so as to sterilize, kill, or prevent emergence of insect pests by damaging their DNA. Three types of ionizing radiation used on foods are gamma rays from radioactive cobalt-60 and cesium-137, high energy electrons, and X-rays [24]. Irradiation with high energy electrons is usually safer and easier to work with because it can be turned on and off while an isotope is always radiating and humans must be shielded from it [25]. The ability of gamma rays to deeply penetrate pallet loads of food makes it one of the most commonly used in postharvest pest control. Sterilization of many species of insects can be accomplished at lower doses. Rusty grain beetles are sterilized at only 0.6 kGy but saw-toothed grain beetles and red flour beetles require a 2.0 kGy dose [26]. The grain mite, however, requires a much higher dose of 4.5 kGy. Though irradiation doses of 3–5 kGy were reported to be effective in controlling insects, they cause damage to product quality [26]. According to Hasan and Khan [27], high dosage of ionizing radiation has a risk of vitamin loss from the food product. The major drawback is substantial initial investment to establish the facility. In order to be economically feasible, the facility must remain in continuous operation. However, the seasonal nature of food produce prevents efficient use of facilities. Besides, this method has not received wide recognition because of high power consumption, large weight, and overall sizes of the installation. Consumers also have concerns over the disposal of radioactive wastes, the safety of the irradiation technology, and its effect on food [28]. Irradiation treatments often lead to live insects found by inspectors or consumers in the treated product because the applied doses do not immediately kill treated insects [29, 30]. Grains will continue to appear to be infested and a grain buyer cannot be certain that the insects are sterilized. Following irradiation with gamma rays at 0.5 kGy, complete insect mortality occurs in 14 days for rusty grain beetles, 28 days for red flour beetles, 70 days for saw-toothed grain beetles, and 200 days for grain mites. To date, irradiation is not accepted by the organic industry. Also approvals for irradiation of some selected food products (e.g., almond) have not been accepted by many countries, such as EU, Japan, and Taiwan [31].

2.3. Controlled Atmosphere Storage. Controlled atmosphere, a disinfestation technology wherein the normal composition of atmospheric air, that is, 21% O_2, 0.03% CO_2, and 78% nitrogen, is altered appropriately for disinfestation process [26]. An atmosphere containing more than 35% CO_2 known as carbon dioxide atmosphere and the atmosphere containing less than 1% oxygen, that is, low-oxygen atmosphere are lethal to insects. Controlled atmospheres are mainly based on the establishment of a low-oxygen environment which kills pests. The oxygen levels vary between 0% and 2%. It can be applied in airtight environments ranging from $1 M^3$ to $1000 M^3$ depending upon food commodity. Insects in all stages are eliminated because of the lack of oxygen which causes the insect to dry out and suffocate. This controlled atmosphere (CA) storage has been shown to be promising

in creating lethal conditions for insects and fungi in stored food commodities. Annis and Morton [36] studied the effect of 15% to 100% CO_2 on developmental stages of insects in wheat at 25°C and 60% RH. They found that pupae were the most tolerant stages for all CO_2 concentrations and eggs were the only stages with 100% mortality at 20% CO_2 for less than 30 days. Gunasekaran and Rajendran [37] also found that the pupae stages were the most tolerant stages when exposed to different concentrations of CO_2. Controlled atmospheres are always being seen as average alternative such as longer treatment time, usability and availability, and being not suitable for dealing with high level of infestation.

2.4. Conventional Heat Treatment. Compared to the use of chemical methods, heat disinfestation has the potential for high market acceptance because of being residue-free. Most research focused on using hot air (80°C to 100°C) for disinfestation of food grains and showed satisfactory results. Thermal treatment methods using hot air/hot water alone or in combination with cold or controlled-atmosphere (CA) storage conditions have been investigated extensively for disinfestation of number of stored commodities [28, 38–55]. The hot air used to increase the temperature of the food product lies above the thermal limits of survival of the pest/insect. Heat disinfestation treatments are relatively easy to apply, leave no chemical residues, and may offer some fungicidal activity. There are a number of factors that affect the mortality of insects when exposed to hot air such as duration of exposure, temperature, species, and stage. Ideal temperatures for growth, reproduction, and movement for most stored product insects are between 25°C and 35°C [56]. Many insect larvae can bore into the center of fruits, nuts, seeds, or kernels, so the center of the commodity must be heated to lethal temperatures. Reported heating times for fruit center to reach the desired maximum temperatures range from 23 min for cherries to 6 h for apples [48, 49]. Conventional heating consists of heat transfer from the heating medium to the fruit surface and then conductive heat transfer from the surface to the center. A common difficulty with hot air or water heating methods is the slow rate of heat transfer resulting in long treatment time [57]. Prolonged heating is proved to be detrimental to the quality of final products such as peel browning, pitting, poor color development, and abnormal softening and may not be practical in industrial applications ([58], cited by [59]). There were reports of damage to grapefruits and mangos when exposed to forced air at 46°C for 3.75 h and 45°C for 1.8 h, respectively ([60, 61] cited by [48]). Flavor and appearance of air-heated grapefruits at 46°C for 3 h were inferior to those of unexposed fruits [62]. Surface browning in avocados was observed when heated with hot air at 43°C for 3.5 h [63] and in apples when exposed to hot water treatment at 46°C for 45 min [64]. A long exposure time requirement also causes alterations to flavor compounds [65]. The low heating rates also may increase the thermotolerance of the few insects [66–68] which was caused by the induction of heat shock proteins in insects during sublethal thermal conditions [69]. It is also important to determine the time-temperature combinations that result in 100% mortality for each insect over relatively large range of temperature. Sometimes temperature and time combinations required to kill the target insects may exceed those that reduce the crop nutrients, germination, or shelf life [70]. The disadvantages associated with conventional heat treatment method stimulated further studies on the possible use of dielectric heating for controlling stored-grain insects.

2.5. Dielectric Heating.

Dielectric heating which covers both radio frequency (RF) and microwave (MW) has been investigated for insect control in foods [71]. Radio frequency (RF) heating is akin to microwave heating but utilizes another part of the electromagnetic spectrum. The frequency used for MW is 2450 MHz or 915 MHz while for RF the frequency is of 13, 27, or 40 MHz [48]. The effects of RF and MW energy are generally believed to be mainly thermal in nature [72]. Most agricultural products that considered dielectric material can store electric energy and convert electric energy into heat [49]. As the wavelength of RF heating frequencies is 22 to 360 times as more than that of the microwave frequencies, this allows RF energy to penetrate dielectric materials more deeply than microwaves. Many studies have explored the use of RF heating for control of insects in agricultural commodities [31, 47, 49, 73–82]. Researchers have reported the acceptable product quality after treating nuts and rice with RF energy to control insect infestation [48, 76–78, 83–85]. Recently, Wang et al. [81] studied postharvest disinfestation treatments for chickpeas and lentils using RF energy and also reported acceptable product quality. But the radio frequency heating is particularly useful when applied to institutional-sized packaged food products because of its deep penetration [86, 87].

Thermal treatment methods involving microwave radiation have extensively been investigated by several researchers as an alternative method of killing insects. Microwave quarantine method seems to have a great potential as an alternative method of killing insects in stored grain because of several advantages such as the control of all developmental stages of storage pests, having no chemical residues on the food product, having minimal impact on the environment, and providing rapid heating [28, 88–90]. Insects are unlikely to develop resistance to this treatment [91]. This electromagnetic energy (MW and RF) interacts directly with the product's interior to quickly raise the center temperature [47–49]. Microwave radiation not only kills insects by the dielectric heat induced within them but also affects the reproduction of survivors [92]. Microwave radiation, with good penetrability, can kill pests existing inside or outside grain kernels [93]. This paper brings the research initiatives especially on microwave (MW) heating for their potential use for disinfestation of stored food grains.

2.5.1. Microwave Heating.

Microwaves are electromagnetic waves with frequencies ranging from about 300 MHz to 300 GHz and corresponding wavelengths from 1 to 0.001 m [94]. In electromagnetic spectrum, microwaves lie between radio frequencies and infrared radiation. Microwave heating is based on the transformation of alternating electromagnetic field energy into thermal energy by affecting polar molecules of a material. Many molecules in food (such as water and fat) are electric dipoles, meaning that they have a positive charge at one end and a negative charge at the other, and therefore they rotate as they try to align themselves with the alternating electric field induced by the microwave beam. The rapid movement of the bipolar molecules creates friction and results in heat dissipation in the material exposed to the microwave radiation. Microwave heating is most efficient on liquid water and much less on fats and sugars (which have less molecular dipole moment) [95]. The most important characteristic of microwave heating is volumetric heating which is different from conventional heating. The electromagnetic energy directly interacts with commodities to raise the interior temperature and significantly reduce treatment times as compared to conventional hot-water immersion and heated air methods. Conventional heating occurs by convection or conduction where heat must diffuse from the surface of the material. Volumetric heating means that materials can absorb microwave energy directly and internally and convert it into heat. The power generated in a material is proportional to the frequency of the source, the dielectric loss of the material, and the square of the field strength within it. The conversion of microwave energy to heat is expressed by the following equation [96]:

$$p = 2\pi E^2 f \varepsilon'' \varepsilon_0 V, \tag{1}$$

where p: power, W, E: the electric field strength, V/m, f: the frequency, Hz, ε_0: the permittivity of free space, F/m, ε'': the dielectric loss factor, and V: volume of the material, m³.

Dielectric properties of food depend on composition, temperature, bulk density, and microwave frequency. Since the influence of a dielectric depends on the amount of mass interacting with the electromagnetic fields, the mass per unit volume, or density, will also have an effect on the dielectric properties. This is especially notable with particulate dielectrics such as food grains. For granular and particulate materials, both the dielectric constant and loss factor tend to increase linearly with increasing bulk density of the materials especially in lower moisture materials such as cereal grains [97].

2.5.2. Microwave Disinfestation.

The use of microwaves for disinfestation is based on the dielectric heating effect produced in grain, which is a relatively poor conductor of electricity. An attractive feature of the insect control using the microwave energy is that the insects are heated at a faster rate than the product they infest because of high moisture content of insects. So, it is possible to heat the insects to a lethal temperature because of their high moisture content while leaving the drier foodstuff unaffected or slightly warm [12]. Raising the temperature of infested materials by any means can be used to control insects if the infested product can tolerate the temperature levels that are necessary to kill the insects. There has been a lot of research on microwave disinfestation of cereals especially wheat [33, 98–104] and of some other food materials such as nuts [28], corn [105], pulses [106, 107], and cherries [89]. Exposure to microwave energy

TABLE 1: Dielectric properties of insects at 20–25°C ([32] cited by [33]).

| Adult insect species | Frequency (GHz) | | | | | | | |
| | 0.2 | | 2.4 | | 9.4 | | 20 | |
	ε'	ε''	ε'	ε''	ε'	ε''	ε'	ε''
S. oryzae	28	12	17	3	17	3	—	—
L. decemlineata	53	81	38	12	30	16	19	17
S. oryzae	42	28	32	9	25	12	18	13
S. oryzae	55	48	42	13	31	16	23	16
T. castaneum	61	56	47	15	34	19	25	19
O. surinamensis	70	68	53	17	40	21	28	22
R. dominica	63	55	43	15	34	19	25	18

ε': dielectric constant, ε'': dielectric loss.

could cause physical injuries and reduced reproduction rates in surviving insects [75].

Feasibility of microwave disinfestation of insect pests has been explored by Andreuccetti et al. [74] for woodworms and by Halverson et al. [108] for wheat, maize, and flour weevils (cited by [89]). Nelson [75] reviewed the susceptibility of various stored grain insect species to radio frequency and microwave treatments. The use of microwave energy to control insects was initiated by Webber et al. [109] and the literature prior to 1980 has been reviewed by Tilton and Vardell [98]. Researchers have reported that microwave treatment is an attractive quarantine treatment. The major advantage of microwave heating is that it interacts directly with food grains and significantly reduce the amount of time required for food grain to reach the lethal temperature for insects as compared with conventional heating methods. Based on microwave heating theory and dissipated power calculations in (1), the absorption of microwave energy is proportional to the dielectric constant and dielectric loss factor of the materials. The dielectric properties of the materials depend on the frequency of the applied electric field and the temperature of the material [110]. If the material is hygroscopic, dielectric properties also depend on the amount of water in the material [111].

The first reported measurements of insect dielectric properties were for rice weevil and flour beetle at 40 MHz frequency. The dielectric constants were 6.6 and 7.8 for rice weevil and flour beetle, respectively, and loss factor was 2.2 for both species [112–114]. Nelson [110] presented data that confirms the dielectric loss factor of the insects was found to be much higher than that of stored-grain insects. The insect permittivity data at 25°C for 2.47 GHz frequency reported by Nelson et al. [32] are shown in Table 1. Alfaifi et al. [115] have reported that the loss factor of insect pests, such as Indian meal moth (Plodia interpunctella) and navel orangeworm (Amyelois transitella), is 26 to 36 times greater than that of dried fruits. Wang et al. [28, 48, 49] have indicated preferential heating of insects when walnut kernels and codling moth were subjected to radio frequency (27 MHz) and microwave frequency (2450 MHz) heating. Similar finding was observed by Ikediala et al. [116] and Raskovan et al. [117] for insect infested grain when subjected to radio frequency heating.

TABLE 2: Dielectric properties of selected food products at 20°C.

| Food product | Dielectric constant | | Dielectric loss | |
	915 MHz	2450 MHz	915 MHz	2450 MHz
Apple	57	54	8	10
Almond	2.1	—	2.6	—
Avocado	47	45	16	12
Banana	64	60	19	18
Carrot	59	56	18	15
Cucumber	71	69	11	12
Dates	12	—	5.7	—
Grape	69	65	15	17
Grapefruit	75	73	14	15
Lemon	73	71	15	14
Lime	72	70	18	15
Mango	64	61	12	14
Onion	61	64	12	14
Orange	73	69	14	16
Papaya	69	67	10	14
Peach	70	67	12	14
Pear	67	64	11	13
Potato	62	57	22	17
Radish	68	67	20	15
Strawberry	73	71	14	14
Walnut	3.2	—	6.4	—

Source: see [34].

They have reported that the dielectric constant and dielectric loss factor for granary weevil were greater than those of host grain in frequency band from 2 to 150 MHz. The reason attributed is that the higher the water content, the higher the values of dielectric properties of a material [118] which consist of dielectric constant (ε') and dielectric loss factor (ε'') described by the complex relative permittivity ε^* [115, 119]:

$$\varepsilon^* = \varepsilon' - j\varepsilon'', \qquad (2)$$

where $j = (-1)^{0.5}$, ε' = real component, a measure of the ability of the material to store electromagnetic energy, and ε'' = Imaginary component, a measure of the ability to dissipate electrical energy into heat.

Table 2 shows the dielectric permittivity of food when subjected to microwave heating. Hallman and Sharp [120] also summarized research on the application of radio frequency and microwave heat (electromagnetic energy) treatments to kill different pests on many postharvest food crops. Power level and exposure time of microwave have been identified as two important parameters to provide 100% insect mortality [101]. Microwave energies have been investigated for a number of food products other than cereal grains such as pulses, nuts, cherries, and dates. Recently, the effect of the microwave-heating method has been studied on disinfestations of the stored green gram seed [107]. The power level and exposure time were optimized based on insect mortality, color, and antinutrient factor with the value of power and exposure time of 800 W and 80 s, respectively. Singh et al. [106] have also studied the disinfestation of pulse beetle (adult stage) in chickpea, pigeon pea, and green gram as a function of exposure time and power level by exposing it continuously to microwave radiation (2450 MHz). The mortality of insect was found to increase with increase in microwave exposure time and power level or both. Though seed viability and germination of all these pulses were found to be affected, the cooking and milling characteristics were not affected by microwave exposure time and power level. Campana et al. [121] and Bhaskara Reddy et al. [122] that though eradication of the insects increased with the increase in microwave energy, but the seed viability, germination capacity, and seedling vigour decreased by exposure to microwave energy. Vadivambal et al. [104] and Vadivambal [33] had studied the mortality of different life stages of three common stored-grain insects, namely, *Tribolium castaneum* (Herbst), *Cryptolestes ferrugineus* (Stephens) and *Sitophilus granarius* (L.) in wheat, barley, and rye, respectively, using industrial microwave dryer operating at 2.45 GHz. Grain samples of 50 g at 14%, 16%, and 18% moisture content (wet basis) were exposed to four different power levels of 200, 300, 400, and 500 W for two exposure times of 28 and 56 s. Complete (100%) mortality was achieved for adults of all three insect species at 500 W for an exposure time of 28 s. The average temperature of wheat, barley, and rye at 500 W and 28 s was around 80°C, 71°C and 82°C, respectively. In all the cases, eggs were found to be most susceptible followed by larvae, and the least susceptible were the pupae and adults. Halverson et al. [103] had also reported that egg and young larva of all the three species were more susceptible than the pupa when microwave energy at 28 GHz frequency. The species tested were *S. granarius, T. castaneum* and *R. dominica*. Germination of seeds was lowered with an increase in power level or exposure time or both but there was no significant difference found in the quality characteristics of microwave-treated and control wheat, barley and rye except for a decrease in the flour yield. There were also reports by several researchers that no significant difference in the quality of grain protein, flour protein, flour yield, and loaf volume of sample was found when treated with microwave energy at which 100% mortality was obtained.

Kaasova et al. [123] studied the chemical and biochemical changes during microwave treatment of wheat and reported that an improvement in the baking quality was found at higher energy doses and higher product temperatures. The decrease in germination capacity/seedling viability was related to the final temperature and the initial moisture content of the grains. Hamid and Boulanger [124] had found that the bread making quality of wheat was affected with increase in treatment temperature when exposed to power of 1.2 kW. They reported that 70% and 100% mortality were obtained when the wheat temperature was 55°C and 65°C, respectively, while, Locatelli and Traversa [125] have reported that temperature of grain has to reach 80°C for achieving complete mortality of insects infected using microwave. In fact, most infesting biological agents do not survive over a certain temperature called lethal temperature, generally between 55°C and 60°C, which can be rapidly reached through microwave irradiation [40]. There are three temperature zones for all insects: optimum, the zone at which highest rate of development can be achieved; suboptimum,

TABLE 3: Response of stored-product insects to temperature [35].

Temperature (°C)	Zone	Effect
50–60	Lethal	Death in minutes
45		Death in hours
35	Suboptimum	Development stops
33–35		Development slows
25–33	Optimum	Maximum rate of development
13–25	Suboptimum	Development slows
13–20		Development stops
5	Lethal	Death in days
−10 to −5		Death in weeks to month
−25 to −15		Death in minutes, insects freeze

a zone below or above optimum during which insects can complete their life cycle; and thirdly lethal zone, above or below suboptimum zones when insects get killed over period of time (Table 3). But the lethal temperatures to kill insects were found to vary considerably not only with species but also with the stage of development, temperature, and relative humidity. When developing effective treatment protocols, it is essential to determine which of the targeted insects is the most heat resistant, as well as the most heat resistant life stage for each species. To accomplish this, information is required on the minimum time-temperature combinations that result in 100% mortality for each insect over a relatively large range of temperature. The thermal death kinetics for fifth-instars of Indian meal moth, codling moth, and navel orange worm have been separately reported [76, 77, 126–129]. The authors of [130] recommended 15 min exposures to microwaves, which increased temperature to 60°C to kill all insects in confectionery walnuts without adversely affecting flavor.

Halverson et al. [103] conducted experiments to determine life stages of the insect that is most susceptible to microwave energy at 28 GHz frequency. The life stages tested were egg, young larva, and pupa. Their results suggested that egg and young larva of all were always more susceptible than the pupa. Zouba et al. [131] had employed a research-scale microwave unit to investigate the mortality of insect in date and the thermal impact on the date quality. The heating characteristics of dates are highly influenced by initial moisture content. Date having various moisture contents gets heated up in a very heterogeneous manner and the energy of the microwaves seems to be preferentially absorbed by soft dates. Similar results were reported during insect control of walnuts using radio frequency treatments. The insect mortality is achieved in less than 90 s during microwave treatment without altering date quality.

2.5.3. Challenges in Microwave Disinfestation. Although microwaves have potential for disinfestation applications in the grain industry, they have not been used widely due to their adverse effects on various quality parameters. The major problems associated with microwave heating are the nonuniform temperature distribution and thereby incomplete kill of microbes [132–137]. Hot spots, produced on products due to the nonuniform heating pattern of microwaves, are one of the important factors for the quality degradation of products during microwave treatment. A hot spot can be defined as a local area of very high temperature that results from the temperature dependence of material properties. Although hot spots are desirable to increase insect mortality, detrimental effects on the medium are not. Manickavasagan et al. [138] had determined the germination percentage of microwave-treated wheat samples collected from hot-spot and normal heating zones. The wheat samples having four different moisture levels (12%, 15%, 18%, and 21% wb) were subjected to different microwave treatments (100, 200, 300, 400, and 500 watt with exposure times of 28 and 56 s) in continuous industrial microwave dryer (2450 MHz). The germination percentage of the sample in the hot-spot region was significantly lower than that of the normal heating region for all moisture and power levels. Apart from germination, the other quality parameters were also found to get affected more in the hot-spot zone than the remaining bulk grain. Moreover, the generation of "hot spots" during industrial microwave heating has recently become of great concern to scientists involved in microwave research [139, 140]. The other major drawback in microwave disinfestation is the poor penetrating power of microwaves. The microwave's intensity diminishes with increased penetration. It has been reported that microwave treatment of bulk grain is not feasible when the depth is greater than 4 inch ([92] cited by [33]). Due to the limited penetration of microwave energy into foodstuff mass, it seems likely that employment of microwave radiation alone could not be considered as a promising insect control measure under field condition. Disinfestations of stored products using microwaves energy coupled with other modes of treatment can be an alternative measure in killing insects effectively, but little work has been reported on combined application. The mechanisms involved in the lethal action of microwave radiation are already understood. Microwave radiation has effects on insects such as reduction of reproductive rate, losing body weight, and malformation [75]. However, application of microwaves radiation could be limited due to insufficient penetration depth. Songping et al., [140] reported that microwaves attenuate exponentially in penetration to foodstuffs.

Combined application of microwaves with hot-air treatment/cold storage/gamma radiation could be considered as a potential measure which can help reduce stored-product insects population. The combined impact of microwave radiation and cold storage on adults was evaluated by several researchers. Ayvaz and Karabörklü [141] had reported that there is a decrease in reproductive ability and number of living adults depending on the length of the cold storage period. In general, the reduction of temperature in the environment stresses the insect, thereby, making it more susceptible to other control measures [48, 89]. The major advantage is that the cold storage can easily be coupled with microwave radiation for pest control measure. There was sufficient indication that longer microwaves energy exposure and cold storage duration could achieve better kill than shorter ones of similar power level. When the insects such as *Tribolium castaneum* H. and *Sitophilus oryzae* L. were

exposed to 2450 MHz at power level of 100 W for exposure time 10 min, continuously and intermittently, the highest rate of mortality was achieved for intermittent exposure time of 10 min and 72 h of cold storage duration [142]. Valizadegan et al. [143] have also evaluated the impact of microwave radiation in conjunction with cold storage on adult insects (saw-toothed grain beetle and cigarette beetle) under laboratory conditions. The insects were exposed to 2450 MHZ at five different power levels of 0, 100, 200, 300, and 400 W for five exposure times of 0, 3, 6, 9, and 12 min. The complete control was achieved at 400 W power levels for exposure time of 12 min and 72 h cold storage period. Similar results, that is, high mortality rate, were also reported by Nasab et al. [5] when Indian meal moth eggs were exposed to power levels of 100, 300, and 500 W with 2, 4, 6, 8, and 10 min exposure duration along and then kept in cold storage conditions (4 ± 1°C) for 24 and 48 h. Combinations of microwave radiation and cold storage were found to be compatible and synergistic which provide an effective and environmentally friendly disinfestation treatment technique.

Researchers have also tried the microwave heating along with gamma irradiation to control insects in stored cereals and cereals products. El-Naggar and Mikhaiel [144] evaluated the biochemical analyses on the samples of wheat grain and flour subjected to combined microwave and gamma irradiation treatment where high mortality was obtained. There were no detectable changes in the quality of protein, fat, fiber, carbohydrates, or ash. Amjad and Akbar Anjum [145] stated that when onion seeds (*Allium cepa* L.) were exposed to various doses of gamma radiation, that is, 0, 20, 40, 80, and 100 krad, there was no significant effect on seed viability except at the highest dose (100 krad) which resulted in reduced viability. But several other researchers reported that microwaves are not suitable for the drying of wheat which is to be used as seeds, even using low power levels, unless some provisions are made to ensure uniform heating [124, 146–148]. They have concluded that bread making quality of wheat and maize get adversely affected by exposure to microwave radiation. Though microwave heating helped in reducing the power consumption in wheat milling process, the textural characteristics of the final products are made from the microwave-treated flour were found not to be acceptable. The viscosity of the flour decreased with increased exposure time because of the alteration in structure of starch and protein when exposed to microwave energy. Campana et al. [121] had also stated that the protein content was not affected but the functionality of gluten altered gradually with increasing time of exposure. The change in the functionality of gluten is because of the absence of elasticity and stretchability of the dough [121, 147].

Several researchers concluded that germination capacity was affected by exposure to microwave energy. The seed viability and germination of chickpea, pigeon pea, and green gram were also reported to be affected by microwave exposure time and power level [106]. Vadivambal et al. [149] found that microwave energy at 2450 MHz with similar heating mechanism to RF treatments might control storage insects in barley and rye but resulted in poor germination due to high sample temperatures. The decrease in germination capacity was related to the final temperature and the initial moisture content of the grains. Nelson and Stetson [150] studied the dielectric heating treatments of rice weevils in wheat at 39 and 2450 MHz and showed that the lower frequency was much more effective in killing the insects. Wang et al. [86] had concluded that differential heating of insects in walnuts does occur at 27 MHz but not at 915 MHz based on the experiments carried out on dried nuts and fruits by exposure to microwave and radio frequencies range.

Bhaskara Reddy et al. [122] studied the effect of microwave treatment on quality of wheat seeds infected with *Fusarium graminearum* (Schwabe). Their results showed that though mortality of insect increased with the microwave power level, but the seed viability and seedling vigor decreased accordingly. It has been concluded that microwave drying of wheat would not be suitable where the final products made out of flour required soft textural characteristics. Another issue with the microwave heating is the large number of factors that affect the microwave heat transfer behavior such as the thickness, the geometry, and the dielectric properties of the food. The heat capacity and the dielectric properties (dielectric constant, loss factor) change with the moisture content and temperature which complicates the microwave heating process. Cost estimates for microwave and radio frequency insect control are estimated to be around three to five times more expensive than chemical control.

3. Conclusion

Insects cause considerable damage to food grains with weight and nutritional losses reducing yields and market values. Postharvest phytosanitary treatments are often required to completely control insect pests before the products are moved through marketing channels to areas where the pests do not occur. Several methods have been suggested to control insect pests in agricultural commodities, including chemical fumigation, thermal treatment, ionizing radiation, cold storage, controlled atmospheres, dielectric heating, and combination treatments. Current technologies involve the use of toxic chemicals which is neither consumer friendly nor environmentally friendly and conventional thermal methods are either undesirable or cause loss of volatile components, browning, and texture change. To date, irradiation is not accepted by the organic industry. Based on the results of the microwave disinfestation studies already conducted, it is considered as safe and competitive alternative method to other quarantine methods and can avoid problems of food safety and environmental pollution. The study conducted on different food products infested with major insects says that complete mortality, that is, 100%, could be achieved using microwave energy. Although microwaves have potential for disinfesting the food products, they have not been used widely due to their adverse effects on various quality parameters. Nonuniformity of heating is one of the important factors that cause quality deterioration of food product. Though several practical means can be used to minimize nonuniform heating such as adding forced hot air to the product surface or sample movement/rotation or mixing during

treatment, and so forth. Also if the microwave disinfestation is made economical, it may serve as a safe and effective alternate method of insect control. The most important factor in the development of an acceptable insect control method using microwave energy is to identify a balance between minimized thermal impact on the product quality and complete killing of the insect population. To achieve a balance between complete eradication of the insects and to maintain the product quality, further research needs to be done on large scale tests, with infested product to confirm the treatment efficacy and product quality after extended storage before this technology would provide an acceptable process for disinfestation.

Acknowledgment

The authors acknowledge the Department of Science and Technology (DST), India, for providing financial support for this study.

References

[1] S. Lal and B. P. Srivastava, "Insect pests of stored wheat of Madhya Pradesh (India)," *Journal of Entomological Research*, vol. 9, pp. 141–148, 1985.

[2] A. K. Sinha and K. K. Sinha, "Insect pests, Aspergillus flavus and aflatoxin contamination in stored wheat: a survey at North Bihar (India)," *Journal of Stored Products Research*, vol. 26, no. 4, pp. 223–226, 1990.

[3] B. Subramanyam and D. W. Hagstrum, *Alternatives to Pesticides in STored-Product IPM*, Kluwer Academic Publishers, Norwell, Mass, USA, 2000.

[4] R. I. Sánchez-MariñEz, M. O. Cortez-Rocha, F. Ortega-Dorame, M. Morales-Valdes, and M. I. Silveira, "End-use quality of flour from Rhyzopertha dominica infested wheat," *Cereal Chemistry*, vol. 74, no. 4, pp. 481–483, 1997.

[5] F. S. Nasab, A. A. Pourmirza, and A. H. Zade, "The effect of microwave radiation with cold storage on the mortality of Indian meal moth (*Plodia Interpunctella* Hub.) Eggs," *Pakistan Journal of Entomology*, vol. 31, no. 2, pp. 111–115, 2009.

[6] J. P. Edwards, J. E. Short, and L. Abraham, "Large-scale evaluation of the insect juvenile hormone analogue fenoxycarb as a long-term protectant of stored wheat," *Journal of Stored Products Research*, vol. 27, no. 1, pp. 31–39, 1991.

[7] Y. Rajashekar, N. Bakthavatsalam, and T. Shivanandappa, "Review article on botanicals as grain protectants," vol. 2012, Article ID 646740, 13 pages, 2012.

[8] S. Rajendran and V. Sriranjini, "Plant products as fumigants for stored-product insect control," *Journal of Stored Products Research*, vol. 44, no. 2, pp. 126–135, 2008.

[9] FDA: Food Drug Administration, *Wheat Flour-Adulteration With Insect Fragments and Rodent Hairs (CPG, 7104. 06). Compliance Policy Guides Manual*, chapter 5, Foods, color and cosmetics, sub chapter 578-processed grain, 1997.

[10] FGIS: Federal Grain Inspection Service, Official United States standards for grain subpart m. United States standards for wheat, 1999.

[11] D. B. Pinniger, M. R. Stubbs, and J. Chambers, "The evaluation of some food attractants for the detection of *Oryzaephilus surinamensis* (L.) and other storage pests," in *Proceedings of*

the 3rd International Working Conference on Stored Product Entomology, pp. 640–650, Kansas, USA, October 1984.

[12] R. Vadivambal, D. S. Jayas, and N. D. White, "Disinfestation of life stages of Tribolium castaneum in wheat using microwave energy," in Proceedings of the CSBE/SCGAB, 2006 Annual Conference Edmonton Alberta, 2006, Paper No. 06-120S.

[13] E. Mitcham, "Quarantine issues in 2000," Acta Horticulture, vol. 553, pp. 451–455, 2001.

[14] J. Gannage, "Pesticides and human health," 2000, http://www .herbs2000.com/articles/pesticides.htm.

[15] R. N. Sinha and F. L. Watters, Insect Pests of Flour Mills, Grain Elevators, and Feed Mills and Their Control, Agriculture Canada, Winnipeg, Canada, 1985.

[16] E. J. Bond, Manual of Fumigation For Insect Control, Food and Agricultural Organization, Rome, Italy, 1984.

[17] J. G. Leesch, G. F. Knapp, and B. E. Mackey, "Methyl bromide adsorption on activated carbon to control emissions from commodity fumigations," Journal of Stored Products Research, vol. 36, no. 1, pp. 65–74, 2000.

[18] UNEP (United Nations Environment Programme), Report of the 9th Meeting of the Parties To the Montreal ProTocol on Substances That Deplete the Ozone Layer, UNEP/OzL.Pro. 9/12, Montreal, Canada, 1997.

[19] S. Rajendran and N. Muralidharan, "Performance of phosphine in fumigation of bagged paddy rice in indoor and outdoor stores," Journal of Stored Products Research, vol. 37, no. 4, pp. 351–358, 2001.

[20] C. H. Bell and S. M. Wilson, "Phosphine tolerance and resistance in Trogoderma granarium Everts (Coleoptera: Dermestidae)," Journal of Stored Products Research, vol. 31, no. 3, pp. 199–205, 1995.

[21] B. H. Subramanyam and D. Hagstrum, "Resistance measurement and management," in Integrated Management of Insects in Stored Products, B. Subramanyam and D. Hagstrum, Eds., pp. 331–398, Marcel Dekker, New York, NY, USA, 1995.

[22] P. Collins, G. Daglish, H. Pavic, T. Lambkin, R. Kopittke, and B. Bridgeman, "Combating strong resistance to phosphine in stored grain pests in Australia," in Proceedings of the 2nd Australian Postharvest Technical Conference, E. J. Wright, H. J. Banks, and E. Highley, Eds., pp. 109–112, Adelaide, South Australia, 2000.

[23] S. J. Langlinais, "Economics of microwave treated rice for controlling weevils," ASAE Paper 893544, ASABE, St. Joseph, Mich, USA, 1989.

[24] E. Shadia and S. E. Abd El-Aziz, "Control strategies of stored product pests," Journal of Entomology, vol. 8, no. 2, pp. 101–122, 2011.

[25] P. G. Fields and W. E. Muir, "Physical control," in Integrated Management of Insects in Stored Products, Marcel Dekker, New York, NY, USA, 1996.

[26] J. Banks and J. B. Fields, "Physical methods for insect control in stored-grain ecosystems," in Stored-Grain Ecosystem, D. S. Jayas, N. D. G. White, and W. E. Muir, Eds., pp. 353–410, Marcel Dekker, New York, NY, USA, 1995.

[27] M. Hasan and A. R. Khan, "Control of stored-product pests by irradiation," Integrated Pest Management Reviews, vol. 3, no. 1, pp. 15–29, 1998.

[28] S. Wang and J. Tang, "Radio frequency and microwave alternative treatments for insect control in nuts: A review," International Agricultural Engineering Journal, vol. 10, no. 3-4, pp. 105–120, 2001.

[29] G. Saour and H. Makee, "Susceptibility of potato tuber moth (Lepidoptera: Gelechiidae) to postharvest gamma irradiation," Journal of Economic Entomology, vol. 97, no. 2, pp. 711–714, 2004.

[30] N. W. Heather, G. Hallman, and N. Heather, Pest Management and Phytosanitary Trade Barriers, CABi Publishing, Cambridge, Mass, USA, 2008.

[31] M. Gao, J. Tang, Y. Wang, J. Powers, and S. Wang, "Almond quality as influenced by radio frequency heat treatments for disinfestation," Postharvest Biology and Technology, vol. 58, no. 3, pp. 225–231, 2010.

[32] S. O. Nelson, P. G. Hartley, and K. C. Lawrence, "RF and microwave dielectric properties of stored-grain insects and their implications for potential insect control," Transactions of the American Society of Agricultural Engineers, vol. 41, no. 3, pp. 685–692, 1998.

[33] R. Vadivambal, Disinfestation of Stored Grain Insects Using Microwave Energy [Ph.D. thesis], University of Manitoba, Winnipeg, Canada, 2009.

[34] M. S. Venkatesh and G. S. V. Raghavan, "An overview of microwave processing and dielectric properties of agri-food materials," Biosystems Engineering, vol. 88, no. 1, pp. 1–18, 2004.

[35] M. P. Fakude, "Eradication of storage insect pests in maize using microwave energy and the effects of the latter on grain quality," 2007, http://upetd.up.ac.za/thesis/available/etd-01292009-131525/unrestricted/dissertation.pdf.

[36] P. C. Annis and R. Morton, "The acute mortality effects of carbon dioxide on various life stages of Sitophilus oryzae," Journal of Stored Products Research, vol. 33, no. 2, pp. 115–124, 1997.

[37] N. Gunasekaran and S. Rajendran, "Toxicity of carbon dioxide to drugstore beetle Stegobium paniceum and cigarette beetle Lasioderma serricorne," Journal of Stored Products Research, vol. 41, no. 3, pp. 283–294, 2005.

[38] D. E. Evans, "Some biological and physical constraints to the use of heat and cold for disinfesting and preserving stored products," in Proceedings of the 4th International Working Conference of Stored Product Protection, E. Donahaye and S. Navarro, Eds., pp. 149–164, Tel-Aviv, Israel, 1986.

[39] J. L. Sharp, J. J. Gaffney, J. I. Moss, and W. P. Gould, "Hot-air treatment device for quarantine research," Journal of Economics Entomology, vol. 84, pp. 520–527, 1991.

[40] P. G. Fields, "The control of stored-product insects and mites with extreme temperatures," Journal of Stored Products Research, vol. 28, no. 2, pp. 89–118, 1992.

[41] V. Y. Yokoyama, G. T. Miller, and R. V. Dowell, "Response of codling Moth (Lepidoptera: Tortricidae) to high temperature, a potential quarantine treatment for exported commodities," Journal of Economics Entomology, vol. 84, pp. 528–531, 1991.

[42] L. G. Neven, "Combined heat treatments and cold storage effects on mortality of fifth-instar codling moth (Lepidoptera: Tortricidae)," Journal of Economic Entomology, vol. 87, no. 5, pp. 1262–1265, 1994.

[43] L. G. Neven and L. M. Rehfield, "Comparison of Pre-Storage Heat Treatments on Fifth-Instar Codling Moth (Lepidoptera: Tortricidae) Mortality," Journal of Economic Entomology, vol. 88, pp. 1371–1375, 1995.

[44] L. J. Mason and C. A. Strait, "Stored product integrated pest management with extreme temperatures," in Temperature Sensitivity in Insects and Application in Integrated Pest Management, G. J. Hallman and D. L. Denlinger, Eds., pp. 141–177, Westview Press, Boulder, Colo, USA, 1998.

[45] A. K. Dowdy, "Heat sterilization as an alternative to methyl bromide fumigation in cereal processing plants," in *Proceedings of 7th International Working Conference for Stored Product Protection*, Z. Jin, Q. Liang, Y. Liang, X. Tan, and L. Guan, Eds., pp. 1089–1095, Beijing, China, 1999.

[46] C. S. Burks, J. A. Johnson, D. E. Maier, and J. W. Heaps, "Temperature.," in *Alternatives to Pesticides in stored Product IPM*, B. Subramanyam and D. W. Hagstrum, Eds., pp. 73–104, Kluwer Academic Publishers, Boston, Mass, USA, 2000.

[47] J. Tang, J. N. Ikediala, S. Wang, J. D. Hansen, and R. P. Cavalieri, "High-temperature-short-time thermal quarantine methods," *Postharvest Biology and Technology*, vol. 21, no. 1, pp. 129–145, 2000.

[48] S. Wang, J. N. Ikediala, J. Tang et al., "Radio frequency treatments to control codling moth in in-shell walnuts," *Postharvest Biology and Technology*, vol. 22, no. 1, pp. 29–38, 2001.

[49] S. Wang, J. Tang, and R. P. Cavalieri, "Modeling fruit internal heating rates for hot air and hot water treatments," *Postharvest Biology and Technology*, vol. 22, no. 3, pp. 257–270, 2001.

[50] S. Lurie, T. Jemric, A. Weksler, R. Akiva, and Y. Gazit, "Heat treatment of 'Oroblanco' citrus fruit to control insect infestation," *Postharvest Biology and Technology*, vol. 34, no. 3, pp. 321–329, 2004.

[51] M. Schirra, M. Mulas, A. Fadda, I. Mignani, and S. Lurie, "Chemical and quality traits of "Olinda" and "Campbell" oranges after heat treatment at 44 or 46∘C for fruit fly disinfestation," *LWT—Food Science and Technology*, vol. 38, no. 5, pp. 519–527, 2005.

[52] O. Dosland, B. Subramanyam, K. Sheppard, and R. Mahroof, "Temperature modification for insect control," in *Insect Management For Food Storage and Processing*, J. Heaps, Ed., pp. 89–103, American Association of Cereal Chemists, St. Paul, Minn, USA, 2005.

[53] S. J. Beckett, P. G. Fields, and B. Subramanyam, "Disinfestation of stored products and associated structures using heat," in *Heat Treatments For Postharvest Pest Control: Theory and Practice*, J. Tang, E. Mitcham, S. Wang, and S. Lurie, Eds., CAB International, Wallingford, UK, 2006.

[54] O. Dosland, B. Subramanyam, K. Sheppard, and R. Mahroof, "Temperature modification for insect control," in *Insect Management For Food Storage and Processing*, J. Heaps, Ed., pp. 89–103, American Association for Clinical Chemistry, St. Paul, Minn, USA, 2nd edition, 2006.

[55] E. J. Mitcham, S. Zhou, and V. Bikoba, "Controlled atmospheres for quarantine control of three pests of table grape," *Journal of Economic Entomology*, vol. 90, no. 5, pp. 1360–1370, 1997.

[56] R. W. Howe, "A summary of estimates of optimal and minimal conditions for population increase of some stored products insects," *Journal of Stored Products Research*, vol. 1, no. 2, pp. 177–184, 1965.

[57] J. Hansen, "Heating curve models of quarantine treatments against insect pests," *Journal of Economics Entomology*, vol. 85, pp. 1846–1854, 1992.

[58] S. Lurie, "Postharvest heat treatments," *Postharvest Biology and Technology*, vol. 14, no. 3, pp. 257–269, 1998.

[59] S. Wang, X. Yin, J. Tang, and J. D. Hansen, "Thermal resistance of different life stages of codling moth (Lepidoptera: Tortricidae)," *Journal of Stored Products Research*, vol. 40, no. 5, pp. 565–574, 2004.

[60] G. J. Hallman, J. J. Gaffney, and J. L. Sharp, "Vapor heat treatment for grapefruit infested with Caribbean fruit fly (Diptera:

[61] D. Ortega-Zaleta and E. M. Yahia, "Tolerance and quality of mango fruit exposed to controlled atmospheres at high temperatures," *Postharvest Biology and Technology*, vol. 20, no. 2, pp. 195–201, 2000.

[62] K. C. Shellie, R. L. Mangan, and S. J. Ingle, "Tolerance of grapefruit and mexican fruit fly larvae to heated controlled atmospheres," *Postharvest Biology and Technology*, vol. 10, no. 2, pp. 179–186, 1997.

[63] E. L. Kerbel, G. Mitchell, and G. Mayer, "Effect of postharvest heat treatment for insect control on the quality and market life of avocados," *Horticultural Science*, vol. 22, pp. 92–94, 1987.

[64] K. J. Smith and M. Lay-Yee, "Response of "Royal Gala" apples to hot water treatment for insect control," *Postharvest Biology and Technology*, vol. 19, no. 2, pp. 111–122, 2000.

[65] D. M. Obenland, M. L. Arpaia, R. K. Austin, and B. E. MacKey, "High-temperature forced-air treatment alters the quantity of flavor-related, volatile constituents present in navel and valencia oranges," *Journal of Agricultural and Food Chemistry*, vol. 47, no. 12, pp. 5184–5188, 1999.

[66] D. B. Thomas and K. C. Shellie, "Heating rate and induced thermotolerance in Mexican fruit fly (diptera: tephritidae) larvae, a quarantine pest of citrus and mangoes," *Journal of Economic Entomology*, vol. 93, no. 4, pp. 1373–1379, 2000.

[67] S. J. Beckett and R. Morton, "Mortality of *Rhyzopertha dominica* (F.) (Coleoptera: Bostrychidae) at grain temperatures ranging from 50∘C to 60∘C obtained at different rates of heating in a spouted bed," *Journal of Stored Products Research*, vol. 39, no. 3, pp. 313–332, 2003.

[68] S. Wang and J. Tang, "Heating uniformity and differential heating of insects in almonds in radio frequency systems," Paper 076019, American Society of Agricultural and Biological Engineers, St. Joseph, Mich, USA, 2007.

[69] X. Yin, S. Wang, J. Tang, J. D. Hansen, and S. Lurie, "Thermal conditioning of fifth-instar *Cydia pomonella* (Lepidoptera: Tortricidae) affects HSP70 accumulation and insect mortality," *Physiological Entomology*, vol. 31, no. 3, pp. 241–247, 2006.

[70] D. E. Evans, G. R. Thorpe, and T. Dermott, "The disinfestation of wheat in a continuous-flow fluidized bed," *Journal of Stored Products Research*, vol. 19, no. 3, pp. 125–137, 1983.

[71] Y. C. Fu, "Fundamentals and industrial applications of microwave and radio frequency in food processing," in *Food Processing: Principles and Applications*, J. S. Smith and Y. H. Hui, Eds., Blackwell, Iowa City, IA, Iowa, USA, 2004.

[72] M. A. Stuchly, "Health effects of exposure to electromagnetic fields," *IEEE Aerospace Applications Conf Proceeding*, vol. 1, pp. 351–368, 1995.

[73] S. O. Nelson and J. A. Payne, "RF dielectric heating for pecan weevil control," *Transactions of the American Society of Agricultural Engineers*, vol. 31, pp. 456–458, 1982.

[74] D. Andreuccetti, M. Bini, A. Ignesti, A. Gambetta, and R. Olmi, "Microwave destruction of woodworms," *Journal of Microwave Power and Electromagnetic Energy*, vol. 29, no. 3, pp. 153–160, 1994.

[75] S. O. Nelson, "Review and assessment of radio-frequency and microwave energy for stored-grain insect control," *Transactions of the American Society of Agricultural Engineers*, vol. 39, no. 4, pp. 1475–1484, 1996.

[76] S. Wang, J. N. Ikediala, J. Tang, and J. D. Hansen, "Thermal death kinetics and heating rate effects for fifth-instar *Cydia*

pomonella (L.) (Lepidoptera: Tortricidae)," *Journal of Stored Products Research*, vol. 38, no. 5, pp. 441–453, 2002.

[77] S. Wang, J. Tang, J. A. Johnson, and J. D. Hansen, "Thermal-death kinetics of fifth-instar *Amyelois transitella* (Walker) (Lepidoptera: Pyralidae)," *Journal of Stored Products Research*, vol. 38, no. 5, pp. 427–440, 2002.

[78] S. Wang, J. Tang, J. A. Johnson et al., "Process protocols based on radio frequency energy to control field and storage pests in in-shell walnuts," *Postharvest Biology and Technology*, vol. 26, no. 3, pp. 265–273, 2002.

[79] E. J. Mitcham, R. H. Veltman, X. Feng et al., "Application of radio frequency treatments to control insects in in-shell walnuts," *Postharvest Biology and Technology*, vol. 33, no. 1, pp. 93–100, 2004.

[80] F. Marra, L. Zhang, and J. G. Lyng, "Radio frequency treatment of foods: review of recent advances," *Journal of Food Engineering*, vol. 91, no. 4, pp. 497–508, 2009.

[81] S. Wang, G. Tiwari, S. Jiao, J. A. Johnson, and J. Tang, "Developing postharvest disinfestations treatments for legumes using radio frequency energy," *Biosystems Engineering*, vol. 105, no. 3, pp. 341–349, 2010.

[82] S. Jiao, J. A. Johnson, J. K. Fellman et al., "Evaluating the storage environment in hypobaric chambers used for disinfesting fresh fruits," *Biosystems Engineering*, vol. 111, no. 3, pp. 271–279, 2012.

[83] S. Wang, M. Monzon, J. A. Johnson, E. J. Mitcham, and J. Tang, "Industrial-scale radio frequency treatments for insect control in walnuts. I: Heating uniformity and energy efficiency," *Postharvest Biology and Technology*, vol. 45, no. 2, pp. 240–246, 2007.

[84] S. Wang, M. Monzon, J. A. Johnson, E. J. Mitcham, and J. Tang, "Industrial-scale radio frequency treatments for insect control in walnuts. II: Insect mortality and product quality," *Postharvest Biology and Technology*, vol. 45, no. 2, pp. 247–253, 2007.

[85] M. C. Lagunas-Solar, Z. Pan, N. X. Zeng, T. D. Truong, R. Khir, and K. S. P. Amaratunga, "Application of radiofrequency power for non-chemical disinfestation of rough rice with full retention of quality attributes," *Applied Engineering in Agriculture*, vol. 23, no. 5, pp. 647–654, 2007.

[86] S. Wang, J. Tang, R. P. Cavalieri, and D. C. Davis, "Differential heating of insects in dried nuts and fruits associated with radio frequency and microwave treatments," *Transactions of the American Society of Agricultural Engineers*, vol. 46, no. 4, pp. 1175–1182, 2003.

[87] S. Wang, J. Tang, J. A. Johnson et al., "Dielectric properties of fruits and insect pests as related to radio frequency and microwave treatments," *Biosystems Engineering*, vol. 85, no. 2, pp. 201–212, 2003.

[88] S. L. Halverson, R. Plarre, T. S. Bigelow, and K. Lieber, "Recent advance in the use of EHF energy for the control of insect in stored products," in *Proceedings of the ASAE Annual International Meeting*, Orlando, Fla, USA, 1998, Paper No. 986052.

[89] J. N. Ikediala, J. Tang, L. G. Neven, and S. R. Drake, "Quarantine treatment of cherries using 915 MHz microwaves: temperature mapping, codling moth mortality and fruit quality," *Postharvest Biology and Technology*, vol. 16, no. 2, pp. 127–1, 1999.

[90] O. A. Karabulut and N. Baykal, "Evaluation of the use of microwave power for the control of postharvest diseases of peaches," *Postharvest Biology and Technology*, vol. 26, no. 2, pp. 237–240, 2002.

[91] F. L. Watters, "Microwave radiation for control of Tribolium confusum in wheat and flour," *Journal of Stored Products Research*, vol. 12, no. 1, pp. 19–25, 1976.

[92] M. A. K. Hamid, C. S. Kashyap, and R. V. Cauwenberghe, "Control of grain insects by microwave power," *Journal of Microwave Power*, vol. 3, no. 3, pp. 126–135, 1968.

[93] S. L. Halverson, T. W. Phillips, T. S. Bigelow, G. N. Mbata, and M. E. Payton, "The control of various species of stored-product insects with EHF energy," in *Proceeding of the Annual International Research Conference on Methyl Bromide Alternatives and Emissions Reductions*, pp. 54-1–54-4, 1999.

[94] R. V. Decareau, *Microwaves in the Food Processing Industry*, Academic Press, Natick, Mass, USA, 1985.

[95] P. P. Sutar and S. Prasad, "Microwave drying technology-recent developments and R&D needs in India," in *proceedings of the 42nd ISAE Annual Convention*, pp. 1–3, Kolkata, India, February 2008.

[96] H. Linn and M. Moller, "Microwave heating," in *Proceedings of the Thermprocess Symposium*, pp. 16–21, Dusseldorf, Germany, June 2003.

[97] S. O. Nelson and S. Trabelsi, "Factors influencing the dielectric properties of agricultural and food products," *Journal of Microwave Power and Electromagnetic Energy*, vol. 46, no. 2, pp. 93–107, 2012.

[98] E. W. Tilton and H. H. Vardell, "Combination of microwaves and partial vacuum for control of four stored-product insects in stored grain," *Georgia Entomological Society*, vol. 17, pp. 96–106, 1982.

[99] E. W. Tilton and J. H. Brower, "Ionizing radiation for insect control in grain and grain products," *Cereal Foods World*, vol. 32, no. 4, pp. 330–335, 1987.

[100] N. Shayesteh and N. N. Barthakur, "Mortality and behaviour of two stored-product insect species during microwave irradiation," *Journal of Stored Products Research*, vol. 32, no. 3, pp. 239–246, 1996.

[101] S. S. Bedi and M. Singh, "Microwaves for control of stored grain insects," *National Academy of Science Letter*, vol. 15, no. 6, pp. 195–197, 1992.

[102] D. Sanchez-Hernandez, J. V. Balbastre, and J. M. Osca, "Microwave energy as a viable alternative to methyl bromide and other pesticides for rice disinfection of industrial processes," in *Proceedings of International Conference on Alternatives to Methyl Bromide*, pp. 159–162, Sevilla, Spain, March 2002.

[103] W. R. Halverson, T. S. Bigelow, and S. L. Halverson, "Design of High-Power Microwave Applicator for the Control of Insects in Stored Products," in *Proceedings of the Annual International Meeting*, pp. 27–30, American Society for Agricultural Engineers, Las Vegas, Nev., USA, July 2003, Paper No. 036156.

[104] R. Vadivambal, D. S. Jayas, and N. D. G. White, "Wheat disinfestation using microwave energy," *Journal of Stored Products Research*, vol. 43, no. 4, pp. 508–514, 2007.

[105] R. Vadivambal, O. F. Deji, D. S. Jayas, and N. D. G. White, "Disinfestation of stored corn using microwave energy," *Agriculture and Biology Journal of North America*, vol. 1, no. 1, pp. 18–26, 2010.

[106] R. Singh, K. K. Singh, and N. Kotwaliwale, "Study on disinfestation of pulses using microwave technique," *Journal of Food Science and Technology*, vol. 49, no. 4, pp. 505–509, 2012.

[107] R. Pande, H. N. Mishra, and M. N. Singh, "Microwave drying for safe storage and improved nutritional quality of green gram seed (Vigna radiata)," *Journal of Agricultural and Food Chemistry*, vol. 60, pp. 3809–3816, 2012.

[108] S. L. Halverson, W. E. Burkholder, T. S. Bigelow, E. V. Norsheim, and M. E. Misenheimer, "High-power microwave radiation

as an alternative insect control method for stored products," *Journal of Economic Entomology*, vol. 89, no. 6, pp. 1638–1648, 1996.

[109] H. H. Webber, R. P. Wangner, and A. G. Pearson, "High frequency electric fields as lethal agents for insects," *Journal of Economics Entomology*, vol. 39, pp. 481–498, 1946.

[110] S. O. Nelson, "Dielectric properties of agricultural products: measurements and applications," *IEEE Transactions on Electrical Insulation*, vol. 26, no. 5, pp. 845–869, 1991.

[111] S. O. Nelson, "Radio-frequency and microwave dielectric properties of insects," *Journal of Microwave Power and Electromagnetic Energy*, vol. 36, no. 1, pp. 47–56, 2001.

[112] S. O. Nelson and W. K. Whitney, "Radiofrequency electric fields for stored grain insect control," *Transactions of the American Society of Agricultural Engineers*, vol. 3, no. 2, pp. 133–137, 1960.

[113] S. O. Nelson, L. E. Stetson, and J. J. Rhine, "Factors Influencing Effectiveness of Radiofrequency Electric Fields for Stored-Grain Insect Control," *Transactions of the American Society of Agricultural Engineers*, vol. 9, pp. 809–815, 1966.

[114] S. O. Nelson, "Electromagnetic Energy," in *Pest Control, Biological, Physical and Selected Chemical Methods*, W. W. Kilgore and R. L. Doutt, Eds., pp. 89–145, Academic Press, New York, NY, USA, 1967.

[115] B. Alfaifi, J. Tang, B. Rasco, S. Sablani, and Y. Jiao, "Radio frequency disinfestation treatments for dried fruit: dielectric properties," *LWT—Food Science and Technology*, vol. 50, no. 2, pp. 746–754, 2013.

[116] J. N. Ikediala, J. Tang, and T. Wig, "A heating block system for studying thermal death kinetics of insect pests," *Transactions of the American Society of Agricultural Engineers*, vol. 43, no. 2, pp. 351–358, 2000.

[117] V. M. Rashkovan, N. A. Khizhnyak, A. V. Basteev, L. A. Bazyma, L. Niño de Rivera, and I. A. Ponomaryova, "Interaction of electromagnetic waves with granular agricultural product and insects," *Journal of Microwave Power and Electromagnetic Energy*, vol. 38, no. 4, pp. 1–12, 2003.

[118] J. Tang, "Dielectric Properties of foods," in *Microwave Processing of Foods*, H. Schubert and M. Regier, Eds., CRC Press, Cambridge UK, 2005.

[119] A. C. Metaxas and R. J. Meredith, *Industrial Microwave Heating*, Peter Peregrinus, 1983.

[120] G. J. Hallman and J. L. Sharp, "Radio Frequency Heat Treatments," in *Quarantine Treatments For Pests of Food Plants*, J. L. Sharp and G. J. Hallman, Eds., pp. 165–170, Westview Press, San Francisco, Calif, USA, 1994.

[121] L. E. Campana, M. E. Sempe, and R. R. Filgueira, "Physical, chemical and baking properties of wheat dried with microwave energy," *Cereal Chemistry*, vol. 70, no. 6, pp. 760–762, 1993.

[122] M. V. Bhaskara Reddy, G. S. V. Raghavan, A. C. Kushalappa, and T. C. Paulitz, "Effect of microwave treatment on quality of wheat seeds infected with *Fusarium graminearum*," *Journal of Agricultural Engineering Research*, vol. 71, no. 2, pp. 113–117, 1998.

[123] J. Kaasova, B. Hubackova, P. Kadlec, J. Prihoda, and Z. O. Bubnik, "Chemical and biochemical changes during microwave treatment of wheat," *Czech Journal of Food Science*, vol. 20, pp. 74–78, 2002.

[124] M. A. K. Hamid and R. J. Boulanger, "A new method for the control of moisture and insect infestations of grain by microwave power," *Journal of Microwave Power*, vol. 4, no. 1, pp. 11–18, 1969.

[125] D. P. Locatelli and S. Traversa, "Microwaves in the Control of Rice Infestants," *Italian Journal of Food Science*, vol. 1, no. 2, p. 62, 1989.

[126] J. L. Sharp and R. G. McGuire, "Control of Caribbean fruit fly (Diptera: Tephritidae) in navel orange by forced hot air," *Journal of Economic Entomology*, vol. 89, no. 5, pp. 1181–1185, 1996.

[127] K. C. Shellie and R. L. Mangan, "Disinfestation: effect of non-chemical treatments on market quality of fruit," in *Postharvest Handling of Tropical Fruits*, B. R. Champ, Ed., ACIAR Proceedings, pp. 304–310, 1994.

[128] J. A. Johnson, K. A. Valero, S. Wang, and J. Tang, "Thermal death kinetics of red flour beetle (Coleoptera: Tenebrionidae)," *Journal of Economic Entomology*, vol. 97, no. 6, pp. 1868–1873, 2004.

[129] K. C. Shellie and R. L. Mangan, "Tolerance of red fleshed grapefruit to a constant or stepped temperature, forced-air quarantine heat treatment," *Postharvest Biology and Technology*, vol. 7, no. 1-2, pp. 151–159, 1996.

[130] D. R. Wilkin and G. Nelson, *Control of Insects in Confectionery Walnuts Using Microwaves*, vol. 87, British Crop Protection Council Monograph, Berks, UK, 1987.

[131] A. Zouba, O. Khoualdia, A. Antoneo Diaferia, V. Valereo Rosito, H. Bouabidi, and B. Chermiti, "Microwave Treatment for Postharvest Control of the Date Moth *Ectomyelois ceratoniae*," *Tunisian Journal of Plant Protection*, vol. 4, no. 2, pp. 173–183, 2012.

[132] S. Ryynanen, H. Tuorila, and L. Hyvonen, "Perceived temperature effects on microwave-heated meals and meal components," *Food Service Technology*, vol. 1, pp. 141–148, 2001.

[133] D. S. Lee, D. Shin, and K. L. Yam, "Improvement of temperature uniformity in microwave-reheated rice by optimizing heat/cold cycle," *Food Service Technology*, vol. 2, no. 2, pp. 87–93, 2002.

[134] N. Sakai, C. Wang, S. Toba, and M. Watanabe, "An analysis of temperature distributions in microwave heating of foods with non-uniform dielectric properties," *Journal of Chemical Engineering of Japan*, vol. 37, no. 7, pp. 858–862, 2004.

[135] A. Manickavasagan, D. S. Jayas, and N. D. G. White, "Non-uniformity of surface temperatures of grain after microwave treatment in an industrial microwave dryer," *Drying Technology*, vol. 24, no. 12, pp. 1559–1567, 2006.

[136] S. Gunasekaran and H. W. Yang, "Effect of experimental parameters on temperature distribution during continuous and pulsed microwave heating," *Journal of Food Engineering*, vol. 78, no. 4, pp. 1452–1456, 2007.

[137] S. S. R. Geedipalli, V. Rakesh, and A. K. Datta, "Modeling the heating uniformity contributed by a rotating turntable in microwave ovens," *Journal of Food Engineering*, vol. 82, no. 3, pp. 359–368, 2007.

[138] A. Manickavasagan, D. S. Jayas, and N. D. G. White, "Germination of wheat grains from uneven microwave heating in an industrial microwave dryer," *Canadian Biosystems Engineering*, vol. 49, pp. 323–327, 2007.

[139] J. M. Hill and M. J. Jennings, "Formulation of model equations for heating by microwave radiation," *Applied Mathematical Modelling*, vol. 11, pp. 369–319, 1993.

[140] Z. Songping, Z. Yinglong, and T. R. Marchant, "A DRBEM model for microwave heating problems," *Applied Mathematical Modelling*, vol. 19, no. 5, pp. 287–297, 1995.

[141] A. Ayvaz and S. Karabörklü, "Effect of cold storage and different diets on *Ephestia kuehniella Zeller*," *Journal of Pest Science*, vol. 81, no. 1, pp. 57–62, 2008.

[142] S. Gasemzadeh, A. Pourmirza, M. Safaralizadeh, and M. Maroufpoor, "Effect of microwave radiation and cold storage on *Tribolium castaneum* Herbst (Coleoptera: Tenebrionidae) and *Sitophilus oryzae* L. (Coleoptera: Curculionidae)," *Journal of Plant Protection Research*, vol. 50, no. 2, pp. 140–145, 2010.

[143] O. Valizadegan, A. A. Pourmirza, and M. H. Safaralizadeh, "The impact of microwaves irradiation and temperature manipulation for control of stored-products insects," *African Journal of Biotechnology*, vol. 10, no. 61, pp. 13256–13262, 2011.

[144] S. M. El-Naggar and A. A. Mikhaiel, "Disinfestation of stored wheat grain and flour using gamma rays and microwave heating," *Journal of Stored Products Research*, vol. 47, no. 3, pp. 191–196, 2011.

[145] M. Amjad and M. Akbar Anjum, "Effect of gamma radiation on onion seed viability, germination potential, seedling growth and morphology," *Pakistan Journal of Agriculture*, vol. 39, pp. 202–206, 2002.

[146] N. C. Doty and C. W. Baker, "Microwave conditioning of hard red spring wheat. I. Effects of wide power range on flour and bread quality," *Cereal Chemistry*, vol. 54, pp. 717–727, 1977.

[147] S. G. Walde, K. Balaswamy, V. Velu, and D. G. Rao, "Microwave drying and grinding characteristics of wheat (*Triticum aestivum*)," *Journal of Food Engineering*, vol. 55, no. 3, pp. 271–276, 2002.

[148] V. Velu, A. Nagender, P. G. Prabhakara Rao, and D. G. Rao, "Dry milling characteristics of microwave dried maize grains (Zea mays L.)," *Journal of Food Engineering*, vol. 74, no. 1, pp. 30–36, 2006.

[149] R. Vadivambal, D. S. Jayas, and N. D. G. White, "Mortality of stored-grain insects exposed to microwave energy," *Transactions of the ASABE*, vol. 51, no. 2, pp. 641–647, 2008.

[150] S. O. Nelson and L. E. Stetson, "Possibilities for vontrolling insects with microwaves and lower frequency RF energy," *IEEE Transactions on Microwave Theory and Techniques*, vol. 22, no. 12, pp. 1303–1305, 1974.

Permissions

The contributors of this book come from diverse backgrounds, making this book a truly international effort. This book will bring forth new frontiers with its revolutionizing research information and detailed analysis of the nascent developments around the world.

We would like to thank all the contributing authors for lending their expertise to make the book truly unique. They have played a crucial role in the development of this book. Without their invaluable contributions this book wouldn't have been possible. They have made vital efforts to compile up to date information on the varied aspects of this subject to make this book a valuable addition to the collection of many professionals and students.

This book was conceptualized with the vision of imparting up-to-date information and advanced data in this field. To ensure the same, a matchless editorial board was set up. Every individual on the board went through rigorous rounds of assessment to prove their worth. After which they invested a large part of their time researching and compiling the most relevant data for our readers. Conferences and sessions were held from time to time between the editorial board and the contributing authors to present the data in the most comprehensible form. The editorial team has worked tirelessly to provide valuable and valid information to help people across the globe.

Every chapter published in this book has been scrutinized by our experts. Their significance has been extensively debated. The topics covered herein carry significant findings which will fuel the growth of the discipline. They may even be implemented as practical applications or may be referred to as a beginning point for another development. Chapters in this book were first published by Hindawi Publishing Corporation; hereby published with permission under the Creative Commons Attribution License or equivalent.

The editorial board has been involved in producing this book since its inception. They have spent rigorous hours researching and exploring the diverse topics which have resulted in the successful publishing of this book. They have passed on their knowledge of decades through this book. To expedite this challenging task, the publisher supported the team at every step. A small team of assistant editors was also appointed to further simplify the editing procedure and attain best results for the readers.

Our editorial team has been hand-picked from every corner of the world. Their multi-ethnicity adds dynamic inputs to the discussions which result in innovative outcomes. These outcomes are then further discussed with the researchers and contributors who give their valuable feedback and opinion regarding the same. The feedback is then collaborated with the researches and they are edited in a comprehensive manner to aid the understanding of the subject.

Apart from the editorial board, the designing team has also invested a significant amount of their time in understanding the subject and creating the most relevant covers. They scrutinized every image to scout for the most suitable representation of the subject and create an appropriate cover for the book.

The publishing team has been involved in this book since its early stages. They were actively engaged in every process, be it collecting the data, connecting with the contributors or procuring relevant information. The team has been an ardent support to the editorial, designing and production team. Their endless efforts to recruit the best for this project, has resulted in the accomplishment of this book. They are a veteran in the field of academics and their pool of knowledge is as vast as their experience in printing. Their expertise and guidance has proved useful at every step. Their uncompromising quality standards have made this book an exceptional effort. Their encouragement from time to time has been an inspiration for everyone.

The publisher and the editorial board hope that this book will prove to be a valuable piece of knowledge for researchers, students, practitioners and scholars across the globe.

List of Contributors

Maira Rubi Segura Campos, Fanny Peralta González, Luis Chel Guerrero and David Betancur Ancona
Facultad de Ingeniería Química, Universidad Autonoma de Yucatan, Periferico Norte. Km 33.5, Tablaje Catastral 13615, Colonia Chuburn´a de Hidalgo Inn, 97203 M´erida, YUC, Mexico

A. M. Maskooki and S. Valibeigi
Food Processing Department, Research Institute of Food Science and Technology (RIFST), Km. 12 Asian Road, Mashhad, Iran

S. H. R. Beheshti and J. Feizi
TESTA Quality Control Laboratory, North-East Food Technology and Biotechnology Zone, Km. 12 Asian Road, Mashhad, Iran

Shelly Coe, Ann Fraser and Lisa Ryan
Functional Food Centre, Oxford Brookes University, Gipsy Lane, Oxford OX3 0BP, UK

Bhawna Chugh and Gurmukh Singh
Department of Food Science & Technology, G. B. Pant University of Agriculture and Technology, Uttarakhand, Pantnagar 263145, India

B. K. Kumbhar
Department of Post-Harvest Process and Food Engineering, G. B. Pant University of Agriculture and Technology, Uttarakhand, Pantnagar 263145, India

Anurag Singh
Department of Food Technology, FET, RBS Engineering Technical Campus, Bichpuri, Agra 283105, India

H. K. Sharma and Sanjay Kumar
Food Engineering and Technology Department, Sant Longowal Institute of Engineering and Technology (SLIET), Longowal, Sangrur, Punjab 148106, India

Ashutosh Upadhyay
National Institute of Food Technology Entrepreneurship and Management, Kundli, Sonipat, Haryana 131028, India

K. P. Mishra
Faculty of Engineering and Technology, Mahatma Gandhi Chitrakoot Gramodaya Vishwavidyalaya, Chitrakoot, Satna, Madhya Pradesh 485331, India

Ulf Svanberg
Department of Chemical and Biological Engineering/Food Science, Chalmers University of Technology, SE-41296 Gothenburg, Sweden

Maria Eduardo
Departamento de Engenharia Química, Faculdade de Engenharia, Universidade Eduardo Mondlane, Maputo, Mozambique
Department of Chemical and Biological Engineering/Food Science, Chalmers University of Technology, SE-41296 Gothenburg, Sweden
SIK, The Swedish Institute for Food and Biotechnology, SE-402 29 Gothenburg, Sweden

Lilia Ahrné
Department of Chemical and Biological Engineering/Food Science, Chalmers University of Technology, SE-41296 Gothenburg, Sweden
SIK, The Swedish Institute for Food and Biotechnology, SE-402 29 Gothenburg, Sweden

Jorge Oliveira
Department of Process and Chemical Engineering, University College Cork, Cork, Ireland

Iulia Movileanu
Department of Neurology, Upstate Medical University, 812 Jacobsen Hall, Syracuse, NY 13210, USA

Máryuri T. Núñez de González
Department of Food Technology, Universidad de Oriente, Nucleo Nueva Esparta, Escuela de Ciencias Aplicadas del Mar, Isla de Margarita 6301, Venezuela

Brian Hafley
Tyson Foods, 1825 Ford Avenue, Springdale, AR 72764, USA

Rhonda K. Miller
Department of Animal Science, Texas A&M University, 338 Kleberg Center, College Station, TX 77843-2471, USA

Jimmy T. Keeton
Department of Nutrition and Food Science, Texas A&M University, 122 Kleberg Center, College Station, TX 77843-2253, USA

Yulia B. Monakhova
Chemisches und Veterinaruntersuchungsamt (CVUA) Karlsruhe, Weissenburger Strasse 3, 76187 Karlsruhe, Germany
Department of Chemistry, Saratov State University, Astrakhanskaya Street 83, 410012 Saratov, Russia
Bruker Biospin GmbH, Silbersteifen, 76287 Rheinstetten, Germany

Dirk W. Lachenmeier
Chemisches und Veterin¨aruntersuchungsamt (CVUA) Karlsruhe, Weissenburger Strasse 3, 76187 Karlsruhe, Germany
Ministry of Rural Affairs and Consumer Protection, Kernerplatz 10, 70182 Stuttgart, Germany

Rolf Godelmann, Claudia Andlauer and Thomas Kuballa
Chemisches und Veterin¨aruntersuchungsamt (CVUA) Karlsruhe, Weissenburger Strasse 3, 76187 Karlsruhe, Germany

Silvia Matiacevich, Daniela Celis Cofré, Patricia Silva and Fernando Osorio
Departamento de Ciencia y Tecnologıa de los Alimentos, Facultad Tecnologica, Universidad de Santiago de Chile, Avenida Libertador Bernardo O'Higgins No. 3363, Estacion Central, 9170022 Santiago, Chile

Javier Enrione
Departamento de Nutricion y Dietetica, Facultad de Medicina, Universidad de los Andes, San Carlos de Apoquindo 2200, Las Condes, 7620001 Santiago, Chile

Adriana R. Weisstaub, Victoria Abdala and Ángela Zuleta
Food Science and Nutritional Department, School of Pharmacy and Biochemistry, Buenos Aires University (UBA), 1114 Buenos Aires, Argentina

Macarena Gonzales Chaves and Susana Zeni
Metabolic Bone Diseases Laboratory, Clinical Hospital, Immunology, Genetic and Metabolism Institute (INIGEM), National Council for Scientific and Technologic Research (CONICET), UBA, 1114 Buenos Aires, Argentina

Patricia Mandalunis
Histologycal and Embryology Department, School of Dentistry, UBA, 1114 Buenos Aires, Argentina

S. Ramesh, R. Muthuvelayudham, R. Rajesh Kannan and T. Viruthagiri
Department of Chemical Engineering, Annamalai University, Annamalainagar 608002, Tamil Nadu, India

Sanguansri Charoenrein and Sunsanee Udomrati
Department of Food Science and Technology, Faculty of Agro-Industry, Kasetsart University, Bangkok 10900, Thailand

John Yew Huat Tang, Bariah Ibrahim Izenty, Ahmad Juanda NurIzzati, Siti Rahmah Masran, Arshad Roslan and Che Abdullah Abu Bakar
Faculty of Food Technology, Universiti Sultan Zainal Abidin, Gong Badak Campus, 21300 Kuala Terengganu, Terengganu, Malaysia

Chew Chieng Yeo
Faculty of Agriculture and Biotechnology, Universiti Sultan Zainal Abidin, Gong Badak Campus,
21300 Kuala Terengganu, Terengganu, Malaysia

Rolando José González, Roberto Luis Torres and Dardo Mario De Greef
Instituto de Tecnolog´ıa de Alimentos, Universidad Nacional del Litoral, 1 de Mayo 3250, 3000 Santa Fe, Argentina

Elena Pastor Cavada and Javier Vioque Peña
Instituto de la Grasa (CSIC), Avenida Padre Garc´ıa Tejero 4, 41012 Seville, Spain

Silvina Rosa Drago
Instituto de Tecnologıa de Alimentos, Universidad Nacional del Litoral, 1 de Mayo 3250, 3000 Santa Fe, Argentina
Consejo Nacional de Investigaciones Cient´ıficas y T´ecnicas (CONICET), Avenida Rivadavia 1917, 1033 Ciudad
Aut´onoma de Buenos Aires, Argentina

Alina Blaszczyk and Aleksandra Augustyniak
Department of General Genetics, Molecular Biology and Plant Biotechnology, Faculty of Biology and
Environmental Protection, University of Łodz, Banacha 12/16, 90-237 Łodz, Poland

Janusz Skolimowski
Department of Organic Chemistry, Faculty of Chemistry, University of Łodz, Tamka 12, 91-403 Łodz, Poland

Nawraj Rummun
Department of Health Sciences, Faculty of Science, University of Mauritius, Reduit, Mauritius

Jhoti Somanah
Department of Biosciences and ANDI Centre of Excellence for Biomedical and Biomaterials Research,
University of Mauritius, Reduit, Mauritius

Srishti Ramsaha and Vidushi S. Neergheen-Bhujun
Department of Health Sciences, Faculty of Science and ANDI Centre of Excellence for Biomedical and Biomaterials
Research, University of Mauritius, Reduit, Mauritius

Theeshan Bahorun
ANDI Centre of Excellence for Biomedical and Biomaterials Research, University of Mauritius, Reduit, Mauritius

Mercedes Pérez-Bonilla, Sofía Salido, Adolfo Sánchez and Joaquín Altarejos
Departamento de Quımica Inorganica y Organica, Facultad de Ciencias Experimentales, Universidad de Jaen, Campus
de Excelencia Internacional Agroalimentario, ceiA3, 23071 Jaen, Spain

Teris A. van Beek
Laboratory of Organic Chemistry, Natural Products Chemistry Group, Wageningen University, Dreijenplein 8, 6703
HB Wageningen, The Netherlands

A. O. Adebayo-Oyetoro
Department of Food Technology, Yaba College of Technology, PMB 2011, Yaba, Lagos 101212, Nigeria

O. B. Oyewole and A. O. Obadina
Department of Food Science and Technology, Federal University of Agriculture, PMB 2240, Ogun State, Abeokuta
110001, Nigeria

M. A. Omemu
Department of Food Service and Tourism, Federal University of Agriculture, PMB 2240, Ogun State, Abeokuta
110001, Nigeria

Laban K. Rutto and Michael Brandt
Alternative Crops Program, Agriculture Research Station, Virginia State University, Petersburg, VA 23806, USA

Yixiang Xu
Food Processing and Engineering Program, Agriculture Research Station, Virginia State University, Petersburg, VA 23806, USA

Elizabeth Ramirez
College of Agriculture and Life Sciences, Virginia Polytechnic Institute and State University, Blacksburg, VA 24061, USA

Michele Proietto Galeano, Monica Scordino, Leonardo Sabatino, Valentina Pantò, Giovanni Morabito, Elena Chiappara, Pasqualino Traulo and Giacomo Gagliano
Dipartimento dell'Ispettorato Centrale della Tutela della Qualita e della Repressione Frodi dei Prodotti Agroalimentari (ICQRF), Laboratory of Catania, Ministero delle Politiche Agricole Alimentari e Forestali (MIPAAF), Via A. Volta 19, 95122 Catania, Italy

Leslaw Juszczak, Dorota Galkowska and Teresa Fortuna
Department of Analysis and Evaluation of Food Quality, University of Agriculture in Krakow, Balicka 122, 30-149 Krakow, Poland

Teresa Witczak
Department of Engineering and Machinery for Food Industry, University of Agriculture in Krakow, Balicka 122, 30-149 Krakow, Poland

Lídia Cedó, Anna Castell-Auví, Victor Pallarès, Mayte Blay, Anna Ardévol and Montserrat Pinent
Nutrigenomics Research Group, Departament de Bioquimica i Biotecnologia, Universitat Rovira i Virgili, Marcelli Domingo s/n, 43007 Tarragona, Spain

Yang Zhou, Xiao-jia Yang, Dan-hua Lin, Yun-fang Gao, Yin-jie Su and Jing-jing Zheng
College of Food Science and Technology, Henan University of Technology, Zhengzhou 450001, China

Sen Yang, Yan-jie Zhang and Jin-shui Wang
College of Bioengineering, Henan University of Technology, Zhengzhou 450001, China

Ipsita Das and Girish Kumar
Department of Electrical Engineering, Indian Institute of Technology, Bombay, Powai, Mumbai 400076, India

Narendra G. Shah
Centre for Technology Alternatives for Rural Areas, Indian Institute of Technology, Bombay, Powai, Mumbai 400076, India

www.ingramcontent.com/pod-product-compliance
Lightning Source LLC
Chambersburg PA
CBHW080659200326
41458CB00013B/4918